STATISTICAL MECHANICS

STATISTICAL MECHANICS

Second Edition

JOSEPH EDWARD MAYER

Department of Chemistry
University of California, San Diego
La Jolla, California

and

the late MARIA GOEPPERT MAYER

A WILEY-INTERSCIENCE PUBLICATION

JOHN WILEY & SONS, New York • London • Sydney • Toronto

***Library of Congress Cataloging in Publication Data*:**

Mayer, Joseph Edward, 1904–
Statistical mechanics.

"A Wiley-Interscience publication."
Includes index.
1. Statistical mechanics. I. Mayer, Maria Goeppert, 1906–1972, joint author. II. Title.
QC174.8.M37 1976 530.1'32 76–20668
ISBN 0–471–57985–8

Printed in the United States of America

10 9 8 7 6 5 4 3 2 1

PREFACE TO THE FIRST EDITION

The rapid increase, in the past few decades, of knowledge concerning the structure of molecules has made the science of statistical mechanics a practical tool for interpreting and correlating experimental data. It is therefore desirable to present this subject in a simple manner in order to make it easily available to scientists whose familiarity with theoretical physics is limited. This book, which grew out of lectures and seminars given to graduate students in chemistry and physics, aims to fulfill this purpose.

The development of quantum mechanics has altered both the axiomatic foundation and the details of the methods of statistical mechanics. Although the results of a large number of statistical calculations are unaffected by the introduction of quantum mechanics, the chemist's interest happens to be largely in fields where quantum effects are important. Consequently, in our presentation, the laws of statistical mechanics are founded on the concepts of both quantum and classical mechanics. The equivalence of the two methods has been stressed, but the quantum-mechanical language has been favored. We believe that this introduction of quantum statistics at the beginning simplifies rather than puts a burden upon the initial concepts. It is to be emphasized that the simpler ideas of quantum mechanics, which are all that is used, are as widely known as the more abstract theorems of classical mechanics which they replace.

Simplicity of presentation rather than brevity and elegance has been our endeavor. However, we have not consciously sacrificed rigor.

Care has been taken to make the book suitable for reference by summarizing and tabulating final equations as well as by an attempt to make individual chapters complete in themselves without too much reference to previous subjects.

All the theorems and results of mechanics and quantum mechanics which are used later have been summarized, largely without proof, in Chapter 2. The last section, 2k, on Einstein-Bose and Fermi-Dirac systems, ties up closely with Chapters 5 and 16 only.

Chapters 3 and 4 contain the derivation of the fundamental statistical laws on which the book is based. Chapter 10 is prerequisite for Chapters 11 to 14. Otherwise, individual subjects may be taken up in different order.

In Chapters 7 to 9 considerable space is devoted to the calculation of thermodynamic functions for perfect gases, which was considered justified by the value of the results for the chemist. These chapters may be omitted by readers uninterested in the subject.

Chapters 13 and 14 on the imperfect gas and condensation theory, respectively, are somewhat more complicated than the remainder, but are included because of our special interest in the subject.

The aim of the book is to give the reader a clear understanding of principles and to prepare him thoroughly for the use of the science and the study of recent papers. Many of the simpler applications are discussed in some detail, but in general language without comparison with experiment. The more complicated subjects have been omitted, as have been those for which at present only partial solutions are obtained. This choice has excluded many of the contemporary developments, especially the interesting work of J. G. Kirkwood, L. Onsager, H. Eyring, and W. F. Giauque.

In conclusion we express our gratitude to Professors Max Born, Karl F. Herzfeld, and Edward Teller, who have read and criticized several parts of the manuscript. We also thank Dr. Elliot Montroll, who aided in reading proof and who made many helpful suggestions.

JOSEPH EDWARD MAYER
MARIA GOEPPERT MAYER

New York City
March 31, 1940

PREFACE TO THE SECOND EDITION

Since the first edition appeared in 1940, the total number of articles and pages in archival journals dealing with statistical mechanics in one form or another exceeds by a factor of more than 10 the quantity published before that date. To cover comprehensively even the limited number of topics treated in the first edition would require many volumes. The treatment of the illustrative examples of applications in this edition is necessarily abbreviated. However, much more attention is given in this volume to both the logic and to the fundamental structure of the subject than in the last edition.

There has also been a considerable change in the extent of mathematical erudition of a large fraction of senior or beginning graduate students in both chemistry and physics since 1940. Nevertheless we still try, in this edition, to present everything in a manner understandable to one whose formal training does not go beyond the calculus of real functions. The concepts of matrix algebra are used extensively only in the last chapter on the density matrix. The necessary mathematics for understanding this is presented without proof in Appendices VII and VIII. There we place special emphasis on certain characteristics of the use in physics of matrix algebra, characteristics that are often not mentioned in mathematical texts.

This book is divided into three parts. Part 1 has an introductory chapter with some historical perspective and a discussion of the use of probabilities and probability densities. Chapter 2 follows with a very elementary presentation of nineteenth century perfect gas kinetic theory. None of formal statistical mechanics is used. Many students will already be familiar with this.

Part 2, which consists of four chapters, concerns the logic and formalism of the Gibbs ensemble method. The first of these, Chapter 3, concludes with the master equations and explanation of their interrelations. These are sufficient for application to all real equilibrium systems. The equations are derived from the mechanical laws in Chapter 4, where it is also shown that the thermodynamic laws follow as a consequence. Some general relations are derived in Chapter 5. Chapter 6 concerns an examination of the assumptions and their logic, particularly that of the ergodic hypothesis. The explanation of various apparent paradoxes is presented. Part 3 then follows with applications to different problems that are much in the style and level of the first edition, understandable to a reader who has not finished Part 2; this part concludes with a more sophisticated discussion of quantum statistics.

Many students, particularly those attracted to the experimental study of science, have the healthy and natural instinct to first ask "What does it do?" and only after knowing the answer, to ask "Why does it work?". The true scientist is never satisfied without also the answer to "why." The reader with this bent might well wish to go to some, or even all of the seven chapters of Part 3, which tell briefly what statistical mechanics does, before studying the last three chapters of Part 2. The book is designed to make this possible.

The individual chapters, and even most of the sections, are, as far as possible, self-contained. Definitions of the symbols, as well as explanation of the notation and statements of the basic starting equations, are given. References to previous sections and equations are dutifully included, but repeated references back to earlier pages should be little needed.

This second edition was planned several years before the death of my wife, Maria Goeppert Mayer. We had discussed and agreed on the outline of the contents. She had been impressed in Chicago by Enrico Fermi's use of statistical mechanical concepts in high energy physics and probably would have included a short chapter on this; I felt incompetent to do so alone.

JOSEPH EDWARD MAYER

La Jolla, California
May 1976

ACKNOWLEDGMENTS

I am much indebted to those who have helped me. Doctors Charles Perrin, Jeff Pressing, Harold Raveché, and Michael Richardson have all read significant portions of the manuscript and offered invaluable criticism and suggestions. Above all, Mrs. Dorothy Prior has prepared a nearly perfectly executed manuscript that I am sure will minimize the usual and unavoidable errors of typography in the finished book.

J. E. M.

CONTENTS

NOTATION

We have employed a conventional notation, P, V, E, S, $A=E-TS$, $G=E+PV-TS$, for most of the thermodynamic quantities, but have generally used N, the number of molecules, rather than n, the number of moles. The quantity $\beta=(kT)^{-1}$ is a "natural" intensive variable for statistical ensembles, and the molecular number density $\rho=N/V$ occurs repeatedly in equations.

Our classical mechanics notation of p for momentum and q for a generalized coordinate with $H(p, q)$ the Hamiltonian is conventional. We use Planck's constant h rather than $\hbar=(2\pi)^{-1}h$ throughout, since $h^{-1}\,dp\,dq$ is the correct dimensionless volume element in phase space to count quantum states.

Indices for single-numbered molecules are usually written i, j, k, whereas a, b, c, . . . or α, γ, . . . generally refer to different chemical species.

Boldface letters are used for vectors or quantities that are intrinsically sets of more than a single number, thus $\mathbf{r}_i = x_i, y_i, z_i$ is the three-dimensional position of molecule i, $\mathbf{p}_i = p_{xi}, p_{yi}, p_{zi}, \ldots$, and dot products are used for $\mathbf{r}_i \cdot \mathbf{p}_i = x_i p_{xi} + y_i p_{yi} + z_i p_{zi}$. However, the same notation, $\boldsymbol{\rho} = \rho_a, \rho_b, \rho_c, \ldots$, $\mathbf{n} = n_a, n_b, \ldots$, with $\boldsymbol{\rho}^{\mathbf{n}} = \rho_a^{n_a} \rho_b^{n_b} \ldots$, and $\mathbf{n}! = n_a!\, n_b! \ldots$, are used for quantities that are in no real sense vectors.

Lowercase letters, n, ν, μ, or $\boldsymbol{\nu}$, $\boldsymbol{\mu}$, $\psi_{\boldsymbol{\nu}}(\mathbf{q})$, and so on, are used for indices, quantum numbers, and so on, or functions of the coordinates of single molecules or even of small numbers $n=1, 2, 3, \ldots$, of molecules. Uppercase capitals are used when the reference is to a macroscopic system of $N \cong 10^{23}$ molecules. Thus $\Psi_{\mathbf{K}}(\mathbf{q}^{(\Gamma)})$, with $\mathbf{q}^{(\Gamma)} = \Pi_{i=1}^{N} \mathbf{q}_i^{(\gamma_i)}$, may be the eigenfunction of the Hamiltonian operator on the whole system of $\Gamma \cong 10^{24}$ degrees of freedom with the intrinsically Γ-dimensional quantum number $\mathbf{K}$.

Script letters $\mathcal{H}$, $\mathcal{K}$, $\mathcal{O}$, are used for quantum-mechanical operators and in Chapter 12 for electric and magnetic fields, $\mathcal{E}$, $\mathcal{H}$, $\mathcal{B}$, $\mathcal{D}$. Boldface uppercase letters are used in Chapter 13 for matrices.

STATISTICAL MECHANICS

PART ONE

This part presents a historical perspective, summary of the needed parts of probability theory, and an elementary account of the successes of nineteenth century perfect gas kinetic theory.

CHAPTER 1

INTRODUCTION

1a. THE TASK OF STATISTICAL MECHANICS

Most of the observational facts that form the basis of physical science, as well as most parts of the theoretical structure used to describe them, fall fairly unambiguously into one of two categories, the macroscopic or the microscopic. Macroscopic laws treat the properties of matter in bulk and use an appropriate set of terms such as density, pressure, volume, and temperature, all of which are meaningless when applied to single atoms and molecules. Molecular behavior is described by microscopic laws involving the concepts of coordinates and momenta, or wave functions and matrix elements. The problem of statistical mechanics is to make a connection between these two points of view. Ordinary matter as observed in the laboratory consists always of atoms and molecules, and these obey in their motions the laws of mechanics: quantum mechanics or in many cases the asymptotically valid laws of classical mechanics. These laws, at least in their statistical aspects, are deterministic in nature and have no need for additional laws to predict molecular behavior. Additional laws must be either in conflict with the assumed quantum mechanics (plus electrodynamics), or redundant and a consequence of the laws of mechanics.

Thus, from this point of view, one of the first tasks of statistical mechanics is to derive from the laws of mechanics all the pertinent macroscopic laws: the laws of thermodynamics, of fluid dynamics, of elasticity, of chemical kinetics, and so on. These, if quantum mechanics is correct, must follow as a consequence of its laws. Were this accomplished satisfactorily, all physical science could be based solely on the fundamental laws of quantum mechanics.

But statistical mechanics goes farther than a derivation of the laws themselves. Macroscopic laws alone do not suffice to predict the behavior of a particular chemical system. Neon, water, iron, and sodium chloride behave differently in their equilibrium properties, although each obeys the thermodynamic laws. The fluid dynamics of glycerine differs from that of water or mercury. In order to use successfully the macroscopic laws for a specified system, a vast amount of experimental information must be available in addition to the bare laws themselves. For instance, the heat-capacity curve for the dilute gas plus the P-V-T equation of state are needed for prediction of the thermodynamic equilibrium properties of a single chemical substance. The task of statistical mechanics includes the computation of these data.

For a given chemical molecule composed of atoms of known mass number and charge, the laws of quantum mechanics presumably suffice to compute the energies and other characteristics of the allowed stationary quantum states. In principle, the forces between pairs of molecules, or even among triplets and quadruplets of molecules, as a function of their relative positions in the different stationary states, should also be calculable. Given this information, the task of statistical mechanics begins, and consists in predicting the behavior of any macroscopic system composed of these molecules, or mixtures of them, under any and all conditions.

Needless to say, this description of statistical mechanical procedure is ridiculously hypothetical. *Ab initio* calculations, even for very simple molecules, of the energies of excited stationary states are extremely difficult and usually imprecise. But, even for some moderately complicated molecules, spectroscopic information, coupled with some theory, yields adequately precise information about the excited states, namely, those of lower energy, needed in the statistical mechanical treatment. Although some attempts have been made to compute *ab initio* the forces between certain molecules, and a few of these have been quite successful, the situation with regard to our knowledge here is lamentable. Most frequently, the forces are assumed to be pairwise additive and to obey a simple two- (or sometimes three-) parameter equation. The parameters are then adjusted to fit experimental data from the macroscopic behavior, for instance, the density and temperature of the gas-liquid critical point.

Given these limitations, even if the purely mathematical complications of the statistical mechanical formalism were completely overcome, which they have not been, the pretentious boast above that the task of statistical mechanics is to explain all the behavior of bulk systems in terms of molecular behavior might seem very empty. Nevertheless, the challenge is there and successful advances are continuously being made.

1b. HISTORICAL PERSPECTIVE

In the development of classical nonrelativistic mechanics an important role is played by the concept of systems having only conservative forces which are the negative derivatives of a potential function dependent on the coordinates alone. For such an ideal system one of the most important of the integrals of motion, which remains constant in time, is the energy. The energy can quite naturally be divided into two additive terms. One of these is the kinetic energy or *vis viva* which is the sum or integral over the system of one-half the product of the mass times the linear velocity magnitude squared. The other is the potential energy. During the motion of such an idealized system, in the absence of time-dependent forces applied from the outside, both kinetic and potential energy generally change, but their sum remains constant. The solar system, consisting of the planets with their moons, moving around the sun, played a great role in the development of Newtonian mechanics. It corresponds quite closely to such an idealized model. However, in all real laboratory systems this idealized model is found to be inexact. In order to obtain precise agreement with experiment it is always necessary to introduce into the theory certain frictional forces which have the property of reducing the kinetic energy without a corresponding increase in the potential energy.

The investigations of Benjamin Thompson (1753–1814) (Count Rumford), in the eighteenth century, followed by the considerations of Julius Robert von Mayer (1814–1878), James Prescott Joule (1818–1889), and others, showed that the decrease in the mechanical energy of the system through frictional forces is always accompanied by a rise in the temperature of the system, or of parts of the system. A new quantity, foreign to mechanics, called heat, may be introduced and defined in such a way that the heat produced in the system is always equal to the mechanical energy lost through friction. By this inclusion of heat as a third form of energy, the mechanical statement that the energy of an isolated system remains constant with time retains its validity, and in this form the law of conservation of energy is known as the first law of thermodynamics.

Observations made by Robert Brown (1773–1858) in 1827 on particles of microscopic size suspended in solution showed that these are in a state of continual random motion, which suggested that the invisible atoms and molecules making up matter in bulk are not at rest. It is immediately obvious that, if this motion is real, the system of atoms and molecules composing bulk matter has associated with it energy in the form of kinetic and potential mechanical energy not different in kind from that associated with a macroscopic system.[(1)]

The assumption that the mysterious disappearance of the mechanical energy of a macroscopic system into the heat of its component parts, as a result of frictional forces, is merely the conversion of macroscopic mechanical energy into the submicroscopic mechanical energy of the

[(1)] Historically, it appears likely that many of the early workers in statistical mechanics, for instance, Boltzmann, were either not aware of or were uninfluenced by Brownian motion.

atoms and molecules is known as the kinetic hypothesis. In this theory thermal energy is no longer essentially distinct in kind from mechanical energy. The theory has been amply confirmed by the remarkable accuracy with which the properties of bulk matter can be predicted by its use.

This, then, is the fundamental step in the kinetic theory: to identify thermal energy with the mechanical energy of the molecule and *to assume that the forces on a molecular level are completely conservative.* Actually, the more important of these forces are the negative gradients of a potential function dependent on the coordinates alone. Additional weak magnetic forces play a role in some problems, but these too lead to conservation of mechanical energy. Thus the first law of thermodynamics is really an assumption of the kinetic theory, namely, that when the mechanics is examined in microscopic detail, all forces are such that mechanical energy is conserved. Thermal energy is the disorganized mechanical energy of the molecules.

The first great successes of the kinetic theory were due to the work of James Clerk Maxwell (1831–1879), who derived the velocity distribution of molecules in gases (1857–1859). The Austrian Ludwig Eduard Boltzmann (1844–1906) generalized the theory, and throughout the nineteenth century the emphasis in statistical mechanical theory was always on the behavior of gases. The American Josiah Willard Gibbs (1839–1903), however, developed at the turn of the century the first completely general treatment for expressing the thermodynamic properties of *any* chemical system in terms of the microscopic mechanical properties of the molecules. Gibbs' basis was of course *classical* mechanics and, as he was well aware, this led to grave discrepancies with experiment, particularly in that many heat capacities were known to be much smaller than could be predicted from any reasonable model of the atoms or molecules (Sec. 5e).

In the meantime, by using the reasoning method of statistical mechanics, it was recognized that the energy density in the electromagnetic radiation field in equilibrium with matter at nonzero temperature, the blackbody radiation, should increase indefinitely with the square of the frequency ν (see Sec. 11d). It actually shows a rapid exponential decrease at high frequencies. Max Planck (1858–1947) cut the Gordian knot in 1901 by assuming that mechanical oscillators could only emit and absorb radiation in units, $h\nu$, of energy. Albert Einstein (1879–1955) in 1907 argued that therefore the oscillators must exist only in energy states differing by discrete energies and showed that a system of "quantized" harmonic oscillators of one frequency had a heat capacity that decreased exponentially with the reciprocal temperature at low temperatures. Peter Debye (1884–1966) in 1912 derived that in a regular crystal of an element the low-frequency distribution of frequencies should be such as to explain the known proportionality of the heat capacity with T^3 rather than an exponential decrease with T^{-1}. The statistical mechanical formalism of Gibbs can readily be adapted to the existence of discrete energy levels. Thus all serious discrepancies between statistical mechanics and the observational facts were removed.

During the 1920s spectroscopists established excellent and precise tables of the stationary energy states of many simple gaseous molecules. Chemists, particularly William F. Giauque (1895–), recognized that these values permitted very precise computation of some thermodynamic properties, for instance, the standard free energies of perfect gases. Coupled with experimental equation-of-state data, and in many cases no more than the vapor-pressure curve was required, the absolute free energies and entropies of condensed phases were then known. These could be compared with values obtained by integration of the heat capacities extrapolated to 0 K and the assumption of the so-called third law of thermodynamics. Thus the third law was experimentally confirmed. In the next few decades a very considerable body of useful thermodynamic data of chemical substances accumulated, much of which depended wholly or in part on statistical mechanical calculations.

In the meantime many other successful applications had been made to the understanding of equilibrium systems. Pierre Curie (1859–1906) investigated the magnetic properties of many substances over a range of temperatures and had observed the abrupt transition (at the Curie point) from ferromagnetic to paramagnetic behavior before the beginning of the century. Pierre-Ernst Weiss (1865–1940) developed a classical theory for this behavior which, with rather minor modification introduced by Louis M. Brillouin attributable to quantum mechanics, is essentially correct, although partly empirical. Louis E. F. Neel (1904–) enlarged on the theory of ferromagnetism and discovered antiferromagnetism. W. F. Giauque described and later applied a method for using magnetic properties to obtain extremely low temperatures. P. Debye and others used statistical theory to explain the properties of materials containing polar molecules. P. Debye[(1)] and W. K. F. Hückel (1895–) formulated a theory in 1923 explaining the square root of concentration dependence of the logarithm of the activity coefficient of strong electrolytes. John Gamble Kirkwood (1907–1959) and his students, and independently J. Yvon (1905–) and later Max Born (1882–1969) and Herbert S. Green (1920–), developed a formal theory of the liquid state.[(2)]

(1) P. Debye and W. K. F. Hückel, *Phys. Z.*, **24,** 185, 305 (1923).

(2) J. E. Mayer contributed to this development.

At midcentury the term statistical mechanics was generally understood to refer to the treatment of equilibrium thermodynamic systems, although the earlier nineteenth century treatment, almost entirely limited to perfect gases, had had its greatest success in the treatment of diffusion, viscosity, and heat conduction. This development, which was greatly extended by the publication in 1939 of the book *Mathematical Theory of Non-Uniform Gases* by Sydney Chapman (1888–) and Thomas George Cowling (1906–) was generally referred to as kinetic theory.

In 1932 Lars Onsager (1903–) published his article on the reciprocal relations in irreversible thermodynamic processes. For small displacements from equilibrium, "fluxes" such as the mutual diffusion of different molecular species in a solution, or the flow of heat, are proportional to thermodynamic "forces" such as gradients in the concentration or temperature. Onsager gave a perfectly general formulation for any set of appropriately defined fluxes by which a set of reciprocal driving forces could be determined. With this prescription the matrix of the proportionality constants relating the fluxes to the various forces is always symmetric.

This stimulated the development in the 1940s of a whole new macroscopic formalism generally called the thermodynamics of irreversible phenomena, developed largely in the low countries, particularly by Ilya Prigogine (1917–) and S. R. de Groot (1916–). A new, rigorous, concise macroscopic theory presented an irresistible challenge to statistical mechanics. Kirkwood and his students and the Brussels school under Prigogine were among the first to work on the problem during the 1950s. A satisfactory formalism, although one that leaves imposing computational difficulties was later developed from the work of Herbert B. Callen (1919–), Melville Green (1922–), and Ryogo Kubo (1920–).

Two other new developments have appeared since midcentury. The first of these, although certainly consisting of the application of quantum mechanics to macroscopic phenomena, has used so little of the older formalism of statistical mechanics that it is seldom regarded as part of the subject. This is the extraordinary advance of solid-state theory, particularly in the realm of semiconductors. The other retains the name statistical mechanics and indeed is sometimes referred to by its practitioners as pure statistical mechanics or statistical mechanics for its own sake. It consists, on the one hand, of rigorous mathematical examination of what restrictions on the forces assumed and other features of the mathematical models of systems are necessary in order to derive results that agree with certain essential features of the behavior of physical systems. For instance, what mathematical limitations must one impose on the assumed mutual forces between molecules in order to obtain a system in which both volume and energy are finite and both increase with the number of molecules linearly in the limit of infinite number? On the other hand, others seek and examine models that show macroscopic phase transitions even when the restriction is made that the treatment remain exact and mathematically rigorous. These latter models are invariably of one- or two-dimensional systems. Most, but not all, of the workers in this branch are mathematicians or applied mathematicians. Some of the results are of interest to physical scientists. This is particularly true of the original example of the model first showing a mathematically rigorous phase transition, the treatment of the two-dimensional Ising model of ferromagnetism by Lars Onsager in 1931.[(1)]

In this book primary emphasis is on the formal methods of treatment for equilibrium real systems and on examples of results that can be obtained by these methods.

1c. PROBABILITY AND PROBABILITY DENSITY

The adjective statistical is used in statistical mechanics in the sense that probability dominates in the theoretical development. The emphasis and the problems, however, have comparatively little resemblance either to those in mathematical probability theory, or to the problems of economics and sociology where probability concepts are also widely used. In so far as the final result of a statistical mechanical problem in calculation of the value of some thermodynamic variable for a specified macroscopic system little is seen of probability in the answer. Because of the large number of molecules and probability that an observation should be made that differs by an experimentally detectable amount from the most probable one is too remotely small to be interesting, except as a mathematical curiosity. Within experimental error, average value, median value, and most probable value are the same; for practical purposes the value is a precise one. Nevertheless, the methods, concepts, and terminology of probability theory appear constantly.

The simplest example of probability is that of the distribution of an integer quantity in distinguishable packages. An example is N cases of oranges in a warehouse. The number n_i of oranges in the ith case, $1 \le i \le N$, can be counted. The *average* number, $\bar{n}$ or $\langle n \rangle$, of oranges per case is

$$\langle n \rangle = N^{-1} \sum_{i=1}^{i=N} n_i. \qquad (1c.1)$$

(1) L. Onsager, *Phys. Rev.*, **37,** 405 (1931).

Alternatively, one may count N_n, the number of cases having n oranges each. Necessarily, the total number of cases is

$$\sum_{n\geq 0} N_n = N, \tag{1c.2}$$

where the upper limit of the sum is infinite or, in practice, that n beyond which all N_n are zero. The probability $W(n)$ that a case selected at random will contain n oranges is

$$W(n) = \frac{N_n}{N}, \qquad \sum_{n\geq 0} W(n) = 1. \tag{1c.3}$$

The total number of oranges in the N cases is

$$M = \sum_{i=1}^{i=N} n_i = \sum_{n\geq 0} nN_n, \tag{1c.4}$$

and, from eqs. (1) and (3),

$$\langle n \rangle = \frac{M}{N} = \sum_{n\geq 0} nW(n). \tag{1c.5}$$

The probability quantity $W(n)$ described above has no cosmic significance. It applies to a very limited specific population, the N cases of oranges in a particular warehouse, and to no other problem. Probability theory gives a method of handling information. Fundamental information must be otherwise supplied. In statistical mechanics there is usually no difficulty in defining the population, and it is a reproducible one. All atoms of a given isotope of a given element are identical. In equilibrium problems the population is most frequently atoms or molecules of a specified species in a macroscopic system whose thermodynamic state is exactly specified. A typical problem is to find and use the probability that a given molecule is in a quantum state n in such a system.

The average of the square of the number n in a given population is not equal to $\bar{n}^2$ in the general case,

$$\langle n^2 \rangle = \sum_{n\geq 0}^{n\leq\infty} n^2 W(n), \tag{1c.6}$$

but is larger. Write $\langle n \rangle^2$ as a double sum over dummy indices n and m,

$$\langle n \rangle^2 = \sum_{n\geq 0} \sum_{m\geq 0} nmW(n)W(m). \tag{1c.7}$$

Since the sum of $W(n)$ or $W(m)$ is unity, we can write eq. (6) as

$$\langle n^2 \rangle = \sum_{n\geq 0} \sum_{m\geq 0} \tfrac{1}{2}(n^2 + m^2)W(n)W(m). \tag{1c.8}$$

Subtract eq. (7) from eq. (8) to obtain

$$\langle n^2\rangle-\langle n\rangle^2=\tfrac{1}{2}\sum_{n\geq 0}\sum_{m\geq 0}(n^2-2nm+m^2)W(n)W(m)\geq 0. \qquad (1c.9)$$

The probabilities $W(n)$ and $W(m)$ are either zero or positive, as is $n^2-2nm+m^2=(n-m)^2$. Every term is zero or positive, so the sum is nonzero positive unless *all* terms are zero. If for two values $k\neq k'$ both $W(k)$ and $W(k')$ are nonzero, then two terms $n=k$, $m=k'$ and $n=k'$, $m=k$ each have the nonzero positive value of $\frac{1}{2}(k-k')^2W(k)W(k')$ and the sum is nonzero. Only if just one value, k, has nonzero $W(k)$, can the sum be zero. Since the sum of all W's is unity, the value of $W(k)$ must be unity. All members of the population have this unique value, which of course is the value of $\langle n\rangle$,

$$W_{\langle n\rangle}=1 \qquad W_n=0 \qquad \text{if } n\neq\langle n\rangle \text{ when } \langle n^2\rangle=\langle n\rangle^2. \qquad (1c.10)$$

Nothing except notation really changes if members of the population are characterized by a set $n_1, n_2, \ldots$ of integers rather than a single integer n. Throughout this book we use boldface letters for vectors and for sets of more than one number. Use $\mathbf{n}$ for the set

$$\mathbf{n}=n_1, n_2, \ldots. \qquad (1c.11)$$

Define $W(\mathbf{n})=W(n_1, n_2, \ldots)$ as the probability that a random member of the population will contain n_1 objects of type 1, n_2 of type 2, and so on, or perhaps be in quantum state n_1 along the first degree of freedom, quantum state n_2 along the second, and so on. The sum over the whole range of $\mathbf{n}$ of the probabilities must be unity,

$$\sum_{\mathbf{n}}W(\mathbf{n})=1, \qquad (1c.12)$$

and the average of any function $F(\mathbf{n})$ of the set n of integers is

$$\langle F\rangle=\sum_{\mathbf{n}}F(\mathbf{n})W(\mathbf{n}). \qquad (1c.13)$$

The probability of observing n_1, independently of the values of all the other members $n_2, n_3, \ldots$, is just the sum of $W(\mathbf{n})$ over all the others with n_1 fixed,

$$W(n_1)=\sum_{n_2}\sum_{n_3}\cdots W(n_1, n_2, n_3, \ldots), \qquad (1c.14)$$

with of course a similar definition for the probability of any specific subset of $\mathbf{n}$. The quantity

$$W(n_1 \mid n_2, n_3, \ldots)=\frac{W(n_1, n_2, n_3, \ldots)}{W(n_2, n_3, \ldots)} \qquad (1c.15)$$

is called a conditional probability. It is the probability of observing the value n_1 when the values $n_2, n_3, \ldots$ *are fixed.* Its sum over n_1 is unity.

Frequently the variable or variables characterizing the members of the population are continuous, rather than an integer number, for instance, the velocity or the momentum of a free particle, or its position in a coordinate system. In this case we use a probability density and substitute integration for summation. If the continuous variables characterizing the population under consideration are $x_1, x_2, \ldots,$ we define the probability density $W(x_1, x_2, \ldots)$ by a limiting process, namely, that

> $W(x_1, x_2, \ldots)$ *is the limit as* $\Delta x_1, \Delta x_2, \ldots \to 0$ *of* $(\Delta x_1, \Delta x_2, \ldots)^{-1}$ *times the probability that the value of the first variable lies between* x_1 *and* $x_1+\Delta x_1$, *that of the second between* x_2 *and* $x_2+\Delta x_2$, *and so on.* (1c.16)

Especially if there is only a single variable x, a shorter definition is practical. Let $W_s(\leq x)$ be the probability that the variable is equal to or less than x. The probability density $W(x)$ is then,

$$W(x)=\frac{dW_s(\leq x)}{dx}. \tag{1c.17}$$

For multiple variables this scheme is often awkward. We prefer the wordier definition of eq. (16).

The equations of use are now essentially identical to those for the integer variables, with the replacement of summation over $\mathbf{n}$ by integration over $\mathbf{x}$. Use

$$\mathbf{x}=x_1,\ x_2,\ \ldots, \tag{1c.18}$$

$$d\mathbf{x}=dx_1\ dx_2\ dx_3\ \ldots, \tag{1c.19}$$

and the condition that the sum of all probabilities be unity is

$$\int d\mathbf{x}\ W(\mathbf{x})=1. \tag{1c.20}$$

The average value of any function $F(\mathbf{x})$ is

$$\langle F\rangle=\int d\mathbf{x}\ F(\mathbf{x})W(\mathbf{x}). \tag{1c.21}$$

The probability density of x_1, independently of the values of $x_2, x_3, \ldots,$ is

$$W(x_1)=\int dx_2\ dx_3 \cdots\ W(x_1, x_2, \ldots). \tag{1c.22}$$

The conditional probability that x_1 have a given value when $x_2, x_3, \ldots$ are fixed is

$$W(x_1 \mid x_2, \ldots) = \frac{W(x_1, x_2, \ldots)}{W(x_2, \ldots)}. \tag{1c.23}$$

A special case of the probability of an integer value occurs when the values of $W(n)$ for consecutive n-values are very close over the whole range of n. The similarity between the probability $W(n)$ and a probability density is even more obvious in this case.

The summation over n can then be asymptotically replaced by integration,

$$\sum_n W(n) \Rightarrow \int dn\, W(n), \qquad \text{if } |W(n+1) - W(n)| \ll 1 \text{ all } n. \tag{1c.24}$$

This is discussed more fully in Appendix II under the Euler–Maclaurin summation formula. However, there is one striking difference between $W(n)$ and $W(x)$. Any naturally occurring quantity n that takes only integer values must be dimensionless, as is the number of oranges, or a quantum number. A continuous quantity x representing an intrinsically continuous variable is most likely to have a physical dimension $[x]$. Since $\int dx\, W(x)$ is a pure number, the dimension of $W(x)$ is that of x^{-1}, or in the general case of more variables,

$$[W(x_1, x_2, \ldots)] = (x_1 x_2 \cdots)^{-1}, \tag{1c.25}$$

so that its numerical value depends on the units used.

If in a case of a single variable x we transform to a new variable $y(x)$, then

$$W(y(x)) = \frac{W(x)}{dy/dx}, \tag{1c.26}$$

which is consistent with eq. (25) and follows directly from eq. (16) or (17). In the case of several variables the factor is the Jacobian.

The average value $\langle x \rangle$ of the variable x is obtained by setting $F(x) = x$ in eq. (21),

$$\langle x \rangle = \int dx\, x W(x). \tag{1c.27}$$

The *most probable* value of x is that for which $W(x)$ is a maximum, which requires that

$$\left[\frac{dW(x)}{dx}\right]_{x = \text{most probable}} = 0 \tag{1c.28}$$

The *median* value of x is that which bisects the area under the $W(x)$ curve,

$$\int^{\text{median } x} dx\, W(x) = \tfrac{1}{2} \tag{1c.29}$$

In general the average value, the most probable value, and the median value are not identical. If, however, the function $W(x)$ is symmetric around its most probable value, they are of course identical. Also if $W(x)$ is not truly symmetric but is extremely sharply peaked at its most probable value, the three values may be numerically equal to high precision, although not identical.

1d. SPECIAL PROBABILITIES AND PROBABILITY DENSITIES

If in every toss of a coin the probability of a head is exactly $\frac{1}{2}$, then in N tosses the probability $W(n)$ of n heads is

$$W(n) = 2^{-N}\binom{N}{n} \equiv 2^{-N}\frac{N!}{(N-n)!\,n!}. \tag{1d.1}$$

This distribution of probabilities is known as a binomial distribution.

If N is very large, the probability $W(n)$ of eq. (1) is sharply peaked at its maximum value of $n=\frac{1}{2}N$. The ratio $R(\Delta n)$ of $W(\frac{1}{2}N\pm\Delta n)$ to the maximum value $W(\frac{1}{2}N)$ is

$$\begin{aligned} R(\Delta n) &= \frac{(\frac{1}{2}N)!}{(\frac{1}{2}N-\Delta n)!}\frac{(\frac{1}{2}N)!}{(\frac{1}{2}N+\Delta n)!} = \frac{\frac{1}{2}N(\frac{1}{2}N-1)(\frac{1}{2}N-2)\cdots(\frac{1}{2}N-(\Delta n-1))}{(\frac{1}{2}N+1)(\frac{1}{2}N+2)(\frac{1}{2}N+3)\cdots(\frac{1}{2}N+\Delta n)} \\ &= \frac{1}{1+2N^{-1}}\cdot\frac{1-2N^{-1}}{1+2(2N^{-1})}\cdot\frac{1-2(2N^{-1})}{1+3(2N^{-1})}\cdots\frac{1-(\Delta n-1)(2N^{-1})}{1+\Delta n(2N^{-1})}. \end{aligned} \tag{1d.2}$$

The logarithm of R is

$$\ln R(\Delta n) = \sum_{\nu=1}^{\nu=\Delta n}\{\ln[1-(\nu-1)(2N^{-1})]-\ln[1+\nu(2N^{-1})]\} \tag{1d.3}$$

Use the approximation for $\Delta n \ll \frac{1}{2}N$ that $\ln(1+x)=x$ to find

$$\begin{aligned} \ln R(\Delta n) &= -(2N^{-1})\left\{\left[2\sum_{\nu=1}^{\nu=\Delta n}\nu\right]-\Delta n\right\} \\ &= -(\Delta n)^2(2N^{-1}) = -\tfrac{1}{2}N\left(\frac{\Delta n}{\frac{1}{2}N}\right)^2 \end{aligned} \tag{1d.4}$$

If $2\Delta nN=\sqrt{2/N}$, the probability has decreased to $1/e$ times its maximum value, and at $\Delta n/N=\sqrt{2/N}$ it is $e^{-4}=0.018\ W_{\text{max}}$. A liter of gas at room temperature and pressure has $N\sim 10^{22}$. If it is in two connected exactly equal half-liter flasks, the probability that the numbers of molecules in the two flasks differ by more than 1 part in 10^{11} is small.

If in the single toss of a weighted coin the probability of a head is

$0<p<1$ (and of a tail $1-p$), then for N tosses,

$$W(n)=p^n(1-p)^{N-n}\binom{N}{n} \tag{1d.5}$$

If the probability p of eq. (5) is extremely small, and the total number of tosses correspondingly large, with pN small, and

$$pN=x \tag{1d.6}$$

held constant, then $W(n)$ of eq. (5) goes over into an asymptotic form known as a Poisson distribution of parameter x (S. D. Poisson, 1781–1840). For very small values of p we can readily see that $W(n)$ has appreciable values only for small values of n/N and, indeed, as we proceed to demonstrate in the limit $p \to 0$, also for $n^2/2N \ll 1$ becomes vanishingly small.

Differentiate the value of $\ln W(n)$ with respect to n:

$$\ln W(n)=n \ln p+(N-n)\ln(1-p)+\ln N!-\ln(N-n)!-\ln n!,$$

using $\ln(n+1)!-\ln n!=\ln(n+1)$. One finds

$$\Delta \ln \frac{W(n)}{\Delta n}=\ln\left[\frac{p(N-n+1)}{n(1-p)}\right], \tag{1d.7}$$

which is zero at $n_{\max}$, the value of n for which $W(n)$ has its maximum value. This occurs when the quantity under the logarithm is unity, or approximately at

$$n_{\max}=pN=x \tag{1d.8}$$

and

$$\lim_{p\to 0}\left[\frac{(n_{\max})^2}{2N}\right]=\tfrac{1}{2}px=0. \tag{1d.9}$$

Now using again $\ln(1-x)\cong -x$, one has

$$\begin{aligned}\ln\frac{N!}{(N-n)!}&=\ln N^n+\sum_{\nu=0}^{\nu=n-1}\ln\left(1-\frac{\nu}{N}\right)\\ &=\ln N^n-\sum_{\nu=0}^{\nu=n-1}\frac{\nu}{N}+0(N^{-2})\\ &\cong \ln N^n-\frac{n(n-1)}{2N}\cong \ln N^n.\end{aligned} \tag{1d.10}$$

Up to values of n well beyond $n_{\max}$, where $W(n)$ becomes negligible, we can approximate $N!/(N-n)!$ by N^n. In addition, use

$$1-p\cong e^{-p}$$

and

$$p(N-n) = x\left(1-\frac{n}{N}\right) \cong x$$

in eq. (2) to find

$$\lim_{p\to 0,\, pN=x} W(n) = \frac{x^n}{n!}e^{-x}. \tag{1d.11}$$

Since

$$\sum_{n\geq 0} \frac{x^n}{n!} = e^x,$$

the normalization $\sum_n W(n) = 1$ has survived the approximation.

The normalized Poisson distribution depends on a parameter x and is

$$W(n) = e^{-x}\frac{x^n}{n!}. \tag{1d.12}$$

Its most probable value is at $n = x$, and its average value is also $\langle n \rangle = x$, but it is not symmetric around this value as is the binomial distribution of eq. (1).

The idealized form for a continuous probability density with a single sharp peak at x_0 is a Gaussian (Karl Friedrich Gauss, 1777–1855),

$$W(x) = (2\pi\sigma^2)^{-1/2} \exp\left[-\frac{1}{2}\left(\frac{x-x_0}{\sigma}\right)^2\right]. \tag{1d.13}$$

For this function the normalization

$$\int_{-\infty}^{+\infty} dx\, W(x) = 1 \tag{1d.14}$$

holds only if the range of x is from $-\infty$ to $+\infty$. If the range is more limited, for instance, if necessarily $x \geq 0$, then still if x_0/σ is sufficiently large the normalizing factor of $(2\pi\sigma^2)^{-1/2}$ is still numerically quite an adequate approximation. The contribution to the integral of eq. (14) from negative x-values will then be negligible; for $x_0/\sigma > 3.85$ the error is less than 10^{-4} in the integral of eq. (14).

For this probability density function the average value, the most probable value, and the median value of x are identical and equal to x_0. The value of $\langle x^2 \rangle - x_0^2$ is σ^2 for the function (13). As σ approaches the zero value, the probability function becomes narrower and narrower and $\langle x^2 \rangle - x_0^2$ approaches the zero value.

The function of eq. (13) has

$$\frac{d^2 \ln W(x)}{dx^2} = -\sigma^{-2} \tag{1d.15}$$

for all values of x. If an empirically obtained $W(x)$ is very steep near its maximum value, one can often satisfactorily approximate it by eq. (13) using the second derivative of its logarithm at the most probable value of x to determine σ. This of course was essentially the procedure used in deriving eq. (4).

The probabilities that atoms or molecules are in a given quantum state are very different from either of the above forms. These probabilities have their maximum values at $n = 0$, the quantum number of the state at lowest energy, and decrease exponentially with energy as the quantum numbers increase.

When the probabilities or probability densities are expressed for a set $\mathbf{n} = n_1, n_2, \ldots,$ of numbers or for a set $\mathbf{x} = x_1, x_2, \ldots,$ of continuous variables, the probabilities are said to be *independent* if the probability of observing each member n_1 or n_2 or ... is independent of the values of all the others. In terms of the conditional probabilities [eq. (1c.15)], this means that $W(n_1 \mid n_2, \ldots)$ does not depend on $n_2, n_3, \ldots$. In this case $W(n_1, n_2, \ldots)$ is a product,

$$W(n_1, n_2, \ldots) = \underset{\text{independent probabilities}}{W_1(n_1) W_2(n_2) \cdots} \tag{1d.16}$$

where of course if the numbers $n_1, n_2, \ldots,$ refer to different kinds of quantum states, the functional form of the dependence of W_1 on n_1 may be very different from that of W_2 on n_2.

In general, the average of a product,

$$\langle F(n_1)G(n_2)\rangle = \sum_{n_1 \geq 0} \sum_{n_2 \geq 0} \cdots F(n_1)G(n_2)W(n_1, n_2, \ldots), \tag{1d.17}$$

is not equal to the product of the averages,

$$\langle F(n_1)G(n_2)\rangle \neq \langle F(n_1)\rangle\langle G(n_2)\rangle, \tag{1d.18}$$

but *is* equal for independent probabilities.

Probabilities and probability densities in sets of an extremely large number of variables $n_1, n_2, \ldots, n_N$ or $x_1, \ldots, x_N$ play an important role in statistical mechanical theory and in the manipulations. Actual use of such probability functions to obtain numerical results is usually completely impractical unless the probabilities are independent in single variables or in small subsets of variables. The simplifications are straight forward. For instance, in this case

$$\begin{aligned}
\langle F(n_1)G(n_2)\rangle &= \sum_{n_1\geq 0}\sum_{n_2\geq 0}\cdots\sum_{n_N\geq 0} F(n_1)G(n_2)W(n_1,\ldots,n_N) \\
&= \sum_{n_1\geq 0}\sum_{n_2\geq 0}\cdots\sum_{n_N\geq 0} [F(n_1)W_1(n_1)] \\
&\qquad\qquad \times [G(n_2)W_2(n_2)]W_3(n_3)\cdots \\
&= \left[\sum_{n_1\geq 0} F(n_1)W_1(n_1)\right] \\
&\qquad\qquad \times\left[\sum_{n_2\geq 0} G(n_2)W_2(n_2)\right]\prod_{\nu\geq 3}^{\nu=N}\left[\sum_{n_\nu} W_\nu(n_\nu)\right] \\
&= \langle F(n_1)\rangle\langle G(n_2)\rangle.
\end{aligned} \tag{1d.19}$$

CHAPTER 2

KINETIC THEORY

2a. PRESSURE OF A PERFECT GAS

Before discussing and using the general methods of statistical mechanics we present some of the simpler conclusions one can derive about the behavior of a perfect monatomic gas. A dilute gas at rather high temperature consists of individual molecules which possess kinetic energy of motion but which, on the average, are so far away from each other that they exert negligible force on each other. Stated somewhat more specifically, only a very small fraction of all the possible instantaneous positions of all the molecules corresponds to a total potential energy which is not infinitesimally small compared with the total kinetic energy of the system.

If the gas is monatomic, it is known from experience that at ordinary temperature the molecules possess no appreciable internal energy. In Chapter 7 it is found that this behavior is to be expected, and certain possible exceptions are noted. For the normal dilute monatomic gas the only important part of the total energy is the kinetic energy of translation of the atoms.

We attempt to predict some of the properties of an idealized system consisting of N identical point particles, each of mass m, exerting no forces on each other, and contained in a vessel of volume V, the walls of which reflect the striking molecules perfectly. N is assumed to be a very large number. Since no forces are operative, the potential energy must be independent of the positions of the particles, and is chosen as zero. The total energy E of the system is the kinetic energy, the sum of $\frac{1}{2}mv^2$ for all the particles. The properties of such a system may be expected to be very close to those of a dilute monatomic gas of an element of atomic weight N_0m, in which N_0 is Avogadro's number.

Later, the properties of this system are calculated without the introduction of any unnatural assumptions concerning the further characteristics of the molecules, but initially the calculation is carried out with the use of two entirely improbable assumptions. It is assumed, namely, that all the N

molecules have exactly the same magnitude of velocity v, and that they move only in the directions of the three principal cartesian axes, one-sixth of the molecules moving in the positive direction, and one-sixth in the negative direction, of each axis. If a molecule hits one of the walls, which are taken to be parallel to the coordinate planes, its velocity changes sign. It so happens that the equations derived under these assumptions are the correct ones and, since the method of derivation is illustrative of the more exact one to come, it appears to be excusable to use these assumptions for preliminary considerations.

The pressure exerted on a wall of the vessel is the force exerted normal to the wall per unit area. This force arises from, and is equal and opposite to, the change in momentum per second suffered by the molecules reflected from the wall. If one considers 1 cm^2 of wall perpendicular to the x-axis, it is clear that this section of the wall is struck in 1 sec only by molecules moving toward the wall along the x-axis and lying, at the beginning of the second, in a rectangular parallelepiped of length v along the x-axis having a cross-sectional area of 1 cm^2. If the density of molecules in all parts of the system is uniformly N/V, the number hitting the wall per square centimeter per second will be $vN/6\,V$. Each molecule striking has a momentum mv normal to the wall and, if the molecules are reflected after striking, the change of momentum per molecule will be $2\,mv$. The total change in momentum per square centimeter per second is the pressure,

$$P = 2mv\frac{vN}{6V},$$
$$PV = \tfrac{2}{3}N(\tfrac{1}{2}mv^2) = \tfrac{2}{3}E, \qquad (2a.1)$$

since the total energy E is $\frac{1}{2}Nmv^2$. The pressure-volume product PV is a constant for constant energy of the system and is proportional to the total energy.

We prefer, however, to relate the pressure-volume product to the more easily measured variable T, the temperature of the system, rather than to the total energy. Here a difficulty is encountered which is not connected with the particular system, but rather with the essentially complicated nature of the function T. It is later shown that two such systems as this, if brought into thermal contact so that energy can flow from one of them to the other, will come into equilibrium in such a way that the average kinetic energy per molecule is equal in the two systems. The qualitative definition of temperature is that it be equal in two systems in thermal equilibrium, and that it be higher in the system from which the energy flows than in the system that gains in energy if the two systems are brought into thermal contact. It follows that the energy per molecule, $\epsilon = E/N$, is a monotonic function of the temperature alone.

Anticipating this result of later considerations, one sees from eq. (1) that the pressure-volume product can be written

$$PV = Nf(T),$$

in which $f(T)$ is a monotonically increasing function of the temperature. Until some more specific definition of the temperature is available, one can proceed no further.

Actually, however, the temperature T has first been defined by just this equation, namely, by setting $f(T) = kT$, where $k = R/N_0$ is the gas constant per molecule, usually called the Boltzmann constant. Its numerical value is $k = 1.3806 \times 10^{-16}$ erg deg^{-1}. The definition of temperature is made by means of the perfect gas equation,

$$PV = \frac{N}{N_0} RT = NkT. \tag{2a.2}$$

By combining eqs. (1) and (2), a relationship between the kinetic energy per molecule and the temperature is found,

$$\tfrac{1}{2}mv^2 = \tfrac{3}{2}kT. \tag{2a.3}$$

We now remove the unnatural assumption of equal velocities and of only six directions of motion, but the justifiable assumption that the gas is completely isotropic, is kept.

Let us again calculate the pressure due to the collisions of the particles on the walls. The pressure is equal to the change in momentum per second of the particles hitting a unit area of wall. The element of wall considered is chosen normal to the x-axis and is of area 1 cm^2.

In general it would be unjustifiable to assume that every particle of velocity v_x, v_y, v_z hitting the wall is perfectly reflected, and leaves with the velocity $-v_x$, v_y, v_z. However, since isotropy has been assumed, and therefore in the assumed stationary state $N(v_x, v_y, v_z) = N(-v_x, v_y, v_z)$, where $N(v_x, v_y, v_z)$ is the number density of molecules in velocity space (see Sec. 1c), it follows that just as many molecules leave the wall with the velocity $-v_x$, v_y, v_z as hit with the velocity v_x, v_y, v_z, and the total change in momentum per second experienced by the molecules due to collision with the wall is the same *as if* the molecules were perfectly reflected. One may therefore, without loss of generality, calculate the total change in momentum per square centimeter per second as being the product $2mv_x$ times the number of molecules of velocity component v_x hitting 1 cm^2 of wall normal to the x-axis, per second, summed by integration over all values of v_x from zero to infinity.

A figure including all the vectors $\mathbf{v}$ with given components v_x, v_y, v_z, whose end points fall in the square centimeter of wall normal to the x-axis, is a parallelepiped of base 1 cm^2 and height v_x, the volume of which is v_x cm^3; see Fig. 2a.1. All the $v_x N(\mathbf{v})/V$ molecules of velocity $\mathbf{v}$ in this figure at any moment will strike the square centimeter of wall within the ensuing second. Integration over all values of v_y, v_z, [see eq. (1c.22)] gives $[v_x N(v_x)/V]\,dv_x$ as the number of molecules of x-components of velocity between v_x and $v_x + dv_x$ striking 1 cm^2 of wall normal to the x-axis per second. Multiplication of this by $2mv_x$, the change in momentum per molecule, and integration, gives for the total change in momentum per square centimeter per second, equal to the pressure P,

$$P = \frac{m}{V} 2 \int_0^\infty v_x^2 N(v_x)\,dv_x. \tag{2a.4}$$

Since $N(v_x) = N(-v_x)$, the integration from zero to infinity is equal to just half the integral from minus infinity to plus infinity. In view of eq. (10′), one obtains

$$PV = 2(\tfrac{1}{2} m \overline{v_x^2}) N, \tag{2a.5}$$

where $\overline{v_x^2}$ is the average value of v_x^2. Since isotropy has been assumed, $\overline{v_x^2} = \overline{v_y^2} = \overline{v_z^2}$, and their sum is $\overline{v^2}$, so that $\overline{v_x^2} = \overline{v^2}/3$, and

$$PV = \tfrac{2}{3}(\tfrac{1}{2} m \overline{v^2}) N = \tfrac{2}{3} E. \tag{2a.6}$$

As before, the pressure-volume product is found to be two-thirds of the total kinetic energy of the system.

Comparison of eq. (13) with the perfect gas equation (2) shows that

$$\langle \tfrac{1}{2} m v^2 \rangle = \tfrac{3}{2} kT, \qquad E = \tfrac{3}{2} NkT, \tag{2a.7}$$

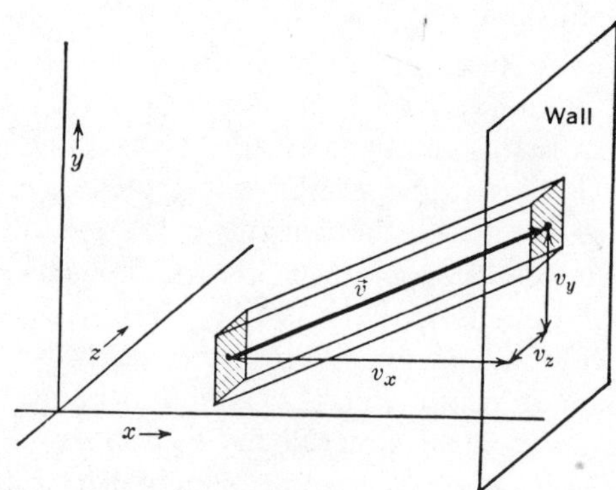

Figure 2a.1

which is similar to eq. (3) except that now the *average* kinetic energy per molecule is used instead of assuming that the kinetic energy is the same for all molecules.

2b. THE MAXWELL-BOLTZMANN DISTRIBUTION LAW

It is interesting to study more closely the distribution of molecules over the velocity ranges, that is, the functional dependence of $N(v_x, v_y, v_z)$ on its arguments. In this section, two proofs of the Maxwell-Boltzmann distribution law are given.

The first derivation, published originally by Maxwell in his first paper on the subject, does not consider the mechanism of collisions between the molecules. This proof, however, is not rigorous, since it is based on an assumption which should first be proved. Maxwell assumed that the distributions of the molecules among the components of velocity in the direction of the three coordinate axes are independent of each other; in other words, the probability that the x-component of the velocity has the value v_x is not influenced by the components in the other two directions. The number of molecules $N(\mathbf{v})$ of velocity $\mathbf{v}$ can then be expressed as a product of three functions of v_x, v_y, and v_z, alone. Since the space is assumed to be isotropic, these three functions must be the same and, moreover, $N(\mathbf{v})$ can depend only on the magnitude of velocity or, if we wish to write it so, on the square of the magnitude, $v_x^2+v_y^2+v_z^2=v^2$.

These two conditions lead to the relation

$$N(\mathbf{v})=f(v_x)f(v_y)f(v_z)=F(v^2). \tag{2b.1}$$

If $v_z=v_y=0$, then $v^2=v_x^2$ and, if the symbol a is used for the value of f when its argument is zero, $a=f(0)$,

$$a^2f(v_x)=F(v_x^2), \tag{2b.2}$$

or, by insertion into eq. (1),

$$F(v^2)=a^{-6}F(v_x^2)F(v_y^2)F(v_z^2). \tag{2b.3}$$

The functional relationship in eq. (3) is satisfied only if F is of the form $Ae^{-\alpha v^2}$. To show this, eq. (3) may be transformed into a differential equation by differentiating both sides with respect to v_y^2 and then setting $v_y=v_z=0$. The symbol α is defined by

$$\alpha=-a^{-3}\left[\frac{\partial F(v_y^2)}{\partial(v_y^2)}\right]_{v_y=0}. \tag{2b.4}$$

Since $F(0)$ is a^3, one obtains

$$\frac{\partial F(v_x^2)}{\partial(v_x^2)} = -\alpha F(v_x^2), \tag{2b.5}$$

or

$$f(v_x^2) = Ae^{-\alpha r_x{}^2}, \qquad N(\mathbf{v}) = Ae^{-\alpha v^2}. \tag{2b.6}$$

The value of the parameters A and α can be determined from the total number of particles N and the total kinetic energy E of the system. Before doing this we derive eq. (6) rigorously without making the assumption that the distribution of the molecules among the components of the velocity is independent.

In order to do this it is necessary to consider the collisions between the molecules in the system. In order to have collisions by which kinetic energy can be transferred from one molecule to another, it must be assumed that there are forces operative in the system. The magnitude of the forces, or the laws governing them, need not be known, but it is essential that they be negligibly small except at distances of approach between the molecules, which are very small compared to the average distances between them. Only under this condition is the potential energy negligible for all probable positions of the molecules in the system. Stated differently, it is important that at any instant an infinitesimal fraction of the molecules is in the process of undergoing a collision.

Consider one particular type of collision process and its reverse, namely, the process by which particles of the vectorial velocities $\mathbf{v}$ and $\mathbf{u}$ collide and emerge with the velocities $\mathbf{v}'$ and $\mathbf{u}'$, respectively. Since the sum of the kinetic energies of the particles must remain unchanged in the collision, the condition

$$v^2 + u^2 = v'^2 + u'^2 \tag{2b.7}$$

must be fulfilled.

The total number of times this process occurs in a second is called the rate of the process and must be proportional to the number of particles of velocity $\mathbf{v}$ and $\mathbf{u}$ present, that is, to the product $N(\mathbf{v})N(\mathbf{u})$. The rate of the reverse process in which particles of velocities $\mathbf{v}'$ and $\mathbf{u}'$ emerge with velocities $\mathbf{v}$ and $\mathbf{u}$ has to be proportional to $N(\mathbf{v}')N(\mathbf{u}')$. At equilibrium the rates of the two processes are equal.[(1)] We show that the proportionality constants entering into the two rates are also identical, from which the relation

$$N(\mathbf{v})N(\mathbf{u}) = N(\mathbf{v}')N(\mathbf{u}') \tag{2b.8}$$

[(1)] We are assuming complete reversibility, namely, that at equilibrium the rate of any process and that of its inverse are equal. The fact that this is generally true is discussed in Secs. 4b and 4c.

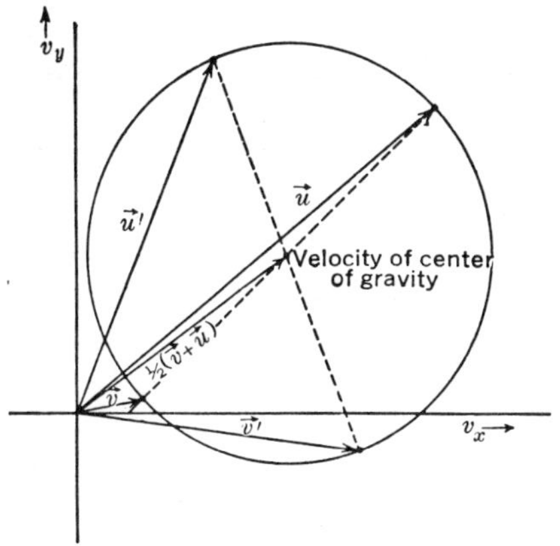

Figure 2b.1

between the equilibrium numbers of molecules of velocities $\mathbf{v}$, $\mathbf{u}$, $\mathbf{v}'$ and $\mathbf{u}'$ results. The only solution of eq. (8) with eq. (7) is eq. (6).

That the proportionality constants of the two rates in question must be equal, if the two particles are unaffected during the course of the collision by the walls or by the other particles of the system, may be demonstrated in the following manner. Take the point of view of an observer moving with the velocity of the center of gravity of the two particles, namely, $\frac{1}{2}(\mathbf{v}+\mathbf{u})=\frac{1}{2}(\mathbf{v}'+\mathbf{u}')$ (conservation of momentum; see Fig. 2b.1). To this observer the two processes, one that converts velocities $\mathbf{v}$ and $\mathbf{u}$ into $\mathbf{v}'$ and $\mathbf{u}'$, and another that converts $\mathbf{v}'$ and $\mathbf{u}'$ into $\mathbf{v}$ and $\mathbf{u}$, are exactly similar. In both cases two molecules moving with equal and opposite velocities of magnitude, $\frac{1}{2}|(\mathbf{v}-\mathbf{u})|=\frac{1}{2}|(\mathbf{v}'-\mathbf{u}')|$, approach, collide, and leave each other with velocities again oppositely directed and of the original magnitude. The angles at which the particles are deflected are the same for both particles and for the two processes. There is no conceivable cause, other than the effect of the other particles or the walls of the system,[(1)] that could make the two absolute rate constants differ. Equation (8) is thereby proved.

The functional relationship (8), together with eq. (7), has eq. (6) as its only solution. This can be shown by taking the special case that $\mathbf{v}'=0$, for which $u'^2=v^2+u^2$. Since the space is isotropic, so that $N(\mathbf{v})$ can depend only on the magnitude and not on the direction of $\mathbf{v}$, one may, as before, write $N(\mathbf{v})=F(v^2)$, obtaining

$$F(0)F(v^2+u^2)=F(v^2)F(u^2). \tag{2b.9}$$

[(1)] The influence of the distribution of the other molecules in the velocity space is the cause of the difference between the results of this consideration and that of Chapter 11 in which quantum mechanics is employed.

This equation has essentially the same nature as eq. (3), and may also be transformed into the differential equation (5) by differentiation with respect to u^2 and subsequent choice of $u = 0$.

It is readily seen that collisions between more than two particles do not change this result.

The number $N(|v|)$ of molecules with magnitude of velocity $|v|$, from eq. (6), is seen to take the form

$$N(|v|) = 4\pi v^2 N(\mathbf{v}) = 4\pi A v^2 e^{-\alpha v^2}. \tag{2b.10}$$

The two constants A and α are calculated from the total number of molecules N and the total kinetic energy E which is related to the temperature by eqs. (13) and (14). In performing these operations two definite integrals $\int_0^\infty v^2 e^{-\alpha v^2}\, dv$ and $\int_0^\infty v^4 e^{-\alpha v^2}\, dv$ are encountered. The transformation to the new variable, $z = \alpha v^2$, $dz = 2\alpha v\, dv$, leads to the forms

$$\int_0^\infty v^2 e^{-\alpha v^2}\, dv = \tfrac{1}{2}\alpha^{-3/2}, \qquad \int_0^\infty z^{1/2} e^{-z}\, dz = \frac{1}{4\alpha}\left(\frac{\pi}{\alpha}\right)^{1/2},$$

$$\int_0^\infty v^4 e^{-\alpha v^2}\, dv = \tfrac{1}{2}\alpha^{-5/2}, \qquad \int_0^\infty z^{3/2} e^{-z}\, dz = \frac{3}{8\alpha^2}\left(\frac{\pi}{\alpha}\right)^{1/2}.$$

The values of integrals of this sort are tabulated in Appendix AI.

The condition

$$N = \int_0^\infty N(v)\, dv = 4\pi A \int_0^\infty v^2 e^{-\alpha v^2}\, dv = A\left(\frac{\pi}{\alpha}\right)^{3/2} \tag{2b.11}$$

leads to

$$A = N\left(\frac{\alpha}{\pi}\right)^{3/2}.$$

or, with eq. (9),

$$N(v) = 4\pi N\left(\frac{\alpha}{\pi}\right)^{3/2} v^2 e^{-\alpha v^2}. \tag{2b.12}$$

The parameter α is necessarily positive, for otherwise the integration of eq. (10) could not have been performed. Indeed, a formula predicting an infinite number of molecules with infinite velocities is obviously nonsensical. For the evaluation of α the total kinetic energy is calculated by the use of eq. (2a.4) and compared with eq. (2a.3) which equates the average kinetic energy per molecule to $3kT/2$. The steps are:

$$E = \int_0^\infty \tfrac{1}{2} m v^2 N(v)\, dv = 2\pi m N\left(\frac{\alpha}{\pi}\right)^{3/2} \int_0^\infty v^4 e^{-\alpha v^2}\, dv = \frac{3mN}{4\alpha} \tag{2b.13}$$

and, combining the above with eq. (2a.7),

$$E = \tfrac{3}{2}NkT = \frac{3mN}{4\alpha},$$

$$\alpha = \frac{m}{2kT}. \tag{2b.14}$$

With this value of α the final form of the Maxwell-Boltzmann distribution law is

$$N(\mathbf{v}) = N(v_x, v_y, v_z) = N\left(\frac{m}{2\pi kT}\right)^{3/2} e^{-mv^2/2kT}, \tag{2b.15}$$

$$N(|v|) = 4\pi N\left(\frac{m}{2\pi kT}\right)^{3/2} v^2 e^{-mv^2/2kT}. \tag{2b.16}$$

The quantity in the exponent, $mv^2/2kT$, is the kinetic energy of the molecule divided by kT. It is found, in general, that, in dealing with molecules having internal energy, the exponential of the energy divided by kT always occurs in the expression for the distribution of molecules with respect to the energy.

If the gas as a whole moves with respect to the observer, that is, if the gas is streaming with the velocity $\mathbf{u}$, the velocities $\mathbf{v}$ of the individual particles will be distributed randomly about this prevalent velocity. In this case eq. (16) has the form

$$N(\mathbf{v}) = N(v_x, v_y, v_z) = N\left(\frac{m}{2\pi kT}\right)^{3/2} e^{-m(\mathbf{v}-\mathbf{u})^2/2kT}. \tag{2b.16'}$$

It is readily seen that then the average value of $\mathbf{v}$ is equal to $\mathbf{u}$.

2c AVERAGES OF VELOCITY

Average values of functions of the velocity vector, or of the velocity magnitude, may be found with the aid of the functions (2b.15) and (2b.16).

The function $N(\mathbf{v}) = N(v_x, v_y, v_z)$ (2b.15) is plotted in Fig. 2c.1 against the magnitude of the velocity. The function has a maximum at $v_x = v_y = v_z = 0$. If v_y and v_z are kept constant and $N(\mathbf{v})$ is plotted as a function of v_x, the resulting curve is proportional to the curve of $N(\mathbf{v})$ plotted against v and is symmetric in $+v_x$ and $-v_x$, decreasing from a maximum at $v_x = 0$ exponentially to zero on both sides. From this fact it is immediately obvious that the average value of v_x, namely,

$$\overline{v_x} = \frac{1}{N}\int_{-\infty}^{+\infty} v_x N(v_x, v_y, v_x)\, dv_x\, dv_y\, dv_z = 0, \tag{2c.1}$$

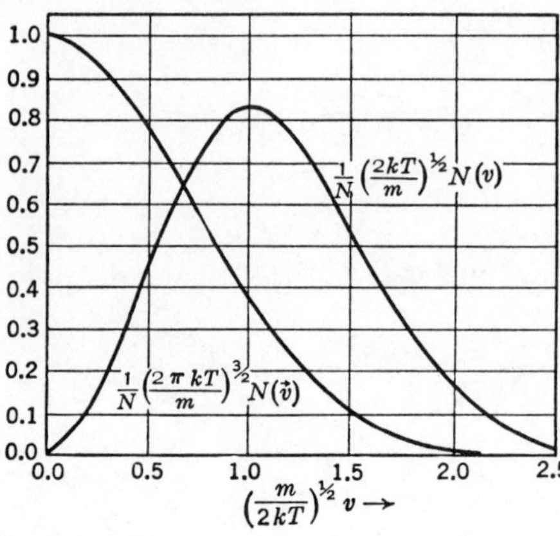

Figure 2c.1

since the integrand is positive for positive values of v_x and antisymmetrically negative for negative values of this variable. Indeed, the average value of any odd power of v_x vanishes. The average value of the velocity vector $\mathbf{v}$ is therefore $\bar{\mathbf{v}} = 0$. This fact is inherent in the assumption of isotropy, the assumption that no preferential direction exists.

The nature of the function $N(|v|)$, (2b.16) is quite different. This function is defined only for positive values of $|v|$, the magnitude of velocity. It, also, is plotted in Fig. 2c.1. It rises from 0 at $v = 0$ to a maximum at $v = v_m$ and goes asymptotically to zero as v goes to infinity. The velocity which corresponds to the maximum v_m is the most probable velocity magnitude or, briefly, the most probable speed. Its value is determined by the condition that

$$\left(\frac{\partial N(v)}{\partial v}\right)_{v=v_m} = 0$$

or

$$1 - \frac{mv_m^2}{2kT} = 0,$$

$$v_m = \left(\frac{2kT}{m}\right)^{1/2}.$$

The kinetic energy corresponding to the most probable velocity is kT.

The function $N(|v|)$ [eq. (2b.16)] may be used to calculate the average of any power of the magnitude of velocity. It is to be noted that the νth root of the average of the νth power of the velocity is *not* the same as the average velocity magnitude. For the average velocity magnitude $\langle |v| \rangle$, or speed,

$$\langle|v|\rangle = \frac{1}{N}\int_0^\infty vN(v)\,dv = 4\pi\left(\frac{m}{2\pi kT}\right)^{3/2}\int_0^\infty v^3 e^{-mv^2/2kT}\,dv$$

$$= 4\pi\left(\frac{m}{2\pi kT}\right)^{3/2}\left(\frac{2kT}{m}\right)^2\int_0^\infty x^3 e^{-x^2}\,dx$$

$$= \frac{4}{\pi^{1/2}}\left(\frac{2kT}{m}\right)^{1/2}\frac{1}{2}\int_0^\infty ye^{-y}\,dy,$$

$$\langle|v|\rangle = \frac{2}{\pi^{1/2}}\left(\frac{2kT}{m}\right)^{1/2} = 1.1283\left(\frac{2kT}{m}\right)^{1/2} \tag{2c.3}$$

is obtained. The average of the velocity squared is

$$\overline{v^2} = \frac{1}{N}\int_0^\infty v^2 N(v)\,dv = \frac{4}{\pi^{1/2}}\left(\frac{2kT}{m}\right)\frac{1}{2}\int_0^\infty y^{3/2}e^{-y}\,dy$$

$$= \frac{3}{2}\left(\frac{2kT}{m}\right). \tag{2c.4}$$

The root mean square velocity is the square root of this:

$$(\overline{v^2})^{1/2} = \left(\frac{3}{2}\right)^{1/2}\left(\frac{2kT}{m}\right)^{1/2} = 1.2247\left(\frac{2kT}{m}\right)^{1/2}. \tag{2c.5}$$

These various averages of the velocity are all proportional to $(2kT/m)^{1/2}$, but differ from each other and from the most probable velocity v_m in numerical factors not greatly different from unity. The $(\nu+1)$th root of the average of the $(\nu+1)$th power of the velocity is always greater than the νth root of the average of the νth power. In calculations into which averages of the velocity enter, care must be taken that the correct average is used. In general, the average value of any function $f(x)$ of the velocity is given by the integral $N^{-1}\int f(v)N(v)\,dv$, [eq. (1c.21)] if the average of a function involving the νth power of the velocity is sought, the average of this νth power, and *not* the νth power of the average velocity, must be taken. For instance, it is the root mean square velocity [eq. (4)] that gives the correct value of the average kinetic energy. Of course, the average velocity square [eq. (3)] may just as well be calculated from $N(v_x, v_y, v_z)$ by $\overline{v^2} = \overline{c_x^2} + \overline{v_y^2} + \overline{v_z^2}$ or, since $\overline{v_x^2} = \overline{v_y^2} = \overline{v_z^2}$, simply as $\overline{v^2} = 3\overline{v_x^2}$.

The numerical values of these average velocities are surprisingly high. One finds, for instance, that the average velocity $\bar{v}$ is 1750 m sec^{-1} for hydrogen at 273 K, and 425 m sec^{-1} for O_2 at the same temperature. These high velocities were once regarded as a severe objection to the theory, since they had to be reconciled with the observed low diffusion velocities. However, these velocities are completely random in direction. A molecule makes frequent collisions with others, so that the macroscopic velocity with which it progresses through the gas is very much smaller. The theory of the mean free path, defined in Sec. 2e, is able to predict the correct diffusion velocities (see Sec. 2h).

2d THE NUMBER OF MOLECULES HITTING A WALL

For the calculation of the pressure in Sec. 2c it was found that the number of molecules with x component of velocity between v_x and $v_x + dv_x$ striking a unit surface of a wall normal to the x-axis is $[v_x N(v_x)/V]\,dv_x$ per second. The total number Z of molecules striking the wall per square centimeter per second is then

$$Z = \frac{1}{V}\int_0^\infty v_x N(v_x)\,dv_x \tag{2d.1}$$

and from eq. (2b.16) giving $N(v_x, v_y, v_z)$,

$$Z = \frac{N}{V}\left(\frac{m}{2\pi kT}\right)^{3/2}\int_0^\infty\int_{-\infty}^\infty\int_{-\infty}^\infty v_x e^{-m(v_x^2+v_y^2+v_z^2)/2kT}\,dv_x\,dv_y\,dv_z, \tag{2d.2}$$

where the integration over dv_y and dv_z is extended from minus to plus infinity, but that over dv_x from only zero to plus infinity.

The transformation to new variables, $\zeta = (m/2kT)^{1/2}v_y$ and an analogously defined variable in place of v_z, changes the integral over dv_y and dv_z to the product of two integrals of the type $\int_{-\infty}^\infty e^{-\zeta^2}\,d\zeta$, each of which has the value $\pi^{1/2}$. Changing to $x = mv_x^2/2kT$, $dx = (mv_x/kT)\,dv_x$ transforms the integral over dv_x into an integral of the type $\int_0^\infty e^{-x}\,dx$, which has the value unity. One then obtains for Z,

$$\begin{aligned} Z &= \frac{N}{V}\frac{1}{\pi}\left(\frac{kT}{2\pi m}\right)^{1/2}\left(\int_{-\infty}^{+\infty} e^{-\zeta^2}\,d\zeta\right)\left(\int_0^\infty e^{-x}\,dx\right) \\ &= \frac{N}{V}\left(\frac{kT}{2\pi m}\right)^{1/2}. \end{aligned} \tag{2d.3}$$

The use of the perfect gas equation, $PV = NkT$, so that $N/V = P/kT$, enables one to express Z in terms of the pressure instead of the number of molecules per unit volume N/V,

$$Z = \frac{P}{(2\pi mkT)^{1/2}}. \tag{2d.4}$$

This, then, is the expression for the number of molecules Z striking a square centimeter of wall per second.

Numerical evaluation of the constants leads to

$$Z = 3.537 \times 10^{22} \frac{P_{\mathrm{mm}}}{(MT)^{1/2}} \qquad \mathrm{sec}^{-1}\ \mathrm{cm}^{-2}, \tag{2d.5}$$

in which P_{mm} is the pressure expressed in millimeters of mercury, and M is the molecular weight of the gas.

The quantity Z is also the number of molecules escaping per second per square centimeter of hole into a vacuum, provided that the pressure is sufficiently low, so that the mean free path of the molecules is much larger than the diameter of the hole.

It is seen from eq. (5) that in hydrogen, of molecular weight $M = 2$, at 1 atm pressure, $P = 760$ mm, and at room temperature, $T = 300$ K, Z has the value 1.1×10^{24} sec^{-1} cm^{-2}. The number of molecules hitting a square centimeter of wall per second corresponds to approximately 1.8 mol of gas.

2e THE MEAN FREE PATH

The average distance traveled by a molecule between collisions cannot be so clearly defined, or so unambiguously measured, as the pressure and average energy.

For a rough calculation we assume the molecules to act as rigid spheres. Two molecules of diameters d_1 and d_2, respectively, collide when the distance between their centers becomes equal to the sum of their radii, $\frac{1}{2}(d_1 + d_2)$. In considering the collisions one specified molecule of diameter d_1 undergoes we may therefore treat that molecule as a point particle, whereas the others are treated as having their diameter increased by d_1, that is, the molecule n as having the diameter $d_1 + d_n$.

Assume the molecule in question, the diameter d_1, moving in the x-direction, to be shot into a gas consisting of molecules of diameter d_2. Each of the gas molecules presents to the approaching one a target of diameter $d_1 + d_2$, and of area $\pi(d_1 + d_2)^2/4$. The number of such targets in a plate normal to the x-direction, of unit area and thickness Δx, is $(N_2/V)\,\Delta x$, where N_2 denotes the number of molecules of kind 2 in the system, and V the total volume. The total area covered by these targets, neglecting possible overlapping, is $[\pi(d_1 + d_2)^2 N_2/4V]\,\Delta x$. The probability that the incoming particle makes a collision in traversing the distance Δx is, then, the ratio of the surface covered by the targets to the total surface, namely,

$$\frac{\pi}{4}(d_1 + d_2)^2 \frac{N_2}{V} \Delta x.$$

The quantity

$$\frac{\pi}{4}(d_1+d_2)^2\frac{N_2}{V}=\frac{1}{l}, \tag{2e.1}$$

having the dimension of a reciprocal length, is abbreviated by the symbol $1/l$.

The significance of l is that it represents the mean free path, or the average distance traversed by a particle (of kind 1) before a collision (with a particle of kind 2). This can be seen in the following manner. Assume that a number n_0 of particles with velocities of approximately equal magnitude and direction, chosen as the x-direction, enter the gas at $x=0$. Each collision removes a particle from the beam, so that the number of particles $n(x)$ arriving at a distance x uniformly decreases. The decrease in n at a place x is equal to the number of molecules reaching that place multiplied by the probability per particle of a collision, namely,

$$-\frac{dn}{dx}\Delta x=\frac{1}{l}n(x)\,\Delta x.$$

This has the solution

$$n(x)=n_0e^{-x/l}. \tag{2e.2}$$

The distance x at which a particle makes a collision is called its free path. The mean free path is obtained by multiplying the path x by the number of particles colliding between x and $x+\Delta x$, summing over all ranges Δx, and dividing by the total number of molecules, namely,

$$-\frac{1}{n_0}\int_0^\infty x\frac{dn}{dx}\,dx=\int_0^\infty\frac{x}{l}e^{-x/l}\,dx=l, \tag{2e.3}$$

which identifies the quantity l with the mean free path.

At the distance $x=l$ the number of particles in the beam has been reduced to the fraction e^{-1} of the initial number, that is, more than half of the molecules have undergone a collision at some smaller value of x. The fact that l is nevertheless the mean free path comes about because, of the molecules reaching the place $x=l$, some go very far; a fraction e^{-2} of them goes further than $x=2l$ before a collision, a fraction e^{-3} further than $x=3l$, and so on.

This formula for the mean free path has been derived by assuming that one molecule is moving, whereas the others are practically at rest. If we use this expression for the motion of one gas molecule among others of the same kind, and therefore of the same average velocity, we obtain a different result. However, this amounts only to a small numerical factor, which is unimportant in view of the much graver assumption of rigid molecular diameters.

We find therefore within the accuracy of this argument, for a gas of one constituent,

$$l=\frac{V}{\pi d^2 N}. \tag{2e.4}$$

In a mixture of two gases the total number of collisions that one particle undergoes is composed additively of the number of collisions it suffers with each kind of particle. We find therefore, for the mean free path l_1 and l_2 of each kind of particle,

$$\begin{aligned} l_1 &= 4V(4\pi d_1^2 N_1 + \pi(d_1+d_2)^2 N_2)^{-1}, \\ l_2 &= 4V(\pi(d_1+d_2)^2 N_1 + 4\pi d_2^2 N_2)^{-1}. \end{aligned} \tag{2e.5}$$

To obtain an idea of the order of magnitude of the mean free path one can use the equation of state of the perfect gas to replace the density N/V in l by P/kT, obtaining

$$l=\frac{kT}{\pi d^2 P}. \tag{2e.6}$$

At room temperature, $T=300$ K, and if the diameter d is measured in ångströms, that is, in 10^{-8} cm, one finds

$$l=132(Pd_{\text{Å}}^2)^{-1}. \tag{2e.7}$$

This relation gives l in centimeters if the pressure is measured in cgs units, namely, in dynes per square centimeter. If P is measured in millimeters of mercury, the relation becomes

$$lP_{\text{mm.}}=0.1(d_{\text{Å}})^{-2}, \tag{2e.7′}$$

with l in centimeters. For atoms and simple molecules the diameter d is a few ångströms, so that the mean free path in millimeters times the pressure in millimeters is about one-tenth. With $d=5$ Å one obtains the following numerical results:

P(dynes cm^{-2})	P(mm of Hg)	P(atm)	l(cm)
1	7.5010×10^{-4}	9.8697×10^{-7}	5.3
1.332×10^{3}	1	1.3158×10^{-3}	3.9×10^{-3}
1.0132×10^{6}	760	1	5.2×10^{-6}

The average time between collisions is obtained by dividing the mean free path by the average velocity. At room temperature the velocity is of the order of 100 m sec^{-1}, so that the time between collisions at atmospheric pressure is about 10^{-10} sec. One molecule of a gas at standard conditions undergoes about 10^{10} collisions per second.

In this development the assumption furthest from the truth is the representation of the molecules as rigid spheres. It is due to this simplification that a mean free path independent of velocity, and therefore independent of temperature, was obtained. Actually, the molecules exert long-range attractive forces and short-range repulsive forces on one another. It is then obviously rather difficult to define a collision and a mean free path, since each particle is at any time interacting with others and is constantly suffering slight deflections of its path. This is borne out by experiments with sharply defined molecular beams[(1)] in an almost perfect vacuum. The effective cross section of the remaining gas particles appears then to be much larger than that calculated from gas kinetic data, since a very small deflection effectively removes a molecule from the beam. However, a very small deflection corresponds to a transfer of only a small amount of momentum and energy and is therefore of no importance for the transfer of heat or the viscosity of gases.

A better approximation for the expression of the mean free path was obtained by Sutherland[(2)] by representing the molecules as hard spheres, of diameter d_0, which, in addition, attract one another. An appreciable deflection of one molecule is obtained only if its sphere touches another one, and only in this case do we speak of a collision. If a fast molecule travels past another one at rest, it will be but slightly deflected. A slow molecule, however, approaching along the same line, may be deflected so much that it touches the other one, that is, it makes a collision. The effective cross section of a molecule in a collision depends then on the relative velocity, and the average cross section on temperature. One obtains

$$d^2 = d_0^2\left(1 + \frac{C}{T}\right), \tag{2e.8}$$

where the quantity C, the Sutherland constant, is determined by the nature of the attractive forces.

2f. VISCOSITY

The mean free path enters into the theories of all phenomena of propagation of physical properties over macroscopic distances. These are notably the transport of momentum, which is connected with the viscosity of gases; the transport of energy, or heat conduction; and the transport of mass, or diffusion. These three processes are treated here in a rather

(1) O. Stern, *Z. Phys.* **39,** 751 (1926).
(2) W. Sutherland, *Phil. Mag.* [V], **36,** 503 (1893).

crude manner. The averaging over different molecules is done somewhat incorrectly, so that numerical factors are quite untrustworthy. A more exact theory, however, becomes very complicated.[1]

The mechanical setup in an experiment for the determination of the viscosity of gases is usually idealized by a model such that the gas is contained between two parallel plane plates a distance a from each other. The plates may be taken to be parallel to the xy-plane and located at the height $z=0$ and $z=a$, respectively. The lower plate is kept at rest, while the upper one moves with a constant velocity u in the x-direction.

If the distance a between the plates is large compared to the mean free path, the gas "sticks" to the plates; near the upper plate, at $z=a$, the average velocity of the molecules is $\bar{v}_x = u$; near the lower one, at $z=0$, $\bar{v}_x = 0$. The average velocity at a height z between the two plates is denoted by $u(z)$ (compare end of Sec. 2b). Since, owing to the random motions of the particles, equally many molecules from above and from below reach the height z, the average velocity $u(z)$ is a linear function of the height, namely, $u(z) = uz/a$. If the mass of the gas particles is denoted by m, there exists a linear drop of average momentum,

$$G(z) = \frac{muz}{a}. \tag{2f.1}$$

Although equally many molecules from above and from below reach the height z during a second, the ones from above, on the average, bring with them a greater value of G than the ones from below. There is therefore a constant flow of momentum through any vertical plane. This flow, through a square centimeter per second, is denoted by $\Gamma(z)$ and will be calculated presently. If the flow in the positive z-direction is calculated, $\Gamma(z)$ will be negative.

The momentum *arriving* at the lower plate $-\Gamma(0)$ represents the force per square centimeter of surface that tends to move the lower plate in the same direction as the upper one. $-\Gamma(a)$ is the momentum lost to the gas by the upper plate per square centimeter of surface per second, or the frictional force counteracting the uniform motion.

The ensuing calculation of $\Gamma(z)$, the flow of the physical quantity G per second through a square centimeter parallel to the xy-plane at the height z, is done without making use of the special form of $G(z)$. The result may then be taken over immediately for cases in which any physical property $G(z)$ varies with height, illustrated in Fig. 2f.1.

[1] See for instance, S. Chapman and T. G. Cowling, *The Mathematical Theory of Non-Uniform Gases*, 2nd ed., Cambridge University Press, New York, 1952; or Ta-You Wu and Takashi Ohmura, *Quantum Theory of Scattering*, Prentice-Hall, Englewood Cliffs, N.J., 1962.

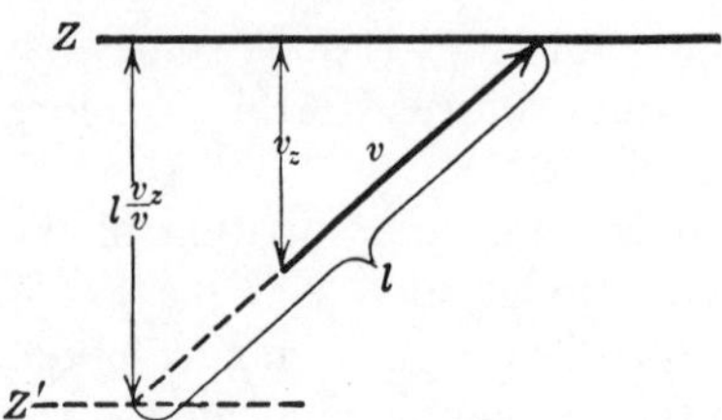

Figure 2f.1

The number of particles of velocity $\mathbf{v}=(v_x, v_y, v_z)$ that pass in a second through the square centimeter in question is, precisely as discussed in the calculation of the pressure in Sec. 2a, equal to the number of particles located at the beginning within a parallelepiped the base of which is the square centimeter and the length of which is $\mathbf{v}$. The height of the figure is therefore $|v_z|$, its volume $|v_z|$, and the average number of particles in it $|v_z| \times N(v_x, v_y, v_z)/V$. If $v_z > 0$, the particles cross the surface from below; otherwise they come from above. The *net* flow of particles through the square centimeter, that is, the surplus of particles going from below to above, is obtained by integrating $v_z \times N(v_x, v_y, v_z)/V$, without the absolute-value sign, over all velocities. In a stationary state the net flow of particles *must* be zero. If the velocity component $+v_z$ occurs just as frequently as $-v_z$, that is, if the variation of G with height does not influence the distribution of the z-component of the velocity, $\int_{-\infty}^{\infty} v_z N(v_x, v_y, v_z)\, dv_z$ is obviously zero.

The particles arriving at the height z have traveled in a straight line since they underwent their last collision. On the average, since that time, they have traversed the distance l, if l signifies the mean free path. The last collision of a particle with velocity $\mathbf{v}$ has therefore, on the average, occurred at a height z' which is given by $z' = z - (v_z l/v)$.

The assumption is now made that at that collision the particle has come into equilibrium with its surroundings. The average value of the quantity G which the particles of velocity $\mathbf{v}$, coming from the height z', bring with them is then

$$G(z') = G\left[z - \frac{v_z l}{v}\right] = G(z) - \frac{v_z l}{v}\frac{dG}{dz}. \tag{2f.2}$$

The net flow of G, namely, the difference of the amount of G carried up through the plane and the amount carried downward, is obtained by multiplying eq. (2) by the number $v_z N(\mathbf{v})/V$ of particles crossing the surface in the positive direction and integrating over all velocities,

$$\Gamma(z) = G(z)\iiint_{-\infty}^{\infty} \frac{N(\mathbf{v})}{V} v_z\, dv_x\, dv_y\, dv_z - l\frac{dG}{dz}\iiint_{-\infty}^{\infty} \frac{v_z^2 N(\mathbf{v})}{vV}\, dv_x\, dv_y\, dv_z. \tag{2f.3}$$

The first term, $G(z)$, times the excess of particles streaming through the element of plane in one direction, is zero in the stationary state. The second term may be simplified by considering that in an isotropic space, on the average, $v_x^2 = v_y^2 = v_z^2 = \frac{1}{3}v^2$. Actually, in the problem treated here, the velocity in the x-direction is a little different. However, the plate velocity u is very small compared to the gas kinetic velocities. One obtains then

$$\Gamma(z) = -\tfrac{1}{3} l \frac{dG}{dz} \frac{N}{V} \bar{v}. \tag{2f.4}$$

The minus sign in the formula shows that the flow takes place in the direction from higher to lower G values. If dG/dz is positive, the flow in the $+z$-direction must be negative.

In the special case of transport of momentum, according to (1), $dG/dz = mu/a$, one finds

$$\Gamma = -\frac{1}{3}\frac{N}{V} m l \bar{v} \frac{u}{a}. \tag{2f.5}$$

The frictional force per square centimeter of surface acting on the upper plate is usually written

$$F = -\frac{\eta u}{a};$$

η, the coefficient of viscosity, has, according to eq. (4), the value

$$\eta = \frac{1}{3}\frac{N}{V} m l \bar{v}. \tag{2f.6}$$

If eq. (2e.4) for the mean free path is inserted in eq. (6), one obtains

$$\eta = \frac{1}{3\pi}\frac{m\bar{v}}{d^2}. \tag{2f.7}$$

The average velocity $\bar{v}$ is, according to eq. (2c.3), $\bar{v} = 2(2kT/\pi m)^{1/2}$. This leads to

$$\eta = \frac{1}{3}\left(\frac{2}{\pi}\right)^{3/2} \frac{(mkT)^{1/2}}{d^2}. \tag{2f.8}$$

This equation predicts that the coefficient of viscosity is independent of the density, or the pressure, a function of temperature only. This result was first deduced theoretically by Maxwell and considered at the time to be rather startling. Subsequent experimentation confirmed the theoretical conclusion over a wide range of pressures. That the viscosity is independent of the pressure has since been regarded as a strong support of the kinetic theory.

Gibson,[1] for instance, measured the viscosity of hydrogen at 25°C in the pressure range from 11 to 295 atm. The viscosity η is 894×10^{-7} p (grams per centimeter per second) at 10.92, 12.66, and 15.28 atm. It then increases gradually to 901×10^{-7} at 60 atm., and 958×10^{-7} at 294.7 atm. The gradual increase does not exceed that expected from deviations from the perfect gas law, which are considerable at the higher pressures.

The fact that the density N/V drops out of eq. (8) comes about in the following manner. The number of particles arriving per second at the height z is proportional to the density. The mean free path, however, is inversely proportional to the density; at increasing density the molecules have made their last collision closer to the z-plane in question and therefore bring with them values of G more nearly equal to $G(z)$.

At very low pressures deviations from eq. (8) are observed; η begins to decrease. This is due to the fact that the assumption that the gas sticks to the plates becomes invalid when the mean free path is comparable to the distance a between the plates.

Equation (8) predicts further that the viscosity increases proportionally to $T^{1/2}$. Actually, a much stronger dependence on temperature has been observed. If, instead of a temperature-independent molecular diameter, the Sutherland approximation [eq. (2e.8)] is used, one obtains

$$\eta = \frac{1}{3}\left(\frac{2}{\pi}\right)^{3/2} \frac{(mkT)^{1/2}}{d_0^2\left(1+\dfrac{C}{T}\right)}, \tag{2f.9}$$

where the constant C, a function of the attractive forces, is unknown. If C is properly adjusted, satisfactory agreement between observed and calculated data is obtained.

2g. HEAT CONDUCTION

If there is a gradient of temperature in the z-direction, the average energy per molecule $\bar{\epsilon}$ will vary with height. We wish to calculate the flow of energy through a plane at the height z and therefore, in the equations of the previous section, have to replace the quantity G by

$$G(z) = \bar{\epsilon}(z), \qquad \frac{dG}{dz} = \frac{d\bar{\epsilon}}{dT}\frac{dT}{dz}. \tag{2g.1}$$

$d\bar{\epsilon}/dT$ is connected with the heat capacity C_V at constant volume. C_V is defined as the increase in energy with increasing temperature for a mole

[1] R. O. Gibson, Dissertation, Amsterdam, 1933.

of substance; that is, for N_0 molecules, if N_0 is Avogadro's number,

$$C_V = \frac{dE}{dT} = N_0 \frac{d\bar{\epsilon}}{dT}. \tag{2g.2}$$

This leads to

$$\frac{dG}{dz} = \frac{C_V}{N_0} \frac{dT}{dz}. \tag{2g.3}$$

By inserting eq. (3) into eq. (2f.4) the flow of heat through a square centimeter parallel to the xy-plane at the height z is found to be

$$\Gamma(z) = -\tfrac{1}{3} l \frac{N}{V} \frac{C_V}{N_0} \bar{v} \frac{dT}{dz}. \tag{2g.4}$$

This is usually written

$$\Gamma(z) = -\kappa \frac{dT}{dz}.$$

The heat conductivity κ is then

$$\kappa = \tfrac{1}{3} l \frac{N}{V} \frac{C_V}{N_0} \bar{v}. \tag{2g.5}$$

Comparison with eq. (2f.6) for the viscosity η, considering that mN_0 is the molecular weight M, gives the relation

$$\kappa = \frac{\eta C_V}{M}. \tag{2g.6}$$

$C_V/M = c_V$ is the specific heat per gram of substance, so that $\eta c_V/\kappa = 1$. A more exact theory still predicts this quotient to be constant but different from unity. In Table 2g.1 the quotient has been calculated for a few monatomic gases[(1)] from the values of κ and η given in Landolt-Börnstein. c_V was calculated from the energy relation (2a.7) and the molecular weight.

Table 2g.1[a]

Substance	$\kappa \times 10^5$	$\eta \times 10^5$	c_V	$\eta c_V/\kappa$
Helium	33.63	19.41	0.745	0.402
Neon	10.92	31.11	0.149	0.424
Argon	4.06	22.17	0.074	0.404

[a] κ in calories per centimeter per second per degree; η in cgs units; c_V in calories per degree.

(1) We have compared here data for monatomic gases only. For these gases the total energy is the kinetic energy of translational motion [eq. (2a.7)]. Polyatomic molecules possess, in addition, internal energy (Chapter 7). It is questionable whether this energy is readily transferred from one molecule to another in every collision.

2h. DIFFUSION

If two vessels connected by a tube with a stopcock are filled with two different gases, and if the stopcock is then opened, molecules will flow from the vessel with higher pressure to that of lower pressure. If pressure and temperature on both sides of the stopcock are equal, there will be no streaming of gas. However, owing to the random heat motion, particles of kind 1 will drift into the vessel which originally contained particles of kind 2 only, and vice versa, until finally both vessels are filled with a uniform mixture of the two gases. This phenomenon is called diffusion.

For calculation of the rate at which this process takes place an idealized experiment is considered. Assume a tube (of infinite diameter) to be filled with a mixture of two gases of kinds 1 and 2. Let the axis of the tube be the z-direction and assume that the composition of the mixture varies along z. The density of molecules of kind 1, that is, the number of molecules of kind 1 per cubic centimeter, which is denoted by n_1, and the density of molecules of kind 2, n_2, are then functions of z. If we stipulate that pressure and temperature are uniform throughout the vessel, the equation of state of the perfect gas demands that the total density of molecules, $n = n_1(z)+n_2(z)$, be constant everywhere. This leads to

$$n = n_1(z)+n_2(z), \qquad \frac{dn_1}{dz}+\frac{dn_2}{dz}=0. \tag{2h.1}$$

Since the distribution of velocities is independent of the density of the gas, the number of particles of each kind per cubic centimeter with a certain velocity $\mathbf{v}$ may be written

$$n_1(\mathbf{v}, z) = n_1(z)f_1(\mathbf{v}), \qquad n_2(\mathbf{v}, z) = n_2(z)f_2(\mathbf{v}), \tag{2h.2}$$

where f_1 and f_2 are independent of z. Indeed, $f_1(\mathbf{v})$ is essentially the Maxwell distribution function, only normalized in such a fashion that $\iiint_{-\infty}^{\infty} f_1(\mathbf{v})\, dv_x\, dv_y\, dv_z = 1$. The average magnitude of velocity of the particles of kind 1 is independent of z and given by

$$\bar{v}_1 = \iiint_{-\infty}^{\infty} |v| f_1(\mathbf{v})\, dv_x\, dv_y\, dv_z. \tag{2h.3}$$

A corresponding equation holds for molecules of kind 2.

The random motion of the molecules tends to bring about a uniform mixture of the gases. If $dn_1/dz > 0$, that is, if the concentration of particles of kind 1 is greater at larger height, there will be an excess of these particles streaming through a plane at the height z in the downward direction. The net flow of particles of kind 1 through an area of 1 cm^2 of the plane perpendicular to the z-axis at the height z in the direction of positive z, $\Gamma_1(z)$, will then be negative. It is usually written

$$\Gamma_1(z) = -D_1 \frac{dn_1}{dz}, \tag{2h.4}$$

where D_1 is the diffusion constant.

The flow is calculated in precisely the same manner as in Sec. 2f. A particle of kind 1 and velocity $\mathbf{v}$ arriving at the height z comes, on the average, from a height $z' = z - (v_z l_1/v)$, where l_1 is the mean free path of the molecules of kind 1. The density of such molecules at that height is

$$n_1(\mathbf{v}, z') = n_1(z') f_1(\mathbf{v}) = n_1(z) f_1(\mathbf{v}) - \frac{v_z}{v} l_1 \frac{dn_1}{dz} f_1(\mathbf{v}). \tag{2h.5}$$

The number of such particles crossing a square centimeter of a plane at the height z in the $+z$-direction per second is then

$$v_z n_1(\mathbf{v}, z') = n_1(z) v_z f_1(\mathbf{v}) - \frac{v_z^2}{v} l_1 \frac{dn_1}{dz} f_1(\mathbf{v}). \tag{2h.6}$$

The expression (6) is positive for all particles coming from below, that is, with $v_z > 0$; it is negative for those coming from above. The excess of particles going in the positive direction Γ_1 is therefore obtained by integrating eq. (6) over all velocities. The first term vanishes, as before, leading to

$$\Gamma_1(z) = -l_1 \frac{dn_1}{dz} \iiint_{-\infty}^{\infty} \frac{v_z^2}{v} f_1(\mathbf{v})\, dv_x\, dv_y\, dv_z. \tag{2h.7}$$

The integral is simply the average value of v_z^2/v. Since all directions of the velocity are equally probable, $\overline{v_z^2/v} = \frac{1}{3}\bar{v}$. The diffusion constant of the molecules of kind 1 is therefore

$$D_1 = \tfrac{1}{3} l_1 \bar{v}_1. \tag{2h.8}$$

Similarly one obtains, for the flow of particles of kind 2,

$$\Gamma_2 = -D_2 \frac{dn_2}{dz} \tag{2h.4′}$$

with

$$D_2 = \tfrac{1}{3} l_2 \bar{v}_2. \tag{2h.8′}$$

By inserting into these equations the value of the mean free path [eq.

(2e.5)], one obtains

$$D_1 = \frac{4}{3\pi}\bar{v}_1[4n_1d_1^2 + n_2(d_1+d_2)^2]^{-1}, \tag{2h.9}$$

$$D_2 = \frac{4}{3\pi}\bar{v}_2[n_1(d_1+d_2)^2 + 4n_2d_2^2]^{-1}. \tag{2h.9$'$}$$

The D's are inversely proportional to the density. If n is expressed with the help of the perfect gas law, $n = N/V = P/kT$, one obtains

$$D_1 = \frac{4}{3\pi}\frac{kT\bar{v}_1}{P}\left[4\frac{n_1}{n}d_1^2 + \frac{n_2}{n}(d_1+d_2)^2\right]^{-1}, \tag{2h.10}$$

$$D_2 = \frac{4}{3\pi}\frac{kT\bar{v}_2}{P}\left[\frac{n_1}{n}(d_1+d_2)^2 + 4\frac{n_2}{n}d_2^2\right]^{-1}. \tag{2h.10$'$}$$

It is seen that D_1 and D_2 depend on the relative concentrations n_1/n, n_2/n but not very strongly, since the diameters of the molecules are not vastly different. If the difference between the diameters is neglected, one obtains a relation between the diffusion constant of a gas and its viscosity coefficient,

$$D \cong \frac{V\eta}{Nm}.$$

This equation can at least be used to obtain the order of magnitude of D.

In general, the two diffusion constants D_1 and D_2 for the two kinds of particles are different, $D_1 \neq D_2$. It is then easily seen that the constants calculated in this manner cannot possibly be those that are observed in a closed tube. The total flow of particles in the $+z$-direction may be calculated by using relation (1),

$$\Gamma = \Gamma_1 + \Gamma_2 = -D_1\frac{dn_1}{dz} - D_2\frac{dn_2}{dz} = -(D_1 - D_2)\frac{dn_1}{dz}, \tag{2h.11}$$

and it is seen that this does not vanish. This means that the density of particles n, and therefore the pressure, do not remain constant throughout the tube but, if Γ is positive, increase in the upper part. This absurd result is usually corrected by assuming that on the calculated diffusion there is superimposed a uniform motion which just counteracts the increase in pressure. This uniform motion corresponds to a velocity $-\Gamma/n$ per particle, and a flow $-(n_1/n)\Gamma$ and $-(n_2/n)\Gamma$ of particles of kinds 1 and 2, respectively, through a square centimeter at the height z in the $+z$-direction.

If this flow is added to the one previously calculated, one obtains the corrected diffusion,

$$\Gamma_1^* = \Gamma_1 - \frac{n_1}{n}\Gamma = \frac{n_2}{n}\Gamma_1 - \frac{n_1}{n}\Gamma_2$$
$$= -\left(\frac{n_2}{n}D_1 + \frac{n_1}{n}D_2\right)\frac{dn_1}{dz} = -D^*\frac{dn_1}{dz}, \tag{2h.12}$$

$$\Gamma_2^* = -\left(\frac{n_2}{n}D_1 + \frac{n_1}{n}D_2\right)\frac{dn_2}{dz} = -D^*\frac{dn_2}{dz}. \tag{2h.12′}$$

The new diffusion constant D^* is equal for the two kinds of particles. The particles of kind 1 diffuse downward just as rapidly as those of type 2 diffuse upward, and the pressure remains constant.

It is seen that D^* depends on the composition of the mixture. The diffusion rate is therefore different at different heights in the tube.

PART TWO

Given in this part are the fundamental equations, the derivation, and some of the more general results of statistical mechanics using the ensemble concept of Willard Gibbs. Various paradoxes and their solution are discussed in Chapter 6.

CHAPTER 3

MECHANICS, PHASE SPACE, THERMODYNAMIC FORMALISM, AND ENSEMBLES

3a. CLASSICAL MECHANICS AND PHASE SPACE

The laws of nonrelativistic classical mechanics are used in statistical mechanics in hamiltonian form (William Rowan Hamilton, 1805–1865). Atoms are adequately assumed to be point masses and, since molecules are composed of atoms, they consist only of point masses fixed in a semirigid geometric structure with forces resisting displacements from an equilibrium configuration. All forces are conservative. There are no restraints that can make the system nonholonomic. These limitations greatly reduce the complexities of general Hamiltonian mechanics. The total number of degrees of freedom Γ is then equal to the total number of coordinates necessary to specify exactly the position of all mass points and, since the real world is three-dimensional, Γ is $3N$, with N the total number of *atoms* in the system. We may always use cartesian coordinates

x_i, y_i, z_i for the ith atom, $1 \le i \le N$, in which case the kinetic energy K is

$$K = \tfrac{1}{2} \sum_{i=1}^{i=N} m_i(\dot{x}_i^2 + \dot{y}_i^2 + \dot{z}_i^2), \tag{3a.1}$$

with m_i the mass of atom i and $\dot{x}_i = dx_i/dt$. If we neglect possible magnetic forces, the potential energy U depends only on the coordinates,

$$U = U(x_1, \ldots, z_N), \tag{3a.2}$$

and the energy E is the sum

$$E = K(\dot{x}_1, \ldots, \dot{z}_1) + U(x_1, \ldots, z_N). \tag{3a.3}$$

The momentum p_{xi} conjugate to the coordinate x_i is given by

$$p_{xi} = \frac{\partial K(\dot{x}_1, \ldots, \dot{z}_N)}{\partial \dot{x}_i} = m_i \dot{x}_i, \tag{3a.4}$$

$$\dot{x}_i = \frac{p_{xi}}{m_i}, \tag{3a.5}$$

so that the kinetic energy K can be written as a function of the momenta,

$$k(p_{x1}, \ldots, p_{zN}) = \tfrac{1}{2} \sum_{i=1}^{i=N} \frac{p_{xi}^2 + p_{yi}^2 + p_{zi}^2}{m_i}. \tag{3a.6}$$

The *Hamiltonian* is now the energy expressed as a function of the *coordinates* and the *momenta,*

$$H(x_1, p_{x1}, \ldots, z_N, p_{zN}) = K(p_{x1}, \ldots, p_{zN}) + U(x_1, \ldots, z_N). \tag{3a.7}$$

The equations of motion, which give the changes in coordinates and momenta with time, are now

$$\dot{x}_i = +\frac{\partial H}{\partial p_{xi}}, \qquad \dot{p}_{xi} = -\frac{\partial H}{\partial x_i}. \tag{3a.8}$$

With the x-component of force acting on point mass i designated by f_{xi},

$$f_{xi} = -\frac{\partial U}{\partial x_i}, \tag{3a.9}$$

the expression for $\dot{p}_{xi}$ is just Newton's law, $\dot{p}_{xi} = f_{xi}$, whereas the expression for $\dot{x}_i$ in eq. (8) is that of eq. (5).

Now the potential function $U(x_1, \ldots, z_{Nz})$ in cartesian coordinates may be inordinately complicated. Sometimes a transformation to a new coordinate set $q_1, \ldots, q_\nu, \ldots, q_\Gamma$ of the *same number,* $\Gamma = 3N$, of coordinates may greatly simplify the form of U. For instance, for a diatomic molecule in free space, transformation to three cartesian coordinates of the center of mass, the distance r between the two atoms, and two angles θ ϕ necessary to define the direction of the vector from atom 1 to atom 2, leaves the potential dependent on r alone. The cartesian coordinates are then, in general, functions of the new set,

$$x_i = x_i(q_1, \ldots, q_\nu, \ldots, q_\Gamma), \tag{3a.10}$$

and

$$\dot{x}_i = \sum_{\nu=1}^{\nu=\Gamma} \frac{\partial x_i}{\partial q_\nu} \dot{q}_\nu. \tag{3a.11}$$

This can be used with eq. (1) for the kinetic energy K to express it as a function of the new coordinates and their time derivatives,

$$K = K(q_1, \dot{q}_1, \ldots, q_\nu, \dot{q}_\nu, \ldots, q_\Gamma, \dot{q}_\Gamma). \tag{3a.12}$$

The momentum p_ν conjugate to the coordinate q_ν is now *defined* as

$$p_\nu = \frac{\partial K(q_1, \ldots, \dot{q}_\Gamma)}{\partial \dot{q}_\nu}. \tag{3a.13}$$

Solve these Γ equations, for $1 \leq \nu \leq \Gamma$, for the $\dot{q}_\nu$'s in terms of the p_ν's and insert in eq. (12) to find K as a function of the coordinates and momenta. The Hamiltonian $H(q_1, p_1, \ldots, p_\Gamma)$ is then the sum of the kinetic plus potential energy *expressed as a function of the coordinates and their conjugated momenta.*

The three important characteristics of this formalism are as follows.

1. The equations of motion retain the same form as eq. (8) for cartesian coordinates, namely,

$$\dot{q}_\nu = \frac{\partial H}{\partial p_\nu}, \qquad \dot{p}_\nu = -\frac{\partial H}{\partial q_\nu}. \tag{3a.14}$$

2. The dimension of the product $p_\nu q_\nu$ is always that of energy times time, ml^2t^{-1}, which is that of Planck's constant h, so that $p_\nu q_\nu/h$ is dimensionless. This dimensionality is readily seen from the definition [eq. (13)] of p_ν. Since K is an energy, p_ν has the dimension of energy times time divided by whatever is the dimension of q_ν.

3. Volume in phase space is unaltered by a transformation of coordinates. The 2Γ-dimensional space formed by the Γ coordinates and their Γ conjugate momenta is called the phase space. The Jacobian that transforms the volume element $dq_1 \cdots dp_\Gamma$ of one set of coordinates q_1, $p_1, \ldots, q_\Gamma$, p_Γ into that of another set q'_1, $p'_1, \ldots, q'_\Gamma$, p'_Γ is always unity. Let a closed $2\Gamma-1$-dimensional surface S include a volume W of dimension h^Γ given by the integral $\int \cdots \int dq_1\, dp_1 \cdots dp_\Gamma$ within the part of phase space enclosed by S. Now transform to a new coordinate system $q'_1, \ldots, q'_\Gamma$ and transform the surface S to the new closed surface S' in the new phase space. The volume W' enclosed by $\int \cdots \int dq'_1 \cdots dp'_\Gamma$ within S' is equal to W, $W' = W$. Thus if g is any physical quantity expressible as a function of coordinates and momenta, the two functions G and G' will be very different,

$$g = G(q_1, p_1, \ldots, q_\Gamma, p_\Gamma) = G'(q'_1, p'_1, \ldots, p'_\Gamma),$$

but the integrals of G and G' within the enclosed volumes will be equal,

$$\iint \underset{CS}{\cdots} \int dq_1 \cdots dp_1 G = \iint \underset{CS}{\cdots} \int dq'_1 \cdots dp'_1 G'. \tag{3a.15}$$

For a macroscopic system with a number N of atoms of the order of 10^{20} or more, the dimensionality of the phase space is larger than one is accustomed to in other fields of physical science. A point in this space, specified by giving the values of $6N$ numbers, gives a precise microscopic state of the system. If no fluctuating forces from outside the system are applied, this state precisely determines the state at any specified later time, according to the laws of classical mechanics, by integration of eq. (14). Paul Ehrenfest (1880–1933) designated this phase space the gamma space (γ-space) for gas space. In contrast, he called the phase space for a single molecule the mu space (μ-space) for molecule space which, if the molecule has n atoms has the more reasonable dimensionality $6n$. For a system consisting of N_m identical molecules of n atoms each, $N_a = nN_m$, the microscopic state of the *system* is determined by giving the position in μ-space of each of the N_m molecules.

An awkward feature is imposed by the nature of the mathematical manipulations required. The various extensive thermodynamic functions such as entropy or free energy are given classically by the logarithm of an integral of a function over the γ-space. To write such a function it is necessary to number the coordinates and momenta of the different atoms differently, x_i, y_i, z_i, p_{xi}, p_{yi}, p_{zi}, $1 \le i \le N_a$, which means effectively that we number the atoms. Now if there are N_α identical atoms of type α in the system, such a numbering will be meaningless in any physical sense. In the γ-space there are $N_\alpha!$ (see Appendix VI) different points, differing only in permutations of identical atoms of species α, which all correspond to the same physical situation. To each of these positions in γ-space there corresponds a diagram in the μ-space with N_α *numbered* points at differing positions, indicating which numbered atom occupied that position. Gibbs found that, in order to obtain results in which the entropy was proportional to the size of the system, as required in thermodynamics, it was necessary to divide the integrals by $N_\alpha!$ for each species α of atom.

The procedure seemed sensible, but arbitrary. In quantum mechanics an equivalent manipulation is found to be required by the generalization of the Pauli exclusion principle (Wolfgang Pauli, 1900–1958).

3b. QUANTUM MECHANICS

In a classical system the microstate is defined by giving the values of a point in γ-space $\mathbf{q}^{(\Gamma)}$, $\mathbf{p}^{(\Gamma)}$. The quantum-mechanical uncertainty principle that for each degree of freedom ν the inequality $\Delta p_\nu \, \Delta q_\nu \geq h$ holds and forbids us to describe the state so closely. Instead the state must be defined by giving a prescription other than the classical one.

A Schrödinger (Erwin Schrödinger, 1887–1961) function $\Psi(q_1, \ldots, q_\Gamma)$ of the Γ coordinates may be used to define the quantum-mechanical state of a system composed of point masses of zero spin. The function Ψ must be normalized, single-valued, and continuous. It may be complex, with the conjugate function designated by Ψ^*. To avoid the continuous use of $q_1, q_2, \ldots, q_\Gamma$ we shorten the notation by introducing a boldface $\mathbf{q}$ for a set of coordinates, sometimes specifying the number of single coordinates Γ by a superscript, $\mathbf{q}^{(\Gamma)}$, or for a molecule, $\mathbf{q}^{(\gamma)}$,

$$\mathbf{q}^{(\Gamma)} \equiv q_1, q_2, \ldots, q_\Gamma, \tag{3b.1}$$

$$d\mathbf{q}^{(\Gamma)} \equiv dq_1\, dq_2 \cdots dq_\Gamma. \tag{3b.2}$$

The real positive product $\Psi^*(\mathbf{q}^{(\Gamma)})\Psi(\mathbf{q}^{(\Gamma)})$ is the probability density in the coordinate space. That is, this product gives the probability density that a system in the quantum state specified by Ψ will be observed to have the set of coordinate values $\mathbf{q}^{(\Gamma)}$ or $\mathbf{q}^{(\gamma)}$.

Two different states, $\boldsymbol{\nu}$ and $\boldsymbol{\mu}$, are said to be *independent quantum states* if their state functions $\Psi_{\boldsymbol{\nu}}(\mathbf{q}^{(\Gamma)})$, $\Psi_{\boldsymbol{\mu}}(\mathbf{q}^{(\Gamma)})$ are orthogonal. This statement can be combined with the normalization condition that the probability density $\Psi^*\Psi$ integrate to unity in one equation. Use the Kronecker delta symbol $\delta(\boldsymbol{\nu} - \boldsymbol{\mu})$ which is defined to be unity for $\boldsymbol{\nu} \equiv \boldsymbol{\mu}$ and zero otherwise. The equation

$$\int d\mathbf{q}^{(\Gamma)} \Psi_{\boldsymbol{\nu}}^*(\mathbf{q}^{(\Gamma)}) \Psi_{\boldsymbol{\mu}}(\mathbf{q}^{(\Gamma)}) = \delta(\boldsymbol{\nu} - \boldsymbol{\mu}) \tag{3b.3}$$

gives the condition that the functions are normalized and orthogonal for sets of numbers $\boldsymbol{\nu}$, $\boldsymbol{\mu}$. These are called quantum numbers and identify independent states. For Γ degrees of freedom each quantum number $\boldsymbol{\nu}$ consists of a set of Γ single numbers,

$$\boldsymbol{\nu} = \nu_1, \nu_2, \ldots, \nu_\Gamma. \tag{3b.4}$$

A normalized linear combination of two independent state functions,

$$\Psi_{\mathbf{m}} = a_{\boldsymbol{\nu}}\Psi_{\boldsymbol{\nu}} + a_{\boldsymbol{\mu}}\Psi_{\boldsymbol{\mu}}, \qquad (a_{\boldsymbol{\nu}}a_{\boldsymbol{\nu}}^* + a_{\boldsymbol{\mu}}^* a_{\boldsymbol{\mu}} = 1), \tag{3b.5}$$

is also a state function. This is in striking contrast to classical mechanics. It makes no sense to say that an atom can be simultaneously at positions x_i and x_i', or partially at both.

To any classical function $F(\mathbf{q}^{(\Gamma)}, \mathbf{p}^{(\Gamma)})$ of the coordinates and momenta there corresponds a quantum-mechanical operator $\mathscr{F}$ obtained from F by replacing p_ν with $(h/2\pi i)(\partial/\partial q_\nu)$, paying respectful attention to the fact that $p_\nu q_\nu$ and $q_\nu p_\nu$ are not identical. One has $q_\nu p_\nu \Psi \to q_\nu (h/2\pi i)(\partial\Psi/\partial q_\nu)$, but $p_\nu q_\nu \Psi \to (h/2\pi i)[\partial(q_\nu\Psi)/\partial q_\nu] = (h/2\pi i)\Psi + q_\nu(h/2\pi i)(\partial\Psi/\partial q_\nu)$. This is the commutation relation that $p_\nu q_\nu - q_\nu p_\nu = h/2\pi i$. Use the symbol $\mathscr{F}^n$ to indicate consecutive operation by $\mathscr{F}$ n-times. The average value $\langle F^n\rangle$ of the nth power of F in a system in the state Ψ is given by the integral,

$$\langle F^n\rangle = \iint\cdots\int d\mathbf{q}^{(\Gamma)}\Psi * \mathscr{F}^n\Psi. \tag{3b.6}$$

If the state $\Psi_{\boldsymbol{\nu}}$ obeys the equation

$$\mathscr{F}\Psi_{\boldsymbol{\nu}} = f_{\boldsymbol{\nu}}\Psi_{\boldsymbol{\nu}}, \tag{3b.7}$$

with $f_{\boldsymbol{\nu}}$ a constant, then Ψ_ν is said to be an eigenfunction of $\mathscr{F}$; $\langle F\rangle = f_{\boldsymbol{\nu}}$ and $\langle F^n\rangle = f_{\boldsymbol{\nu}}^n$ so that $\langle F\rangle$ is a precise value. In the general case $\langle F^n\rangle \neq \langle F\rangle^n$ and $\langle F\rangle$ is an average value, unlike the classical case in which the microscopic state of precisely determined $q_1, p_1, \ldots, q_\Gamma, p_\Gamma$ always defines a precise value of $F(\mathbf{q}^{(\Gamma)}, \mathbf{p}^{(\Gamma)})$.

The operator $\mathscr{H}$ corresponding to the classical Hamiltonian in *cartesian* coordinates,

$$\mathscr{H} = \sum_{i=1}^{i=N} -\frac{h^2}{8\pi^2 m_i}\left(\frac{\partial^2}{\partial x_i^2} + \frac{\partial^2}{\partial y_i^2} + \frac{\partial^2}{\partial z_i^2}\right) + U(x_1, \ldots, z_N), \tag{3b.8}$$

is unique, having no ambiguous $q_\nu p_\nu$ or $p_\nu q_\nu$ terms in its classical function. If generalized coordinates are used, the classical $H(\mathbf{q}^{(\Gamma)}, \mathbf{p}^{(\Gamma)})$ may contain such terms. In this case the Hamiltonian operator $\mathscr{H}$ in the generalized coordinates can always be found by applying the coordinate transformation to eq. (8). The result is usually equivalent to writing the operator by using $\frac{1}{2}(p_\nu q_\nu + q_\nu p_\nu)$ in place of the ambiguous $p_\nu q_\nu$ or $q_\nu p_\nu$. The time dependence of the state $\Psi(t, q^{(\Gamma)})$ is given by the equation[1]

$$\frac{\partial\Psi(t, \mathbf{q}^{(\Gamma)})}{\partial t} = \dot{\Psi} = \frac{2\pi i}{h}\mathscr{H}\Psi. \tag{3b.9}$$

[1] Quantum mechanical equations are usually simplified by using the symbol $\hbar$ instead of h, $\hbar = h/2\pi$. In statistical mechanics the quantity h occurs naturally more often than $h/2\pi$. In the interest of uniformity we use h consistently.

Equation (9) can be integrated formally to write

$$\Psi(t, \mathbf{q}^{(\Gamma)}) = \left(\exp \frac{2\pi i}{h} \mathcal{H} t\right) \Psi(0, \mathbf{q}^{(\Gamma)})$$

$$\equiv \sum_{n=0}^{\infty} \frac{t^n}{n!} \left(\frac{2\pi i}{h} \mathcal{H}\right)^n \Psi(t=0, \mathbf{q}^{(n)}), \tag{3b.10}$$

so that, if the exact microscopic state function $\Psi(\mathbf{q}^{(\Gamma)})$ is known at a time t arbitrarily chosen as zero time, then, in the absence of fluctuating external forces, its precise state function is determined at any later time. In this sense quantum mechanics is deterministic.

The eigenstates of the Hamiltonian play a peculiarly important role in most statistical mechanical treatments. If for a state $\Psi_{\mathbf{K}}$

$$\mathcal{H}\Psi_{\mathbf{K}} = E_{\mathbf{K}}\Psi_{\mathbf{K}}, \tag{3b.11}$$

with $E_{\mathbf{K}}$ a number of dimension energy, then the state $\Psi_{\mathbf{K}}$ has a precise energy $E_{\mathbf{K}}$. Equation (11) is known as the time-independent Schrödinger equation, and eq. (9) as the time-dependent Schrödinger equation. From eq. (10), with eq. (11), we find that

$$\Psi_{\mathbf{K}}(t, \mathbf{q}^{(\Gamma)}) = \left(\exp \frac{2\pi E_{\mathbf{K}}}{h} it\right) \Psi_{\mathbf{K}}(0, \mathbf{q}^{(\Gamma)}), \tag{3b.12}$$

$$\Psi^*_{\mathbf{K}}(t, \mathbf{q}^{(\Gamma)}) = \left[\exp\left(-\frac{2\pi E_{\mathbf{K}}}{h} it\right)\right] \Psi^*_{\mathbf{K}}(0, \mathbf{q}^{(\Gamma)}). \tag{3b.12′}$$

In eq. (6) for the average $\langle F \rangle$ of any physical quantity the product of the two time exponentials in $\Psi * \Psi$ gives unity, and $\langle F \rangle$ is independent of time. These states are said to be stationary.

In Chapter 13, on quantum statistics, we discuss a formalism in which no restriction is made on energy eigenstates. Otherwise, when we refer to a quantum state, we usually implicitly assume that an eigenstate of a Hamiltonian is meant. The Hamiltonian may be, and usually is, if it is specified, an approximation to the true one for the system. The states are not truly stationary but transitions between different quantum states $\mathbf{K}$, $\mathbf{K}'$, ... of the same or very nearly the same energies occur. However, even if one were to use the exact Hamiltonian, the prediction that the state be stationary depends on the complete absence of any fluctuating force from the outside. This is obviously a fiction. Any laboratory system must be in a container. Even if the container is called part of the system and the surroundings are evacuated, there will be photon exchange with the surroundings. We then use eigenstates of a Hamiltonian, approximating as best we can that of the real system, to describe states labeled by a set $\mathbf{K}$

of Γ quantum numbers as possible states of a single system, knowing that transitions between different states occur in the real system.

The method of specification of the quantum numbers, and calculation of the corresponding energies in a macroscopic system, compose one of the main subjects of the various chapters of this book. If we refer to a small system of, say, $n = 1$, 2, or 20 or 30 atoms, and if they are confined to a volume of the order of a molecular volume, the spectrum of the allowed energy levels will be discrete, sometimes with rather large gaps of unallowed energies in the stationary states. Because of symmetries in the Hamiltonian there is frequently a number $g > 1$ of independent mutually orthogonal solutions to eq. (11) at one energy. At this energy there is an allowed energy *level* consisting of g states. The level has a degeneracy g. When the term state is used in this book we refer to a single *nondegenerate* state in which every quantum number is specified. For a macroscopic system the degeneracy or number of states at a given energy is found to have fantastic values like $10^{10^{20}}$, far larger than the numbers appearing elsewhere in any physical science.

In addition to the three coordinates necessary to describe the position of a point mass in space, the fundamental particles composing atoms, protons, neutrons, and electrons have a completely nonclassical degree of freedom, the spin. Quantum numbers of angular momentum take integer or half-integer values j. The angular momentum squared is $(h/2\pi)^2 j(j+1)$, and the projection of the angular momentum measured in units $h/2\pi$ along any unique axis in space can take values $-j, -j+1, \ldots, +j-1, +j$; in all, $2j+1$ values. The spin of a proton, neutron, or electron is an intrinsic angular momentum of quantum number $\frac{1}{2}$, which can then take two values, $+\frac{1}{2}$ and $-\frac{1}{2}$. An atomic nucleus of atomic mass A has a sum of protons plus neutrons equal to A, and an angular momentum of which at least part is due to their spin. It has a total angular momentum $(h/2\pi)S_n$. The value of S_n, if A is odd, is a half-integer, hence never zero; but if A is even, the nuclear spin is an integer and quite often zero. The energies of the $2S_n + 1$ different nuclear spin orientations may differ because of the magnetic interaction with the magnetic field of the electrons, but this direct effect, called the hyperfine splitting, is extremely small. For most problems the energies of the differing orientations can be considered equal.

The electronic states of an atom or molecule are characterized by angular momentum quantum numbers j, which are vector sums of the electron spins and electron angular momenta. Very frequently, but by no means always, this electronic angular momentum is zero in chemically stable molecules in their lowest energy level. If it is not the degeneracy, or number of states in this level, is $2j+1$.

Although various complications of considerable difficulty can and do occur, a very considerable fraction of systems composed of stable molecules can be treated adequately by considering only the degrees of freedom in which the atoms are treated as point masses. The total number of these degrees of freedom is $3N$, with N the total number of atoms composing the molecules of the system. A brief summary of the situation follows in the next section.

3c. THE MODEL SYSTEMS

In the last analysis whenever a detailed numerical calculation is made in statistical mechanics, the treatment is made of a mathematical model which is seldom if ever exactly a true replica of the real system it is intended to mimic. Sometimes the imitation is so close that there are no discrepancies of which we are aware. In other cases the model only imitates certain features of the physical system whose influence on the system's behavior we wish to understand. In some cases, indeed, the model is so far from reality that its behavior must be different in essential details from that of anything in the physical world. Nevertheless, from a study of the behavior of the model we may sometimes hope to obtain some insight into what happens in the real system. This is particularly true of many two-dimensional models.

Most frequently, we are concerned with atoms whose structure and excited quantum states are assumed to be known. The atoms of course consist of electrons and a nucleus, the latter made up of protons and neutrons. The nuclear structure is complex, and excited levels exist but, except in the treatment of stellar interiors or atomic bombs, the energy of excitation is so high that their existence can be ignored. The nucleus is then regarded as a single point mass sometimes possessing spin S_n and consisting of $2S_n+1$ states. Even this degeneracy can frequently be disregarded with impunity in the model, adding only a constant term to the absolute entropy. Except in metals, electronic excitation of the atoms again requires so high an energy that often only the lowest energy electronic level need be considered. Less often only one or two excited levels whose energy and angular momenta are taken from spectroscopic data need enter the model.

In the case of metals the outer valence electrons exist in a band of continuous energy, and low-energy excitation plays a role. The simplest model then treats the electrons alone, moving in the constant potential energy field of a fixed lattice of ions. Even so crude an approximation as assuming this potential to be constant in the volume of the metal explains many properties of metals. A more sophisticated model includes the interaction of the electrons with the vibrational modes of the ion lattice.

In a considerable number of cases where atoms have low-lying excited electronic energy levels, or even when the lowest level has nonzero angular momentum, the atoms do not occur free under normal laboratory conditions but form stable molecules or ionic crystals. The model used then treats the molecules or ions as the basis units. Molecules consisting of two or more atoms always have rotational energy levels, and often vibrational levels, of low enough energy to be excited at laboratory temperatures. Low-lying excited electronic levels are more rare. The allowed energy levels due to rotational and vibrational modes are generally taken from spectroscopic information and are often sufficiently precise to give excellent thermodynamic data for the gases. For condensed states, liquid or crystalline, the model usually used assumes that the mutual potential energies between different molecules are independent of the internal state and is therefore rather crude.

Many features of real systems can be handled piecemeal fairly satisfactorily by simple models. Ionic solution theory treats charged ions in water as though the water were a continuous medium of uniform dielectric constant only modifying the forces between the ions. The transition elements have shielded electronic levels with nonzero angular momenta and a consequent magnetic moment. The magnetic properties can be fairly well explained by assuming strong or weak empirical forces tending to align nearby magnets parallel or antiparallel. A large number of similar cases exist.

The models used seldom or never treat protons, neutrons, and electrons as the fundamental units. Nevertheless, to give results that could be expected to be near the behavior of real systems they must, particularly with regard to the symmetry properties discussed in Sec. 3e, be tailored to agree with the behavior of atoms or molecules composed of these units.

3d. SEPARABILITY OF THE HAMILTONIAN

If the classical Hamiltonian of a system consists of a sum of terms each dependent *only* on a subset of phase space coordinate momentum pairs none of which occur in another subset, the Hamiltonian is said to be separable. The classical motion in each of the subsets of the phase space is independent of the others. The quantum-mechanical stationary state functions are then a product of functions of the coordinate subsets. The quantum number **K** is given by the set of quantum numbers of the subset functions, and the energy is a sum of the energies of the subsets. In detail separability requires that we can write

$$\begin{aligned}\mathbf{q}^{(\Gamma)} &= \mathbf{q}_\alpha^{(\gamma\alpha)}, \mathbf{q}_\beta^{(\gamma\beta)}, \ldots, \\ \mathbf{p}^{(\Gamma)} &= \mathbf{p}_\alpha^{(\gamma\alpha)}, \mathbf{p}_\beta^{(\gamma\beta)}, \ldots,\end{aligned} \qquad \sum_\alpha \gamma_\alpha = \Gamma, \tag{3d.1}$$

in such a way that

$$H(\mathbf{q}^{(\Gamma)}, \mathbf{q}^{(\Gamma)}) = \sum_\alpha H_\alpha(\mathbf{q}_\alpha^{(\gamma\alpha)}, \mathbf{p}_\alpha^{(\gamma\alpha)}). \tag{3d.2}$$

The quantum-mechanical operator $\mathcal{H}_\Gamma$ is then obviously also a sum,

$$\mathcal{H}_\Gamma = \sum_\alpha \mathcal{H}_\alpha. \tag{3d.3}$$

The solutions, for each α, to the equations,

$$\mathcal{H}_\alpha \psi_{\mathbf{m}\alpha}(\mathbf{q}_\alpha^{(\gamma\alpha)}) = \epsilon_{\mathbf{m}\alpha} \psi_{\mathbf{m}\alpha}(\mathbf{q}_\alpha^{(\gamma\alpha)}), \tag{3d.4}$$

form a complete orthonormal set in the q_α-coordinate space. Their products,

$$\Psi_{\mathbf{K}}(\mathbf{q}^{(\Gamma)}) = \prod_\alpha \psi_{\mathbf{m}\alpha}(\mathbf{q}^{(\gamma\alpha)}). \tag{3d.5}$$

$$\mathbf{K} = \mathbf{m}_\alpha, \mathbf{m}_\beta, \ldots, \tag{3d.6}$$

form a complete set in the $\mathbf{q}^{(\Gamma)}$-space. Any function in this space can be written as a linear combination of these products. The simple product is a solution of the time-independent Schrödinger equation (3b.11) with energy

$$E_{\mathbf{K}} = \sum_\alpha \epsilon_{\mathbf{m}\alpha}, \tag{3d.7}$$

since $\mathcal{H}_\alpha$ operates only on the member $\psi_{\mathbf{m}\alpha}$ of the product in eq. (5), $\mathcal{H}_\alpha \Psi_{\mathbf{K}} = \epsilon_{\mathbf{m}\alpha} \Psi_{\mathbf{K}}$.

The simplest and most obvious example of a separable Hamiltonian is a perfect gas, in which one *assumes* there to be no mutual interaction potential between different molecules. The Hamiltonian of the system is the sum of the Hamiltonians of the individual free molecules. The α's of eq. (1) are the indices 1, 2, ..., N numbering the individual molecules, and the $\mathbf{q}_\alpha$, $\mathbf{p}_\alpha$ form the phase space of the individual molecule α, namely, its μ-space of $3n$ degrees of freedom with n the number of constituent atoms.

The Hamiltonian of the single free molecule is further separable. If the cartesian coordinates x, y, z of the center of mass are used, there are three translational terms H_x, H_y, H_z,

$$H_x = (\tfrac{1}{2}m)p_x^2, \qquad \mathcal{H}_x = -\frac{1}{8\pi^2 m}\frac{\partial^2}{\partial x^2}, \tag{3d.8}$$

and an internal Hamiltonian H_i of $3n-3$ coordinates. The latter in turn is at least approximately separable into rotational and vibrational terms for each electronic state if excited electronic states play a role.

As discussed in Sec. 3e, the simple products of eq. (5) in which each numbered molecule is assigned a quantum number $\mathbf{m}_\alpha$ are not allowed states of a real physical system. The allowed states are a linear combination of all those having the same number of identical molecules in a given quantum state $\mathbf{m}\alpha$. This follows from a generalization of the Pauli exclusion principle, as discussed in the next section.

3e. FERMI–DIRAC AND BOSE–EINSTEIN SYSTEMS

Pauli postulated the exclusion principle in order to explain the buildup of electronic states in the atoms of the periodic system. With the later understanding of the electron spin quantum number it takes the form that no two electrons can occupy the same fully defined quantum state. A more general statement is that the state functions of any system composed of fermions (Enrico Fermi, 1901–1954) is antisymmetric, that is, changes sign only in the coordinate exchange of any two identical fermions. The fermions with which we are concerned are protons, neutrons, and electrons all having spin quantum number $\frac{1}{2}$, which compose all ordinary laboratory matter.

To simplify the discussion consider a gas of N neutrons treated as a perfect gas. Number the neutrons using an index i, $1 \le i \le N$. In a rectangular box the quantum state $\mathbf{m}_i$ of the ith neutron is given by three cartesian translational quantum numbers k_x, k_y, k_z and one spin number s which can be $+\frac{1}{2}$ or $-\frac{1}{2}$. The state function for the gas corresponding to one of the products of (3d.5) is

$$\Psi_{\mathbf{K}\dagger}^{(0)} = \prod_{i=1}^{i=N} \psi_{\mathbf{m}i}(\mathbf{q}_i). \tag{3e.1}$$

If two neutrons i and j are in different quantum states $\mathbf{m}_i \neq \mathbf{m}_j$, $\Psi_{\mathbf{K}\dagger}$ has no symmetry in exchange of the pair i and j. We can make one and only one totally antisymmetric state from $\Psi_{K\dagger}^{(0)}$. There are, in all, $N!$ possible permutations of the coordinates. Use a symbol $\mathcal{P}_p$, with $0 \le p \le N! - 1$, for an operator which carries out one of the permutations of coordinates. The operator $\mathcal{P}_0 = 1$ is the identity operator that permutes nothing. The index number p is odd if an odd number of pairs is permuted and is even if an even number are exchanged. The function

$$\Psi_{\mathbf{K}}^{(A)} = A \sum_{p=0}^{p=N!-1} (-1)^p \mathscr{P}_p \Psi_{\mathbf{K}\dagger}^{(0)}, \tag{3e.2}$$

if it is nonzero, is antisymmetric in exchange of the coordinates of any pair. The number A is a normalization constant. For the product $\Psi_{\mathbf{K}\dagger}^{(0)}$ of eq. (1) it is convenient to write eq. (2) as a Slater determinant (John C. Slater, 1900–   ),

$$\Psi_{\mathbf{K}}^{(A)} = (N!)^{-1/2} \begin{vmatrix} \psi_{\mathbf{m}1}(\mathbf{q}_1) & \psi_{\mathbf{m}1}(\mathbf{q}_2) & \cdots & \psi_{\mathbf{m}1}(\mathbf{q}_N) \\ \psi_{\mathbf{m}2}(\mathbf{q}_1) & \psi_{\mathbf{m}2}(\mathbf{q}_2) & \cdots & \psi_{\mathbf{m}2}(\mathbf{q}_N) \\ \cdot & \cdot & \cdots & \cdot \\ \cdot & \cdot & \cdots & \cdot \\ \cdot & \cdot & \cdots & \cdot \\ \psi_{\mathbf{m}N}(\mathbf{q}_1) & \psi_{\mathbf{m}N}(\mathbf{q}_2) & \cdots & \psi_{\mathbf{m}N}(\mathbf{q}_N) \end{vmatrix}. \tag{3e.2'}$$

The single particle functions $\psi_{\mathbf{m}}$, $\psi_{\mathbf{m}'}$ are orthogonal if $\mathbf{m} \neq \mathbf{m}'$, so that, if no two neutrons have the same quantum state, all $N!$ permutations $\mathscr{P}_p \Psi_{\mathbf{K}\dagger}^{(0)}$ will be orthogonal, and the normalization factor A is $(N!)^{-1/2}$ as written in eq. (2′). If, however, two neutrons have identical quantum numbers, including spin, $\mathbf{m}_i = \mathbf{m}_j$, $i \neq j$, then, since the determinant has two equal rows, $\Psi_{\mathbf{K}}^{(A)}$ will be zero. The state of the system is forbidden.

In the more general case that the fermions interact and do not form a perfect gas one frequently transforms to coordinates that are linear combinations of the individual particle coordinates. For example, in the treatment of a molecule we use rotational and vibrational coordinates. The state function $\Psi_{\mathbf{K}\dagger}^{(0)}$ is no longer the simple product of eq. (1). Nevertheless, the permutation operators $\mathscr{P}_p$ carry out some definite transformation of the coordinates $\mathscr{P}_p \mathbf{q}_0^{(\Gamma)} = \mathbf{q}_p^{(\Gamma)}$, and

$$\mathscr{P}_p \Psi_{\mathbf{K}\dagger}^{(0)}(\mathbf{q}_0^{(\Gamma)}) = \Psi_{\mathbf{K}\dagger}^{(0)}(\mathbf{q}_p^{(\Gamma)}). \tag{3e.3}$$

The function $\Psi_{\mathbf{K}}^{(A)}$ defined by eq. (2) is still antisymmetric in all pair exchanges if it is nonzero.

In such a case $\Psi_{\mathbf{K}\dagger}^{(0)}$ may already be antisymmetric in some pair exchanges so that fewer than $N!$ independent states $\mathbf{K}\dagger$ are involved in the summation of eq. (2). The normalization factor A is not always $(N!)^{-1/2}$. If, however, $\Psi_{\mathbf{K}\dagger}^{(0)}$ is symmetric in any pair exchange of identical fermions, then $\Psi_{\mathbf{K}}^{(A)}$ of eq. (2) is zero. In this case for every even permutation there exists an odd permutation of opposite sign which is the same function. The sum of eq. (2) cancels pairwise to zero.

Now consider a perfect gas composed of identical atoms or of molecules. These in turn consist of neutrons, protons, and electrons, all of which are fermions. The state of each molecule is given by translational quantum numbers k_x, k_y, k_z and by a set of internal quantum numbers $\boldsymbol{\nu}$. The single molecule function $\psi_{\mathbf{m}i}(\mathbf{q}_i)$ is a product,

$$\psi_{\mathbf{m}i}(\mathbf{q}_i) = \psi_{kxi}(x_i)\psi_{kyi}(y_i)\psi_{kzi}(z_i)\psi_{\nu i}(\mathbf{q}_i^{\text{internal}}). \tag{3e.4}$$

The permutation of identical fermions within the molecule does not change the center-of-mass coordinates x_i, y_i, z_i, so the translational factor in eq. (4) is necessarily totally symmetric in all such permutations. The internal state $\psi_{\nu i}(\boldsymbol{q}_i^{\text{internal}})$ must therefore be antisymmetric in all pair exchanges of identical constituents.

The product of eq. (1) is now a solution of the time-independent Schrödinger equation but must be made antisymmetric in *all* pair exchanges of identical fermions. Suppose that the total number of fermions in each molecule is μ,

$$\mu = \text{number of electrons plus neutrons plus protons.} \tag{3e.5}$$

When we exchange two identical molecules we exchange μ pairs of identical fermions. The allowed gas functions must be multiplied by $(-1)^\mu$. If μ is odd, the function must be antisymmetric. Equation (2) or (2′) gives the allowed function $\Psi_{\mathbf{K}}^{(A)}$. The system is a Fermi–Dirac system (P. A. M. Dirac, 1902–   ). Molecules for which μ is even, however, require a function that is totally symmetric in all pair exchanges, namely,

$$\Psi_{\mathbf{K}}^{(S)} = A \sum_{p=0}^{p=N!-1} \mathscr{P}_p \Psi_{\mathbf{K}\dagger}^{(0)}. \tag{3e.6}$$

The system is a Bose–Einstein system. The molecules are said to be bosons (S. N. Bose, 1894–   ).

For neutral molecules the numbers of protons and electrons are equal, so that the system is Fermi–Dirac if the total number of neutrons is odd, and Bose–Einstein otherwise. The particles that form Fermi–Dirac systems have always a half-integer total angular momentum quantum number, including the nuclear spin. Bosons have integer angular momenta. In addition to atoms and molecules of even μ-value, photons form a Bose–Einstein system.

In each perfect gas the total quantum number $\mathbf{K}$ of the allowed states of the gas is given by specifying the number $n(\mathbf{m})$ of molecules in each of the completely specified quantum states $\mathbf{m}$ of the individual molecules. For Fermi–Dirac systems only $n(\mathbf{m})$ equal to 0 or 1 is allowed. For Bose–Einstein systems $n(\mathbf{m})$ can have any integer value 0, 1, 2, ... For those states $\mathbf{K}$ for which $n(\mathbf{m})$ is 0 or 1 for all m, exactly $N!$ independent products [eq. (1)] form a *single* allowed state $\mathbf{K}$ of the correct symmetry. For Fermi–Dirac systems those with any $n(\mathbf{m})>1$ are excluded, but for Bose–Einstein systems fewer of these than $N!$ are necessary to form an allowed symmetric state. Thus these states occur more frequently than $1/N!$ times the number of solutions [eq. (1)] of the time-independent Schrödinger equation.

For systems in which the number of molecules per molecular quantum state is small,

$$\langle n(\mathbf{m})\rangle \ll 1, \qquad \text{all } \mathbf{m}, \tag{3e.7}$$

it is very closely correct to count the number of products [eq. (1)] and divide by $N!$ to find the number of allowed states in a given energy range. In the general case of different species α the division is by $\prod_\alpha N_\alpha!$. This system of counting is known as Boltzmann counting. It falls between that for Fermi–Dirac and Bose–Einstein systems. Fermi–Dirac systems have fewer, and Bose–Einstein systems more, allowed states. However, only for gases in helium at temperatures near 0 K is the error appreciable. Two other cases in which Boltzmann counting is invalid are the treatment of electrons in metals and blackbody radiation or equilibrium photon gas. In these cases the correct quantum counting is necessary.

In the case in which the quantum numbers $\mathbf{K}^\dagger$ are not even approximately given by a set of single particle states $\mathbf{m}_i$, the correct symmetrization can be complicated in detail. One can, however, show that, asymptotically in an energy range containing a sufficient number of solutions $\Psi_{\mathbf{K}^\dagger}$ of the time-independent Schrödinger equation, only the fraction $(\prod_\alpha N_\alpha!)^{-1}$ has the correct symmetry. This is discussed briefly under symmetry number in Chapter 7, Sec. 7m, and more extensively but more abstractly in Chapter 13 on quantum statistics.

Several remarks are quite obvious but very important. First, all meaningful physically measurable functions of the coordinates and momenta F and their corresponding quantum-mechanical operators $\mathscr{F}$ must be symmetric in exchange of identical units. It is permissible to ask the average momentum of the particles of type α, or even those in some reasonable-sized region of the system, but it is meaningless to ask the momentum of a particle of type α number 21 when it is identical to $N_\alpha - 1$ others. It follows that in eq. (3b.6) the average value $\langle F\rangle$ for all meaningful functions F is the same for all $\prod_\alpha N_\alpha!$ permutations $\mathscr{P}_p\Psi^{(0)}_{\mathbf{K}^\dagger}$. *All* functions $\mathscr{P}_p\Psi^{(0)}_{\mathbf{K}^\dagger}$ have the same energy $E_{\mathbf{K}}$.

Second, as a corollary of the first remark, any perturbation Hamiltonian that may cause transitions between eigenstates of an approximate Hamiltonian, even a random fluctuating time-dependent Hamiltonian, is symmetric in identical particle exchange. Only transitions between states of the same symmetry character occur. If our universe is once in a state totally antisymmetric in exchange of identical fermions, it will remain so.

Third, any linear combination of totally symmetric (antisymmetric) functions is symmetric (antisymmetric). The rule that any linear combination of allowed states is an allowed state remains valid.

3f. QUANTUM STATES IN PHASE SPACE

For a system of Γ degrees of freedom let w be the phase volume of dimension h^Γ between two numerical energy values E and $E+\Delta E$,

$$w = \int\int_{H(\mathbf{q}^{(\Gamma)},\, \mathbf{p}^{(\Gamma)})\geq E}^{H(\mathbf{q}^{(\Gamma)},\, \mathbf{p}^{(\Gamma)})\leq E+\Delta E} \cdots \int d\mathbf{q}^{(\Gamma)}\, d\mathbf{p}^{(\Gamma)}. \tag{3f.1}$$

Suppose that the size and shape of the region w is such that for *every* degree of freedom ν, $1\leq\nu\leq\Gamma$, the average cross-sectional area is large compared to h,

$$\left\langle \int dq_\nu\, dp_\nu \right\rangle \text{ in } w \gg h, \qquad 1\leq\nu\leq\Gamma. \tag{3f.2}$$

In this case, asymptotically as eq. (2) is obeyed more and more strongly, the number $\Omega\dagger$ of independent solutions of the time-independent Schrödinger equations having energies between E and $E+\Delta E$ approaches the value

$$\Omega\dagger \Rightarrow h^{-\Gamma} w. \tag{3f.3}$$

Actually, the more mathematically respectable limiting process is to fix ΔE in eq. (1) and solve for the number of eigenstates $\Omega\dagger$ of energies between E and $E+\Delta E$ for quantum-mechanical equations with a variable h. In the limit that h approaches zero value,

$$\lim_{h\to 0} (h^\Gamma \Omega\dagger) = w, \tag{3f.3'}$$

The validity of eq. (3) or (3′) is more general than for phase space volumes w defined by limiting the energy values. If w is defined by limiting the integral of eq. (1) to values of any physical function $F(\mathbf{q}^{(\Gamma)}, \mathbf{p}^{(\Gamma)})$ between f and $f+\Delta f$ and eq. (2) is satisfied, the number $\Omega\dagger$ of eigenstates Ψ_ν of the Hermitian operator $\mathscr{F}$ satisfying eq. (3b.7) with f_ν lying between f and $f+\Delta f$ also obeys eq. (3).

The relation between number of quantum states and phase volume is of course related to the uncertainty principle that for every degree of freedom ν there is an uncertainty in coordinate value Δq_ν, and one in the momentum Δp_ν such that[1]

$$\Delta q_\nu \, \Delta p_\nu \geq h. \tag{3f.4}$$

Eigenfunctions of the Hamiltonians, or eigenfunctions of physical operators $\mathcal{F}$ do not usually correspond to compact rectangular volume elements in phase space,

$$w = \prod_\nu \Delta q_\nu \, \Delta p_\nu. \tag{3f.5}$$

However nonstationary time-dependent functions can be constructed having appreciable amplitudes only within such a volume element, and the number of orthogonal functions that do so is given by eq. (3).

Thus there is a correspondence between quantum states and volume elements in phase space. To a sufficiently large dimensionless volume element $h^{-\Gamma}w$ in a phase space of Γ degrees of freedom there correspond asymptotically $\Omega\dagger = h^{-\Gamma}w$ solutions of any eigenfunction equation, and if $\Omega = h^{-\Gamma}(\prod_\alpha N_\alpha!)^{-1}w$ is much larger than unity, $\Omega \gg 1$, then Ω is the number of allowed states of the proper symmetry, symmetric or antisymmetric. Instead of eq. (3′) we have,

$$\lim_{h \to 0} (h^\Gamma \Omega) = \frac{w}{\prod_\alpha N_\alpha!}. \tag{3f.6}$$

Agreement is approached if two independent conditions are satisfied in the real world of nonzero value of h.

1. The dimensionless volume element $h^{-\Gamma}w$ must have a cross section considerably greater than unity

$$\left\langle \frac{\int dq_\nu \, dp_\nu}{h} \right\rangle \gg 1, \qquad 1 \leq \nu \leq \Gamma,$$

for every degree of freedom ν.

2. The density of points describing identical particles, electrons, atoms or molecules in the dimensionless μ-space must be much lower than unity throughout all of $h^{-\Gamma}w$. If this is so, a negligible portion of w will correspond to having two or more identical particles in the same quantum state.

[1] Modern quantum-mechanical texts usually write $\hbar = h/2\pi$ in eq. (4). Since uncertainty is imprecisely defined, the factor 2π is unimportant in eq. (4). The statement of eq. (3), however, is definitely correct with h and *not* $\hbar$. It is for this reason that we prefer the consistent use of h in this book.

The foregoing considerations apply to any Hamiltonian with any number of degrees of freedom in phase space. If the Hamiltonian of the system is separable, it applies to the phase space of each of the subsets $\mathbf{q}_\alpha^{(\gamma\alpha)}$, $\mathbf{q}_\beta^{(\gamma\beta)}$ of coordinates of eqs. (3c.1) and (3c.2). Frequently, in the subspace of some sets $\mathbf{q}_\alpha^{(\gamma\alpha)}$, $\mathbf{q}_\beta^{(\gamma\beta)}, \ldots$ the quantum state energies lie densely spaced, but not so in other subspaces. For instance, in a perfect gas the Hamiltonians of the three center-of-mass coordinates of each molecule are separable. The $\Delta q_\nu = \Delta x_i$ of each such coordinate in a range of constant energy is the macroscopic size of the system, millimeters or greater. The spacings in translational momenta space, and energy, are extremely dense, whereas, for instance, the states of the vibrational degree of freedom of a diatomic molecule are widely spaced in energy compared to kT ($T \sim 300$ K).

The computational procedures of statistical mechanics involve summation of a probability function over all quantum states of a system. If the Hamiltonian is separable, the probabilities are independent in the subsets of the separable degrees of freedom and, as in eq. (1d.12), the summation becomes a product of simpler sums. In the degrees of freedom for which the quantum-mechanical state density is such that there are many states in a range kT of energy, the summation can be replaced by integration over phase space. In other degrees of freedom, such as vibration, summation over quantum states must be used.

Thus for a gas integration over the classical translational phase, space is always employed. If, in addition, for each internal quantum state the density of points in the translational μ-space is low compared to h^{-3}, Boltzmann counting by dividing with $\prod_\alpha N_\alpha!$ is valid.

3g. TIME AVERAGE AND ENSEMBLE AVERAGE

Consider a simple system such as Avogadro's number of argon atoms in a volume of 25 liters with energy (above that of the gas at 0 K) of 900 cal, which corresponds to 1 mol of argon at room temperature and approximately atmospheric pressure. The macroscopic thermodynamic state is precisely determined by the three numbers N, E, V. Consistent with this macroscopic prescription there is an infinite continuum of points in a multidimensional phase space that represent possible classical microscopic states. In the quantum-mechanical description the situation is much better. The number of quantum states is finite but, although mathematicians assure us that any finite number is much less than infinity, the number in this case is somewhat appalling, as great as $10^{10^{20}}$.

Among these states are many that have rather unusual characteristics, for instance, that all or most of the atoms are clustered together in approximately one-thousandth of the total volume, or that all atoms are moving along the x-axis, exerting pressure only on the walls perpendicular to this axis. True these states might be expected to be rather evanescent in time but, for an only slightly more complicated system, quite long-lasting exceptional states are consistent with the thermodynamic prescription. For instance, let the system contain half Avogadro's number of neon atoms and the same number of argon atoms, both in 25 liters with 900 cal of energy. Among the enormous number of microscopic states are many with all, or nearly all the neon in the top half of the vessel and the argon in the bottom. Diffusion is not particularly rapid, and the excess of neon in the upper half persists for a long time.

Clearly, some sort of statistical averaging is needed. It was customary in the early development for theoreticians to define the averaging process as being the *time average* behavior of a single completely isolated system. Gibbs introduced ensemble averaging,[1] which seems to us to correspond far more closely to the experimental processes by which the properties of a macroscopic system are determined. The Gibbs ensemble consists of an infinite number of systems which are *macroscopically* identical. This means that the systems are prepared according to the same sufficient macroscopic prescription. For equilibrium systems the requirement is that they all be in the same completely defined *thermodynamic state.*

Gibbs then derived the probability density in the γ-space of the single systems that a randomly selected member of the ensemble will be at any point in this γ-space. Gibbs of course had no knowledge of quantum mechanics. We ask now for the probability that the system will be in each completely defined allowed quantum state. Given this probability, we can compute the average value of any function F of coordinates and momenta, since its average $\langle F_{\mathbf{K}} \rangle$ is known in principle [eq. (3b.6)] for each quantum state **K**.

One can then further compute, again in principle at least, the probability of observing a value of F different from the average. In agreement with experimental fact one finds for the completely defined thermodynamic state that this probability is negligibly small for any experimentally observable deviation from the equilibrium average value.

The Gibbs formalism then gives a probability density function $W(\mathbf{q}^{(\Gamma)}, \mathbf{p}^{(\Gamma)})$ for finding a system having Γ degrees of freedom at the γ-space position $\mathbf{q}^{(\Gamma)}, \mathbf{p}^{(\Gamma)}$. The system is a randomly selected one from an

[1] Boltzmann discussed ensemble averaging before Gibbs, but the Gibbs procedure was more general.

infinite ensemble of systems all in the same thermodynamic state. Actually, there are two ways of defining such a probability density. They differ only slightly in the wording of the definition but enormously in numerical value, namely, by a factor that may be $10^{10^{20}}$ or greater. The definition we prefer, and use, is:

> $W(\mathbf{q}^{(\Gamma)}, \mathbf{p}^{(\Gamma)})$ *is the probability density that molecules (of the appropriate species) occupy the numbered positions* $\mathbf{q}_i^{(\gamma)}$, $\mathbf{p}_i^{(\gamma)}$, $i = 1, 2, \ldots, N$, *in the* μ*-space of the individual molecules.* (3g.1)

An alternative function $W^*(\mathbf{q}^{(\Gamma)}, \mathbf{p}^{(\Gamma)})$ can be defined as the probability density that numbered molecules $i = 1, 2, \ldots, N$ occupy the positions $\mathbf{q}_1^{(\gamma)}$, $\mathbf{p}_1^{(\gamma)}$, $\mathbf{q}_2^{(\gamma)}, \ldots, \mathbf{p}_N^{(\gamma)}$, respectively. If there are N identical molecules of one species, the two functions differ by a factor $N!$. If several species α are present, and N_α molecules of each α,

$$W^*(\mathbf{q}^{(\Gamma)}, \mathbf{p}^{(\Gamma)}) = \left(\prod_\alpha N_\alpha!\right)^{-1} W(q^{(\Gamma)}, p^{(\Gamma)}), \tag{3g.2}$$

since there are $\prod_\alpha N_\alpha!$ different permutations of the positions of the identical molecules in W^* corresponding to the same statement of molecules occupying numbered positions in the definition of W.

However, in integrating W over the γ-space there are always $\prod_\alpha N_\alpha!$ different points corresponding to the same probability density statement. Where the condition that the sum of probabilities over all possible microstates requires that the integral of W^* be unity, we have for W,

$$\iint \cdots \int \frac{d\mathbf{q}^{(\Gamma)}\, d\mathbf{p}^{(\Gamma)}\, W(\mathbf{q}^{(\Gamma)}, \mathbf{p}^{(\Gamma)})}{\prod_\alpha N_\alpha!} = 1. \tag{3g.3}$$

The initial quantum hypothesis of Planck in 1900 introduced into physical science a fundamental constant of nature h of the dimension energy times time, or action (ml^2t^{-1}). This is the dimension of $q_\nu p_\nu$ for any degree of freedom. Since $W(q^{(\Gamma)}, p^{(\Gamma)})$ has the dimension of action to the power $-\Gamma$ and therefore depends on the units used for mass, length, and time, it might seem natural to measure it in such units that h has a unit value. Define $W^{\text{abs}}(\mathbf{q}^{(\Gamma)}, \mathbf{p}^{(\Gamma)})$ as a dimensionless probability density in these units, so that, if $W(\mathbf{q}^{(\Gamma)}, \mathbf{p}^{(\Gamma)}$ is given in arbitrary units such as cgs units,

$$W(\mathbf{q}^{(\Gamma)}, \mathbf{p}^{(\Gamma)}) = h^{-\Gamma} W^{\text{abs}}(\mathbf{q}^{(\Gamma)}, \mathbf{p}^{(\Gamma)}), \tag{3g.4}$$

and eq. (3) becomes,

$$\iint \cdots \int \frac{d\mathbf{q}^{(\Gamma)}\, d\mathbf{p}^{(\Gamma)}\, W^{\text{abs}}(\mathbf{q}^{(\Gamma)}, \mathbf{p}^{(\Gamma)})}{h^\Gamma \prod_\alpha N_\alpha!} = 1 \tag{3g.5}$$

Essentially, this step was taken by Sackur and Tetrode in the second decade of this century. When the Gibbs expressions were used with W^{abs}, the resulting equations led to an absolute thermodynamic entropy instead of one defined only to an arbitrary constant. This absolute entropy was in agreement with the third-law value, S at $T = 0$ K equal to zero.

The significance of the dimensionless W^{abs} of eq. (4) is clear from the discussion in Sec. 3f. For quantum-mechanical systems we define $W(\mathbf{K})$ as the probability that in an infinite ensemble of systems a randomly selected system will be in the fully defined allowable quantum state $\mathbf{K}$ of the correct symmetry. The condition that the sum of all probabilities be unity is

$$\sum_{\mathbf{K}} W(\mathbf{K}) = 1. \tag{3g.6}$$

If eq. (3f.6) is used, we see that we can replace summation over the quantum numbers $\mathbf{K}$ in the limit that $h \to 0$ by integration over phase space,

$$\lim_{h\to 0}\left(\sum_{\mathbf{K}}\right) = \iint \cdots \frac{d\mathbf{q}^{(\Gamma)}\, d\mathbf{p}^{(\Gamma)}}{h^{\Gamma}(\prod_{\alpha} N_{\alpha}!)}. \tag{3g.7}$$

Comparing eq. (5) with eq. (6) we see that for quantum states $\mathbf{K}$ corresponding to a given value $\mathbf{q}^{(\Gamma)}$, $\mathbf{p}^{(\Gamma)}$ of the classical phase space,

$$\lim_{h\to 0} W(\mathbf{K}) = W^{\text{abs}}(\mathbf{q}^{(\Gamma)}, \mathbf{p}^{(\Gamma)}). \tag{3g.8}$$

One characteristic of the Gibbs ensembles is mentioned here, but is discussed in greater detail in Sec. 3i. A thermodynamic state of a system having a single chemical component is fully defined by specifying the volume V, energy E, and number of molecules N. Alternatively, the same classical thermodynamic state is defined by giving V, T, and N, or P, T, and N, or V, T, and μ, with μ the chemical potential. One of the "extensive" variables V, E, or N must be retained, or the size of the system is undefined. The Gibbs ensembles for the different descriptions are *not* the same. The *microcanonical* ensemble is one of the systems of fixed V, E, N. The *canonical* or *petite canonical* ensemble of systems of fixed V, T, N consists of an infinite number of systems each of fixed V and N but in thermal contact with an infinite reservoir of temperature T. Energy can flow in and out through the walls and fluctuates. Only its average value is fixed. The *grand canonical* ensemble of systems of fixed V, T, μ consists of open systems of fixed volume with walls permeable to energy and to molecules. Each system is in contact with an infinite reservoir of fixed T and μ. Both energy and number of molecules can fluctuate.

3h. THERMODYNAMIC FORMALISM

The conjugate extensive-intensive variable pairs V, $-P$; S, T; N, μ are defined by a quite general type of mathematical relationship. An alternative choice based on the same mathematics is a more natural choice for statistical mechanics.

Let $F(X_1, X_2, \ldots, X_n)$ be a finite real function of the n real variables $X_1, \ldots, X_n$. We limit ourselves to homogeneous linear finite functions F, which means that, for a constant λ,

$$F(\lambda X_1, \lambda X_2, \ldots, \lambda X_n) = \lambda F(X_1, X_2, \ldots, X_n). \tag{3h.1}$$

It follows that

$$F(0, 0, \ldots, 0) = 0, \tag{3h.2}$$

and that the quantities x_i defined by

$$x_i = \frac{\partial F}{\partial X_i}, \qquad 1 \le i \le n, \tag{3h.3}$$

are independent of λ. For fixed $X_1^{(0)}, X_2^{(0)}, \ldots, X_n^{(0)}$ let $X_i = \lambda X_i^{(0)}$ and integrate $F(\lambda)$ from $\lambda = 0$ to $\lambda = 1$. Since

$$\frac{dF}{d\lambda} = \sum_{i=1}^{i=n} x_i X_i^{(0)} \tag{3h.4}$$

is independent of λ, we have

$$F(X_1, X_2, \ldots, X_n) = \sum_{i=1}^{i=n} x_i X_i \tag{3h.5}$$

for any set of variables $X_i = X_i^{(0)}$, and so on. Since, from eq. (3),

$$dF = \sum_{i=1}^{i=n} x_i \, dX_i, \tag{3h.6}$$

whereas from eq. (5)

$$dF = \sum_{i=1}^{i=n} (x_i \, dX_i + X_i \, dx_i), \tag{3h.7}$$

we have the Gibbs–Duhem relation that

$$\sum_{i=1}^{i=N} X_i \, dx_i = 0. \tag{3h.8}$$

Now further assume that the diagonal second derivatives of F are all positive,

$$\left(\frac{\partial x_i}{\partial X_i}\right)_{X,\ldots} = \frac{\partial^2 F}{\partial X_i^2} > 0. \tag{3h.9}$$

More strictly, we assume that, along any line L in the $(n-1)$-dimensional space of any of $(n-1)$ X_i's, $dX_i = l_i\,dL$, the second derivative with respect to L is positive, say, in the space of $X_1, \ldots, X_{n-1}$, that

$$\frac{d^2F}{dL^2} = \sum_{i=1}^{i=n-1} \sum_{j=1}^{j=n-1} l_i l_j \left(\frac{\partial^2 F}{\partial X_i\,\partial X_j}\right)_{X_k,\ldots,X_n} > 0. \qquad (3h.9')$$

This requires that the determinant of the second derivatives $d^2F/\partial X_i\,\partial X_j$ be positive,

$$\left|\left(\frac{\partial^2 F}{\partial X_i X_j}\right)_{X_k,\ldots,X_n}\right| > 0, \qquad 1 \le i,\ j \le n-1. \qquad (3h.10)$$

This assures us that we can make a one-to-one unique correspondence of X_i to x_i as long as X_k, say X_n, is held fixed. From eq. (8) the determinant of *all* second derivatives is zero.

Now introduce a *quantity* having the same physical dimensions as the quantity represented by the function F, namely,

$$Q = F - x_1 X_1. \qquad (3h.11)$$

We have then that,

$$\begin{aligned} dQ &= dF - x_1\,dX_1 - X_1\,dx_1 \\ &= -X_1\,dx_1 + \sum_{i=2}^{i=n} x_i\,dX_i, \end{aligned} \qquad (3h.12)$$

and

$$\left(\frac{\partial Q}{\partial x_1}\right)_{X_2,\ldots,X_n} = -X_1, \qquad \left(\frac{\partial Q}{\partial X_i}\right)_{x_1,X_2,\ldots} = x_i, \qquad i \ne 1. \qquad (3h.13)$$

One says that the *quantity* Q can be expressed as a *natural function* of the variables $x_1, X_2, \ldots, X_n$, in that its partial derivatives in that set of variables are simply the conjugate variable, paying proper attention to the sign. Similarly, of course if

$$R = F - x_1 X_1 - x_2 X_2, \qquad (3h.14)$$

then

$$dR(x_1, x_2, X_3 \ldots, X_n) = -X_1\,dx_1 - X_2\,dx_2 + \sum_{i\ge 3}^{i\le n} x_i\,dX_i, \ldots. \qquad (3h.15)$$

The mathematics of conventional thermodynamics follows this scheme. The minimum set of "extensive" coordinates of a one-component system is V, S, N. For more chemical components $\alpha, \beta, \ldots$, we need V, S, N_α, $N_\beta, \ldots$. However, many other extensive variables X_i may play a role. Most of these are normally zero in the absence of forces, other than a uniform pressure, acting from outside on the system. For instance, one may need three vector components of electrical polarization, three of magnetic polarization, and possibly the elements of a strain tensor, normalized to be proportional to the system size. Their conjugate forces are electric and magnetic field vectors and the elements of the stress tensor.

The internal energy E is a natural function $E(V, S, N_\alpha, \ldots, X_1, \ldots)$ of these variables corresponding to the mathematical function F. It obeys the requirements of eq. (1) of being linear homogeneous in all variables.[(1)] The quantities x_i of eq. (3) are the negatives of the generalized thermodynamic forces f_i conjugate to X_i. The reversible work done *on* the system is $-f_i\,dX_i$. They are given in Table 3h.1.

Table 3h.1 Variables X_i and Their Conjugate Negative Forces,
$-f_i = (\partial E/\partial X_i)_{X_j, \ldots}$

$V,\ S,\ N_\alpha,\ N_\beta, \ldots, X_i, \ldots$
$-P,\ T,\ \mu_\alpha,\ \mu_\beta, \ldots, -f_i, \ldots$

We have that

$$dE = (-P)\,dV + T\,dS + \sum_\alpha \mu_\alpha\,dN_\alpha + \sum_i (-f_i)\,dX_i. \qquad (3h.16)$$

The requirement in eq. (9) on the single second derivatives is met by E, and the more general requirements of eqs. (9′) and (10) if the system is in a single thermodynamic phase. It is not if the state of the system is such that two or more phases are in equilibrium. For instance P, T, N does not define V if the P, T pair is on a vapor-pressure curve. If p phases are in equilibrium p different X_i's are required to define the complete state. The determinant of the second derivatives of the remaining $n-p$ is positive. Such minor vexations as this may cause are never serious and are discussed in Sec. 4j.

The common energy quantities corresponding to Q and R of eqs. (11) and (14) are:

enthalpy, $H = E - (-PV)$,

$$dH = -V\,d(-P) + T\,dS + \sum_\alpha \mu_\alpha\,dN_\alpha + \sum_i (-f_i)\,dX_i, \qquad (3h.17)$$

Helmholtz free energy, $A = E - TS$,

$$dA = (-P)\,dV - S\,dT + \sum_\alpha \mu_\alpha\,dN_\alpha + \sum_i (-f_i)\,dX_i, \qquad (3h.18)$$

(1) Were it not, normal thermodynamics would fail. A system in which gravitational forces between units i, j proportional to r_{ij}^{-2} are important does not obey normal thermodynamics. For charged particles the condition of electric neutrality saves thermodynamics (Sec. 8n).

Gibbs free energy, $G = E - (-PV) - TS$,

$$dG = -V\,d(-P) - S\,dT + \sum_\alpha \mu_\alpha\,dN_\alpha + \sum_i (-f_i)\,dX_i. \qquad (3h.19)$$

Obviously, many more such functions can be constructed and are occasionally useful, especially if the variables listed here as f_i, X_i are not all zero. Since the energy E is a minimum at fixed $V, S, N_\alpha, \ldots, X_i, \ldots$ at equilibrium, the forces f_i are zero at equilibrium in the absence of an externally applied field such as an electric field or a stress. The X_i's may have arbitrary zeros, but these are most often chosen so that $X_i = 0$ if $f_i = 0$. If the f_i, X_i are all zero, one useful example not commonly listed in this group is the negative pressure volume product $(-PV) = E - TS - \sum_\alpha \mu_\alpha N_\alpha$,

$$d(-PV) = (-P)\,dV - S\,dT - \sum_\alpha N_\alpha\,d\mu_\alpha. \qquad (3h.20)$$

All these quantities have the property of having minimum values at equilibrium if their "natural" variables are held constant.

One comment on notation might clarify some difficulties. Mathematicians commonly use a symbol like $F(X_1, \ldots, X_n)$ for a function of the variables $X_1, \ldots, X_n$. If a transformation to new variables $X_1, \ldots, X_{n-1}$ and $x_n(X_1, \ldots, X_n)$ is made, then a *new function* $\Phi(X_1, \ldots, X_{n-1}, x_n)$ can be used to express the same quantity at corresponding values x_n and X_n, $\Phi[X_1, \ldots, X_{n-1}, x_n(X_1, \ldots, X_n)] = F(X_1, \ldots, X_n)$. There is then no ambiguity about partial derivatives; in general, $\partial\Phi/\partial X_1 \neq \partial F/\partial X_1$. Not having the Chinese advantage of several thousand symbols, thermodynamicists use one letter, E, H, A, G, and so on, for a *quantity*. Since one may want to know the behavior of E when the state is given by V, T, N, one distinguishes the variables held constant in the differentiation, and $(\partial E/\partial V)_{T,N} \neq (\partial E/\partial V)_{S,N}$.

An alternative scheme of thermodynamic equations and conjugated variables utilizes the same mathematical formalism. In this formulation the equations of statistical mechanics take a very simple form.

Choose as extensive variables $V, E, N_\alpha, N_\beta, \ldots, X_i, \ldots$, replacing S in the conventional systematics by E. A dimensionless quantity is S/k with k Boltzmann's constant.[1] The quantity S/k is a *maximum* at equilibrium.

[1] Boltzmann's constant k is the gas constant per molecule, $k = R/N_0$, with N_0 Avogadro's number and R the usual gas constant per mole. With n the number of moles and N the number of molecules, $N = nN_0$, the perfect gas equation is $PV = nRT = NkT$. The dimension of k, like that of R, is thus energy divided by the scale of T, and its numerical value depends on the arbitrary Celsius scale,

$$R = 8.3143\ \mathrm{J\,deg^{-1}\,mol^{-1}},$$
$$k = 1.38066 \times 10^{-16}\ \mathrm{erg\,deg^{-1}\,mol^{-1}}.$$

Solve eq. (16) for $d(S/k)$ to write

$$d\frac{S}{k}=\frac{P}{kT}dV+\frac{1}{kT}dE+\sum_{\alpha}\frac{-\mu_{\alpha}}{kT}dN_{\alpha}+\sum_{i}\frac{f_i}{kT}dX_i. \tag{3h.21}$$

The reciprocal energy quantity

$$\beta=\frac{1}{kT} \tag{3h.22}$$

occurs so constantly in statistical mechanics that it seems worthwhile to introduce it and use it. Extensive intensive variables are shown in Table 3h.2.

Table 3h.2 Variables X_i and Their Conjugate $x_i=[\partial(S/k)/\partial X_i]_{X_j},\ldots$

V	E	N_{α}	N_{γ}	$\cdots$	X_i
βP	β	$-\beta\mu_{\alpha}$	$-\beta\mu_{\gamma}$	$\cdots$	$x_i=\beta f_i$

We have that

$$\frac{S}{k}=\beta PV+\beta E-\sum_{\alpha}\beta\mu_{\alpha}N_{\alpha}+\sum_{i}x_iX_i, \tag{3h.23}$$

where again the terms x_iX_i are zero in the absence of externally applied forces other than pressure.

Since we chose a quantity that is maximum at equilibrium rather than minimum, the second derivatives [eq. (3)] and determinant [eq. (10)] for a single phase state are all negative, which does not alter the uniqueness theorem. The quantities that are now maximum for other variable choices are, omitting the x_iX_i terms:

$$Q=\frac{S}{k}-\beta PV,$$

$$dQ=-V\,d(\beta P)+\beta\,dE-\sum_{\alpha}\beta\mu_{\alpha}\,dN_{\alpha}; \tag{3h.24}$$

$$-\beta A=\frac{S}{k}-\beta E,$$

$$d(-\beta A)=\beta P\,dV-E\,d\beta-\sum_{\alpha}\beta\mu_{\alpha}\,dN_{\alpha}; \tag{3h.25}$$

$$-\beta G=\frac{S}{k}-\beta PV-\beta E,$$

$$d(-\beta G)=-V\,d(\beta P)-E\,d\beta-\sum_{\alpha}\beta\mu_{\alpha}\,dN_{\alpha}; \tag{3h.26}$$

$$\beta PV = \frac{S}{k} - \beta E + \sum_{\alpha} \beta\mu_\alpha N_\sigma,$$

$$d(\beta PV) = \beta P\, dV - E\, d\beta + \sum_{\alpha} N_\alpha\, d(\beta\mu_\alpha); \tag{3h.27}$$

of which the first one, Q, is hardly of use, but the others are all familiar.

3i. ENSEMBLE PROBABILITY EQUATIONS

The probabilities $W(\mathbf{K})$ that a randomly selected member system from any specified ensemble will be in quantum state $\mathbf{K}$ are derived in Chapter 4. We state the equations in this section and discuss certain features of them. Readers who find abstract derivations difficult unless the use and characteristics of the equations derived are already familiar to them are encouraged to omit Chapters 4 to 6 until later. The book is consciously so organized that this omission should cause no difficulty in comprehending later chapters.

All the equilibrium ensembles are defined as consisting of an infinite number of member macroscopic systems which are all prepared in the same thermodynamic state. For the *microcanonical* ensemble the thermodynamic state is specified by giving all the necessary extensive variables. The simplest system is that of one chemical component which requires at least the three variables V, E, N to define the state.

The probabilities $W(\mathbf{K})$ in this microcanonical ensemble are all equal for all Ω quantum states of the system consistent with the specified values of V, E, N and possibly of other necessary extensive variables. These are all the states of N molecules in V having energies $E(\mathbf{K}) = E$, more strictly, those states $\mathbf{K}$ for which $E \leq E(\mathbf{K}) \leq E + \Delta E$. The value of ΔE is discussed later, as is the effect of requiring fixed values of other variables. With S the entropy and k Boltzmann's constant the value of $W_{\mathbf{K}}$ is

$$W(\mathbf{K};\, V,\, E,\, N) = \exp\left(\frac{-S}{k}\right). \tag{3i.1}$$

Since

$$\sum_{\mathbf{K}=1}^{\mathbf{K}=\Omega} W(\mathbf{K}) = \Omega \exp\left(\frac{-S}{k}\right) = 1, \tag{3i.2}$$

we have that

$$\frac{S}{k} = \ln \Omega \tag{3i.3}$$

The *petite canonical* ensemble consists of systems of specified V, T, N. The systems are maintained at T by thermal contact with an infinite heat reservoir at that temperature. The energy can fluctuate. The probability is given for *all* states $\mathbf{K}$ of N molecules in V whatever their energy $E_{\mathbf{K}}$. The probabilities are

$$W(\mathbf{K};\ V,\ \beta,\ N) = \exp \beta[A - E(\mathbf{K})], \tag{3i.4}$$

with $\beta = 1/kT$ and $A = E - TS$ the Helmholtz free energy. The requirement that the sum of all probabilities be unity gives an expression for $-\beta A$,

$$-\beta A = \ln \sum_{\mathbf{K} \geq 1} \exp[-\beta E(\mathbf{K})]. \tag{3i.5}$$

Use $\Omega(V,\ E,\ N)$ as the number density of system quantum states $\mathbf{K}$ per unit energy range, and we can rewrite eq. (5) as

$$-\beta A = \ln \int_0^\infty dE\ \Omega(V, E, N) \exp(-\beta E) \tag{3i.6}$$

The *grand canonical* ensemble is one of systems of specified V, β, $-\beta\mu$, with μ the chemical potential. The systems have thermally conducting walls which are also permeable to molecules and are in contact with an infinite reservoir of the specified T and μ. Both energy and numbers of molecules can fluctuate. This is frequently referred to as an ensemble of open systems. Indeed, the walls need be no more than mathematical abstractions defining boundaries of a volume V. The probabilities are

$$W(\mathbf{K},\ N;\ V,\ \beta,\ -\beta\mu) = \exp \beta[-PV + N\mu - E(\mathbf{K})]. \tag{3i.7}$$

The sum of W must now be made over all numbers N of molecules and all quantum states $\mathbf{K}$ for each number N of molecules in V. Setting this sum equal to unity, we have

$$\beta PV = \ln \sum_{N} \sum_{\mathbf{K}(N)} \exp \beta[N\mu - E(\mathbf{K})]. \tag{3i.8}$$

These are the three most used ensembles. We discuss more general ensembles later, but before doing so some points should be cleared up and some definitions made of terms in common usage. The number density $\Omega(V,\ E,\ N)$ of quantum states $\mathbf{K}$ in the energy coordinate can be defined by

$$\Omega(V,\ E,\ N) = \lim_{\Delta E \to 0} \left(\Delta E^{-1} \sum_{K, (E \leq E(\mathbf{K}) \leq E + \Delta E)} 1 \right). \tag{3i.9}$$

The Ω of eq. (3) is

$$\Omega = \Omega(V,\ E,\ N) \Delta E = \sum_{\mathbf{K}(E \leq E(\mathbf{K}) \leq E + \Delta E)} 1, \tag{3i.10}$$

with ΔE the uncertainty in the energy of the ensemble systems. The sum on the right of eq. (10) can be called the partition function of the microcanonical ensemble, although the term is less frequently used for this ensemble than for the others. A partition function is a sum over quantum states or, in the classical approximation, an integral over the phase space divided by h^Γ, of a function. In eq. (10) the function is the trivial one of unity for all states. The partition function for the petite canonical ensemble is

$$Q(V, \beta, N) = \sum_{\mathbf{K}} \exp[-\beta E(\mathbf{K})] \tag{3i.11}$$

$$= \int_0^\infty dE\, \Omega(V, E, N) \exp(-\beta E), \tag{3i.11'}$$

$$-\beta A = \ln Q(V, \beta, N). \tag{3i.11''}$$

For the grand canonical ensemble the partition function is,

$$\mathcal{Q}(V, \beta, -\beta\mu) = \sum_N \sum_{\mathbf{K}(N)} \exp \beta[N\mu - E(K)]. \tag{3i.12}$$

$$= \sum_N Q(V, \beta, N) \exp \beta N\mu, \tag{3i.12'}$$

$$\beta PV = \ln \mathcal{Q}(V, \beta, -\beta\mu). \tag{3i.12''}$$

The argument under the sum in the partition function is always proportional to the probability of the quantum state. The logarithm of the partition function is always the negative of the thermodynamic function, $-S/k$, βA, $-\beta PV$, that takes minimum values at equilibrium when the variables of the ensemble are fixed.

With eq. (10) for Ω the entropy of eq. (3) is

$$\frac{S}{k} = \ln \Omega(V, E, N) + \ln \Delta E \tag{3i.13}$$

and appears to depend on the measurement accuracy ΔE with which the energies of the member systems of the ensemble were determined. This is logically the case. In the past many authors defined the Ω of eq. (3) as the total number of states $\mathbf{K}$ with $E(K) \le E$, which removes this awkwardness. This is numerically possible for most large systems, since $\Omega(V, E, N)$ normally increases rapidly with E so that, as we discuss shortly, $\ln \int_0^E dE'\, \Omega(V, E', N)$ is numerically the same as $\ln \Omega(V, E, N)\, \Delta E$. Several arguments militate against this usage. First, there is a considerable number of other cases in which we are forced to conclude that the entropy S is dependent on our knowledge of the system beyond what looks like an adequate thermodynamic description of the state (see

Chapter 6). Second, there exist spin systems that behave like independent systems for which $\Omega(V, E, N)$ has a maximum value at some E' and then decreases rapidly. In these cases $\ln(V, E, N)\,\Delta E$ and the logarithm of the integral are utterly different in value. Only the use of the former in eq. (13) gives behavior that conforms to observations.

Actually, the term $\ln \Delta E$ in eq. (13) can be ignored with impunity for all truly macroscopic systems. Entropies per mole are usually the gas constant R times a factor of order unity or greater. Since R/k is Avogadro's number S/k for a mole of material is close to 10^{24} and usually greater. The error of a very large factor f in the value of $\Omega(V, E, N)\,\Delta E$ adds only $\ln f$ to S/k. Even if f were as large as 10^{100}, the additive $\ln f = 230$ changes $S/k \sim 10^{24}$ by 2 parts in 10^{22}. For a micromole with $S/k \sim 10^{18}$ the error of 2 parts in 10^{16} is percentually 10^6-fold greater but can still rather obviously be ignored. If $\partial\Omega(V, E', N)/\partial E'$ is positive for all $E' < E$, then $\int_0^E dE'\,\Omega(V, E', N) < E\Omega(V, E, N)$. Unless we assume that the uncertainty $\Delta E/E$ in E is very much less than 10^{-100}, the integral can be substituted for Ω without noticeable numerical error.

One might of course ask why should not ΔE be identically zero? The uncertainty principle for energy, $\Delta E \geq h/t$, with t the time of isolation of the system, forbids this. Set t the lifetime of our universe $t = 10^{10}$ yr $=$ 3×10^{17} sec, and $\Delta E > 2 \times 10^{-44}$ ergs, whereas E per mole seldom exceeds 10^4 J $= 10^{11}$ ergs, $E/\Delta E < 10^{55}$.

The extremely large numerical values of all the partition functions for systems of micromoles or more play a crucial role in statistical mechanics. The partition functions are all sums of positive numbers. The theorem that follows is used often. Let a quantity σ be a finite sum of positive terms T_n,

$$\sigma = \sum_{n\geq 1}^{n\leq K} T_n, \qquad \text{with } T_n \geq 0,\ 1 \leq n \leq K. \tag{3i.14}$$

Define T_m as the largest term in the sum,

$$T_m \geq T_n, \qquad 1 \leq n \leq K. \tag{3i.15}$$

We can bracket the value of the sum between T_m and KT_m so that

$$\ln T_m \leq \ln \sigma \leq \ln T_m + \ln K, \tag{3i.16}$$

since the sum cannot exceed KT_m. Now, for instance, the grand canonical partition function [eq. (12′)] is a sum over N. There is no specified upper limit to the sum. However, at the cost of a little more mathematics one can prove that, for N greater than some K, which is a reasonable number a greater than unity times the average number $\langle N\rangle$, $K = a\langle N\rangle$, the sum for greater than K has a negligible logarithm. However $\ln T_m$ is some constant b of order near unity times $\langle N\rangle$. We bracket $\ln\sigma$, then, by

$$\ln T_m = b\langle N\rangle \le \ln\sigma \le \langle N\rangle[1+(B\langle N\rangle)^{-1}\ln(a\langle N\rangle)] \tag{3i.17}$$

Similarly, by choosing a sufficiently small ΔE we can convert eq. (11′) for the macrocanonical ensemble into a sum with

$$T_n = \Delta E\,\Omega(V, n\,\Delta E, N)\exp(-\beta n\,\Delta E) \tag{3i.18}$$

and sum to some effective upper limit K,

$$K\,\Delta E = a\langle E\rangle > \langle E\rangle, \tag{3i.19}$$

with $\langle E\rangle$ the average energy.

The conventional mathematical procedure at this stage is to write

$$\lim_{\langle N\rangle\to\infty}\left(\frac{\ln\sigma}{\ln T_m}\right) = 1 \tag{3i.19}$$

The limit $N\to\infty$, V/N and E/N constant, is often referred to as the thermodynamic limit. Its frequent use occasionally leads to such statements as, "Thermodynamics is valid only for infinite systems," or "Statistical mechanics is correct only for infinite systems." Laboratories are not of infinite size, and we prefer to adapt our terminology to laboratory systems. We use the partition functions for any average $\langle N\rangle$, or exact N, as defining the corresponding thermodynamic function. The thermodynamic relations $A = E - TS$, $G = A + PV$, and so on, are numerically correct to fractional errors of order $N^{-1}\ln N$. Only when computer calculations are made with N of order 10^3, or experiments are made on truly microscopic crystals or droplets, need such errors concern us. In such cases there can be differences between average quantities $\langle N\rangle$, $\langle E\rangle, \ldots$ and most probable values of N or E. In most cases we ignore these differences.

Actually, in cases in which numerically valid results are sought we usually ignore much larger fractional errors of order $N^{-1/3}$ due to surface effects. These surface tension terms can be, and in some cases are, calculated, but the normal procedure is to ask for and calculate the bulk values of the thermodynamic functions E, A, S, and so on, namely, the terms proportional to N.

3j. THE GENERAL ENSEMBLE PROBABILITY

In Sec. 3i we discussed the three most used ensembles. Obviously, many more are possible, especially if we have several chemical components and if we concern ourselves with electric or magnetic polarization, and so on. It is possible to write a single probability equation for all cases, the sole difference being in the quantum numbers for which the probability is nonzero. We limit ourselves first to the cases in which we always keep V as one variable. We can then replace any or all of the remaining extensive variables $X_i = E, N_\alpha, N_\gamma, \ldots, X_j, \ldots$ by its conjugate intensive variable. We use the set $x_i = \partial(S/k)/\partial X_i$ of Table 3h.2 as the intensive variables, namely, $\beta, -\beta\mu_\alpha, -\beta\mu_\gamma, \ldots, x_j = \beta f_j, \ldots$, respectively.

Of the extensive variables $E, N_\alpha, N_\gamma, \ldots$ are conservative. In an isolated system with walls impervious to heat or molecules they cannot change in value. Most of the other extensive variables are capable of altering their values in a real laboratory situation, but not necessarily all. Usually, they may be given a specified *average* value by some force field acting on them, such as an electric field for electrical polarization, or a magnetic field for magnetic polarization. However, we can imagine a crystal clamped in a position fixing the strain tensor. For a sufficiently slow chemical reaction we can introduce an extensive variable ΔN, giving the displacement in numbers of molecules from equilibrium. A system can be prepared with a fixed value of ΔN. If measurements can then be made on its properties before ΔN changes appreciably, it will be legitimate to treat it as an equilibrium system with a new conservative extensive variable ΔN. Alternatively, in principle at least, an average ΔN can be maintained by keeping the system in contact through walls permeable to one kind of molecule only, with reservoirs of fixed $-\beta\mu_\alpha$ for all species α of reactants and products. In general, the concept of semipermeable membranes permeable to some species, impermeable to others, permits us to imagine ensembles of systems with N_α fixed for some species $\alpha, \ldots$, but $-\beta\mu_\gamma, \ldots$ fixed for others. Not all the variable choices correspond to practical laboratory situations. The theoretician, however, need not hamper his imagination by the fact that perfect semipermeable membranes are not on all laboratory shelves.

We have employed the terms extensive and intensive in describing the paired variables X_i, x_i as though no possible ambiguity could arise. Every extensive variable can be made intensive by division with N: $v = V/N$, the volume per molecule, or $\epsilon = E/N$, the energy per molecule. We can equally well convert every intensive variable into an extensive one by multiplication with N. Indeed, for a one-component system, $N\mu = G$ is the extensive Gibbs free energy and is a most important function. From a purely operational point of view limited to macroscopic experimental concepts, it is difficult to give a general prescription for decision as to the proper choice of what we call extensive variables, which are often called thermodynamic coordinates.

In terms of the microstate description of coordinates and momenta a simple rule helps. The volume V is a boundary condition on the molecular coordinates. In classical mechanics all other extensive variables are sums over the atoms or molecules of functions of their phase space coordinates and momenta. For instance E is the Hamiltonian of the system. The number N_α of molecules of type α is the sum over all molecules of this species of the trivial function unity. For neutral molecules the electrical polarization is the vector sum of their dipole moments. For ionic constituents it is the sum of their vector positions times their charge. The magnetic polarization is the vector sum of the molecular magnetic moments. The strain tensor elements are expressible in terms of the displacements of molecular coordinates from equilibrium positions.

Corresponding to each such function of phase space there is a quantum-mechanical operator. In each state **K** of the systems there is an average value $\langle X_i \rangle(\mathbf{K})$ [eq. (3b.6)] of X_i. For all ensembles a *single* equation $W(\mathbf{K})$ can be written. The difference in the different ensembles consists only in the range of quantum state **K** for which it is nonzero. Outside this range the probability is arbitrarily set equal to zero. The single equation is

$$W(\mathbf{K}) = \exp\left\{-\frac{S}{k} + \sum_i x_i [X_i - \langle X_i \rangle(\mathbf{K})]\right\}. \tag{3j.1}$$

If this is spelled out explicitly for the more common variables X_i, x_i, it is

$$\begin{aligned} W(\mathbf{K}) = \exp\Big\{ -\frac{S}{k} &+ \beta[E - E(\mathbf{K})] \\ &+ \sum_\alpha (-\beta\mu_\alpha)[N_\alpha - N_\alpha(\mathbf{K}) + \sum_i (\beta f_i)[X_i - \langle X_i \rangle(\mathbf{K})]\Big\}. \end{aligned} \tag{3j.1'}$$

If all the extensive variables are fixed, $W(\mathbf{K})$ is set equal to zero except for quantum states **K** for which $E(\mathbf{K}) = E$, $N_\alpha(\mathbf{K}) = N_\alpha$ and $X_i(\mathbf{K}) = X_i$. All terms in the exponent of eq. (1) except the constant $-S/k$ vanish. The equation reproduces the microcanonical probability of eq. (3i.1). If β is fixed, but the extensive variables are not, we include all **K** of any energy $E(\mathbf{K})$, but with $N_\alpha(\mathbf{K}) = N_\alpha$, $\langle X_i \rangle(\mathbf{K}) = X_i$, and so on. With $-(S/k) + \beta E = \beta A$ we have the petite canonical probability of eq. (3i.4). Fix β and $-\beta\mu_\alpha, \ldots$ and note that

$$-\frac{S}{k} + \beta E - \sum_\alpha \beta\mu_\alpha N_\alpha = \beta(A - G) = -\beta PV. \tag{3j.2}$$

The equation is valid for all energies $E(\mathbf{K})$ and molecule numbers. The probability is that of the grand canonical ensemble for any number of chemical components, analogous to eq. (3i.7) for one component.

For the nonconservative variables written as X_i in eq. (1′) the conjugates βf_i are zero unless special forces f_i such as an electric or a magnetic field or stress is applied. If such forces are not present, we can completely forget the existence of the variable X_i, as we now show. From the very definition of probability [see eq. (1c.13)] the average value $\bar{X}_i$ of X_i in the member systems of the ensemble is

$$\langle X_i \rangle = \sum_{\mathbf{K}} \bar{X}_i(\mathbf{K}) W(\mathbf{K}), \tag{3j.3}$$

where $\bar{X}_i(\mathbf{K})$ is the expectation value in the state **K**. Suppose that in eq. (1′) we choose an ensemble for which we explicitly consider the thermodynamic state to be given by $x_i = \beta f_i$ rather than X_i, but set $f_i = 0$. The term $\beta f_i[X_i - \bar{X}_i(\mathbf{K})]$ is then zero. The rule is that in eq. (2) we now sum over states **K** with all values of $X_i(\mathbf{K})$ to find the thermodynamic value of X_i at zero thermodynamic force f_i, $X_i(f_i = 0) = X_i^{(0)}$.

Had we wished to do so, we could alternatively have chosen to define an ensemble with X_i specified as $X_i^{(0)}$ and restrict the **K** values to those for which $\bar{X}_i(\mathbf{K}) = X_i^{(0)}$. This would have been a much more awkward procedure. It is far easier to forget X_i and x_i completely and include the allowed K values independently of their $\bar{X}_i(\mathbf{K})$ values. The behavior discussed in Sec. 3i, that the logarithm of a sum of positive terms can safely be replaced by the logarithm of the largest one, means that differences in the thermodynamic quantities computed are negligible. Unless we explicitly wish to investigate the properties of a system in an electric or magnetic field, or under stress, we can completely ignore the existence of the corresponding X_i, x_i as possible thermodynamic variables. Only the volume and the conservative variables E, $N_\alpha, \ldots$ or their conjugates are always necessary to specify the thermodynamic state.

Employ the notation of eq. (1) in which we use X_i for all extensive variables. Consider the ensemble for which $V, x_1, \ldots, x_\nu, X_{\nu+1}, \ldots, X_n$ are specified. With

$$\Theta = -\frac{S}{k} + \sum_{i=1}^{i=\nu} x_i X_i, \tag{3j.4}$$

the probability $W(K)$ is,

$$W(\mathbf{K}) = \exp\left[\Theta - \sum_{i=1}^{i=\nu} x_i \langle X_i \rangle(\mathbf{K})\right]. \tag{3j.5}$$

With eq. (5) for $W(\mathbf{K})$ in the relation

$$\sum_{\mathbf{K}} W(\mathbf{K}) = 1, \tag{3j.6}$$

differentiate both sides with respect to x_i for any $1 \leq i \leq \nu$. One finds

$$\sum_{\mathbf{K}} W(\mathbf{K})\left[\left(\frac{\partial\Theta}{\partial x_i}\right)_{x_1,\ldots,x,X_{\nu+1},\ldots,X_n} - \langle X_i\rangle_{(\mathbf{K})}\right] = 0 \tag{3j.7}$$

or, with eqs. (3) and (6), that

$$\left(\frac{\partial\Theta}{\partial x_i}\right)_{x_1,\ldots,x_\nu,X_{\nu+1},\ldots,X_n} = \langle X_i\rangle. \tag{3j.8}$$

Now the quantity Θ of eq. (4) is that dimensionless quantity which is a minimum at equilibrium when the variables $x_1, \ldots, x_\nu X_{\nu+1}, \ldots, X_n$ are held fixed. Its partial derivative with respect to x_i is the thermodynamic variable X_i. Thus the general probability equation (1) reproduces the thermodynamic relations.

The case in which βP is held fixed and the sum runs over all volume V is slightly different, since V is a boundary condition rather than a sum of functions of the phase space variables. At least one of the other extensive variables must now be held fixed, or the thermodynamic state is undefined. If all of them are fixed, the function corresponding to Θ of eq. (4) is $-S/k+\beta PV$ which is not dignified with a name in conventional thermodynamics. If, however, E is replaced by β and all other extensive variables held fixed, we have $-S/k+\beta E+\beta PV=\beta G$ for Θ, and

$$W(\mathbf{K}) = \exp[\beta G - \beta PV - \beta E(\mathbf{K})], \tag{3j.9}$$

which is now valid for all quantum states $\mathbf{K}$ of any energy $E(\mathbf{K})$ and at all volumes V.

The condition that $W(\mathbf{K})$ summed over all K at all volumes shall be unity allows us to equate $-\beta G$ to the logarithm of a partition function $Q(\beta P, \beta, N)$. This partition function appears to be the $Q(V, \beta, N)$ of eq. (3i.11) multiplied by $\exp(-\beta PV)$ and summed by integration over all volumes. But this would give us a partition function of the dimension of volume, since $Q(V, \beta, N)\exp(-\beta PV)$ is a number. Now it is obvious from the discussion following eq. (3i.13), in which we justify the numerical validity of omitting the ΔE in $\ln[\Omega(V, E, N)\,\Delta E]$ that here too we are not likely to cause numerical error by using the simple integration. However, it is equally obvious that some logical error has occurred in the argument.

The error is obvious if we consider carefully the meaning of the number $\Omega=\Omega(V, E, N)\,\Delta E$, and particularly obvious if we remember that in the classical limit this is a normalized integral over $d\mathbf{q}^{(\Gamma)}\,d\mathbf{p}^{(\Gamma)}h^{-\Gamma}$ between two energy limits. The number Ω is the number of states in the correct energy range in which the molecules occupy the volume V *or less.* This is logically the correct procedure for a system constrained by rigid walls of volume V to remain within V. It includes a negligible fraction of the $10^{10^{20}}$ states for which the molecules occupy a lesser volume. But for a system of constant pressure we should count only the states occupying V and pushing on the walls, namely, the derivative with respect to V of $\Omega(V, E, N)$.

We therefore write

$$-\beta G = \ln \int_0^\infty dV \frac{\partial Q(V, \beta, N)}{\partial V} \exp(-\beta PV), \tag{3j.10}$$

which is dimensionally correct. Partial integration now gives

$$\begin{aligned} -\beta G &= \ln \int_0^\infty d(\beta PV)\, Q(V, \beta, N) \exp(-\beta PV) \\ &= \ln Q(\beta P, N). \end{aligned} \tag{3j.11}$$

For all ensembles when any extensive variable X_i is *not* fixed, but only its conjugate x_i, we identify the *thermodynamic* value of X_i with the average $\langle X_i \rangle$ of eq. (3). From eq. (1) it is seen then that

$$\sum_{\mathbf{K}} W(\mathbf{K}) \ln W(\mathbf{r}) = \sum_K W(\mathbf{K}) \left\{ -\frac{S}{k} + \sum_i x_i [X_i - \langle X_i \rangle (K)] \right\} = -\frac{S}{k} \tag{3j.12}$$

for all ensembles. We later discuss in Sec. 6g the use of $\sum W \ln W$ for $-S/k$ in ensembles of nonequilibrium systems. Whenever the entropy S is definable through a thermodynamic reversible cycle, the identification is valid. We can use eq. (12) as the general statistical mechanical definition of entropy.

For any special equilibrium system we have a variety of ensembles available. For each ensemble, defined by the choice X_i or x_i of variables for each i, the logarithm of the partition function gives the thermodynamic function S/k, $-\beta A$, βPV, $-\beta G, \ldots$ that is a maximum at equilibrium if the ensemble variables are held fixed. Partial differentiation gives the missing conjugate variables. We can readily compute all the other thermodynamic functions from any one. The choice of ensemble need be dictated only by the ease with which it promises an answer. The simplest looking of the partition functions $\Omega(V, E, N)$ is seldom the easiest. The petite canonical $Q(V, \beta, N)$ is probably still the most used, although the grand canonical partition function $Q(V, \beta, -\beta\mu)$ is often easier and seldom more difficult to use. The partition function $Q(\beta P, \beta, N)$ for the Gibbs free energy G is very seldom used, since the integral over a three-dimensional volume in terms of molecular coordinates is impractical. It has been used for a one-dimensional model in which the volume integral is easy. Recently, Kohn and Onffroy[(1)] showed that it has advantages of giving exponential convergence with N. This may make its employment in model computer calculations with relatively small N-values, $N < 10^3$, very useful.

[(1)] W. Kohn and J. R. Onffroy, *Phys. Rev.*, **B8,** 2485, (1973).

CHAPTER 4

DERIVATION OF THE ENSEMBLE EQUATIONS

4a. INTRODUCTION

In this chapter we undertake the derivation of the ensemble probability equations. Various workers have different preferences, deriving the validity of the equation for different ensembles first. It is then relatively easy (see Sec. 4g and 4h) to show that those for the other ensembles follow. We prefer to show first the validity of the expression $W(\mathbf{K}, V, E, N)$ of eq. (3i.1) as $\exp(-S/k)$.

From the Liouville equation (Sec. 4b) it follows that for an ensemble of systems all following the same closed path in phase space the probability density along all portions of the path must be uniform at equilibrium. In Sec. 4c the equivalent quantum-mechanical derivation is made. For an ensemble of equilibrium systems of fixed V, E, N having the same spectrum of Ω states $\mathbf{K}$ available to them, the probabilities of all states are equal. A dimensionless quantity σ is defined in Sec. 4d to be $\sigma = \ln \Omega$. It is shown that σ for a system composed of two independent parts is the sum of their individual σ-values, and also that σ is extensive, proportional to the size of the system.[1] It is demonstrated that σ increases in any spontaneous change in an isolated system. In Sec. 4e the derivative $(\partial\sigma/\partial E)_{V,N}$ is shown to be qualitatively a reciprocal temperature, and $(\partial\sigma/\partial V)_{E,N}/(\partial\sigma/\partial E)_{V,N}$ is shown to be proportional to a pressure. Thus σ has the properties of a dimensionless entropy.

[1] The mathematical statement is that σ is linear homogeneous in the extensive variables V, E, N (Sec. 3h).

The demonstration that σ is actually equal to S/k is completed in Sec. 4f. Thus the equation for the microcanonical ensemble is proved. The derivation of the other ensemble probability equations from $\sigma = S/k = \ln \Omega$ is made in Sec. 4g. Section 4h completes the other half of the necessary and sufficient statement, namely, that if any of the other ensembles is proved to be correct, that for the microcanonical ensemble follows.

4b CONSERVATION OF PHASE VOLUME, THE LIOUVILLE EQUATION

If $F(\mathbf{q}^{(\Gamma)}, \mathbf{p}^{(\Gamma)})$ is any function of Γ coordinates, $\mathbf{q}^{(\Gamma)} = q_1, \ldots, q_\Gamma$ and their conjugate momenta $\mathbf{p}^{(\Gamma)}$, then the total time derivative of F, *measured along the path of motion in the phase space* is

$$\frac{dF}{dt} = \sum_{\nu \geq 1}^{\nu \leq \Gamma} \left[\left(\frac{\partial F}{\partial q_\nu} \right) \dot{q}_\nu + \left(\frac{\partial F}{\partial p_\nu} \right) \dot{p}_\nu \right]. \tag{4b.1}$$

Use the Hamiltonian form [eq. (3a.14)] of the equations of motion in this to write it as

$$\frac{dF}{dt} = \sum_{\nu \geq 1}^{\nu \leq \Gamma} \left[\left(\frac{\partial H}{\partial p_\nu} \right) \left(\frac{\partial F}{\partial q_\nu} \right) - \left(\frac{\partial H}{\partial q_\nu} \right) \left(\frac{\partial F}{\partial p_\nu} \right) \right]. \tag{4b.2}$$

The Liouville operator (Joseph Liouville, 1809–1882) may be defined as[1]

$$\mathscr{L} = \sum_{\nu \geq 1}^{\nu \leq \Gamma} \left[\left(\frac{\partial H}{\partial p_\nu} \right) \left(\frac{\partial}{\partial q_\nu} \right) - \left(\frac{\partial H}{\partial q_\nu} \right) \left(\frac{\partial}{\partial p_\nu} \right) \right]. \tag{4b.3}$$

With its use eq. (2) can be written symbolically as

$$\frac{dF}{dt} = \mathscr{L} F. \tag{4b.4}$$

Now with sufficiently small $\Delta q_\nu\ \Delta p_\nu$,

$$\begin{aligned} \Delta q_\nu &= q_\nu^{(2)} - q_\nu^{(1)}, \\ \Delta p_\nu &= p_\nu^{(2)} - p_\nu^{(1)}, \end{aligned} \tag{4b.5}$$

we define a volume Δw in the phase space Γ degrees of freedom,

$$\Delta w = \prod_{\nu \geq 1}^{\nu \leq \Gamma} \Delta q_\nu \Delta p_\nu. \tag{4b.6}$$

[1] Sometimes the Liouville operator is defined as the negative of the expression on the right of eq. (3) and occasionally as i times this, since $i\mathscr{L}$ with $i = \sqrt{-1}$ is a Hermitian operator.

The change $\overline{\Delta w}^{\cdot}$ in volume with time is then

$$\frac{d\,\Delta w}{dt}=\dot{\overline{\Delta w}}=\Delta w\sum_{\nu\geq 1}^{\nu\leq\Gamma}\left[\left(\frac{\dot{\overline{\Delta q_\nu}}}{\Delta q_\nu}\right)+\left(\frac{\dot{\overline{\Delta p_\nu}}}{\Delta p_\nu}\right)\right]. \tag{4b.7}$$

But now, for small enough $\Delta q_\nu = q^{(2)} - q^{(1)}$,

$$\dot{\overline{\Delta q}}=\dot{q}_\nu^{(2)}-\dot{q}_\nu^{(1)}=\left(\frac{\partial H}{\partial p_\nu}\right)_{q_\nu=q_\nu^{(2)}}-\left(\frac{\partial H}{\partial p_\nu}\right)_{q_\nu=q_\nu^{(1)}}$$

$$=\Delta q_\nu\left(\frac{\partial^2 H}{\partial p_\nu\,\partial q_\nu}\right). \tag{4b.8}$$

Similarly, since $\dot{p}_\nu = -\partial H/\partial q_\nu$

$$\dot{\overline{\Delta p_\nu}}=-\Delta p_\nu\left(\frac{\partial^2 H}{\partial p_\nu\,\partial q_\nu}\right) \tag{4b.9}$$

and

$$\left(\frac{\dot{\overline{\Delta q_\nu}}}{\Delta q_\nu}\right)+\left(\frac{\dot{\overline{\Delta p_\nu}}}{\Delta p_\nu}\right)=0. \tag{4b.10}$$

We then have, from eq. (7), that

$$\dot{\overline{\Delta w}}=0; \tag{4b.11}$$

the infinitesimal elements of volume in phase space move through the system to new positions but retain the same magnitude of volume. The motion through phase space is like the motion of a flowing incompressible fluid.

Now suppose that an ensemble has $n(\mathbf{q}^{(\Gamma)}, \mathbf{p}^{(\Gamma)})\,\Delta w$ systems in such a volume element at $\mathbf{q}^{(\Gamma)}, \mathbf{p}^{(\Gamma)}$ at a given time t. As time changes, the phase space positions of the systems change, remaining in the volume element as it moves through the phase space. We have that

$$\frac{d}{dt}[n(q^{(\Gamma)}p^{(\Gamma)})\,\Delta w]=0=n\dot{\overline{\Delta w}}+\dot{n}\Delta w \tag{4b.12}$$

or, since $\dot{\overline{\Delta w}}=0$ from eq. (11), the change in density n in phase space is zero,

$$\dot{n}=0. \tag{4b.13}$$

If there are a total of N systems in the ensemble, the probability density $W(\mathbf{q}^{(\Gamma)}, \mathbf{p}^{(\Gamma)})$ is

$$W(\mathbf{q}^{(\Gamma)}, \mathbf{p}^{(\Gamma)})=\frac{n(q^{(\Gamma)}, p^{(\Gamma)})}{N} \tag{4b.14}$$

and, since N is constant,

$$\dot{W}(\mathbf{q}^{(\Gamma)}, \mathbf{p}^{(\Gamma)}) = 0. \tag{4b.15}$$

The probability densities along the paths on which the systems move through phase space move with the systems. If a system at time $t=0$ is at $\mathbf{q}_0^{(\Gamma)}$, $\mathbf{p}_0^{(\Gamma)}$, it will at a later time t be at some position $q_t^{(\Gamma)}$, $p_t^{(\Gamma)}$, and

$$W(t, \mathbf{q}_t^{(\Gamma)}, \mathbf{p}_t^{(\Gamma)}) = W(t=0, \mathbf{q}_0^{(\Gamma)}, \mathbf{p}_0^{(\Gamma)}). \tag{4b.16}$$

At equilibrium the probability densities at each $\mathbf{q}^{(\Gamma)}$, $\mathbf{p}^{(\Gamma)}$ must be constant in time. This can be true only if $W(\mathbf{q}^{(\Gamma)}, \mathbf{p}^{(\Gamma)})$ is constant for every phase point on the allowed path.

From eq. (15) one can immediately write the equation for the *partial derivative* of W with respect to t at constant $q^{(\Gamma)}$, $p^{(\Gamma)}$. Since

$$\begin{aligned}\dot{W}(t, q^{(\Gamma)}, p^{(\Gamma)}) &= \frac{d}{dt} W(t, \mathbf{q}^{(\Gamma)}, \mathbf{p}^{(\Gamma)}) \\ &= \frac{\partial}{\partial t} W(t, \mathbf{q}^{(\Gamma)}, \mathbf{p}^{(\Gamma)}) + \sum_{\nu=1}^{\nu=\Gamma}\left[\dot{q}_\nu \frac{\partial W}{\partial q_\nu} + \dot{p}_\nu \frac{\partial W}{\partial p_\nu}\right] = 0,\end{aligned} \tag{4b.17}$$

with eq. (3a.14) for $\dot{q}_\nu$, $\dot{p}_\nu$, we have

$$\frac{\partial W(t, \mathbf{q}^{(\Gamma)}, \mathbf{p}^{(\Gamma)})}{\partial t} = -\sum_{\nu=1}^{\nu=\Gamma}\left[\frac{\partial H}{\partial p_\nu}\frac{\partial W}{\partial q_\nu} - \frac{\partial H}{\partial q_\nu}\frac{\partial W}{\partial p_\nu}\right]. \tag{4b.18}$$

With the symbol $\mathscr{L}$ for the Liouville operator [eq. (3)], one can write this as

$$\frac{\partial W}{\partial t} = -\mathscr{L}W. \tag{4b.19}$$

If W is a constant everywhere, then $\mathscr{L}W=0$ and $\partial W/\partial t$ is everywhere zero.

We notice also that, if $W(\mathbf{q}^{(\Gamma)}, \mathbf{p}^{(\Gamma)})$ is not constant everywhere but is a function only of the value H of the Hamiltonian $H(\mathbf{q}^{(\Gamma)}, \mathbf{p}^{(\Gamma)})$, then

$$\frac{\partial W}{\partial q^\nu} = \frac{\partial H}{\partial q^\nu}\frac{dW}{dH}, \tag{4b.20}$$

and

$$\frac{\partial W}{\partial p_\nu} = \frac{\partial H}{\partial p_\nu}\frac{dW}{dH}, \tag{4b.20$'$}$$

so that again,

$$\frac{\partial W}{\partial t} = \frac{dW}{dH}\sum_{\nu=1}^{\nu=\Gamma}\left[\frac{\partial H}{\partial p_\nu}\frac{\partial H}{\partial q_\nu} - \frac{\partial H}{\partial q_\nu}\frac{\partial H}{\partial p_\nu}\right] = 0. \tag{4b.21}$$

The individual paths of each system follow lines of constant $H(\mathbf{q}^{(\Gamma)}, \mathbf{p}^{(\Gamma)})$ in the phase space when uninfluenced by forces from outside the system.

An isolated system is a convenient thermodynamic fiction which can be approached experimentally but never completely attained. In an isolated classical system the energy would be absolutely constant. In a real system there are always some random fluctuations at the container walls, which alter the detailed paths of the systems in the phase space. If these perturbations from outside the system are minimal, no appreciable change in energy occurs. The allowed paths are effectively at constant energy. The ergodic hypothesis, which is discussed in more detail in Sec. 4i, is that the complete description of the macroscopic state is adequate to inform us of what portions of phase space are available to the systems, and that all systems can traverse all parts of this phase space. With this assumption it follows that an equilibrium ensemble of isolated systems has equal probability density in all parts of the allowed phase space.

4c. EQUAL A PRIORI PROBABILITIES OF QUANTUM STATES

The quantum-mechanical statement equivalent to the classical one of equal probability density in phase space is that all quantum states $\mathbf{K}$ of equal energies $E(\mathbf{K})$ are equally probable in an isolated system.

Assume that our state functions $\Psi_{\mathbf{K}}, \Psi_{\mathbf{K}'}, \ldots$ are eigenfunctions of a Hamiltonian operator $\mathcal{H}^{(0)}$ of the same energy $E(K)$,

$$\mathcal{H}^{(0)}\Psi_{\mathbf{K}} = E(\mathbf{K})\Psi_{\mathbf{K}}, \qquad \mathcal{H}^{(0)}\Psi_{\mathbf{K}'} = E(K)\Psi_{\mathbf{K}'}, \qquad \ldots . \tag{4c.1}$$

Let the true Hamiltonian operator $\mathcal{H}$ be

$$\mathcal{H} = \mathcal{H}^{(0)} + \Delta\mathcal{H}, \tag{4c.2}$$

with $\Delta\mathcal{H}$ a sufficiently small perturbation such that all diagonal elements which give the change ΔE in the energies,

$$\Delta\mathcal{H}(\mathbf{K}, \mathbf{K}) = \int dq^{(\Gamma)}\, \Psi^*_{\mathbf{K}}\, \Delta\mathcal{H}\Psi_{\mathbf{K}} \qquad \text{for all } \mathbf{K}, \tag{4c.3}$$

are zero or completely negligible compared to $E(\mathbf{K})$. Nevertheless, the perturbation causes transitions between pairs of states $\mathbf{K}$ and $\mathbf{K}'$ of the same energy. Let $w(\mathbf{K}, \mathbf{K}')\, dt$ be the probability that if a system is in the state $\mathbf{K}$ at time t it will be in state $\mathbf{K}'$ at time $t + dt$. The probability of the reverse happening per unit time is $w(\mathbf{K}', \mathbf{K})$. It is a fundamental theorem of quantum mechanics for all forces that are derivatives of a potential that these two transition probabilities are equal,

$$w(\mathbf{K}, \mathbf{K}') = w(\mathbf{K}', \mathbf{K}). \tag{4c.4}$$

If now $W(\mathbf{K})$, $W(\mathbf{K}'),\ldots$ are the probabilities in an ensemble of systems of one energy $E(\mathbf{K})$, then the time change $\dot{W}(\mathbf{K})$ of $W(\mathbf{K})$ will be,

$$\dot{W}(\mathbf{K})=\sum_{K'}[w(\mathbf{K}',\mathbf{K})W(\mathbf{K}')-w(\mathbf{K},\mathbf{K}')W(\mathbf{K})], \tag{4c.5}$$

or, with eq. (4),

$$\dot{W}(\mathbf{K})=\sum_{K'}w(\mathbf{K},\mathbf{K}')[W(\mathbf{K}')-W(\mathbf{K})]. \tag{4c.6}$$

Just as we could formally integrate the time-dependent Schrödinger equation [see eq. (3b.10)], so we can write, for a time-independent Liouville operator $\mathscr{L}$,

$$W(t,\mathbf{q}^{(\Gamma)},\mathbf{p}^{(\Gamma)})=e^{-\mathscr{L}(t-t_0)}W(t_0,\mathbf{q}^{(\Gamma)},\mathbf{p}^{(\Gamma)}).$$

The condition of equilibrium is that all $W(\mathbf{K})$ are independent of time; $\dot{W}(\mathbf{K})=0$ for all states $\mathbf{K}$. Obviously, this is satisfied by having all $W(\mathbf{K})$ equal. With the ergodic hypothesis this is not only a sufficient but also a necessary condition.

The condition that the systems be ergodic is now that we cannot break the states $\mathbf{K}$, $\mathbf{K}',\ldots$ into blocks such that transitions between those in one block may occur, but that *all* $w(\mathbf{K},\mathbf{K}')$ are zero when $\mathbf{K}$ and $\mathbf{K}'$ are in different blocks. If this condition is satisfied, then the equality of *all* $W(\mathbf{K})$ is the *only* solution to $\dot{W}(\mathbf{K})=0$ (all $\mathbf{K}$). This can be shown as follows. If not all the $E(\mathbf{K})$ are equal, then there must be *at least one* that is as large or larger than all others. Assume there are $N\geq 1$ states $\mathbf{K}''$ for which $W(\mathbf{K}'')$ has this maximum value. Of these at least one, which we call $\mathbf{K}$, has a transition probability $w(\mathbf{K},\mathbf{K}')$ greater than zero to at least one of the states K' of lesser value of $W(K')$. Were this not so, the ergodic condition would be violated. For this state $\mathbf{K}$ we then have from eq. (6) that $\dot{W}(\mathbf{K})<0$, since all $w(\mathbf{K},\mathbf{K}')$ are positive or zero, and all $[W(\mathbf{K}')-W(K)]$ are zero or negative, and both are not zero for at least one term.

The argument of this section and that of the last are startingly different when one considers that they prove completely corresponding theorems in quantum and classical mechanics, respectively. The stationary states or eigenstates of the Hamiltonian do not at all correspond to the compact rectangular volume Δw of eq. (4b.6). Indeed, were the exact Hamiltonian of the system completely time-independent, the eigenstates of that Hamiltonian would show no transitions to other states. They would correspond to closed paths in phase space, with changing phase along one of the integrals of motion in the classical phase space. It has long been known that the simply stated ergodic theorem is not rigorously true for an absolutely isolated classical system. There is an enormous number of integrals of motion defining closed paths in phase space, and a system once in such a path does not leave it. Instead the hypothesis of *quasi ergodicity* is used, namely, that each, or more strictly all but a quite negligible fraction of these paths, come asymptotically close to each point in phase space before continuing on their way.

For any real physical system, even were we to use the exact time-independent Hamiltonian operator $\mathcal{H}$ in eq. (1) to select the eigenfunctions $\Psi_{\mathbf{K}}$ of states **K**, the small time-dependent fluctuations due to the surroundings would cause transitions. These still necessarily obey relation (4). The difficulties with the ergodic hypothesis are discussed more fully in Secs. 4i and 6b.

4d. THE QUANTITY $\sigma = \ln \Omega$

For an ensemble of fixed volume V all having energies between[1] E and $E+\Delta E$ and with the same set $\mathbf{N} = N_\alpha, N_\gamma, \ldots$ of molecules of species α, $\gamma, \ldots$, let Ω be the number of quantum states available to the systems Define a dimensionless number

$$\sigma = \ln \Omega, \tag{4d.1}$$

which we intend to identify with S/k.

Suppose each system consists of two completely independent parts a and b. Designate by Ω_a and Ω_b the number of states available to the two parts, respectively, and by Ω_{a+b} the number of states available to the system a plus b. Since each quantum state of a is available to the a-part with each state of b the number Ω_{a+b} is $\Omega_a\Omega_b$, and

$$\sigma_{a+b} = \ln \Omega_{a+b} = \ln \Omega_a + \ln \Omega_b = \sigma_a + \sigma_b. \tag{4d.2}$$

It follows that, if the system consists of n thermodynamically identical but independent and *unconnected* parts, the σ-value is n-fold larger than that of the individual parts.

This is not enough to show that σ is extensive. If n thermodynamically identical systems are fused into a single system, new states are created. These are due to the fact that with the *total* energy constant some of the originally independent and equal regions can gain energy and others lose. Fluctuations are now possible. Similarly, in a fluid system fluctuations in numbers of molecules in the different parts can occur. As we show, this increase in the number of states Ω contributes negligibly to the logarithm as long as we are always talking in terms of bulk systems, $N \geq 10^{19}$.

The argument that this is so follows similarly to the discussion in Sec. 3i for eq. (3i.16), in which the numerical equivalence of the logarithm of a

[1] That the numerical value of ΔE is unimportant was discussed in Sec. 3i.

limited sum of very large positive terms and the logarithm of the largest of these terms were discussed. We first discuss the general case in which an isolated system of fixed total V, E, and N initially not at equilibrium may attain equilibrium. In every case we can describe the initial nonequilibrium state as being maintained by an inhibition, and the attainment of equilibrium by the lifting of the inhibition, which makes available to the system a limited number D of distributions. To each distribution d, $1 \leq d \leq D$, there exist a number Ω_d of quantum states and a value $\sigma_d = \ln \Omega_d$ of the quantity σ. The uninhibited system has available to it quantum states of all values of the distribution variable d,

$$\Omega = \sum_{d=1}^{d=D} \Omega_d \tag{4d.3}$$

Since all quantum states have equal probability, the probability $W(d)$ of observing the distribution d in an ensemble of uninhibited systems is

$$W(d) = \frac{\Omega_d}{\Omega}, \tag{4d.4}$$

and is a maximum for that distribution d_0 for which Ω_d is maximum,

$$\Omega(d_0) = \Omega_0 \geq \Omega_d \qquad d \neq d_0. \tag{4d.5}$$

Now since Ω_d is necessarily nonnegative, we can, as discussed in Sec. 3i, bracket σ in value between the values

$$\sigma_0 \leq \sigma \leq \sigma_0(1 + \sigma_0^{-1} \ln D), \tag{4d.6}$$

where σ_0 is the value of σ for the most probable distribution d_0.

We now must inquire as to the nature of the inhibitions that can be applied in a real physical situation, the characteristics of the distributions d, and the order of magnitude of $\sigma_0^{-1} \ln D$. We choose first the simplest possible example to discuss, namely, a volume V consisting of two flasks a and b of equal volume $\frac{1}{2}V$ connected through a stopcock, containing N molecules of a perfect gas having a total energy E. A distribution d is defined by giving the number N_a of molecules in flask a, and the total number D of distributions is $N+1$ since N_a can vary from zero to N. As we later find, σ_0 is of order N, so that $\sigma_0^{-1} \ln D$ for $N = 10^{22}$ is $\cong 10^{-22} \ln 10^{22}$, of order 10^{-20}; indeed, truly negligible. The probability $W(N_a)$ of the distribution at equilibrium with an open stopcock is that of the binomial function of eq. (1d.1) with N_a replacing n. As discussed in Sec. 1d for $N = 10^{22}$, the probability that N_a differs from $\frac{1}{2}N$ by a given ΔN, $N_a = \frac{1}{2}N(1 \pm \Delta N / \frac{1}{2}N)$, is already small for $\Delta N / \frac{1}{2}N = 10^{-11}$.

Having analyzed this one example in detail, two different but related questions remain. One of them is whether all thermodynamic examples of attainment of equilibrium can be analyzed with an analogous model in which $\sigma_0^{-1} \ln D$ is negligible for a macroscopic system. The other involves showing that therefore the quantity σ is extensive, namely, linearly proportional to the size for systems of the same intensive properties.

The first question is answered only by examining a number of examples. Consider first a simple example differing in only one important respect from that of the distribution of numbers of molecules between two flasks of gas. This example is that of two equal systems connected, or not connected, through a heat conductor which can transfer energy from part a to part b, or vice versa. The distribution parameter E_a, which can take values between zero and the total energy E, is now continuous. However, there is a limit ΔE to the accuracy with which the energy can be measured. We can define an integer distribution parameter d by setting $E_a = d\,\Delta E$, or d the nearest integer to $E_a/\Delta E$, $0 \le d \le E/\Delta E \equiv D$. We again find the bracket of eq. (6) for σ as differing from the most probable distribution σ_0 of $E_a = \frac{1}{2}E$ by no more than one part in $\sigma_0^{-1}[\ln(E/\Delta E)]$.

Now one may ask, Why not let ΔE approach zero and $E/\Delta E$ approach infinite value? But there is a theoretical limit to the smallness of ΔE, namely, that given by the uncertainty principle that $\Delta E \ge ht^{-1}$ with t a time of measurement. Even with the accepted life[(1)] of the universe $t = 10^{10}$ yr $= 3 \times 10^{17}$ sec and $h = 6 \times 10^{-27}$ erg sec, we find $\Delta E \ge 2 \times 10^{-44}$ erg $= 2 \times 10^{-51}$ J. The total energy E will be of order $aNkT$ with a of order unity or perhaps 10 or even 50, not more than 2×10^3 J for $N = 10^{22}$ at room temperature. Even in this absurd limit for ΔE we have $D \cong 10^{54}$ and $\sigma_0^{-1} \ln D \cong 120/\sigma_0$, still of order 10^{-20} if σ_0 is of order $N = 10^{22}$. If we increase D by choosing a larger system with larger E, we increase $\ln D$ only logarithmically with the size, but σ_0 increases linearly and $\sigma_0^{-1} \ln D$ decreases.

In both the above examples we imagined the existence of a mechanism, in one case opening or shutting a stopcock, in the other the establishment or breaking of a thermal connection, by which without doing work on the system, or changing V or the numbers of molecules, we could inhibit or permit a change in distribution. Other examples are obvious. The insertion or removal of a (minute) catalyst can permit or prevent a chemical reaction. The distribution d is then the number of molecules that have reacted in one direction. Opening or closing an electric switch may permit or prohibit the flow of electrons between two battery poles; the distribution d is the number of electrons that pass between the poles.

[(1)] The accepted life of the universe makes quantum jumps of considerable magnitude at roughly decade intervals. Since this was first written it seems to have increased 60%.

Consider the case in which the distribution d is measured by several independent variables with different limiting numbers $D_1, D_2, \ldots$, for instance, if d is determined by the number $N_{\alpha a}$ of molecules of type α in flask a, the number $N_{\gamma a}$ of molecules of type $\gamma, \ldots$, the energy E_a in flask a, ... The total number D of distributions is the product $D = D_1 \times D_2 \ldots$, and $\ln D$ is the sum $\ln D_1 + \ln D_2 + \cdots$. As long as the number of such independent distribution variables $d_1, d_2, \ldots$ is limited to the number of independent *macroscopic* variables that can reasonably be actually measured in a system (including a few that we may imagine measuring although it would be difficult to carry out), we can still safely neglect $\sigma_0^{-1} \ln D$.

One example of a multiple distribution description is that the system, instead of being divided into two parts, a and b, is divided into n equal parts $1, 2, \ldots, \nu, \ldots, n$, and molecules or energy or both are allowed to fluctuate within each part. Consider the case of only molecular exchange. The distributions are defined by the set $d_1 = N_1$, $d_2 = N_2$, $d_n = N_n$, of numbers, but are not independent, since

$$\sum_{\nu=1}^{\nu=n} d_\nu = N, \tag{4d.7}$$

with N the total number of molecules. Since any one d_ν is limited to values from zero to N, it is obvious that $D < (N+1)^n$. With more careful examination one finds the asymptotic relation for large n and N, with e the Naperian e,

$$\ln D \cong n \ln \frac{eN}{n}. \tag{4d.8}$$

Let $\bar{N}_p = N/n$ be the average number of molecules in each part, which is the number in the most probable distribution, and σ_p the value of $\ln \Omega_p(\bar{N}_p)$. The value of σ_0 for the n parts inhibited, but to equilibrium, is $n\sigma_p$. The value of $\sigma_0^{-1} \ln D$ is then

$$\sigma_0^{-1} \ln D = \sigma_p^{-1} \ln e\bar{N}_p, \tag{4d.9}$$

independent of n, the number of parts. For macroscopic values of $\bar{N}_p$, $\bar{N}_p \geq 10^{19}$, the value of $\sigma_p^{-1} \ln e\bar{N}_p$ is negligible. It follows that, for systems of total N greater than some such number, the uninhibited value of σ is extensive in the size of the system.

We note, however, a necessary caution. If we compute σ from a model requiring a uniform density of molecules or of energy, in volume elements containing only a small number of molecules, the result may be, and usually is, in serious error. Uniform number density and energy density can be assumed without error only in volume elements large enough to contain a macroscopic number of molecules.

One note of historical interest belongs here. As discussed earlier, the counting of states Ω is often very accurately proportional to an integral over phase space, and this integral was the classical substitute for Ω used by Gibbs and others. Division of the integral by h^Γ then gives the number of solutions of the time-independent Schrödinger equation. The logarithm of the integral is not proportional to N, but contains an additive contribution of $N \ln N$. This arises from counting as independent the $N!$ positions differing only in permutations of the numbered identical molecules. Gibbs divided the integral, somewhat arbitrarily, by $N!$. With the Stirling approximation that $\ln N! \cong N(\ln N - 1)$ this cancels the unwanted term. The logarithm of $(h^\Gamma N!)^{-1}$ times the integral is thus extensive. The quantum-mechanical requirement that allowed states be described by totally symmetric or totally antisymmetric functions introduces the division by $N!$ with no arbitrariness.[(1)]

In terms of our analysis of the distribution of identical molecules between two equal volume flasks, the opening of the stopcock, even if limited to keeping exactly $\frac{1}{2}N$ in each flask, gives $N!/(\frac{1}{2}N!)^2$ new distributions of the numbered molecules in the two flasks, each of which contributes the same to the classical integral. With Stirling's approximation the logarithm of the integral then increases by $N \ln 2$ when the inhibition is lifted by opening the stopcock. This is indeed correct if initially the molecules in the two different flasks are of different species, but not if they are identical.

Thus the dimensionless quantity σ of eq. (1) has all the qualitative properties of a dimensionless entropy. It depends on the thermodynamic state variables of the system V, E, and N, and is fully defined if these and any constraints are given. For two or more independent systems the value for the sum regarded as a single system is the sum of their individual values. It is an extensive quantity,

$$\alpha(nV, nE, nN_\alpha, nN_\gamma, \ldots) = n\sigma(V, E, N_\alpha, N_\gamma, \ldots), \qquad (4d.10)$$

(1) Gibbs worried about this division by $N!$. If there were several kinds of molecules $\alpha, \gamma, \ldots$ with N_α of kind α, N_γ of kind γ, and so on, the division should be by $\prod_\alpha N_\alpha!$. With $N = \sum_\alpha N_\alpha$, $x_\alpha = N_\alpha/N$ the mole fraction of species α, the value of $N^{-1} \ln N!/\prod N_\alpha!$ is $-\sum x_\alpha \ln x_\alpha$ with use of the Stirling approximation (Appendix III). This leads to the familiar entropy of mixing expression $S_{\text{mix}}/kN = -\sum x_\alpha \ln x_\alpha$. The Gibbs paradox was that this term goes discontinuously to zero if two species of molecules are considered to become asymptotically identical in physical properties. Present concepts that atoms are characterized by integers, numbers Z_α of protons and A_α of neutrons plus protons, disallows a *continuous* transition to identity.

for any $n > 1$ as long as V, E, $\mathbf{N}$ correspond truly to a bulk macroscopic system. If for any system of fixed V, E, $\mathbf{N}$ inhibited from a definite distribution d of the thermodynamic variables the inhibition is lifted and the system proceeds to equilibrium with fixed total V, E, $\mathbf{N}$, the value of σ increases or stays constant. It stays constant only if no macroscopic change occurs in the distribution in going to equilibrium. Thus σ, like entropy, increases for any naturally occurring process in an isolated system and is a maximum at equilibrium.

We note from eq. (4) that, with eq. (1) for σ, the ratio of probabilities for an uninhibited system at equilibrium for a distribution of d and d_0, is

$$\frac{W(d)}{W(d_0)} = \exp[-(\sigma_0 - \sigma(d)]. \tag{4d.11}$$

Again we invoke the large magnitude of σ_0 for a bulk system, say, of order 10^{20}. The probability of trapping an uninhibited equilibrium system in a state d of reduced $\sigma(d)$ by 1 part in 10^{10} less than σ_0 is $\exp(-10^{10})$.

Finally, we remark that much of the discussion of this section could have been eliminated by always going to the thermodynamic limit of infinite size, $N \to \infty$. In this limit all the quantities we show to be negligible for $N \geq 10^{20}$ go strictly to zero. However, our value of $\sigma_0^{-1} \ln D$ for the fractional error is strictly an upper limit. Normally, only a very, very small fraction of the total number D of possible distributions has appreciable probabilities.

4e. THE DERIVATIVES OF σ

Define two quantities τ and Π by the equations

$$\tau^{-1} = \left(\frac{\partial \sigma}{\partial E}\right)_{V,N}, \tag{4e.1}$$

$$\Pi = \tau\left(\frac{\partial \sigma}{\partial V}\right)_{E,N}. \tag{4e.2}$$

Since σ is dimensionless, τ has the dimension of energy, and Π of energy divided by volume which is the dimension of pressure. We proceed to show that these are kT and P, respectively.

Consider a system consisting of two parts, a and b, each of fixed volume, energy, and number of molecules, V_a, E_a, N_a and V_b, E_b, N_b, respectively. The number Ω_{a+b} of quantum states in the combined system is $\Omega_a\Omega_b$ and the value of σ_{a+b} is

$$\sigma_{a+b}(V_a, E_a, N_a; V_b, E_b, N_b) = \sigma_a(V_a, E_a, N_a) + \sigma_b(V_b, E_b, N_b). \tag{4e.3}$$

Now permit thermal contact so that with $E = E_a + E_b$ constant, $\delta E_a = -\delta E_b$, the energies of the two parts can change,

$$\begin{aligned}\delta\sigma_{a+b} = \delta\sigma_a + \delta\sigma_b &= \frac{\partial\sigma_a}{\partial E}\,\delta E_a + \frac{\partial\sigma_b}{\partial E}\,\delta E_b \\ &= (\tau_a^{-1} - \tau_b^{-1})\,\delta E_a.\end{aligned} \tag{4e.4}$$

In the transition to equilibrium the change will be in such a direction that σ_{a+b} increases. The energy flow δE_a into a will be positive if $(\tau_a^{-1} - \tau_b^{-1}) > 0$, that is, if $\tau_a < \tau_b$. Equilibrium is reached with maximum σ_{a+b} when $\tau_a = \tau_b$. The quantity τ is some scale of temperature having the dimension of energy.

Again consider this two-part system with thermal equilibrium established so that $\tau_a = \tau_b = \tau$. Keep the total volume $V = V_a + V_b$ fixed, but let a movable wall separate the two parts so that $\delta V_a = -\delta V_b$. With eq. (2) for Π we now have

$$\begin{aligned}\delta\sigma_{a+b} &= \frac{\partial\sigma_a}{\partial V}\,\delta V_a + \frac{\partial\sigma_b}{\partial V}\,\delta V_b \\ &= \tau^{-1}(\Pi_a - \Pi_b)\,\delta V_a.\end{aligned} \tag{4e.5}$$

The spontaneous process goes in the direction that the volume of a increases at the expense of that of b if $\Pi_a > \Pi_b$, and equilibrium is reached when $\Pi_a = \Pi_b$. Two systems at equilibrium with respect to such a volume exchange have equal mechanical pressure.

We have demonstrated the following. Two systems in thermal equilibrium, hence having the same temperature, have equal values of τ defined by eq. (1). Two systems having the same pressure have equal values of Π.

4f. THE IDENTIFICATION $\sigma = S/k$

The simplest procedure now is to consider a perfect monatomic gas for which we know experimentally that, for N atoms,

$$PV = NkT, \qquad E = \tfrac{3}{2}NkT, \tag{4f.1}$$

with k Boltzmann's constant and T the perfect gas temperature which is also the thermodynamic absolute temperature.

But for this gas the value of Ω is given by the classical phase volume divided by $N!\,h^{3N}$. The phase volume has one factor V^N due to integration over the coordinates. The factor, from the momentum integral with

$$\sum_{i=1}^{i=N} (p_{xi}^2 + p_{yi}^2 + p_{zi}^2) = 2mE, \tag{4f.2}$$

is the surface area of a $3N$-dimensional sphere of radius $\sqrt{2mE}$ times the energy uncertainty ΔE. The logarithm of this has an additive term $\frac{3}{2}N \ln E$ plus terms dependent on m and N only. We have, for N atoms, with Q some constant involving π, h, and e,

$$\sigma = \ln \Omega = N(\ln V + \tfrac{3}{2} \ln E + Q) \tag{4f.3}$$

With this and eq. (4e.1) for τ, we find

$$\tau = \left[\left(\frac{\partial \sigma}{\partial E}\right)_{V,N}\right]^{-1} = \frac{2}{3}\frac{E}{N}, \tag{4f.4}$$

and from eq. (4e.2) for Π,

$$\Pi = \tau\left(\frac{\partial \sigma}{\partial V}\right)_{E,N} = \tau\frac{N}{V}. \tag{4f.5}$$

From eq. (1) we then have

$$\tau = kT, \tag{4f.6}$$

$$\Pi = \frac{NkT}{V} = P. \tag{4f.7}$$

Now in the last section we found that, if two systems have the same temperature, they have the same values of τ. Therefore $\tau = kT$ for *all* systems. We also found that two systems of equal pressure have equal Π-values. Therefore Π is the pressure for *all* systems. We have, then, at constant N,

$$(d\sigma)_N = (kT)^{-1}[(dE)_N + P(dV)_N], \tag{4f.8}$$

$$(dE)_N = -P(dV)_N + Td(k\sigma)_N. \tag{4f.9}$$

The reversible work done on the system is $-PdV$, and $Td(k\sigma)$ is the reversible heat flow into the system. We have demonstrated that

$$S = k\sigma = k \ln \Omega. \tag{4f.10}$$

A somewhat more elegant argument avoids invoking any particular system such as a perfect gas. In general, for any thermodynamic system and any extensive thermodynamic coordinate X_i, the quantity

$$\left(\frac{\partial E}{\partial X_i}\right)_{S,X_j,\ldots} = -f_i \tag{4f.11}$$

is the thermodynamic force conjugate to X_i. The reversible work done *on* the system when the other coordinates X_j are held fixed and there is no heat flow (since the work is reversible $T\,dS=0$) is

$$(dE)_{S,X_j,\ldots} = -f_i\,dX_i. \tag{4f.12}$$

Ehrenfest's adiabatic principle is that, for a change in any *continuous* variable X_i, the individual quantum states $\mathbf{K}$ keep their identity but in general change their energy $E(\mathbf{K})$. The force $f_i(\mathbf{K})$ conjugate to X_i for a system in the quantum state $\mathbf{K}$ is

$$f_i(\mathbf{K}) = -\left[\frac{\partial E(\mathbf{K})}{\partial X_i}\right]_{X_j}. \tag{4f.13}$$

The average work done by a change dX_i on systems of an ensemble in the equally probable Ω quantum states $\mathbf{K}$ is $\langle \partial E/dX_i\rangle_{\text{av}}\,dX_i$,

$$dX_i\left\langle\frac{\partial E}{\partial X_i}\right\rangle_{\text{av}} = \Omega^{-1}\sum_{\mathbf{K}=1}^{\mathbf{K}=\Omega}\left[\frac{\partial E(\mathbf{K})}{\partial X_i}\right]_{X_j} dX_i \tag{4f.14}$$

This average mechanical work is just the average change in energy in the fixed Ω quantum states,

$$\left\langle\frac{\partial E}{\partial X_i}\right\rangle_{\text{av.}} = \left(\frac{\partial E}{\partial X_i}\right)_{\sigma,X_j,\ldots} = -f_i. \tag{4f.15}$$

When X_i is the volume V, the force is $f_i=+P$. The chemical potential μ_α of species α of molecules can be defined by

$$\mu_\alpha = \left(\frac{\partial E}{\partial N_\alpha}\right)_{V,\sigma,N_\gamma,\ldots}, \tag{4f.16}$$

and corresponds to the negative force conjugate to N_α. We have

$$dE = -P\,dV + \tau\,d\sigma + \sum_\alpha \mu_\alpha\,dN_\alpha - \sum_i f_i\,dX_i, \tag{4f.17}$$

where τ is defined by

$$\tau^{-1} = \left(\frac{\partial\sigma}{\partial E}\right)_{V,\mathbf{N},X_i,\ldots}. \tag{4f.18}$$

The reversible work done on a system with a constant number of molecules is

$$\text{Reversible work} = -P\,dV - \sum_i f_i\,dX_i, \tag{4f.19}$$

so that from eq. (17) the quantity $\tau\,d\sigma$ is the reversible heat flow. The dimensionless quantity σ is a function of the state of the system, and τ is the integrating denominator for the heat flow. With k an arbitrary constant of dimension energy divided by a temperature scale we have eqs. (6) and (10).

4g. DERIVATION OF THE GENERAL ENSEMBLE PROBABILITY

In the last section we completed the derivation of the microcanonical ensemble probability as $W(\mathbf{K}) = \exp(-S/k)$. We also showed that, if the character of the system is analyzed by a distribution variable d which can take D values, $1 \le d \le D$, then the equilibrium distribution d_0 is that for which $\Omega(d_0)$ and $\sigma(d_0)$ are maxima, $\Omega(d_0) \ge \Omega(d)$, $1 \le d \le D$.

We now consider an ensemble of M systems plus a reservoir. The ensemble has fixed total volume V_t, fixed energy E_t, and fixed number set $N_t = N_{\alpha t}, N_{\gamma t}, \ldots$ of molecules. Each of these M systems is in contact with the reservoir in a way as yet unspecified, hence in indirect contact with each other. All M systems have the same set of quantum states available to them. Let $\mathbf{K}$ be one of these states. Consider now a distribution d in this ensemble in which d is given by $M(\mathbf{K})$, the number of systems of the ensemble in the single quantum state $\mathbf{K}$. The systems in this quantum state each have an energy $E(\mathbf{K})$, a volume $V(\mathbf{K})$, and a number set $\mathbf{N}(\mathbf{K})$ of molecules. The remaining $M - M(K)$ systems, plus the reservoir, have volume $V_t - M(\mathbf{K})V(\mathbf{K})$, energy $E_t - M(\mathbf{K})E(\mathbf{K})$, and a number set $\mathbf{N}_t - M(\mathbf{K})\mathbf{N}(\mathbf{K})$.

For given $M(\mathbf{K})$ the total number of quantum states $\Omega_{En}[M(\mathbf{K})]$ available to the ensemble consists of three factors. The first is the number of states available to the $M(\mathbf{K})$ systems, but this is just unity. The second factor is the number of ways we can select $M(\mathbf{K})$ out of the M distinguishable systems, which is

$$A[M(\mathbf{K})] = \frac{M!}{[M - M(\mathbf{K})]!\,M(\mathbf{K})!}. \tag{4g.1}$$

The third factor we call Ω_{rest}, and it is the number of quantum states available to the rest of the ensemble, the reservoir plus $M - M(\mathbf{K})$ systems. This depends on its volume, energy, and number set of molecules. Its logarithm is

$$\ln \Omega_{\text{rest}} = k^{-1} S_{\text{rest}}[V_t - M(\mathbf{K})V(\mathbf{K}),\ E_t - M(\mathbf{K})E(\mathbf{K}),\ \mathbf{N}_t - M(\mathbf{K})\mathbf{N}(\mathbf{K})]. \tag{4g.2}$$

The value of k^{-1} times the entropy $S_{\text{en}}[M(\mathbf{K})]$ of the ensemble in the distribution $M(K)$ is then

$$k^{-1} S_{\text{en}}[M(\mathbf{K})] = \ln A[M(\mathbf{K})] + \ln \Omega_{\text{rest}}[M(\mathbf{K})]. \tag{4g.3}$$

We now take two more steps. The first is purely formal. We consider an ensemble of such ensembles so that we can talk about a probability and define, at equilibrium when S_{en} is a maximum,

$$W(\mathbf{K}) = \left[\frac{M(\mathbf{K})}{M}\right]_{S_{\text{en}}} = \max. \tag{4g.4}$$

The other step is to go to the limit $M \to \infty$, keeping V_t/M, E_t/M, $N_{\alpha t}/M, \ldots$ constant. This permits us to treat $M(\mathbf{K})$ as a continuous variable. Differentiate $k^{-1}S_{\text{en}}$ with respect to $M(\mathbf{K})$, setting $dS_{\text{en}}/dM(\mathbf{K}) = 0$ at the maximum. From eq. (1) for A we have, from eq. (4) for $W(K)$,

$$\frac{d \ln A}{dM(\mathbf{K})} = \ln \frac{M - M(\mathbf{K})}{M(\mathbf{K})} = \ln \frac{M}{M(\mathbf{K})} = -\ln W(\mathbf{K}), \tag{4g.5}$$

since $M(\mathbf{K})$ is completely negligible compared to M. For sufficiently large M the ratios $V(\mathbf{K})/V_t$, $E(\mathbf{K})/E_t$, $N_\alpha(K)/N_{\alpha t}, \ldots$ approach zero, and the derivatives of S_{rest} remain constant for the finite increments $V(\mathbf{K})$, $E(\mathbf{K}), \ldots$. Use the thermodynamic expressions for the partial derivatives of $k^{-1}S$,

$$d(k^{-1}S) = \beta P\, dV + \beta\, dE + \sum_\alpha (-\beta\mu_\alpha)\, dN_\alpha, \tag{4g.6}$$

and one finds, from eq. (2),

$$\frac{d \ln \Omega_{\text{rest}} M(\mathbf{K})}{dM(\mathbf{K})} = -\beta\left[PV(\mathbf{K}) + E(\mathbf{K}) - \sum_\alpha \mu_\alpha N_\alpha(\mathbf{K})\right]. \tag{4g.7}$$

With eqs. (3), (4), (5), and (7) one then has

$$W(\mathbf{K}) = \exp \beta[-PV(\mathbf{K}) - E(\mathbf{K}) + \sum_\alpha \mu_\alpha N_\alpha(\mathbf{K})]. \tag{4g.8}$$

The thermodynamic equation

$$-PV - E + \sum_\alpha \mu_\alpha N_\alpha = -PV - E + G = -TS \tag{4g.9}$$

permits us to write eq. (8) in the form

$$W(\mathbf{K}) = \exp\left\{-\frac{S}{k} + \beta P[V - V(\mathbf{K})] + \beta[E - E(\mathbf{K})] + \sum (-\beta\mu_\alpha) \times [N - N_\alpha(\mathbf{K})]\right\}, \tag{4g.10}$$

which is displayed in Sec. 3j [eq. (3j.1′)] as the general equation for any ensemble, omitting only the nonconservative extensive variables X_i.

At the beginning of this section we discussed the ensemble as having systems in contact with a reservoir in an unspecified manner. It is now clear from eq. (10) and the discussion in Sec. 3j what the choices are and what the results are. If there is no contact and all the systems have the same fixed V, E, N, then only states $\mathbf{K}$ with $V(\mathbf{K}) = V$, $E(\mathbf{K}) = E, \ldots$ are available, and we recover the microcanonical ensemble equation from eq. (10). If there is thermal contact only all $E(\mathbf{K})$ are available and eq. (10) becomes the equation for the petite canonical ensemble. If the walls are also permeable to all molecules and only V, β, $\mu_\alpha, \ldots$ are fixed, we have the grand canonical ensemble.

We must, however, retain at least one extensive variable. If we do not, we would include states $\mathbf{K}$ for which volume, energy, and number set increase indefinitely in size. From eq. (8) the probability for each *single* state decreases with size drastically, but the number of states in each range of size increases equally rapidly. The necessary condition for the use of eq. (7) is that, for all states of nonnegligible probability, the variables $V(\mathbf{K})$, $E(\mathbf{K}), \ldots$ shall be negligible in size compared to V_t, $E_t, \ldots$, respectively. This condition is violated unless one of them is assumed to be fixed in size.

We can include any or several of the nonconservative extensive variables X_i in the discussion, describing its average value in the quantum state $\mathbf{K}$ by $X_i(\mathbf{K})$. One may require that all M systems be somehow clamped in such a way that only states of $X_i(\mathbf{K}) = X_i$ are available to them. Considerable imagination may be required to perform this feat in such a way that no work is done, for instance, in the case of electric polarization. More realistically, one might include in the reservoir a machine that produces a specified thermodynamic force f_i acting on all M systems, for instance, a battery bank that puts all of them in a specified electric field. In any case such an apparatus included in the term reservoir must be imagined to be in the ensemble whose total volume, energy, and molecular content are fixed.

The variables X_i are extensive, which means that for the ensemble the total value $X_{i,\text{total}}$ of X_i is the sum of that in all M systems plus the reservoir. If there are $M(\mathbf{K})$ systems in quantum state $\mathbf{K}$ for which the average value of X_i is $X_i(\mathbf{K})$, then the value of X_i in the rest of the ensemble is $X_{i,\text{total}} - M(\mathbf{K})X_i(\mathbf{K})$. The change in $k^{-1}S_{\text{rest}}$ due to unit change in $M(K)$ is then $x_i X_i(K)$, with

$$x_i = \beta f_i = \left(\frac{\partial k^{-1}S}{\partial X_i}\right)_{X_i}.$$

The terms in eq. (3j.1′) with the nonconservative variables appear in the same way as the corresponding ones for E and $N_\alpha, \ldots$.

4h. EQUIVALENCE OF THE ENSEMBLES

In this section we wish to show that any one of the ensemble probability equations can be used to derive all the others. We derived all of them from the microcanonical in the last section. In this section we derive the microcanonical from any of the others. The method used consists essentially of substituting the logarithm of the largest term for the logarithm of a sum of positive terms [eq. (3i.16)]. More correctly stated, use the logarithm of the maximum value of a positive function $f(x)$ times a width Δx for the logarithm of the integral of the function over x. This is legitimate only if the logarithm of the function at its maximum is very great, circa 10^{20}, a condition satisfied by all the partition functions.

Start with the ensemble of fixed V, β, N, for which [eq. (3i.10)]

$$-\beta A = \ln Q(V, \beta, N), \tag{4h.1}$$

$$Q(V, \beta, N) = \int_0^\infty dE\, \Omega(V, E, N) \exp(-\beta E), \tag{4h.2}$$

where $\Omega(V, E, N)$ is the number density of states along the energy axis E. From this we seek to derive eq. (3i.12), that

$$k^{-1} S(V, E, N) = \ln[\Omega(V, E, N)\, \Delta E]. \tag{4h.3}$$

Define a quantity σ as

$$\sigma(V, E, N) = \ln \Omega(V, E, N), \tag{4h.4}$$

so that eqs. (1) and (2) become

$$-\beta A = \ln \int_0^\infty dE \, \exp[\sigma(V, E, N) - \beta E]. \tag{4h.5}$$

The integrand of eq. (5) has an extremum when E has a value $E(\beta)$ such that

$$\left(\frac{\partial \sigma}{\partial E}\right)_{V,N,E=E(\beta)} = \beta, \tag{4h.6}$$

and this extremum is a maximum since[1] always $(\partial\beta/\partial E)_{V,N} \leq 0$, and therefore

$$\frac{\partial^2 \ln \Omega(V, E, N)}{\partial E^2} \leq 0. \tag{4h.7}$$

Substituting the logarithm of an appropriate ΔE times the integrand at this maximum for the logarithm of the integral in eq. (5), one has

$$-\beta A(V, \beta, N) = \sigma(V, E(\beta), N) + \ln \Delta E - \beta E(\beta), \tag{4h.8}$$

where $E(\beta)$ is the *most probable* value of the energy at fixed V, β, and N. But since $E - A = TS$, we have derived eq. (3).

[1] This is proved in Sec. 5a.

The same argument shows that, if the grand canonical partition function is valid, that for the petite canonical follows or, in general, if that for a set with a particular intensive variable x_i is valid, that with X_i in place of x_i and no other change is also correct. We then have two thermodynamic functions,

$$\beta Y(\ldots, X_i, \ldots) = \beta Z(\ldots, x_i, \ldots) + x_i X_i, \tag{4h.9}$$

in which βY is a minimum at equilibrium when X_i and the other variables are fixed, and βZ is a minimum at fixed x_i. An equation

$$-\beta Z(x_i) = \ln \int dX_i\, Q(\ldots, X_i, \ldots)\exp(-xX_i) \tag{4h.10}$$

is assumed valid. We wish to prove that

$$-\beta Y(\ldots, X_i, \ldots) = \ln Q(\ldots, X_i, \ldots). \tag{4h.11}$$

The demonstration follows exactly the sequence of eqs. (4) to (8), with $Q(\ldots, X_i, \ldots)$ replacing $\Omega(V, E, N)$, $-\beta Y$ replacing $k^{-1}S$, and βZ replacing βA. We can thus proceed from any one of the other ensembles to derive the microcanonical equation as a consequence, and from the microcanonical to derive all the others.

4i. THE THIRD LAW OF THERMODYNAMICS

In classical statistical mechanics the probability density $W(\mathbf{q}^{(\Gamma)}, \mathbf{p}^{(\Gamma)})$ of a system of Γ degrees of freedom has the dimension of action to the power $-\Gamma$, $(ml^2/t)^{-\Gamma}$. The integral of $W \ln W$ has a numerical value dependent on the units of m, l, and t. There is no absolute value of the entropy; it is defined only to an arbitrary additive constant. The same is true in classical thermodynamics. Only entropy differences of systems containing the same number of atoms of each species are defined.

Early in the twentieth century W. H. Nernst proposed a theorem stating that that the entropy of all crystalline pure materials at 0 K was zero. This is experimentally verifiable. For two elements, A and B, which react at T and P to form a compound AB we measure ΔG and ΔH of the reaction $A+B=AB$, thus determining $\Delta S(T)=(\Delta H-\Delta G)/T$,

$$\Delta S(T) = S_{AB}(T) - S_A(T) - S_B(T). \tag{4i.1}$$

But if C_p is the heat capacity at constant pressure, the entropy difference of AB at T minus that at $T=0$ K is

$$S_{AB}(T) - S_{AB}(0) = \int_0^T dT' \left(\frac{C_p}{T'}\right)_{AB}. \tag{4i.2}$$

We then have

$$\Delta S(T=0) = S_{AB}(0) - S_A(0) - S_B(0)$$
$$= \Delta S(T) - \int_0^T dT' \frac{C_P^{AB} - C_P^A - C_P^B}{T'}, \tag{4i.3}$$

and if the Nernst theorem is correct and all three, AB, A, and B, are perfect crystals at 0 K, then $\Delta S(T=0)=0$. It has been verified in many cases, first by W. F. Giauque.

The theorem has since been expanded and is known as the third law. It states that *all* pure materials, in *complete thermodynamic equilibrium*, approach zero entropy as 0 K is approached. The statement applies equally to hypothetical perfect gases, liquids, or crystals. Indeed, if *complete thermodynamic equilibrium* is taken rigorously, the limitation to pure materials is presumably unnecessary, since in equilibrium only pure phases or ordered solid solutions, which are then really pure compounds, are in true equilibrium. Even a liquid solution of the two isotopes, ^{3}He and ^{4}He, of helium separates into two pure phases as 0 K is approached.

The hitch in the statement of the third law is the limitation to complete equilibrium. In purely thermodynamic terms one is eventually driven to the circular statement that one recognizes complete equilibrium by having a zero entropy at 0 K. We discuss, in Secs. 6b to 6d, some of the more usual apparent exceptions to the third-law rule.

The introduction of quantum mechanics, even in its earlier form, gave meaning to an *absolute* entropy derived from statistical mechanics. The fundamental physical constant h has the dimension of action, energy times time $= ml^2/t$. Divide the 2Γ-dimensional phase space $\mathbf{q}^{(\Gamma)}\mathbf{p}^{(\Gamma)}$ by h^Γ, and one has a space that is physically dimensionless. If one normalizes $W(\mathbf{q}^{(\Gamma)}, \mathbf{p}^{(\Gamma)})$ by making the integral over $d\mathbf{q}^{(\Gamma)}\, d\mathbf{p}^{(\Gamma)}\, h^{-\Gamma}$ equal to unity, then W is dimensionless. Further division by $\mathbf{N}!$ is still necessary to make the integral of $W \ln W$ proportional to N, but the difficulty in using the logarithm of a number with physical dimensions has been removed. The entropy now has a natural absolute value. The further understanding that this integral operator of $d\mathbf{q}^{(\Gamma)}\, d\mathbf{p}^{(\Gamma)}\, h^{-\Gamma}(N!)^{-1}$ just (approximately) counts quantum states gives a simple meaning to the operation. For an isolated system, $S = k \ln \Omega$, and Ω is the total number of quantum states available to the systems of the ensemble. Since Ω is a positive integer, S is always positive or zero.

The law that S approaches zero as T goes to zero now has a simple interpretation. For any system of N molecules there is one, or perhaps a number g, of quantum states of minimum energy, and the limit $N^{-1} \ln g$ always goes to zero for infinite N. Only if, for some reason, the system is frozen into a band of states above this minimum energy will the entropy be other than zero as T approaches 0 K (see Secs. 6b to 6d).

CHAPTER 5

GENERAL RELATIONS

5a. RELATIONS BETWEEN $\partial X_i/\partial x_j$ AND MOMENTS

In Sec. 3j we showed that the general ensemble probability equation in which the ensemble was defined with at least one intensive variable x_i, such as β, βP, or $-\beta\mu_\alpha$, gave a correct thermodynamic relation. If $\Theta(x_1, \ldots, x_i, \ldots, x_\nu, X_{\nu+1}, X_{\nu+2}, \ldots)$ is the dimensionless minimum function for fixed values $x_1, \ldots, x_\nu, X_{\nu+1}, \ldots$, then by differentiation of both sides of the normalization condition $\sum W(\mathbf{K}) = 1$ for $W(\mathbf{K})$ of the corresponding ensemble with respect to x_i we found the required thermodynamic relation

$$\frac{\partial\Theta}{\partial x_i} = X_i. \tag{5a.1}$$

Take the explicit case of the petite canonical ensemble,

$$W(\mathbf{K};\, V, \beta, \mathbf{N}) = \exp[\beta A - \beta E(\mathbf{K})], \tag{5a.2}$$

and differentiate with respect to β both sides of the equation,

$$\sum_{\mathbf{K}} W(\mathbf{K};\, V, \beta, \mathbf{N}) = 1. \tag{5a.3}$$

One finds

$$\sum_{\mathbf{K}} \left(\frac{\partial}{\partial\beta}\right)_{V,\mathbf{N}} W(\mathbf{K}) = 0 = \sum_{\mathbf{K}} W(\mathbf{K}) \left\{ \left[\frac{\partial(\beta A)}{\partial\beta}\right]_{V,\mathbf{N}} - E(\mathbf{K}) \right\}. \tag{5a.4}$$

Since $\sum W(\mathbf{K})E(\mathbf{K}) = E$, and $\partial(\beta A)\partial\partial\beta$ is a constant independent of $\mathbf{K}$, one finds, with eq. (3), that

$$\left[\frac{\partial(\beta A)}{\partial\beta}\right]_{V,\mathbf{N}} = \langle E \rangle, \tag{5a.5}$$

as a special case of eq. (1). Since with $\beta = 1/kT$ the thermodynamic relation is

$$\left[\frac{\partial(\beta A)}{\partial\beta}\right]_{V,N} = A + \beta\left(\frac{\partial A}{\partial\beta}\right)_{V,N} = A - T\left(\frac{\partial A}{\partial T}\right)_{V,N} = A + TS = E, \quad (5a.6)$$

we find that the average energy $\langle E\rangle$ of the ensemble satisfies the thermodynamic equation for energy.

Now go one step further and take the second derivative of both sides of eq. (3). We use eq. (5) to find

$$\frac{\partial^2 W(\mathbf{K}, V, \beta, \mathbf{N})}{\partial\beta^2} = W\left[\left(\frac{\partial\langle E\rangle}{\partial\beta}\right)_{V,N} + (\langle E\rangle - E(\mathbf{K}))^2\right]. \quad (5a.7)$$

Since $[\langle E\rangle - E(\mathbf{K})]^2 = E^2(\mathbf{K}) - 2\langle E\rangle E(\mathbf{K}) + \langle E\rangle^2$ and the sum of W times this is $\langle E^2\rangle - 2\langle E\rangle\langle E\rangle + \langle E\rangle^2$, we find from the condition $\sum \partial^2 W(\mathbf{K})/\partial\beta^2 = 0$ that

$$\left(\frac{\partial E}{\partial\beta}\right)_{V,N} = -(\langle E^2\rangle - \langle E\rangle^2)_{V,\beta,\mathbf{N}}. \quad (5a.8)$$

The subscripts V, β, $\mathbf{N}$ are placed after the combination of averages to emphasize that the averages are those in the petite canonical ensemble of fixed V, β, $\mathbf{N}$. Identify $\langle E\rangle$ with the thermodynamic energy and write C_V for the constant-volume heat capacity,

$$C_V = \left(\frac{\partial E}{\partial T}\right)_{V,N}. \quad (5a.9)$$

The result, from eq. (8), is that

$$kT^2 C_V = (\langle E^2\rangle - \langle E\rangle^2)_{V,\beta,\mathbf{N}}. \quad (5a.10)$$

Since, as shown in Sec. 1c [eq. (1c.9)], $\langle E^2\rangle - \langle E\rangle^2 \geq 0$ we have proved that C_V can never be negative.

Now let us use eq. (10) to examine the probability of fluctuations of energy in the ensemble of fixed V, β, $\mathbf{N}$, that is, of energy fluctuations in systems of constant volume and fixed molecular content in contact with a large heat bath of fixed temperature. Introduce a dimensionless quantity ζ which gives the fractional deviation of the energy E from its average value $\langle E\rangle$,

$$\zeta = \frac{(E - \langle E\rangle)}{\langle E\rangle}, \quad (5a.11)$$

and approximate the probability density of ζ by a Gaussian [eq. (1d.6)],

$$W(\zeta) = (2\pi\alpha^2)^{-1/2} \exp -\frac{1}{2}\left(\frac{\zeta}{\alpha}\right)^2, \quad (5a.12)$$

for which

$$\langle \zeta^2 \rangle = \alpha^2. \tag{5a.13}$$

From eqs. (10) and (11) we find, with $E = E$, that

$$\alpha^2 = \langle \zeta^2 \rangle = \frac{kT^2 C_V}{E^2}. \tag{5a.14}$$

Now with E measured from the energy of the system at 0 K the value of TC_V is of nearly the same order as E, except near a critical point, in any system existing in no more than a single thermodynamic phase. The quantity E/kT is within a few orders of magnitude equal to N, so that α does not differ greatly from $N^{-1/2}$. The probability density $W(\zeta)$ [eq. (12)] of a fractional fluctuation of energy by an amount greatly exceeding 1 part in $N^{1/2}$ is very, very small.

Had we chosen to use the grand canonical ensemble of fixed values V, β, $-\beta\boldsymbol{\mu}$, with

$$W(\mathbf{K}, V, \beta, -\beta\mu_\alpha, -\beta\mu_\gamma, \ldots) = \exp -[\beta PV + \beta E(\mathbf{K}) + \sum_\alpha (-\beta\mu_\alpha) N_\alpha(\mathbf{K})], \tag{5a.15}$$

instead of eq. (2), and gone through the equivalent steps of eqs. (3) and (4), we would have found, instead of eq. (5), that

$$\left[\frac{\partial(\beta PV)}{\partial\beta}\right]_{V,\beta\boldsymbol{\mu}} = \langle E \rangle. \tag{5a.16}$$

Going on to the steps in $\sum \partial^2 W(\mathbf{K})/\partial\beta^2 = 0$ one would find, instead of eq. (8) that

$$-\left(\frac{\partial E}{\partial \beta}\right)_{V,\beta\boldsymbol{\mu}} = kT^2 C^\dagger [\langle E^2 \rangle - \langle E \rangle^2]_{V,\beta,-\beta\boldsymbol{\mu}}, \tag{5a.17}$$

in which $C^\dagger$ is a very strange heat capacity[(1)] at constant volume, but a constant value of $\beta\mu_\alpha$ for all species of molecules α. In this change in temperature at constant V, $\beta\boldsymbol{\mu}$, the numbers N_α of molecules of each species α change. The fluctuation quantity $\langle E^2 \rangle - \langle E \rangle^2$ is in general very different in the open systems of fixed V, β, $-\beta\boldsymbol{\mu}$, within which the numbers of molecules $\mathbf{N}$ can also fluctuate, than in the systems of fixed V, β, $\mathbf{N}$.

Quite generally, if we go through the steps equivalent to those of eqs. (2) to (8) for any ensemble in which the ith variable is specified by the

(1) J. E. Mayer, *J. Phys. Chem.*, **66,** 591 (1962). The details are discussed in Sec. 9i.

intensive partner x_i, the equation equivalent to eq. (8) is found to be

$$\left(\frac{\partial X_i}{\partial x_i}\right)_{\mathbf{x}_k,\mathbf{X}_l} = -[\langle X_i^2\rangle - \langle X_i^2\rangle]_{x_i\mathbf{x}_k\mathbf{X}_l} \leq 0, \tag{5a.18}$$

whatever other variable set $\mathbf{x}_k\mathbf{X}_l$ with x_i characterizes the ensemble. It follows that $\partial X_i/\partial x_i$ is always negative. This is a generalization of Le Chatelier's principle.

Similarly, for an ensemble specified by x_i, x_j and any other set $\mathbf{x}_k$, $\mathbf{X}_l$ with minimum function $\Theta(x_i, x_j, \mathbf{x}_k, \mathbf{X}_l)$, the condition that the sum over $\mathbf{K}$ of $\partial^2 W(\mathbf{K})/\partial x_i\, \partial x_j$ be zero leads to

$$\sum_{\mathbf{K}} W(\mathbf{K})\left\{\left(\frac{\partial^2\Theta}{\partial x_i\, \partial x_j}\right)_{\mathbf{x}_k,\mathbf{X}_l} + [X_i - X_i(\mathbf{K})][X_j - X_j(\mathbf{K})]\right\} = 0 \tag{5a.19}$$

or

$$\left(\frac{\partial^2\Theta}{\partial x_i\, \partial x_j}\right)_{\mathbf{x}_k,\mathbf{X}_l} = \left(\frac{\partial X_i}{\partial x_j}\right)_{\mathbf{x}_k,\mathbf{X}_l} = \left(\frac{\partial X_j}{\partial x_i}\right)_{\mathbf{x}_k,\mathbf{X}_l}$$
$$= -[\langle X_iX_j\rangle - \langle X_i\rangle\langle X_j\rangle]_{x_ix_j\mathbf{x}_k\mathbf{X}_l}. \tag{5a.20}$$

The Cauchy–Schwarz inequality applies here, and one can prove, as follows, that

$$\left(\frac{\partial X_i}{\partial x_j}\right)^2 = \left(\frac{\partial X_j}{\partial x_i}\right)^2 \leq \frac{\partial X_i}{\partial x_i}\frac{\partial X_j}{\partial x_j}, \tag{5a.21}$$

whatever other variables are held constant. Let

$$\Delta X_i(\mathbf{K}) = X_i(\mathbf{K}) - \langle X_i\rangle, \tag{5a.22}$$

so that with a similarly defined $\Delta X_j(\mathbf{K})$ we have

$$\sum_{\mathbf{K}} W(\mathbf{K})\,\Delta X_i^2(\mathbf{K}) = \langle \Delta X_i^2\rangle = \langle X_i^2\rangle - \langle X_i^2\rangle = -\frac{\partial X_i}{\partial x_i}, \tag{5a.23}$$

and

$$\sum_{\mathbf{K}} W(\mathbf{K})\,\Delta X_i(\mathbf{K})\,\Delta X_j(\mathbf{K}) = \langle \Delta X_i\, \Delta X_j\rangle = \langle X_iX_j\rangle - \langle X_i\rangle\langle X_j\rangle = \frac{\partial X_i}{\partial x_j} = \frac{\partial X_j}{\partial x_i}. \tag{5a.24}$$

Now form the necessarily positive quantity (remember that $\langle \Delta X_j^2\rangle > 0$),

$$F(\mathbf{K}) = \langle \Delta X_j^2\rangle[\Delta X_i(\mathbf{K}) - \Delta X_j(K)\langle \Delta X_i\, \Delta X_j\rangle\langle \Delta X_j^2\rangle^{-1}]^2$$
$$= \langle \Delta X_j^2\rangle\,\Delta X_i^2(\mathbf{K}) - 2[\Delta X_i(\mathbf{K})\,\Delta X_j(\mathbf{K})]\langle \Delta X_i\, \Delta X_j\rangle$$
$$+ [\Delta X_j^2(K)\langle \Delta X_j^2\rangle^{-1}]\langle \Delta X_i\, \Delta X_j\rangle^2 \geq 0. \tag{5a.25}$$

The sum over $\mathbf{K}$ of $W(\mathbf{K})F(\mathbf{K})$ is necessarily equal to or greater than zero and is seen to be

$$\langle \Delta X_j^2 \rangle \langle \Delta X_i^2 \rangle - \langle \Delta X_i \, \Delta X_j \rangle^2 \geq 0, \tag{5a.26}$$

which, with eqs. (20), (23), and (25) proves (21).

Obviously, relations for higher derivatives of the X_j's with respect to the intensive variables $x_i, \ldots$ can be obtained in terms of averages of greater products. We do not display them.

5b. CHOICES OF VARIABLES

Usually, but not always, there is an obvious simplest choice of the extensive variables for a given system, for instance V, E and the numbers N_a, N_b, . . . of molecules of species a, b, ... In the case in which there is a chemical reaction, $2a + 3b \leftrightarrows c$, at complete equilibrium, the choice is noı quite so obvious. One might specify, in addition to V and E, the numbers $N_a^{(0)}$ and $N_b^{(0)}$ of species a and b, were there no molecules of c formed. There would then be an equilibrium number N_c^{eq} of molecules of type c present, and the actual numbers of a and b would be $N_a^{\text{eq}} = N_a^{(0)} - 2N_c^{\text{eq}}$ and $N_b^{\text{eq}} = N_b^{(0)} - 3N_c^{\text{eq}}$. Alternatively, preparation of the system by adding $K_c^{(0)}$ molecules of pure c to $K_a^{(0)}$ molecules of a,

$$K_a^{(0)} = N_a^{(0)} - \tfrac{2}{3} N_b^{(0)} \tag{5b.1}$$

and

$$K_c^{(0)} = \tfrac{1}{3} N_b^{(0)} \tag{5b.1$'$}$$

might seem an obvious choice. One would then have

$$N_b^{(\text{eq})} = 3(K_c^{(0)} - N_c^{\text{eq}}), \tag{5b.2}$$

$$N_a^{(\text{eq})} = K_a^{(0)} + 2(K_c^{(0)} - N_c^{\text{eq}}). \tag{5b.2$'$}$$

This choice could even be used with $K_a^{(0)}$ a negative number (but always greater than $-2K_c^{(0)}$).

In the above example the linear combination is of two quantities of the same physical dimensions, namely, pure numbers. In general, different variables have different physical dimensions, and a linear combination of two appears to be meaningless. However, one example that arises naturally should be mentioned. The zero from which the energy is measured is arbitrary. It is often customary in statistical mechanical calculations to choose zero energy as that of the system in its lowest energy quantum state, so that the energy is always positive and equal to the energy of the same system in excess of that at absolute zero. Sometimes, especially if calculations of isotopic equilibria are contemplated, the energy zero is chosen as that of the system at its lowest potential energy. If there are vibrational degrees of freedom, the energy of the lowest quantum state is greater by the sum of $\frac{1}{2}h\nu$ for all frequencies. There are then two (and indeed many more) alternative choices E and, for instance,

$$E' = E + N\,\Delta\epsilon, \tag{5b.3}$$

in a one-component system. If V and N are the other variables, we have

$$\beta = \left[\frac{\partial(S/k)}{\partial E'}\right]_{V,N} = \left[\frac{\partial(S/k)}{\partial E}\right]_{V,N}, \tag{5b.4}$$

unchanged by the choice. However,

$$-\beta\mu' = \left[\frac{\partial(S/k)}{\partial N}\right]_{V,E'} = \left[\frac{\partial(S/k)}{\partial N}\right]_{V,E} + \left[\frac{\partial(S/k)}{\partial E}\right]_{V,N}\left(\frac{\partial E}{\partial N}\right)_{V,E'}$$

$$= -\beta\mu - \beta\,\Delta\epsilon, \tag{5b.5}$$

or

$$\mu' = \mu + \Delta\epsilon. \tag{5b.5'}$$

The chemical potential is shifted in such a way that it is always measured from the chosen zero of energy per molecule.

More exotic choices of variables are possible but are seldom used in discussing purely equilibrium phenomena. For a treatment of the decay of a nonequilibrium system to its equilibrium thermodynamic state, a more general set of variables has some significance. We discuss here a general choice.

To make the discussion explicit we consider the grand canonical ensemble for which the thermodynamic state is specified by one extensive variable V, and all others are intensive, $x_1, \ldots, x_i, \ldots, x_n$. The minimum function is $\Theta(V, x_1, \ldots, x_n)$,

$$\Theta(V, x_1, \ldots) = -\frac{S}{k} + \sum_{i=1}^{i=n} x_i X_i = -\beta PV. \tag{5b.6}$$

Define, for $X_i \neq V$, $X_j \neq V$,

$$\Theta_{ij} = \left(\frac{\partial^2\Theta}{\partial x_i\,\partial x_j}\right)_{V,x_k\ldots} = \left(\frac{\partial X_i}{\partial x_j}\right)_{V,x_i} = \left(\frac{\partial X_j}{\partial x_i}\right)_{V,x_j,x_k,\ldots} \tag{5b.7}$$

so that, at constant V,

$$(dX_i)_V = \sum_j \Theta_{ij}\,dx_j. \tag{5b.8}$$

Now on the other hand,

$$\left[\frac{\partial(S/k)}{\partial X_i}\right]_{V,X_j,\ldots} = x_i, \tag{5b.9}$$

so that if one defines

$$\sigma_{ij} = \left[\frac{\partial^2(S/k)}{\partial X_i\,\partial X_j}\right]_{V,X_k,\dots}$$
$$= \left(\frac{\partial x_i}{\partial X_j}\right)_{V,X_i,X_k,\dots} = \left(\frac{\partial x_j}{\partial X_i}\right)_{V,X_j,\dots}, \tag{5b.10}$$

then

$$(dx_i)_V = \sum_j \sigma_{ij}\, dX_j. \tag{5b.11}$$

From eqs. (8) and (11) we have that

$$\sum_j \sigma_{ij}\theta_{jk} = \sum_j \theta_{ij}\sigma_{jk} = \delta(i-k) \tag{5b.12}$$

is zero if $i \neq k$, and unity if $i = k$.

The quantities $\theta_{ij} = \theta_{ji}$ of eq. (7) and $\sigma_{ij} = \sigma_{ji}$ of eq. (10) can both be regarded as elements of two square $n \times n$ real symmetric matrices, $\boldsymbol{\theta}$ and $\boldsymbol{\sigma}$, respectively. Expression (12) gives the rule for the i, k elements of the two products $\boldsymbol{\sigma\theta}$ and $\boldsymbol{\theta\sigma}$ as unity on the diagonal and zero for all off-diagonal terms, namely, the elements of the unit matrix. In matrix notation we have (see Appendix VII)

$$\boldsymbol{\sigma\theta} = \boldsymbol{\theta\sigma} = \mathbf{1}, \tag{5b.13}$$

with $\mathbf{1}$ the unit matrix. The matrices $\boldsymbol{\sigma}$ and $\boldsymbol{\theta}$ are said to be reciprocal matrices, $\boldsymbol{\sigma} = \boldsymbol{\theta}^{-1}$.

Now suppose we introduce n new independent extensive variables which are linear combinations of the X_i's,

$$Y_\nu = \sum_i r_{\nu i} X_i, \qquad 1 \leq \nu \leq n, \tag{5b.14}$$

and for convenience we choose the Y_ν's to be dimensionless by requiring each $r_{\nu i}$ to have physical dimensions reciprocal to X_i. The X_i's can be expressed as combinations of the Y_ν's,

$$X_i = \sum_\nu s_{i\nu} Y_\nu, \tag{5b.15}$$

where $s_{i\nu}$ has the physical dimensions of X_i. Necessarily, if the X_i's are independent,

$$\sum_i r_{\nu i} s_{i\mu} = \delta(\nu - \mu),$$
$$\sum_\nu s_{i\nu} r_{\nu j} = \delta(i-j), \tag{5b.16}$$

so that the matrices **r** and **s** of elements $r_{\nu i}$ and $s_{i\nu}$ are reciprocal real $n \times n$ matrices,

$$\mathbf{rs} = \mathbf{sr} = \mathbf{1}. \tag{5b.17}$$

By definition the intensive variable y_ν, conjugate to Y_ν is,

$$y_\nu = \left[\frac{\partial(S/k)}{\partial Y_\nu}\right]_{V,Y_\mu,\ldots} = \sum \left[\frac{\partial(S/k)}{\partial X_i}\right]_{V,X_j\cdots} \left(\frac{\partial X_i}{\partial Y_\nu}\right)_{V,Y_\mu}, \tag{5b.18}$$

or, since $\partial(S/k)/\partial X_i = x_i$, and with eq. (15) for X_i,

$$y_\nu = \sum_i x_i s_i, \qquad x_i = \sum y_\nu r_{\nu i}. \tag{5b.19}$$

This, then, is the general rule for transforming the thermodynamic variables. We can in general pick any independent set Y_ν given by coefficients $r_{\nu i}$ [eq. (14)], although we certainly must have the dimensions of $r_{\nu i}X_i$ the same for all i with any given ν-value. The matrix **s** of elements $s_{i\nu}$ is then the reciprocal of **r** and is unique if the Y_ν's are independent, so that the determinant of **r** is nonzero. The reciprocal intensive variables are given by eq. (19).

We seek now the transformations for $(\partial y_\nu/\partial Y_\mu)_{V,Y_\lambda,\ldots}$ and $(\partial Y_\nu/\partial y_\mu)_{V,y_\lambda,\ldots}$. Introduce matrices $\mathbf{r}^{(t)}$ and $\mathbf{s}^{(t)}$, which are the transposes of **r** and **s** having elements

$$r^{(t)}_{i\nu} = r_{\nu i}, \qquad s^{(t)}_{\nu i} = s_{i\nu}. \tag{5b.20}$$

These matrices are, from eq. (16), also reciprocal,

$$\mathbf{r}^{(t)}\mathbf{s}^{(t)} = \mathbf{s}^{(t)}\mathbf{r}^{(t)} = \mathbf{1}. \tag{5b.21}$$

Now, using $\sigma_{ij} = (\partial x_j/\partial X_i)_{V,X_k,\ldots}$ calculate $(\partial y_\nu/\partial Y_\mu)_{V,Y_\lambda,\ldots}$. Using eqs. (19) and (15) and defining $\sigma_{\mu\nu}$ as

$$\begin{aligned}\sigma_{\mu\nu} &= \left(\frac{\partial y_\nu}{\partial Y_\mu}\right)_{V,Y_\lambda} = \sum_i \left(\frac{\partial}{\partial X_i}\right)_{V,X_k} \left(\sum_j x_j s_{j\nu}\right)\left(\frac{\partial X_i}{\partial Y_\mu}\right)_{V,Y_\lambda,\ldots} \\ &= \sum_i \sum_j \sigma_{ij} s_{j\nu} s_{i\mu},\end{aligned}$$

or, using eq. (20), $s^{(t)}_{\mu i} = s_{i\mu}$,

$$\left(\frac{\partial y_\nu}{\partial Y_\mu}\right)_{V,Y_\nu,Y_\lambda,\ldots} = \left(\frac{\partial y_\mu}{\partial Y_\nu}\right)_{V,Y_u,Y_\lambda} = \left(\frac{\partial^2(S/k)}{\partial Y_\nu\,\partial Y_\mu}\right)_{V,Y_\lambda} = \sum_i \sum_j s^{(t)}_{\mu i}\sigma_{ij}S_{j\nu} = \sigma_{\mu\nu}. \tag{5b.22}$$

This is the μ, ν element of the product $\mathbf{s}^{(t)}\boldsymbol{\sigma}\mathbf{s}$. Quite similarly, using

$\theta_{ij} = (\partial x_j/\partial x_i)_{V,s_j,x_k}$ and eqs. (14) and (19) one finds that

$$\left(\frac{\partial Y_\nu}{\partial y_\mu}\right)_{V,y_\nu,y_\lambda,\ldots} = \theta_{\nu\mu} = \theta_{\mu\nu} = \sum_i \sum_j r_{\nu i}\theta_{ij} r^{(t)}_{j\mu}, \tag{5b.23}$$

which is the ν, μ element of $\mathbf{r\theta r}^{(t)}$.

We can check that

$$\sum_\mu \sigma_{\nu\mu}\theta_{\mu\lambda} = \sum_\mu \theta_{\nu\mu}\sigma_{\mu\lambda} = \delta(\nu-\lambda), \tag{5b.24}$$

as it must if we have made no error. The check is most easily made by the matrix product

$$\mathbf{s}^{(t)}\boldsymbol{\sigma}\mathbf{s}\mathbf{r}\boldsymbol{\theta}\mathbf{r}^{(t)} = \mathbf{1} = \mathbf{r}\boldsymbol{\theta}\mathbf{r}^{(t)}\mathbf{s}^{(t)}\boldsymbol{\sigma}\mathbf{s}, \tag{5b.25}$$

which follows from eqs. (13), (17), and (21) by consecutively multiplying the middle pair in order in each expression (remembering that for any matrix **a**, **b** that $\mathbf{a1b} = \mathbf{ab}$).

The extensive variables X_i, $X_j, \ldots$, with the exception of the volume V, are all ensemble average values of some functions $\chi_i(\mathbf{q}^{(\Gamma)}, \mathbf{p}^{(\Gamma)})$, $\chi_j(\mathbf{p}^{(\Gamma)}, \mathbf{p}^{(\Gamma)}), \ldots$ in classical mechanics and of operators $\boldsymbol{\chi}_i$, $\boldsymbol{\chi}_j$ in quantum mechanics, such that

$$X_i(\mathbf{K}) = \int dq^{(\Gamma)} \Psi^*_{\mathbf{K}}(q^{(\Gamma)}) \boldsymbol{\chi}_i \Psi_{\mathbf{K}}(q^{(\Gamma)}) \tag{5b.26}$$

is the expectation value of the classical quantity χ_i in the state **K**. The Y_ν of eq. (14) is the ensemble average of the classical function

$$Y_\nu(\mathbf{q}^{(\Gamma)}\mathbf{p}^{(\Gamma)}) = \sum_i r_{\nu i}\chi_i(\mathbf{q}^{(\Gamma)}\mathbf{p}^{(\Gamma)}) \tag{5b.27}$$

or operator

$$\mathbf{Y}_\nu = \sum_i r_{\nu i}\ \boldsymbol{\chi}_i, \tag{5b.28}$$

having expectation value $Y(\mathbf{K})$ in the quantum state **K**,

$$Y_\nu(\mathbf{K}) = \sum_i r_{\nu i} X_i(\mathbf{K}). \tag{5b.29}$$

All the relations of Sec. 5a for derivatives $\partial X_i/\partial x_j$ apply equally, replacing X_i's and x_j's by Y_ν's and y_μ's, in particular that of eq. (5a.18),

$$\left(\frac{\partial Y_\nu}{\partial y_\nu}\right)_{V,y_\mu,\ldots} = -(\langle Y^2\rangle - \langle Y\rangle^2) \le 0, \tag{5b.30}$$

and that of eq. (5a.21),

$$\left(\frac{\partial Y_\nu}{\partial y_\mu}\right)^2 \le \frac{\partial Y_\nu}{\partial y_\nu}\,\frac{\partial Y_\mu}{\partial y_\mu}, \tag{5b.31}$$

namely, in our matrix element notation that

$$\theta_{\nu\nu} \le 0 \qquad \text{and} \qquad \theta_{\nu\mu}^2 \le \theta_{\nu\nu}\theta_{\mu\mu}. \tag{5b.32}$$

The conditions of eq. (32) suffice to require that for the matrix $\boldsymbol{\sigma}$ reciprocal to $\boldsymbol{\theta}$ we also have

$$\left[\frac{\partial^2(S/k)}{\partial Y_\nu^2}\right]_{V,Y_\mu,\ldots} = \sigma_{\nu\nu} \le 0, \qquad \sigma_{\nu\mu}^2 \le \sigma_{\nu\nu}\sigma_{\mu\mu}, \tag{5b.33}$$

The requirement that the second derivative of the entropy with respect to any displacement of the extensive variables never be positive is a familiar thermodynamic requirement. It follows from the facts that entropy is extensive and is a maximum at equilibrium, and that for sufficiently large systems surface terms between two different thermodynamic phases can be neglected. The argument is given in the caption for Figure 5b.1.

The discussion in the caption appears to imply that the curve below the two-phase straight line of constant y value has some real significance.

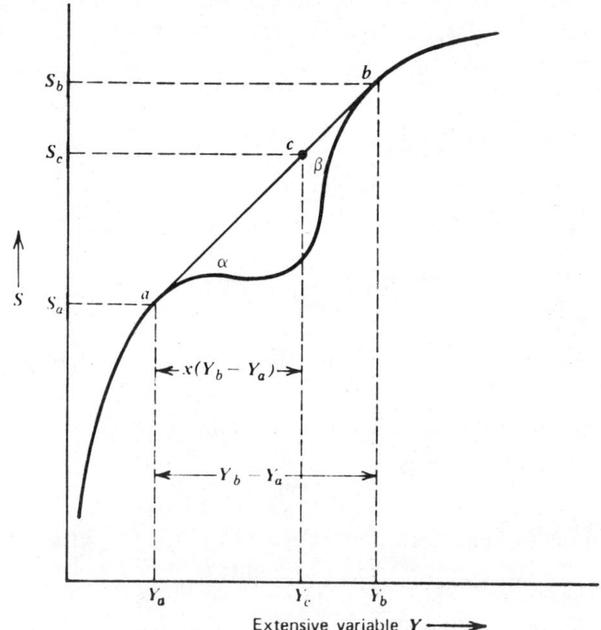

Figure 5b.1 Plot of entropy S versus any extensive variable Y for the hypothetical case that $\partial^2 S/\partial Y^2 > 0$ in a region between α and β. The two-phase system of fraction $(1-x)$ of phase a of S_a and Y_a and fraction x of phase b of S_b and Y_b has $Y_c = (1-x)Y_a + xY_b = Y_a + x(Y_b - Y_a)$ and $S_c = S_a + x(S_b - S_a)$. Its entropy lies on the straight line connecting points a and b above the hypothetical unallowed region and will be thermodynamically stable.

This is not necessarily so, and probably is never so for the region between α and β for which $\partial^2 S/\partial Y^2$ is positive. Since, however, metastable phases exist, the regions a to α and β to b for which $\partial^2 S/\partial Y^2 < 0$ correspond to metastable phase a and phase b, respectively.

The statistical mechanical calculation, if all possible quantum states, or, in a classical approximation, all parts of the phase space are included, follows the two-phase line. Only if a restriction of some sort is imposed which counts only the one-phase quantum states (or integrates only that part of phase space) will the metastable region be computed. It is very doubtful that any even reasonably realistic model or restriction can reproduce the curve through the portion where $d^2 S/\partial Y^2$ is positive.

5c. THE MAGNITUDE OF FLUCTUATIONS

In Sec. 5a we discussed the magnitude of fluctuations of energy in a system of constant V and N in thermal contact with a thermostat. The dimensionless quantity $\zeta = (E - \langle E \rangle)/\langle E \rangle$ giving the fractional deviation of the energy from its average value $\langle E \rangle$ had an average square value [eq. (5a.14)] of $\langle \zeta^2 \rangle = kT^2 C_V/E^2$ of the order of N^{-1} if C_V is finite of the order of $TC_V \cong E$. A fractional fluctuation of E much greater than 1 part in $N^{1/2}$ is extremely improbable. The more general case [eq. (5a.18)] that $-(\partial X_i/\partial x_i) = \langle X_i^2 \rangle - \langle X_i^2 \rangle$ permits us to write for a dimensionless ζ_i,

$$\zeta_i = \frac{(X_i - \langle X_i \rangle)}{\langle X_i \rangle}, \tag{5c.1}$$

that

$$\langle \zeta_i^2 \rangle = -\frac{(\partial X_i/\partial x_i)}{\langle X_i \rangle^2}, \tag{5c.2}$$

and normally, since X_i is extensive and proportional to N as is $\partial X_i/\partial x_i$, we have $\langle \zeta_i^2 \rangle$ not many orders of magnitude different from N^{-1}. The above statement is subject to two caveats.

The first of these is a purely formal correction which does not change the fact that the fluctuations are negligible for large numbers N of molecules. As long as X_i is one of the conservative variables E, N_α, $N_\gamma, \ldots$ and measured from a normal zero, the above statement holds for a system in a single thermodynamic phase. However, for variables such as electric or magnetic moment, or the elements of the strain tensor, the natural zero for X_i is usually such that $\langle X_i \rangle$ is zero in the absence of an external force. The quantity ζ_i defined by eq. (1) has no realistic meaning. If we choose ΔX_i to be some value of X_i giving the accuracy within which the variable is normally measured for a mole of material, we can redefine, for a system of n moles,

$$\zeta_i = \frac{X_i - \langle X_i \rangle}{n\,\Delta X_i}. \tag{5c.3}$$

We still find $\langle \zeta_i^2 \rangle$ proportional to N^{-1} and fluctuations many orders of magnitude greater than $N^{-1/2}\,\Delta X_i$ very rare in one-phase systems.

The second limitation to the statement that fluctuations are small is a real one but is consistent with a limitation on the thermodynamic choice of variables. For any thermodynamic system at least one extensive variable must be specified to define the full thermodynamic state, since the intensive variables never define the extent of the system. If the state is one for which two thermodynamic phases coexist at equilibrium, *two* extensive variables are necessary for a complete specification in which the amount of each phase is detailed. For instance, with one chemical component P, T, and N fully define the state for a single thermodynamic phase but, on the vapor-pressure line $P(T)$, the volume V may take any value between V_l with all N molecules in a liquid, and V_g with all in the gas phase, $V_l \leq V \leq V_g$. Similarly, with V, β, $-\beta\mu$ specified with $\mu(\beta)$ having the value along the liquid-vapor coexistence line, N can take all values between N_g and N_l. These are the variables of the grand canonical ensemble. At any point on the coexistence line,

$$-\left[\frac{\partial N}{\partial(-\beta\mu)}\right]_{V,\beta} = \left(\frac{\partial N}{\partial \mu}\right)_{V,T} = (\langle E^2\rangle - \langle E\rangle^2)_{V,\beta,-\beta\mu} = \infty.^{(1)} \tag{5c.4}$$

In general, if the thermodynamic state is such that n phases coexist and less than n extensive variables are specified, there will be at least one line Y_ν, a "tie line," along which $(\partial Y_\nu/dy_\nu)$ is negatively infinite, and the fluctuations are unlimited in the corresponding ensemble. On the tie line y_ν is constant. It corresponds to the straight-line region of Fig. (5b.1).

On the coexistence line, if the grand canonical partition function $Q(V, \beta, -\beta\mu)$ of eq. (3i.11′) is written as

$$Q(V, \beta, -\beta\mu) = \sum_N T(N), \tag{5c.5}$$

$$T(N) = Q(V, \beta, N) \exp N\beta\mu, \tag{5c.6}$$

the values of ln $T(N)$ will be effectively constant for all values of N between N_{gas} and N_{liq}. The large contribution to the N values between these two limits does not come from states corresponding to uniform density, but from states with part of the volume having gas density and the rest liquid density.

(1) Actually, $\langle E^2\rangle - \langle E\rangle^2$ is infinite only in the limit $N \to \infty$, but it is no longer of the order of $\langle E\rangle$ but rather of the order $\langle E\rangle^2$.

The statistical mechanical partition function with only one extensive variable gives the same information in the two-phase region that the thermodynamic variables do. It does not completely define the state of the system. Large macroscopic fluctuations in at least one extensive variable are predicted. The use of the grand canonical ensemble is not erroneous, but the information it gives is incomplete. Similarly, in the one-phase region very close to a critical point, the fluctuations are large and approach infinite values as the critical point is approached.

This constancy of $\ln T(N)$ over a considerable range, $N_g \le N \le N_l$, for particular paired values of β, $\beta\mu$ and in the range where $\ln T(N)$ has its maximum value does not invalidate the use of this maximum value as the value of the logarithm of the sum $\ln \sum T(N)$. This is evident from the nature of the proof for eq. (3i.15). However, now the effective number of distributions D (effective) discussed at the end of Sec. 4d, namely, the number of nonnegligible distributions, no longer needs to be many orders of magnitude smaller than the total number D of possible distributions.

5d. HEAT CAPACITY WITH SEPARABLE HAMILTONIAN

As discussed in Sec. 3d, if the classical Hamiltonian $H(\mathbf{q}^{(\Gamma)}, \mathbf{p}^{(\Gamma)})$ of the system for all Γ degrees of freedom is separable as a sum $\sum_i H_i(\mathbf{q}^{(\gamma)}, \mathbf{p}^{(\gamma)})$, with $\sum_i \gamma_i = \Gamma$ and no coordinate or momentum in the subset i appearing in any other subset $j \neq i$, then the quantum-mechanical Hamiltonian is also separable, and the solutions $\Psi_{\mathbf{K}'}$ of the time-independent Schrödinger equation are products and the energies are sums,

$$E(\mathbf{K}') = \sum_i \epsilon_i(\mathbf{n}_{\gamma i}). \tag{5d.1}$$

The quantum state index $\mathbf{K}'$ for the system is the set of quantum states $\mathbf{n}_{\gamma i}$ for all subsets, and the probabilities $W(K')$ are products,

$$W(K') = \prod_i W_i(\mathbf{n}_{\gamma i}). \tag{5d.2}$$

With partition functions Q_i defined by

$$Q_i = \sum_{\mathbf{n}_{\gamma i}} \exp\{-[\beta\epsilon_i(\mathbf{n}_{\gamma i})]\}, \tag{5d.3}$$

we have

$$W_i(\mathbf{n}_{\gamma i}) = Q_i^{-1} \exp\{-[\beta\epsilon_i(\mathbf{n}_{\gamma i})]\}, \tag{5d.4}$$

and[1]

$$-Q_i^{-1}\left(\frac{\partial Q_i}{\partial \beta}\right)_V = \sum_{\mathbf{n}_{\gamma i}} \epsilon_i(\mathbf{n}_{\gamma i}) W_i(\mathbf{n}_{\gamma i}) = \langle \epsilon_i \rangle. \tag{5d.5}$$

[1] For some degrees of freedom, for instance, for translational degrees bounded by the walls, the energies $\epsilon(\mathbf{n}_{\gamma i})$ depend on V. The differentiation in eq. (5) is at constant volume.

The total average energy, from eqs. (1) and (2), is

$$\langle E\rangle = \sum_{K'} E(\mathbf{K}')\,W(\mathbf{K}') = \sum_i \sum_{\mathbf{n}_{\gamma i}} \epsilon_i(\mathbf{n}_{\gamma i})\,W_i(\mathbf{n}_{\gamma i}) = \sum_i \langle \epsilon_i\rangle, \tag{5d.6}$$

and

$$C_V = \left(\frac{\partial\langle E\rangle}{\partial T}\right)_V = \sum_i \left(\frac{\langle\partial\epsilon_i\rangle}{\partial T}\right)_V = -k\beta^2 \sum_i \left(\frac{\partial\langle\epsilon_i\rangle}{\partial\beta}\right)_V. \tag{5d.7}$$

With eqs. (4) and (5) we have that

$$\begin{aligned} -\left(\frac{\partial\epsilon_i}{\partial\beta}\right)_V &= -\sum_{\mathbf{n}_{\gamma i}} \epsilon_i(\mathbf{n}_{\gamma i}) \left[\frac{\partial W_i(\mathbf{n}_{\gamma i})}{\partial\beta}\right]_V \\ &= \sum_{\mathbf{n}_{\gamma i}} \epsilon_i(\mathbf{n}_{\gamma i})[\epsilon_i(\mathbf{n}_{\gamma i}) - \langle\epsilon_i\rangle]\,W_i(\mathbf{n}_{\gamma i}) \\ &= \langle\epsilon_i^2\rangle - \langle\epsilon_i\rangle^2, \end{aligned} \tag{5d.8}$$

so that

$$C_V = k\beta^2 \sum_i [\langle\epsilon_i^2\rangle - \langle\epsilon_i\rangle^2]. \tag{5d.9}$$

As a check that eq. (9) is consistent with the general relation that

$$C_V = k\beta^2(\langle E^2\rangle - \langle E\rangle^2) \tag{5d.10}$$

of eq. (5a.10), we use eq. (6) for $\langle E^2\rangle$ to write

$$\begin{aligned} \langle E^2\rangle &= \sum_{K'} \left|\sum_i \epsilon_i(\mathbf{n}_{\gamma i})\right|^2 W_{K'} \\ &= \sum_i \sum_{\mathbf{n}_{\gamma i}} [\in_i(\mathbf{n}_{\gamma i})]^2\,W_i(\mathbf{n}_{\gamma i}) + 2\sum_{i>j} \sum \sum_{\mathbf{n}_{\gamma i}} \sum_{\mathbf{n}_{\gamma j}} \in_i(n_{\gamma i})_{\in j}(n_{\gamma j})\,W_i(n_{\gamma i})\,W_n(n_{\gamma j}) \\ &= \sum_i \langle\epsilon_i^2\rangle + 2\sum_i \sum_j \langle\epsilon_i\rangle\langle\epsilon_j\rangle, \end{aligned} \tag{5d.11}$$

and, from eq. (6),

$$\langle E\rangle^2 = \sum_i (\langle\epsilon_i\rangle)^2 + 2\sum_{i>j} \sum \langle\epsilon_i\rangle\langle\epsilon_j\rangle, \tag{5d.12}$$

so that eq. (10) does lead to eq. (9).

One very important caveat must be noted about the use of eq. (9). We have discussed probabilities and sums over the quantum states $\mathbf{K}'$ which are quantum numbers of states of solutions of the time-independent Schrödinger equation. With Boltzmann counting these are all equally weighted, the assumption being that a linear combination of $N!$ (or of $\prod_\alpha N'_\alpha$ if there are N_α molecules of species α) of them can be used to form a state of the correct allowed symmetry either symmetric in all pair exchanges of identical molecules or antisymmetric in such exchanges. If the Boltzmann approximation is inadequate, eq. (9) is wrong. The symmetry requirement either excludes certain combinations of the quantum numbers $\mathbf{n}_{\gamma i}$ or gives them an abnormal weight. For instance, in the case of a perfect gas the subsets i are the coordinates and momenta of single-numbered molecules, and $\mathbf{n}_{\gamma i}$ the quantum numbers of the individual molecules. For Fermi–Dirac systems, no two molecules may have the same fully defined quantum numbers $\mathbf{n}_{\gamma i}$. For Bose–Einstein systems, states with two or more molecules in the same state, $\mathbf{n}_{\gamma i} = \mathbf{n}_{\gamma j}$, $i \neq j$, have an extra weight, since fewer than $N!$ states K' are required to make a symmetric allowed state. There is thus a mutual statistical interaction between the molecules, which makes the probabilities of quantum state occupancy nonindependent.

5e. AVERAGE CLASSICAL KINETIC ENERGY

For any system composed of molecules obeying the laws of *classical* mechanics it is simple to derive that the total average energy at equilibrium must be *at least* $\frac{1}{2}kT$ times the number of degrees of freedom plus another term with an arbitrary zero that has a positive temperature coefficient, so that the heat capacity at constant volume can never be less than $\frac{3}{2}N_a k$, with N_a the number of *atoms* in the system. For diatomic molecules with two atoms per molecule a mole of gas containing Avogadro's number N_0 of molecules and $2N_0$ atoms the heat capacity C_V is at least $3N_0 k = 3R$, with R the gas constant per mole. The measured value of C_V near room temperature of most diatomic molecules is very close to $\frac{5}{2}R$, less than the required minimum value. This discrepancy between experiment and a rigorously proved mathematical theorem led many of the most eminent scientists of the late nineteenth and early twentieth century to doubt the reality of the atomic molecular hypothesis. The discrepancy is increased when one recognizes that the existence of atomic and molecular spectra shows that atoms and molecules must have a complex structure, hence many more degrees of freedom than the three translational coordinates per point mass atom.

The solution lay not in the rejection of the atomic hypothesis but in the then unthinkable failure of classical mechanics to apply to motions of particles with the low masses of atoms and electrons in a potential that varies in the short distances within molecules. Quantum mechanics predicts that many of the degrees of freedom, particularly those of electrons, are frozen and inactive up to temperatures very considerably above room temperature, in some cases several thousands of degrees.

We proceed to demonstrate the classical theorem. The theorem rests on the fact that, if cartesian coordinates and momenta are used, the classical Hamiltonian is separable in a kinetic and potential energy term, and that the kinetic Hamiltonian is separable as a sum of independent terms for each degree of freedom. Namely, for N point masses, $1 \leq i \leq N$, we have

$$H(\mathbf{x}^{(3N)}, \mathbf{p}^{(3N)}) = \sum_{i=1}^{i=N} \sum_{\alpha=x,y,z} \frac{1}{2m_i} p_{\alpha i}^2 + U(x_1, \ldots, z_N). \tag{5e.1}$$

The potential energy function U, which depends only on the coordinates in this cartesian representation, may be inordinately complicated, but for our present theorem this does not concern us. The probability density $W(\mathbf{x}^{(3N)}, \mathbf{p}^{(3N)})$ is proportional to $\exp(-\beta H)$ and is a product of $3N+1$ independent probabilities,

$$W(\mathbf{x}^{(3N)}, \mathbf{p}^{(3N)}) = \prod_{i=1}^{i=3N} \prod_{\alpha=x,y,z} W_{\alpha i}(p_{\alpha i}) W_{\mathbf{x}}(x_1, \ldots, z_N). \tag{5e.2}$$

The $3N$ probability densities $W_{\alpha i}(p_{\alpha i})$ are all of the same analytic form,

$$W_{\alpha i}(p_{\alpha i}) = \left(\frac{\beta}{2\pi m_i}\right)^{1/2} \exp(-\beta \tfrac{1}{2} m_i^{-1} p_{\alpha i}^2), \tag{5e.3}$$

normalized to unity

$$\int_{-\infty}^{+\infty} dp_\alpha W_{\alpha i}(p_{\alpha i}) = 1. \tag{5e.4}$$

The average kinetic energy for each of the $3N$ degrees of freedom is

$$\begin{aligned}
&\int_{-\infty}^{+\infty} dp_{\alpha i} (\tfrac{1}{2} m_i p_{\alpha i}^2) W_{\alpha i}(p_{\alpha i}) \\
&\quad = \pi^{-1/2} kT \int_{-\infty}^{+\infty} \left(\frac{\beta}{2m_i}\right)^{1/2} dp_{\alpha i} \left[\left(\frac{\beta}{2m_i}\right)^{1/2} p_{\alpha i}\right]^2 \exp\left\{-\left[\left(\frac{\beta}{2m_i}\right)^{1/2} p_{\alpha i}\right]^2\right\} \\
&\quad = \tfrac{1}{2} kT,
\end{aligned} \tag{5e.5}$$

since $\int_{-\infty}^{+\infty} dz\, z^2 e^{-z} = \frac{1}{2}\pi^{1/2}$. Thus the contribution of the kinetic energy of each degree of freedom is $\frac{1}{2}kT$ and to the heat capacity at constant volume is $\frac{1}{2}k$. The probability density in the coordinate part of the phase space is, in this cartesian choice of coordinates and momenta, independent of that in the momenta part. As discussed in Sec. 5d, the contribution of this independent part of the Hamiltonian, $U(x_1, \ldots, z_N)$, to the heat capacity is independently additive and is $k\beta^2(\langle U^2 \rangle - \bar{U}^2)$ which can never be negative.

We have thus completed the classical proof that, for a system containing N atoms,

$$C_V = k\tfrac{3}{2}N + \beta^2(\langle \bar{U}^2 \rangle - U^2) \geq \tfrac{3}{2}Nk. \tag{5e.6}$$

For a monatomic perfect gas, for which the potential energy is zero, the heat capacity at constant volume is indeed $\frac{3}{2}R$ per mole. This was known from experimental measurements on mercury vapor, the only monatomic gas then known, in the last quarter of the nineteenth century and constituted a triumph of the theory, but measurements on diatomic gases proved disastrous.

One comment belongs here. In a suitable choice of coordinates and conjugate momenta, for instance, the cartesian, the Hamiltonian is separable as kinetic plus potential energy, dependent on the momenta only and coordinates only, respectively. This separation never occurs in the quantum-mechanical Hamiltonian, since each $p_{\alpha i}$ in the classical function is replaced by $(h/2\pi i)\partial/\partial q_{\alpha i}$ in the quantum-mechanical operator.

CHAPTER 6

CRITIQUE

6a. INTRODUCTION

There is probably no other well-established field of theoretical physical science that is as much plagued by paradox and criticism of its fundamental logic as is statistical mechanics. In other realms such as classical or quantum mechanics, electromagnetism, thermodynamics, and so on, the basis is a generalization of observed experimental behavior. The generalization is reduced to one, or more often a few, fundamental laws cast in mathematical form. These laws permit the prediction, in quantitative form, of the results of a vast body of experiments. The validity of the laws is proved by the enormous variety of facts that are correctly predicted, and the absence of failures other than those due merely to the difficulty of exact numerical solution of the mathematical equations.

The same criterion can be applied to the theory of the statistical mechanics of equilibrium systems. The equation for the probability distribution of the grand canonical ensemble can be stated as a dogmatic law and used to compute the thermodynamic behavior of any arbitrary molecular system at equilibrium, subject only to the usual limitation that, using present techniques, for sufficiently complicated cases the exact numerical solution may be too difficult. The results agree with observational facts and are never in contradiction with them. This proves the validity of the method in the same sense that the validity of molecular quantum mechanics is proven by its empirical validity. This method is that adopted in Part 3 of this book.

But statistical mechanics can legitimately be required to answer a far more stringent requirement, and here much more difficulty is encountered. We accept the laws of quantum mechanics, or often the asymptotically valid laws of classical mechanics, as applying to all the laboratory systems for which we use equilibrium statistical mechanics. These laws are deterministic in nature in the following sense. Given the initial microscopic state of a system at time $t=0$ by giving all coordinates $\mathbf{q}_0^{(\Gamma)}$ and all momenta $\mathbf{p}_0^{(\Gamma)}$, the classical laws predict, for a truly isolated system, the values $\mathbf{q}^{(\Gamma)}(t)$ and $\mathbf{p}^{(\Gamma)}(t)$, hence the values of all observable properties at a later time t.

Determinism in quantum mechanics is more sophisticated in nature, but a similar predictability follows from its laws. The initial quantum state at time $t=0$ can be defined in various ways, of which one is to say that it is described by a function $\Psi(\mathbf{q}^{(\Gamma)}, t=0)$ of the coordinates $\mathbf{q}^{(\Gamma)}$. This function changes in time by an equation known as the time-dependent Schrödinger equation,

$$\frac{\partial\Psi(\mathbf{q}^{(\Gamma)}, t)}{\partial t}=\frac{2\pi i}{h}\mathscr{H}\Psi(\mathbf{q}^{(\Gamma)}, t), \tag{6a.1}$$

in which $\mathscr{H}$ is the quantum mechanical Hamiltonian operator. This equation can even be formally integrated, so wonderful is the formalism of mathematics, to write

$$\Psi(\mathbf{q}^{(\Gamma)}, t)=\left(\exp\frac{2\pi i}{h}\mathscr{H}t\right)\Psi(\mathbf{q}^{(\Gamma)}, t=0), \tag{6a.2}$$

where, by definition,

$$\exp\frac{2\pi i}{h}\mathscr{H}\equiv\sum_{n=0}^{\infty}\frac{1}{n!}\left(\frac{2\pi i}{h}\mathscr{H}\right)^n, \tag{6a.3}$$

and $\mathscr{H}^n$ means n-fold repeated operation. With $\Psi(\mathbf{q}^{(\Gamma)}, t)$ given, the average or expectation value of any physical property at time t is determined.

It does not matter that we may have infinite difficulty in determining the initial state. Our present wisdom assumes that the system knows its own state and, in the complete absence of externally applied perturbation, evolves in a predetermined manner.

Now in this case it follows that *no other independent laws exist* influencing the evolution of an isolated system. Any other law either invalidates our assumption of the correctness of the laws of mechanics, or it is consistent with them, hence derivable from mechanics. There exists in macroscopic systems a set of other laws, those of thermodynamics. If we believe the laws of mechanics and those of thermodynamics, then the latter must be derivable from the former. We have formulated such a derivation in Chapter 4.

The criticism of the derivation often takes an unjustified, although intuitively quite natural, form. Since the laws of mechanics and those of thermodynamics can be expressed in concise mathematical expressions, one might expect that a derivation could follow the accepted formalism of a rigorous mathematical proof, in short, that we should demand mathematical rigor as though no more wordy logic could be compelling. Unfortunately, an examination of the behavior of experimental systems shows quite clearly that this is an ephemeral ideal.

There are two lacunae in the derivation of Chapter 4. The first of these is in the assumption called the ergodic hypothesis. In Chapter 4 we identified S/k with the natural logarithm of the number Ω of quantum states available to a defined system of fixed $V, E, N_\alpha, \ldots$ and any other set of extensive variables defining the thermodynamic state. We tacitly assumed that this number is clearly defined, and in deriving the equal probability of all quantum states we assumed that there exists at least one chain of transitions of nonzero probability between all pairs of these states. This is the ergodic hypothesis. In classical mechanics the equivalent statement is that the phase volume is uniquely defined, and that all (except possibly in cases of vanishingly small phase volume) systems traverse this whole phase volume before returning (infinitesimally close) to their initial phase point. This assumption is discussed in the next sections.

The discussion there completely ignores a great body of mathematical literature going back to the nineteenth century, which treates the rigorous mathematical problem, usually with the assumptions of classical mechanics. This literature is imposing and has been contributed to by many of the best mathematicians of the past century. The ground rule of the treatments is to assume a *completely* isolated system with time-independent forces. One can readily prove[(1)] that strict ergodicity cannot exist under these assumptions. Phase paths passing through a point P in the phase space cannot pass through every point Q. The Ehrenfests[(2)] introduced the quasi-ergodic hypothesis, that for all points P the paths through phase space pass arbitrarily close to every point Q, except for points P of zero measure. A good summary of this mathematical literature is given by Munster.[(3)] One might be inclined to assume that quantum mechanics, with the uncertainty principle, would remove the difficulties inherent in the classical treatment. This is by no means so. For one thing, with strictly time-independent forces at the walls, the states described by solutions to the time-independent Schrödinger equation, $\mathcal{H}\Psi_{\mathbf{K}} = E\Psi_{\mathbf{K}}$, are truly stationary; there are no transitions between states

(1) See, for instance, A. Rosenthal, *Ann. Phys.*, **42,** 796 (1913); M. Plancherel, *Ann. Phys.*, **42,** 1061 (1913).

(2) P. Ehrenfest and T. Ehrenfest, *Encyclopädi der Mathematischen Wissenschaften*, Vol. IV, Pt. 32, B. G. Teubner, Leipzig, 1911.

(3) Arnold Munster, *Statistical Thermodynamics*, Springer-Verlag, New York, 1969, p. 16 ff.

of different quantum numbers K, K'. It is only when we use an approximate Hamiltonian $\mathcal{H}^{(0)}$ to find the states that we can assume the existence of a perturbation, $\Delta\mathcal{H} = \mathcal{H} - \mathcal{H}^{(0)}$, which causes transitions between the approximately stationary states. Tolman[4] has discussed the axiomatics in the quantum-mechanical case.

We adopt the point of view that we are concerned with applications to real laboratory systems. These are never completely isolated. Randomly fluctuating forces are always exerted by the walls. If the walls are part of the system then these systems are subjected to fluctuations due to emitted or impinging blackbody radiation. The problems are different than those of the mathematical idealization.

The second hole in our argument of the preceding chapter does not invalidate the proof that eventual equilibrium is that of maximum entropy. It is that we have not proved that, in the transition of an isolated nonequilibrium system from a state of low entropy to the highest entropy state of equilibrium, the path can never *reproducibly* pass through a thermodynamic state of still lower entropy than the initial one. This would permit a clever experimentalist to manipulate the conditions in such a way as to violate the second law. We discuss this, and certain spin echo experiments that appear to be exceptions to this rule, in Sec. 6f.

This latter problem of the uniform increase in entropy in all systems is intimately connected with a large number of paradoxes which have been raised. All of these in turn originate in the fact that the laws of mechanics are reversible with respect to time, whereas in thermodynamics time's arrow has a unique direction. Time enters the laws of mechanics in essentially the same way that a coordinate does. It makes no difference to the fundamental laws whether we measure the coordinate x as increasing from left to right or from right to left, nor is any fundamental change made in the laws by replacing t with $-t$. For every state of a given set of coordinates and momenta there exists another of the same coordinates but reversed momenta which evolves with positive t in the same way that the first one does with negative t. Little wonder that there are paradoxes connected with the derivation of a law that a quantity S, given by an integral over coordinates and momenta, must increase only along the positive axis of time. Asked how one could tell whether a given phenomenon involved entropy production or not Gilbert N. Lewis replied, "Take a moving picture of it, and run the film backwards. If it looks natural entropy production plays no role, if it looks ridiculous entropy is produced."

(4) Richard Tolman, *The Principles of Statistical Mechanics,* Oxford University Press, London, 1938.

In recent years the behavior of some meson reactions in high-energy phenomena have led to doubt as to whether *all* purely mechanical behavior is indeed completely invariant with respect to time reversal. If firmly established, this may have a real effect on our philosophical understanding of cosmology. It does not alter the well-established fact that the parts of the laws of mechanics that influence the motion of molecules for reasonable-temperature laboratory systems are invariant under time reversal.

The actual increase in entropy in one direction is a consequence of mechanical laws which are themselves ignorant of the direction of time. Probably the most instructive understanding of entropy increase is that it is no more mysterious than the necessity for shuffling a new deck of cards before use. A later shuffling of a disordered deck may produce a perfectly ordered one, but only with a probability of 1 part in $52! \cong 8 \times 10^{67}$. Nevertheless, the paradoxes raised by the time reversibility of mechanics seem to be worthy of discussion.

6b. THE ERGODIC HYPOTHESIS

It is always assumed that the macroscopic state prescription, for instance, in the simplest one-chemical-component system V, E, and N, completely determines a set of $\Omega(V, E, N)$ quantum states $\mathbf{K}$, $1 \leq \mathbf{K} \leq \Omega$, and that between any two states $\mathbf{K}$, $\mathbf{K}'$ there is at least one chain of allowed transitions of nonzero transition probability. This is the ergodic hypothesis. No mathematical proof of the general case can ever be made, for the very simple reason that it is not always true. We can, however, discuss the consequences of its failure to be correct.

We first briefly mention a trivial difficulty connected with an ambiguity in the choice of thermodynamic variables. Consider a system of fixed V and E containing three kinds, a, b, and c, of molecules for which a chemical reaction $\mathrm{a}+\mathrm{b}=\mathrm{c}$ can be written. If the reaction proceeds only infinitely slowly, we need the numbers N_a, N_b, and N_c of the three molecular species in addition to V and E to describe the state. If the reaction proceeds rapidly, we require only $N_a^{(0)}$ and $N_b^{(0)}$, the numbers of a and b that would be present were no c formed. The decision as to fast or slow reaction depends not only on its natural half-time under the conditions of the state, but also on the time with which we assume observations are made. If the half-time is, say, about a minute, that is, 10^2 sec, then for measurements that take only 10^{-3} sec or less, and there are many such, we should require the specification of all three numbers in a nonequilibrium system. For measurements requiring an hour the numbers $N_a^{(0)}$ and $N_b^{(0)}$ suffice.

The appropriate number of quantum states depends on this choice. Only by a complete theoretical analysis using nonequilibrium statistical mechanics can we decide a priori on the reaction rate. Usually, we are forced to resort to experimental evidence to make the correct choice.

A far less trivial case is offered by a number of known and understood examples in which the systems are truly nonergodic. This means that for an ensemble of given V, E, $N_a, \ldots$ there is a total of $\Omega_t(V, E, \mathbf{N})$ quantum states, but that these fall into Q blocks, labeled with i, $1 \leq i \leq Q$, having $\Omega_i(V, E, \mathbf{N})$ states in block i, and that no transitions in a single system at the energy E occur between states $\mathbf{K}^{(i)}$ and $\mathbf{K}'^{(j)}$ in different blocks $i \neq j$.

Two typical examples occur, frozen crystalline solid solutions and frozen disordered configurational arrangements of molecules at lattice sites, as in CO. In both cases the same analysis applies.

There are many examples of crystalline solid solutions in which, at sufficiently high temperatures, the lattice site occupancy of molecules or atoms of type a or of type b is random. Presumably, as T approaches 0 K, these are never at true thermodynamic equilibrium. The equilibrium state would be either a two-phase mixture of pure a and pure b, or an ordered one-phase lattice having some regularly defined periodic occupancy by alternating a and b species. Sometimes at a temperature T_c a sharp observable transition occurs to the ordered phase for $T < T_c$. Lattice diffusion is sufficiently rapid to establish equilibrium if dT/dt is not too great. In other cases diffusion at and below T_c is so slow that the equilibrium ordered phase is hard to obtain, or even completely unknown. In many of the former cases the disordered supercooled phase can be obtained by rapid quenching from a temperature above T_c. We discuss as an example the case of an equimolar binary solution for which the equilibrium low-temperature phase is ordered. The behavior of crystalline CO is mathematically equivalent. The CO molecule is isoelectronic with N_2 and has an extremely small dipole moment. Molecules orient randomly on lattice sites with "carbon up" or "carbon down." An ordered arrangement must be thermodynamically stable below some temperature T_c, but rotation of the molecules is frozen.

In both cases a crystal of N lattice sites has $Q = 2^N$ possible configurations i, $1 \leq i \leq Q$. Of these, one is ordered and has a lower potential energy ΔE, proportional to N, than the disordered configurations have. Actually, of course, there is presumably a small number n of these, but with $\lim_{N\to\infty}[N^{-1}(\ln n)] = 0$. Similarly, there must be semiordered configurations with intermediate energies. In the solid solution example, if a petite canonical ensemble is used, the limitation to exactly $N_a = N_b = \frac{1}{2}N$ requires Q to be less than 2^N, but by a factor whose logarithm is negligible compared to N. In order to clarify the situation we discuss an approximate model for which $Q=2^N$ and there is only one ordered configuration. To each configuration i there are $\Omega_{\text{vib}}(T)$ quantum states of vibration, which we assume to be the same for all configurations including the ordered one. The total number of states in the ordered phase is then $\Omega_{\text{vib}}(T)$, but in the disordered phase is $2^N\,\Omega_{\text{vib}}(T)$, so that

$$S_{\text{ord}} = k \ln \Omega_{\text{vib}}(T), \tag{6b.1}$$

$$S_{\text{dis}} = k[\ln \Omega_{\text{vib}}(T) + N \ln 2]. \tag{6b.2}$$

We can relate ΔE to T_c by the condition that the Gibbs free energy G is equal for the two phases at $T=T_c$ or, since $\Delta G = G_{\text{dis}} - G_{\text{ord}} \cong \Delta A = E_{\text{dis}} - E_{\text{ord}} - (TS_{\text{dis}} - TS_{\text{ord}})$, that for $\Delta E = E_{\text{dis}} - E_{\text{ord}} > 0$ we have

$$\Delta E = NkT_c \ln 2. \tag{6b.3}$$

For true equilibrium at temperature T with all quantum states attainable by lattice site diffusion from all others, the relative probability $W_{\text{ord}}/W_{\text{dis}}$ of the two phases is

$$\frac{W_{\text{ord}}}{W_{\text{dis}}} = 2^{-N} \exp \frac{\Delta E}{kT}. \tag{6b.4}$$

With eq. (3), since $W_{\text{ord}} + W_{\text{dis}} = 1$, we have

$$W_{\text{ord}} = 2^{N(T_c/T)}(2^N + 2^{N(T_c/T)})^{-1}, \tag{6b.5}$$

which for very large N values is zero for $T > T_c$ and unity for $T < T_c$, crossing 0.5 at $T \equiv T_c$. The model is overly simple. More sophisticated models give a more complex description of the behavior at the transition temperature. This model, however, is adequate to clarify the behavior of this ensemble of nonergodic systems and to set a framework for the more general case.

The transition probabilities $r_{\mathbf{K},\mathbf{K}'} = r_{\mathbf{K}',\mathbf{K}}$ between quantum states $\mathbf{K}$ and $\mathbf{K}'$ of the same energy in the single systems can be regarded as the elements of a square real symmetric matrix with Ω_t rows and Ω_t columns. The states can be ordered, giving consecutively the $\Omega_i = \Omega_{\text{vib}}(E)$ vibrational states of the $Q=2^N$ different configurations, $i=1, 2, 3, \ldots, 2^N$. The matrix elements fall into blocks of Ω_i rows and Ω_i columns along the diagonal. Transitions within each block are between states differing only in vibrational quantum number of the same energy. These are very rapid, and $r_{\mathbf{k},\mathbf{k}'}$ is large. The elements further off the diagonal giving transitions due to diffusion change the state from one configuration i to another of configuration j. These are slow, and $r_{\mathbf{K}(i),\mathbf{K}'(j)}$ is small. Presumably, for configurations i, j differing in more than one neighboring pair exchange, they are zero at all temperatures. The numerical values of transition rates between different configurations decrease rapidly for low values of E. If $T(E)$ is the temperature of a system of total energy E, the values vary approximately as $\exp-[\epsilon/kT(E)]$, with ϵ some activation energy. Below some loosely defined temperature T_d all transitions between differing configurations may be regarded as zero.

At temperatures well above T_d the systems are truly ergodic. At a given total energy all quantum states must have equal probability at equilibrium. Those of the ordered configuration have more vibrational energy, by an amount ΔE, than those in disordered configurations. The number ratio $(\Omega_{\text{vib}}^{\text{ord}}/\Omega_{\text{vib}}^{\text{dis}})$ just equals 2^N at $T=T_c$, exceeds it at lower temperatures, and is less at higher temperatures. The fraction of systems in the ordered phase follows eq. (5), being only little more than 2^{-N} for $T>T_c$ but essentially unity for $T<T_c$. If $T_d<T_c$, the systems will follow the equilibrium condition, although perhaps slowly.

If, however, $T_d>T_c$, or if the systems are quenched sufficiently rapidly from a temperature $T>T_d$ so that no ordering occurs, then we freeze each system in the configuration it is in at T_d or T, respectively. The ordered configuration has such negligible probability that we can ignore it. The probability, in an ensemble of such systems, is the same per vibrational quantum state for all disordered configurations. The entropy S is that of eq. (2) and, as $T\to 0$ K, it approaches the value $kN \ln 2$. This can be, and is, checked experimentally by integration of C_P/T to room temperature and comparison, through some reversible cycle, with an absolute entropy scale.

It makes no difference if transitions, or chains of nonzero transitions, connect all quantum states or not, as long as all quantum states of the ensemble are equally probable. The ensemble is defined as consisting of an infinite number of systems in the same thermodynamic state, sufficiently described to ensure macroscopically reproducible behavior of all systems. If the systems are ergodic, we readily proved in Secs. 4b and 4c that equal probability of all quantum states is a requirement at equilibrium. No such universal proof can be made for nonergodic systems, but the method of preparation of the systems in the ensemble may satisfy the equal probability condition. The method of calculation is then the same as for ergodic systems. Indeed, as we discuss in Sec. 6e on Poincaré times, even for perfectly ergodic systems the individual system can sample only a fraction of zero weight of all $\Omega(V, E, N)$ states. This is also true if the course of an observation extends over days or years.

One may legitimately inquire what one does with systems that are partially annealed at or below T_c so that partial ordering has occurred but not complete equilibrium. In such a case the systems usually are not reproducible; they are not sufficiently described to give a unique answer to all macroscopic properties. If the annealing process can be completely controlled in temperature and time, one must seek one or more new thermodynamic variables to describe the exact state. If, for instance, the state can be adequately described as a two-phase mixture with fraction $x<1$ in an ordered phase and $1-x$ disordered, we would know how to proceed, but it is unlikely that such a simple description suffices.

One must remember that many examples, particularly among the hard metals and their alloys, exist in which the state and properties depend critically on previous history of annealing and working. The exact state is not simply described by the straightforward thermodynamic variables V, E, N_a, $N_b, \ldots$. In other cases the state is completely reproducible but requires a specification in addition to the usual minimum number of variables. The element carbon exists in two well-defined crystalline forms at room temperature and atmospheric pressure, graphite and diamond. Graphite is the thermodynamically stable phase but, as we are told by the jeweler's advertisements, "Diamonds are forever." In each case we should not count *all* quantum states consistent with V, E, N, even if we knew how to do so, but only those corresponding to small vibrational displacements from the pertinent lattice sites. Of course, since graphite is presumably *the* stable form of carbon at room temperature and pressure, counting all states would give us zero weight to the diamond configuration, as well as to all other nongraphitic arrangements.

One interesting example of nonergodicity in which the state requires an extra continuous variable that is well understood is that of hydrogen as gas or liquid at very low temperature. It is discussed in detail in Sec. 7k. The molecule H_2 exists in two forms. One is ortho hydrogen with three states of unit total nuclear spin which are states symmetric in nuclear exchange, and can exist only with the antisymmetric molecular rotational levels of odd quantum number $j=1, 3, \ldots$. The other, para hydrogen, with one antisymmetric nuclear spin state of zero spin exists only in even $j=0, 2, \ldots$ symmetric rotational states. At room temperature and above, equilibrium exists with three-fourths of the hydrogen ortho and one-fourth as para hydrogen. The level $j=1$ has an energy $\Delta\epsilon$, which, measured in temperature units, $\Delta\epsilon = kT$, is about 150° above that of the zero rotational state, so that at low temperatures the equilibrium ratio is almost pure para. However, in the absence of paramagnetic impurities the rate of transition is very low, taking several days to approach equilibrium. The behavior is that of a two-component solution described by N_{ortho} and N_{para}, rather than $N_{H_2} = N_{\text{ortho}} + N_{\text{para}}$.

6c. THE NONERGODIC PARADOX

The assignment of entropy equal to $kN \ln 2$ to disordered nonergodic crystals such as CO at 0 K appears to raise a paradox. Any single bridge hand of 13 cards, no matter how ordinary looking to the player, is just as improbable as 13 spades. Similarly, any single fixed arrangement of carbon up or carbon down is as improbable as the ordered arrangement whatever that is. The distribution of a single frozen crystal near 0 K, no matter how disorderly it may appear to us, might be considered to be of singular unique beauty by that crystal itself; why should it be assigned an entropy $kN \ln 2$ merely because we poor mortals cannot distinguish it from among the $2^N - 1$ other disordered distributions?

The assignment of greater entropy can indeed be justified. The thermodynamic entropy can be measured only by a reversible cycle to some standard entropy state. A convenient one is to measure the equilibrium vapor pressure. The assignment of an extra $kN \ln 2$ to the entropy of the crystal means that the predicted vapor pressure is just half that of a crystal of the lower entropy and *of the same energy*, measured from that of the gas. Now the rate, per unit area, that molecules leave the crystal surface for the gas phase is independent of the order, except for the truly unique ordered crystal of slightly lower energy. If, however, one of the disordered crystals is to maintain its particular distribution, that is, if it is to keep its individuality in equilibrium with the vapor, there must be some mechanism such as a "Maxwell demon" by which it rejects just half of the vapor molecules that alight on the surface with carbon in the wrong orientation. An ensemble of such crystals that keep any *one* of the 2^N configurations can only be in equilibrium with the vapor if the pressure is double that of an ensemble of random orientation.

Of course, crystals in equilibrium with their own vapor are no longer strictly nonergodic. There exists a mechanism by which transitions can occur, converting one configuration into others, but the example serves to clarify the apparently arbitrary assignment of k times the logarithm of the total number of states of the ensemble for the entropy of each crystal.

6d. CONCLUSIONS ABOUT ERGODICITY

From the discussion of the last two sections we can draw some rather general conclusions.

First, it is clear that systems exist that are nonergodic, namely, in which the real symmetric matrix of transition probabilities between all of the possible quantum states consistent with the minimum thermodynamic state specification V, E, N_a, N_b, . . . , fall into blocks, with zero transition rates between states in different blocks.

Suppose the systems trapped in the quantum states of the different blocks are physically distinguishable by macroscopic measurements. In this case different methods of preparation of the systems lead to ensembles of differing fractions in the various types of systems. The ensemble is not macroscopically defined and reproducible; one or more new thermodynamic specifications are necessary to define the state. The simplest examples are those of two or more thermodynamic phases, all but one of which are metastable. Other examples include cases that can be described as solutions of differing composition, such as the ortho and para hydrogen example.

Alternatively, there are examples in which a number Q of such blocks corresponds to quantum states of systems which are physically identical in all macroscopic behavior. Just by the definition of physical indistinguishability, any process of preparation of the members of an ensemble of such systems will be equally probable for all quantum states of equal energy, independently of the block index i, $1 \leq i \leq Q$.

Since entropy itself is measurable, the systems of different i values will be indistinguishable only if the number Ω_i of quantum states ergodically available to them are equal or, more precisely if $N^{-1} \ln \Omega_i$ is the same, to terms of vanishing order as $N \to \infty$, for all i-values. In such cases the entropy is given by $S = k \ln \Omega_t$, with $\Omega_t = Q\Omega_i$, so that

$$S = k(\ln Q + \ln \Omega_i). \tag{6d.1}$$

The statistical mechanical treatment is *as if* the systems were in fact ergodic.

If $N^{-1} \ln Q$ vanishes for large enough N values, the case is trivial. Probably, for most systems at sufficiently low temperatures this kind of nonergodicity occurs. The third-law assumption that only *one* quantum state of lowest energy is reached at 0 K is probably a fictitious statement for most crystalline materials. The more real case is simply that $N^{-1} \ln Q$ vanishes and the entropy is immeasurably small at $T \to 0$ K.

If $N^{-1} \ln Q$ remains finite, as it does for the specific examples we have discussed, the entropy, as T approaches zero, approaches a finite extensive value. In such cases there is presumably at least one block in the transition probability matrix, which is unique and distinguishable from the others, with a lower energy of the lowest quantum state or states, but which occurs with vanishing probability in the preparation of the system of the ensemble.

6e. THE POINCARÉ RECURSION TIME

One of the objections to uniform entropy increase associated with the reversibility of the laws of mechanics is due to the rather obvious fact that a system starting at time $t=0$ from a point P in phase space must return to P, or infinitesimally close to it, in a time t_P, known as the Poincaré recursion time. If, then, the system is truly undisturbed, it repeats on the same path.

Some idea of the magnitude of this time is most easily obtained using the concepts of a quantum-mechanical system. Neon at atmospheric pressure and room temperature has an S/k per mole of about 10^{25}, or $\exp 10^{25}$ quantum states for a mole of gas. Each molecule makes about 10^{12} collisions per second, so that 10^{24} molecules make $10^{36} \cong \exp 100$ changes in quantum state of the gas per second. This sounds impressively large, but the time required to sample all $\exp 10^{25}$ quantum states is $\exp(10^{25}-100)$, which is essentially $\exp 10^{25}$ sec. The commonly assumed life of the universe is 10^{10} yr, which is $\pi \times 10^{17}$ sec.

One may well object that this is a nonsense calculation. More interesting is the time required to return to a given nonequilibrium macroscopically defined state, consistent with a very large number of quantum states. Suppose the macroscopically defined state has an entropy lower than the equilibrium value by 1 part in 1000, $S/k = 0.999 \times 10^{25}$, so that there are $\exp(0.999 \times 10^{25})$ quantum states available to it. The time above must be divided by this number, but this still requires $\exp 10^{22}$ sec.

This numbers game makes clear an assertion made in the discussion of the ensemble of nonergodic systems, namely, that even during the course of a rather long drawn-out observation on a system, the number of quantum states the system samples is always a completely negligible fraction of those available to the ensemble. In an observation extended over several hours, say 10^4 sec, the system may sample 10^{40} quantum states. This is a decently large statistical sample, but it is a quite negligible fraction, about $\exp(-10^{25})$, of all those of the ensemble. Of course, the fact that 10^{40} is a large number does not preclude that they are all states corresponding to a nonequilibrium macroscopic distribution which relaxes slowly to equilibrium. The average of the ensemble is equilibrium.

6f. THE LOSCHMIDT PARADOX AND SPIN ECHO

If a classical completely isolated system starts at time $t=0$ far from equilibrium, say, much hotter at the upper end than at the lower, it will gradually approach equilibrium by heat conduction. Suppose that at a later time t_0 an extraordinarily clever experimentalist succeeds in exactly

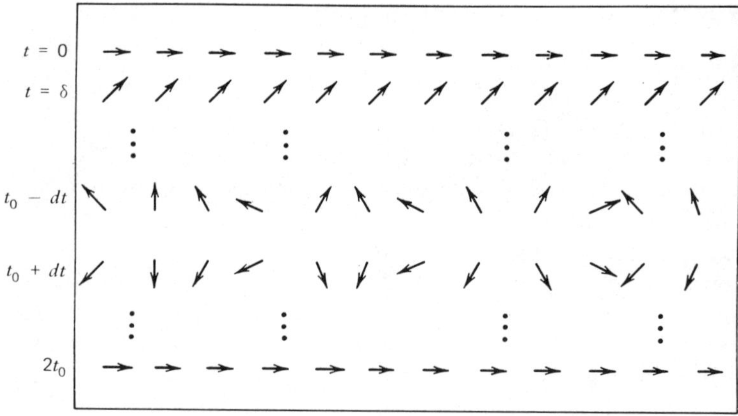

Figure 6f.1 Spins in echo.

reversing the sign of all molecular momenta. The molecules will then exactly retrace their previous paths, and at $2t_0$ the system will again be hotter at the top than at the bottom. During the period from t_0 to $2t_0$ it appears that the entropy is decreasing. Since no obvious simple criterion seems to preclude the reversed path as inherently less probable than the normal one, there appears to be a serious paradox.

There is a story, probably quite apocryphal, that when Boltzmann was presented with this paradox by Loschmidt, he pointed his finger at the latter and said, "You reverse the momenta!" Since then Hahn[(1)] has accomplished this miracle in the demonstration of spin echo. By now spin echo experiments and spin echo spectroscopy have become a common field of experimentation. We describe an overly simplified cariacature of the experiment[(2)] and answer the heresy that it demonstrates a violation of the second law.

At sufficiently low temperatures a strong magnetic field parallel to the x-axis can be used to line up the nuclear spins of a crystal along the x-axis (Fig. 6f.1, line $t=0$). If the magnetic field is removed, the spins remain aligned and the crystal has a magnetic moment of some magnitude M parallel to x. A magnetic field along the z-axis, perpendicular to the plane of the paper, is now applied at $t=0$. The moments precess in the xy-plane with an angular velocity ω and, a very short time later, $t=\delta$, have the position shown in the figure. The angular velocity ω_i of the ith

(1) Erwin L. Hahn, *Phys. Rev.* **80,** 580 (1950).

(2) This particular model is due to John Blatt who first called attention to the fact that spin echo was essentially a realization of the Loschmidt paradox.

spin is proportional to the local field H_{zi} and, since there are unavoidable inhomogeneities, these are not absolutely identical. The macroscopic magnetic moment **M** rotates in the xy-plane with the average angular velocity $\bar{\omega}$ but, since the individual directions, $\phi_i(t) = \omega_i t$, get slowly out of phase, the magnitude $|\mathbf{M}|$ decreases in time (Fig. 6f.2).

Now at some arbitrary time t_0, a magnetic π-pulse along the x-axis is applied without changing the field parallel to z. The pulse is exactly of magnitude enough to rotate each moment $\mathbf{m}_i$ by an amount π in the zy-plane, bringing its direction back into the xy-plane but changing $\phi(t_0) = \omega_i t_0$ to its negative (lines $t = t_0 - dt$ and $t_0 + dt$ in the figure). The individual moments continue to precess at the rate $d\phi_i/dt = \omega_i$, so that for each i at $t = 2t_0$ we have $\phi_i(2t_0) = (\omega_i - \omega_i)t_0 = 0$. The magnets are all aligned again, and $|\mathbf{M}|$ is back at its original value M, or nearly so. Actually, of course, there is some random interaction with the lattice vibrations, and $|\mathbf{M}(2t_0)|$ is less than $|\mathbf{M}(t = 0)|$ (Fig. 6f.2).

Now the state of the random spin orientation contributes an additive entropy term $\Delta S_M = kN \ln(2s + 1)$, with S the quantum number of nuclear spin, and since the aligned spins are all in one state, $\Delta S = 0$. The high value of $|\mathbf{M}|$ corresponds to a more negative entropy than the low values in any normal system. One therefore intuitively assumes that the entropy *decreases* between t_0 and $2t_0$ in the experiment, and this is an anathema to a thermodynamicist.

But a really careful classical thermodynamicist, without any reference to the mechanical nature of the microscopic state, can readily show that

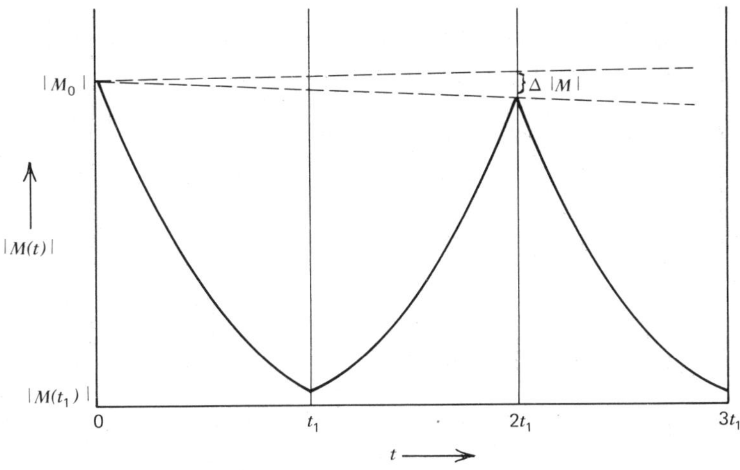

Figure 6f.2 $|M(t)|$ in spin echo.

the entropy decrease is not present. To measure the entropy of the state at t_0 one must remove it from the magnetic apparatus and by some *reversible* path, involving heat flow into or out of the system, measure the entropy difference between it and some standard state of known entropy. If one uses the conventional techniques of measuring the entropy of a system with nonzero magnetic moment, one indeed finds that the heat flow corresponds to the entropy normally associated with the moment $|\mathbf{M}(t_0)|$. But one must assure oneself that the heat path is indeed reversible. To do this one reverses the path and brings the system to $\mathbf{M}(t_0)$. But this system is *not* macroscopically identical to the system left in the magnetic field at t_0; it has lost its ability to climb up to $|\mathbf{M}(2t_o)|$.

The spin echo experimentalist explains this readily; the individual spins have lost the memory of their exact angles $\phi_i(t_0)$. Only the average value $\langle\phi(t_0)\rangle$ is reproduced. The classical thermodynamicist is not concerned with this. He only knows that the measurement did not follow a reversible path, and the original entropy at t_0 in the apparatus was more negative than the measurement indicated.

The situation is not yet clarified. A reversible path must be found to measure the entropy. There is one. Leave the system in the magnetic field, after having π-pulsed it, until $2t_0$ with moment $\mathbf{M}(2t_0)$. Then measure its entropy, finding the conventional value corresponding to a moment $\mathbf{M}(2t_0)$. The path is now reversible for, when returned to the moment $\mathbf{M}(2t_0)$ and replaced in the magnetic field, it repeats the path from $2t_0$ on, and at $3t_0$ is (nearly) in the same state as at t_0. The entropy at t_0 is the value of a conventional system of moment between $\mathbf{M}(\mathrm{t}=0)$ and $\mathbf{M}(2t_0)$.

By pulsing at various times one finds an always slowly increasing entropy following that of a system whose magnetic moment decreases slowly along the line connecting the peaks at the echo.

The example is clearly one that simulates that of the original Loschmidt paradox and, as far as entropy decrease is concerned, the resolution is always the same. A measurement of entropy by a thermodynamic heat flow path always erases the memory and prevents a return to the initial state. The measurement is not by a reversible path. Clever experimentalists may possibly find other cases in which the microscopic path can be reversed, and indeed there are very many more sophisticated examples in nuclear spin echo.

One feature of the discussion above is unpleasant to the classical thermodynamicist, but is by no means without precedent. One normally assumes that the thermodynamic state of a system at a given time t is fully defined, or can be defined, by its macroscopic properties at that time, without recourse to its history. The variation in ω_i in our hypothetical example is due primarily to uncontrollable inhomogeneities in the magnets used and possibly also to variations in the local magnetic susceptibility of the system itself due to crystal imperfections. These are not reproducible. A particular crystal removed from one apparatus at time t_0 does not show an echo if placed in a similar apparatus. The different systems are unique, unless the complete magnetic apparatus is included as part of the system. One must, however, remember that other examples have been known for a long time in which the properties of a system are best described by its history, rather than by simple numerical values of measurable macroscopic variables. Such examples include most of the harder metals, and alloys of them, which depend in almost all of their properties on previous histories of annealing and cold-working.

Two comments concerning the likelihood of experimental manifestations of the Loschmidt paradox other than in nuclear magnetic resonance are called for. The times t_0 in these experiments are generally short, measurable in units of seconds or less, much shorter than the relaxation times of 10^3 sec or more characteristic of the approach to equilibrium of an ordinary laboratory system in which heat diffusion or molecular diffusion is the mechanism by which equilibrium is established. Second, the coupling between nuclear spins and their surroundings is notoriously weak. The pertinent surroundings in this case are the molecular vibrations or phonons which interact very little with the nuclear spins. In room-temperature systems and in relaxation times of, say, 10^3 sec each molecule makes about 10^{15} collisions. To expect that it completely retraces its path on the exact reversal of all momenta under laboratory conditions is scarcely credible. Each random disturbance from outside the system, no matter how small, propagates itself throughout the system with the velocity of sound. It is very likely, even apart from the difficulty in making an equivalent-to-momentum reversal in other than a magnetic system, that similar echo effects will remain pretty well limited to nuclear spin systems.

6g. THE STATISTICAL MECHANICAL ENTROPY

As discussed in Sec. 3j, the expression for $-S/k$ for all the equilibrium ensembles is

$$\frac{-S}{k}=\sum_{\mathbf{N}}\sum_{\mathbf{K}(\mathbf{N})} W_{\mathbf{K}} \ln W_{K}, \tag{6g.1}$$

which, in the classical approximation is

$$\frac{-S}{k}=\sum_{\mathbf{N}}\int\int_{V}\cdots\int \frac{d\mathbf{q}^{(\Gamma)}\,d\mathbf{p}^{(\Gamma)}}{h^{\Gamma}\mathbf{N}!} W_{\mathbf{n}}(\mathbf{q}^{(\Gamma)},\mathbf{p}^{(\Gamma)}) \ln W_{\mathbf{N}}(\mathbf{q}^{(\Gamma)},\mathbf{p}^{(\Gamma)}), \tag{6g.1'}$$

where for some ensembles such as the microcanonical or petite canonical the sum over $\mathbf{N} = N_a, N_b, \ldots$ involves only one nonzero member.

Now the thermodynamic entropy of a nonequilibrium system is not always definable. One must at least imagine a "Gedankenexperiment" by which, through a process involving only reversible steps, the system can be brought into some state for which the entropy is known and the difference ΔS of the standard state minus the entropy in the original state is measured by $\int dQ/T$, where dQ is the element of heat removed reversibly from the system, and T is the temperature of a reservoir. For the process to be reversible the reservoir must be in temperature equilibrium with at least that part of the system from which the heat comes.

This requirement still permits us to assign entropies to a wide variety of nonequilibrium systems. As long as all regions of the system containing a sufficient number of molecules to be regarded as macroscopic are in thermodynamic equilibrium within the region, we can imagine insulating partitions inserted and measure and sum the entropies of the individual regions.

There are even more exotic conditions under which we can define an entropy meaningfully. We mentioned in the last section that nuclear spins and lattice vibrations in a crystal exchange energy very slowly. One can create experimentally a situation in which a crystal having a very low lattice vibration temperature contains nuclear spins in an external magnetic field having the equilibrium distribution of spin directions corresponding to a very high temperature. Indeed, one can even bring the nuclear spin temperature to transinfinite negative values. These are values for which $\beta = 1/kT$ crosses zero to become negative. Heat flows very slowly, then, from the spin system of negative β (low β) to the vibration system of high positive β-value.

Another example of two coexistent systems of different temperatures in the same volume element occurs often in plasmas. A neutral dilute gaseous plasma of hydrogen ions and electrons can exist in which both the protons and the electrons have an essentially equilibrium Gaussian distribution of momentum components, but with very different temperatures, the electrons having much higher temperature than the protons. This is possible because momentum exchange between particles of the same mass is much faster than that between the two kinds of vastly different masses.

In all such cases one can define a quantity $-S/k$ as the sum or integral of $W \ln W$, and the macroscopic behavior of the system is in agreement with the thermodynamic expectations one would deduce if this quantity S, so defined, were the entropy.

We give an example of a classical probability density $W(r_1, r_2, \ldots, r_N,$

$p_1, \ldots, p_N$) for a grand canonical ensemble of nonequilibrium multicomponent systems of monatomic molecules. Assume the temperature T is a function of position $\mathbf{r}$ in the systems, $T = T(\mathbf{r})$, and the chemical potential μ_a of species a is also a function of $\mathbf{r}$, $\mu_a(\mathbf{r})$. We use

$$\beta(\mathbf{r}) = \frac{1}{kT(\mathbf{r})}, \tag{6g.2}$$

$$\nu_a(\mathbf{r}) = \frac{-\mu_a(\mathbf{r})}{kT(\mathbf{r})}. \tag{6g.3}$$

For simplicity assume that the potential energy $U_N(\mathbf{r}_1, \ldots, \mathbf{r}_n)$ is a sum of pair terms,

$$U_N = \sum_{N \geq i > j \geq 1}\sum u_{ij}(|\mathbf{r}_i - \mathbf{r}_j|), \tag{6g.4}$$

where $u_{ij}(|\mathbf{r}_i - \mathbf{r}_j|) = u_{ij}(r_{ij})$ depends on the species of molecules i and j. Then define a Hamiltonian H_i for each molecule i, which is a function of the positions of all others, but which for any given position $\mathbf{r}^{(N)} \equiv \mathbf{r}_1, \ldots, \mathbf{r}_N$ depends only on the positions $\mathbf{r}_j$ of molecules j for which r_{ij} is small and $u_{ij}(r_{ij})$ is nonzero,

$$H_i(\mathbf{r}_i, \mathbf{r}^{(N-i)}) = \frac{\mathbf{p}_i \cdot \mathbf{p}_i}{2m_i} + \frac{1}{2}\sum_{j=i} u_{ij}(r_{ij}). \tag{6g.5}$$

With this the total Hamiltonian H_N is

$$H(\mathbf{r}^{(N)}, \mathbf{p}^{(N)}) = \sum_{i=1}^{i=N} H_i(\mathbf{r}_i, \mathbf{r}^{(N-i)}). \tag{6g.6}$$

We now write the probability density as

$$W_\mathbf{N} = \exp -\left\{\bar{\phi}V + \sum_{i=1}^{i=N}\left[\nu_i(\mathbf{r}_i) + \beta(\mathbf{r}_i)H_i(\mathbf{r}_i, \mathbf{r}^{(N-i)})\right]\right\}, \tag{6g.7}$$

as a rather obvious intuitive guess. The quantity $\bar{\phi}$ is to be evaluated by the condition

$$\sum_N \iint \cdots \int \frac{d\mathbf{r}^{(N)}\, d\mathbf{p}^{(N)}}{h^{3N} N_a! N_b! \cdots} W_\mathbf{N}(\mathbf{r}^{(N)}, \mathbf{p}^{(N)} \equiv \mathcal{I} W_N = 1, \tag{6g.8}$$

where we use $\mathcal{I}$ for the summation and integration operator,

$$\mathcal{I} \equiv \sum_N \iint \cdots \int \frac{d\mathbf{r}^{(N)}\, d\mathbf{p}^{(N)}}{h^{3N} N_a! N_b! \cdots}. \tag{6g.9}$$

Now the proof that the guess of eq. (7) for $W_{\mathbf{N}}$ is legitimate is as follows. First, we require that temperature and chemical potential have a meaning at each $\mathbf{r}$-value, which demands that they vary negligibly near each $\mathbf{r}$ in distances that contain a statistical number of molecules. This requires that the gradients be small, namely, that

$$|\nabla \nu_a(\mathbf{r})| \ll [\rho_a(\mathbf{r})]^{1/3}, \qquad \text{for all } a \text{ and } \mathbf{r}, \tag{6g.10}$$

$$|\nabla \ln \beta(\mathbf{r})| \ll \left[\sum_a \rho_a(\mathbf{r})\right]^{1/3}, \qquad \text{for all } \mathbf{r}, \tag{6g.11}$$

where $\rho_a(\mathbf{r})$ is the number density of species a at $\mathbf{r}$. This in turn is given by

$$\rho_a(\mathbf{r}) = \mathscr{I}_N \sum_{ia=1a}^{ia=N_a} \delta(\mathbf{r}-\mathbf{r}_{ia}) W_N, \tag{6g.12}$$

where $\delta(\mathbf{r}-\mathbf{r}_{ia})$ is the Dirac delta function,

$$\delta(\mathbf{r}-\mathbf{r}_{ia}) = 0, \qquad \text{if } \mathbf{r}_{ia} \neq \mathbf{r}_i, \qquad \int d\mathbf{r}_{ia}\, \delta(\mathbf{r}-\mathbf{r}_{ia}) = 1. \qquad \text{6g.13)}$$

The energy density $\epsilon_V(\mathbf{r})$ at $\mathbf{r}$ is

$$\epsilon_V(\mathbf{r}) = \mathscr{I}_N \sum_{i=1}^{i=N} \delta(\mathbf{r}-\mathbf{r}_i) H_i(\mathbf{r}, \mathbf{r}^{(N-i)}) W_N. \tag{6g.14}$$

In any small volume v near r, still containing a macroscopic number of molecules of each species, the total number N_a of molecules of species a is $N_a^{(v)} = v\rho_a(r)$, and the energy is $E^{(v)} = v\epsilon_V(\mathbf{r})$, so that the local thermodynamic state is well defined by v, $E^{(v)}$, $N_a^{(v)}$, $N_b^{(v)}, \ldots$. For such an *equilibrium* system we can calculate the pressure, temperature, and chemical potential of each molecular species. It is comparatively easy, then, to prove that to order $\{|\nabla \ln \beta|/\sum_a \rho_a(r)]^{1/3}\}^2$ and $\{|\nabla \nu_a|/[\rho_a(r)]^{1/3}\}^2$ these values of temperature and chemical potential agree, through eqs. (2) and (3) with those of $\beta(\mathbf{r})$ and $\nu_a(\mathbf{r})$ used in $W_{\mathbf{N}}$ of eq. (7). We further find that the constant $\bar{\phi}$ in eq. (7) is

$$\bar{\phi} = \left\langle \frac{P(\mathbf{r})}{kT(\mathbf{r})} \right\rangle_{\text{space average}} \tag{6g.15}$$

The quantity

$$\frac{-S}{k} = \mathscr{I} W_{\mathbf{N}} \ln W_N, \tag{6g.16}$$

with eq. (7) for $W_{\mathbf{N}}$ and eq. (9) for $\mathscr{I}$, and with eqs. (12), (14), and (15) for $\rho_a(\mathbf{r})$, $\varepsilon_V(r)$, and $\bar{\phi}$, respectively, gives

$$\frac{-S}{k} = \int d\mathbf{r} \left[\beta(\mathbf{r})P(\mathbf{r}) + \sum_a \rho_a(\mathbf{r})\nu_a(\mathbf{r}) + \beta(\mathbf{r})\epsilon_V(\mathbf{r})\right]. \tag{6g.17}$$

But the thermodynamic relationship for the local entropy per unit volume $S_V(\mathbf{r})$ is given by the equilibrium equation,

$$k^{-1}S_V(\mathbf{r}) = \beta(\mathbf{r})P(\mathbf{r}) + \sum_a \rho_a(\mathbf{r})\nu_a(\mathbf{r}) + \beta(\mathbf{r})\epsilon(\mathbf{r}), \tag{6g.18}$$

so that, from eq. (17),

$$S = \int d\mathbf{r}\, S_V(\mathbf{r}). \tag{6g.19}$$

From this example, and others, we find that, whenever a meaningful thermodynamic entropy can be assigned to a nonequilibrium system, the quantity defined by eq. (16) is always the negative entropy divided by Boltzmann's constant k. We thus find it useful to define the S of eq. (16) as entropy, even for systems so far distorted from equilibrium that no thermodynamic measurement is possible.

6h. A DIFFERENT DERIVATION OF W(q, p)

We remarked at the beginning of Sec. 4a that there are alternate methods of deriving the equations for the classical probability density $W(\mathbf{q}^{(\Gamma)}, \mathbf{p}^{(\Gamma)})$ for the various ensembles. One of these consists of defining the entropy by the equation

$$\frac{-S}{k} = \mathscr{I}W(\mathbf{q}^{(\Gamma)}, \mathbf{p}^{(\Gamma)}) \ln W_{\mathbf{N}}(\mathbf{q}^{(\Gamma)}, \mathbf{p}^{(\Gamma)}), \tag{6h.1}$$

$$\mathscr{I} \equiv \sum_{\mathbf{N}} \int\int\int \cdots \int \left(h^{\Gamma} \prod_{\alpha} N_\alpha!\right)^{-1} d\mathbf{q}^{(\Gamma)}\, d\mathbf{p}^{(\Gamma)}, \tag{6h.2}$$

and then, by a variational method, to require $W_{\mathbf{N}}$, subject to whatever constraints are appropriate, to be such that $-S/k$ takes a minimum value.

Consider first the microcanonical ensemble. We require that $W_{\mathbf{N}}$ be nonzero only for a fixed set $\mathbf{N} = N_\alpha, N_\gamma, \ldots$ of numbers of N_α of molecules of type α, and so on, and only for those values of the phase space variables $\mathbf{q}^{(\Gamma)}$, $\mathbf{p}^{(\Gamma)}$ such that $E \le H(\mathbf{q}^{(\Gamma)}, \mathbf{p}^{(\Gamma)} \le E + \Delta E$, namely, that the energy lie between E and $E + \Delta E$. The one constraint is now that of the normalization,

$$\mathscr{I}W_{\mathbf{N}} = 1. \tag{6h.3}$$

Use the method of undetermined multipliers (Appendix V). Multiply eq. (3) by an arbitrary constant a and subtract it from eq. (1). The condition that $-S/k$ be an extremum is now that, for any arbitrary variation $\delta W(\mathbf{q}^{(\Gamma)}, \mathbf{p}^{(\Gamma)})$,

$$\begin{aligned} \delta\frac{-S}{k} = 0 &= \mathscr{I}\delta\{W_{\mathbf{N}}(\mathbf{q}^{(\Gamma)}, \mathbf{p}^{(\Gamma)})[\ln W_{\mathbf{N}}(\mathbf{q}^{(\Gamma)}, \mathbf{p}^{(\Gamma)}) - a]\} \\ &= \mathscr{I}[(1-a) + \ln W_{\mathbf{N}}(\mathbf{q}^{(\Gamma)}, \mathbf{p}^{(\Gamma)})]\, \delta W_{\mathbf{N}}(\mathbf{q}^{(\Gamma)}, \mathbf{q}^{(\Gamma)}). \end{aligned} \tag{6h.4}$$

The only solution that the right-hand side of eq. (4) be equal to zero for *any* arbitrary variation $\delta W_{\mathbf{N}}$ is that

$$\ln W_{\mathbf{N}}(\mathbf{q}^{(\Gamma)}, \mathbf{p}^{(\Gamma)}) = (a-1) \tag{6h.5}$$

wherever it is nonzero, namely, for the given set $\mathbf{N}$ of molecules and energy E. The arbitrary constant a is found by using eq. (3) with eq. (5) in eq. (1),

$$\frac{-S}{k} = (a-1),$$

$$\ln W(\mathbf{q}^{(\Gamma)}, \mathbf{p}^{(\Gamma)}) = \frac{-S}{k}. \tag{6h.6}$$

That the extremum of eq. (4) is a minimum is readily checked by taking the second derivative,

$$\frac{\delta(W \ln W)}{\delta W^2} = W^{-1} > 0.$$

For the petite canonical ensemble we require $W_{\mathbf{N}}$ to be nonzero for only one set $\mathbf{N}$ of molecules, but for all values of the coordinates and momenta. However, we require that the *average* energy $\langle E \rangle$, of the members of the ensemble be fixed,

$$\mathscr{I} H(\mathbf{q}^{(\Gamma)}, \mathbf{p}^{(\Gamma)}) W(\mathbf{q}^{(\Gamma)}, \mathbf{p}^{(\Gamma)}) = \langle E \rangle. \tag{6h.7}$$

If $-\beta$ times the left side of eq. (7) is now also subtracted from eq. (1), we find

$$\delta \frac{-S}{k} = 0 = \mathscr{I}[(1-a) + \beta H(\mathbf{q}^{(\Gamma)}, \mathbf{p}^{(\Gamma)}) + \ln \mathrm{W}_{\mathbf{N}}] \delta \mathrm{W}_{\mathbf{N}}, \tag{6h.8}$$

$$\ln W_{\mathbf{N}} = (a-1) - \beta H(\mathbf{q}^{(\Gamma)}, \mathbf{p}^{(\Gamma)}). \tag{6h.8$'$}$$

Again we use the normalization condition (3) in eq. (1) with eq. (7) for $\bar{E}$ to find that, with A the Helmholtz free energy,

$$\frac{-S}{k} = (a-1) - \beta \langle E \rangle, \tag{6h.9}$$

$$\beta = \frac{\partial(-S/k)}{\partial \langle E \rangle} = \frac{1}{kT}, \tag*{(6h.9$'$)[1]}$$

$$(a-1) = \frac{1}{kT}(\langle E \rangle - TS) = \frac{A}{kT}, \tag{6h.9$''$}$$

[1] To derive eq. (9′) by differentiation of eq. (9) may look far too facile. A careful examination of the derivation of the method of undetermined multipliers (Appendix V) shows that, if one seeks an extremum to $Q = \int dx\, F(x)$ subject to $R = \int dx\, G(x)$ by an arbitrary variation of $\int dx\, (F + \alpha G)$, then $\alpha = \partial Q / \partial R$ even if other constraints are required.

or, finally,

$$\ln W = \beta[A - H(\mathbf{q}^{(\Gamma)}, \mathbf{p}^{(\Gamma)})]. \tag{6h.10}$$

If we permit W to be nonzero for all sets $\mathbf{N}$ of numbers of molecules, we must impose the constraint

$$\mathscr{I} N_\alpha W_{\mathbf{N}}(\mathbf{q}, \mathbf{p}) = \langle N_\alpha \rangle \tag{6h.11}$$

for each species α of molecules. With

$$-\nu_\alpha = \beta\mu_\alpha = \left(\partial \frac{(-S/k)}{\partial N_\alpha}\right)_{V,E,N_\gamma,\ldots}, \tag{6h.12}$$

we add $-\nu_\alpha$ times the left-hand side of eq. (11) for each species α of molecule before the variation of the integrand. One finds

$$\ln W = (a-1) - \beta H(\mathbf{q}^{(\Gamma)}, \mathbf{p}^{(\Gamma)}) + \sum_\alpha \beta\mu_\alpha, \tag{6h.13}$$

and again using eq. (3) in eq. (1) with this,

$$(a-1) = \beta\left(-\sum_\alpha N_\alpha\mu_\alpha + \beta\langle E\rangle - TS\right) = -\beta PV, \tag{6h.14}$$

$$\ln W = -\beta\left[PV - \sum_\alpha \mu_\alpha N_\alpha + H(\mathbf{q}^{(\Gamma)}, \mathbf{p}^{(\Gamma)})\right]. \tag{6h.15}$$

The above is all presented in a scheme appropriate for classical systems. Had we so chosen, we could as well have used eq. (6g.1) in which $-S/k$ is the sum of $W_{\mathbf{K}} \ln W_{\mathbf{K}}$. By defining the operator $\mathscr{I}$ to be

$$\mathscr{I} \equiv \sum_{\mathbf{N}} \sum_{\mathbf{K(N)}} \tag{6h.16}$$

instead of the classical equivalent [eq. (2)], and replacing $W_{\mathbf{N}}(\mathbf{q}^{(\Gamma)}, \mathbf{p}^{(\Gamma)})$ by $W_{\mathbf{K}}$ and $H(\mathbf{q}^{(\Gamma)}, \mathbf{p}^{(\Gamma)})$ by $E_{\mathbf{K}}$, every step from eq. (3) to eq. (15) remains exactly valid.

Thus by the single assumption that the dimensionless quantity $-S/k$ defined by eq. (6g.1) or (6g.1′) for quantum or classical systems, respectively, and the requirement that $-S/k$ be a minimum at equilibrium, we can derive the form of $W_{\mathbf{K}}$ or $W(\mathbf{q}^{(\Gamma)}, \mathbf{p}^{(\Gamma)})$ for all of the Gibbs ensembles.

The argument by which one identifies the sum or integral of $W \ln W$ with $-S/k$ appears to us to be less compelling than the method used by us in Chapter 4 to derive that $S/k = \ln \Omega$. Since the predictions are the same, the point is not worth pursuing. However, one qualitative feature of the argument is worthy of emphasis. The integral of $W \ln W$ has minimum value when W is "smoothest," consistent with any other constraints. For the sole constraint of the normalization the solution for minimum value is that W is a constant. If $\langle E\rangle$ is fixed (and the integration is extended to all energies), ln W acquires an additive term $-\beta H(\mathbf{q}^{(\Gamma)}, \mathbf{p}^{(\Gamma)})$. If β decreases, so that T increases, S/k increases, which means that the integral of $W \ln W$ has decreased.

Let us examine the effect of adding a small perturbation to the equilibrium function ln $W_{\mathbf{N}}$. Set

$$W_{\mathbf{N}}(\mathbf{q}^{(\Gamma)}, \mathbf{p}^{(\Gamma)}) = W_{\mathbf{N}}^{\text{eq}}(\mathbf{q}^{(\Gamma)}, \mathbf{p}^{(\Gamma)})[1+\lambda \Psi_{\mathbf{N}}(\mathbf{p}^{(\Gamma)}, \mathbf{q}^{(\Gamma)})], \tag{6h.17}$$

with λ a small perturbation parameter. Since W can never be negative, and $W^{(\text{eq})}$ is positive or zero, we must limit λ for given ψ to values such that $\lambda\Psi \geq -1$ at all $\mathbf{q}^{(\Gamma)}$, $\mathbf{p}^{(\Gamma)}$. For given perturbed W we choose W^{eq} to be the equilibrium probability density of an ensemble of systems having the same energy $\langle E\rangle$, and number set average $\langle \mathbf{N}\rangle$ as the ensemble of the perturbed nonequilibrium systems. This means that, including the normalization requirement, we demand that

$$\mathscr{I} W_{\mathbf{N}} = \mathscr{I} W_{\mathbf{N}}^{\text{eq}}(1+\lambda\Psi_{\mathbf{N}}) = 1 = \mathscr{I} W_{\mathbf{N}}^{\text{eq}}, \tag{6h.18}$$

$$\mathscr{I} H W_{\mathbf{N}} = \mathscr{I} H W_{\mathbf{N}}^{\text{eq}}(1+\lambda\Psi_{\mathbf{N}}) = \langle E\rangle = \mathscr{I} H W_{\mathbf{N}}^{\text{eq}}, \tag{6h.19}$$

$$\mathscr{I} N_\alpha W_{\mathbf{N}} = \mathscr{I} N_\alpha W_{\mathbf{N}}^{\text{eq}}(1+\lambda\Psi_{\mathbf{N}}) = \langle N_\alpha\rangle = \mathscr{I} N_\alpha W_{\mathbf{N}}^{\text{eq}}, \tag{6h.20}$$

or

$$\mathscr{I} W_{\mathbf{N}}^{\text{eq}} \Psi_{\mathbf{N}} = 0, \tag{6h.21}$$

$$\mathscr{I} W_{\mathbf{N}}^{\text{eq}} H \Psi_{\mathbf{N}} = 0, \tag{6h.22}$$

$$\mathscr{I} W_{\mathbf{N}}^{\text{eq}} N_\alpha \Psi_{\mathbf{N}} = 0 \qquad \text{for all } \alpha. \tag{6h.23}$$

Since ln $W_{\mathbf{N}}^{\text{eq}}$ contains only additive terms which are constants plus terms of N_α times constants, plus $-\beta H$, we have that

$$\mathscr{I} W_{\mathbf{N}}^{\text{eq}} \Psi_{\mathbf{N}} \ln W_{\mathbf{N}}^{\text{eq}} = 0. \tag{6h.24}$$

We might parenthetically remark that, if the ensemble is such that for some other thermodynamic coordinate X_i we assign a fixed average value $\langle X_i\rangle$ so that ln W^{eq} contains a term $-x_i\chi(q^{(\Gamma)}, p^{(\Gamma)})$, we should also require $\mathscr{I} W_{\mathbf{N}}^{\text{eq}} \Psi_{\mathbf{N}} \chi_{\mathbf{N}}$ to be zero, so that eq. (24) would still be valid.

If now we develop ln W to quadratic terms in λ, $\ln(1+x) = x - \frac{1}{2}x^2 - \cdots$,

$$\begin{aligned} \ln W &= \ln W^{\text{eq}} + \ln(1+\lambda\Psi) \\ &= \ln W^{\text{eq}} + \lambda\Psi - \tfrac{1}{2}\lambda^2\Psi^2 + 0(\lambda^3), \end{aligned} \tag{6h.25}$$

and

$$W \ln W = W^{\text{eq}}(1+\lambda\Psi)(\ln W^{\text{eq}} + \lambda\Psi - \tfrac{1}{2}\lambda^2\Psi^2 + \cdots)$$
$$= W^{\text{eq}}[\ln W^{\text{eq}} + \lambda\Psi(1+\ln W^{\text{eq}}) + \tfrac{1}{2}\lambda^2\Psi^2 + \cdots]. \qquad (6h.26)$$

The difference $\Delta(-S/k)$ is now

$$\frac{-S}{k} = \frac{-S}{k} - \frac{-S^{\text{eq}}}{k}$$
$$= \mathscr{I} W_{\mathbf{N}} \ln W_{\mathbf{N}} - W_{\mathbf{N}}^{\text{eq}} \ln W_{\mathbf{N}}^{\text{eq}}$$
$$= \mathscr{I}\, W_{\mathbf{N}}^{\text{eq}}[\lambda\Psi_{\mathbf{N}}(1+\ln W_{\mathbf{N}}^{\text{eq}}) + \tfrac{1}{2}\lambda^2\Psi_{\mathbf{N}}^2$$
$$= \frac{\lambda^2}{2}\, \mathscr{I} W_{\mathbf{N}}^{\text{eq}} \Psi_{\mathbf{N}}^2, \qquad (6h.27)$$

since from eqs. (21) and (24) the linear terms and λ integrate to zero. The increase in $-S/k$ above that of an equilibrium system of the same conservative set of thermodynamic variables, E and $\mathbf{N}$, is always positive and quadratic in the magnitude of the perturbation. This of course is required if $-S/k$ is to be a minimum at equilibrium.

6i. THE CONSTANCY OF ENTROPY PARADOX

An unfortunate apparent disaster arises in using eq. (6g.1) or (6h.16) or their classical equivalents eqs. (6g.1′) and (6g.1) for S/k. This is that with S/k defined by these equations one can very readily prove that, in any isolated system, whether in equilibrium or not, $dS/dt = 0$; the entropy so defined never increases.

We demonstrate this for an ensemble of systems obeying classical mechanics. The Liouville operator is, from eq. (4b.3),

$$\mathscr{L}^{(\Gamma)} = \sum_{i=1}^{i=\Gamma} \left(\frac{\partial H}{\partial p_i}\frac{\partial}{\partial q_i} - \frac{\partial H}{\partial q_i}\frac{\partial}{\partial p_i} \right) \qquad (6i.1)$$

and, from eq. (4b.19), that $\partial W/\partial t = -\mathscr{L}W$,

$$\frac{d(S/k)}{dt} = -\mathscr{I}\frac{\partial}{\partial t}\, W_{\mathbf{N}}(t, \mathbf{q}^{(\Gamma)}, \mathbf{p}^{(\Gamma)}) \ln W_{\mathbf{N}}(t, \mathbf{q}^{(\Gamma)}, \mathbf{p}^{(\Gamma)})$$
$$= \mathscr{I}\mathscr{L}^{(\Gamma)}[W_{\mathbf{N}}(t, \mathbf{q}^{(\Gamma)}, \mathbf{p}^{(\Gamma)}) \ln W_{\mathbf{N}}(t, \mathbf{q}^{(\Gamma)}, \mathbf{p}^{(\Gamma)})]. \qquad (6i.2)$$

Now $\mathscr{L}W_N \ln W_N$ is a sum of terms involving $\partial/\partial q_i(W \ln W)$ or $\partial/\partial p_i(W \ln W)$, and the summation integration operator $\mathscr{I}$ contains an integral over all coordinates and momenta.[1] The integral $\int_a^b dx\,(dF/dx)$ is

[1] If cartesian coordinates are used, and this is always permissible, $\partial H/\partial q_i$ is independent of p_i, and $\partial H/\partial p_i$ is independent of q_i.

$F(b)-F(a)$, so that the integral of each term is the difference of $W \ln W$ at the extreme boundaries of the coordinate of momentum integration. For the momenta integrals the boundaries are plus and minus infinity for which W, hence $W \ln W$, must be zero for a system of finite energy. For the coordinates of center of mass of the molecules in a fluid the boundaries are the walls, and for closed systems W and $W \ln W$ are again zero. For the internal molecular coordinates either of two cases arises. For periodic angular coordinates such as the Eulerian ϕ, $0 \le \phi \le 2\pi$, the condition that W be single-valued requires $W \ln W$ to be the same at $\phi=0$ and at $\phi=2\pi$. For vibrational coordinates the model used has infinite energy at infinite extensions. If one objects that the model is erroneous, that the energies of bond breaking are not infinite, then the boundaries are the walls with $W \ln W$ equal to zero. In any case eq. (2) gives zero change in entropy with time in an isolated closed system.

The example of the isolated system is disaster enough. If one assumes an open system having walls permeable to molecules, or closed but with fluctuating forces at the walls simulating molecular impact, one does indeed have the possibility of entropy increase. However, the important terms correspond exactly to the net integrated flux of S/k through the spatial boundaries of the systems. Thus again entropy can flow into and out of the systems, but none is produced within the systems.

6j. SOLUTION OF THE CONSTANT-ENTROPY PARADOX

The qualitative clarification of the paradox of the apparent constancy of $-S/k$ if defined by $\mathscr{I}\, W \ln W$ has long been understood. It is connected with the fact that the smoothest probability density W has the lowest value of $\mathscr{I}\, W \ln W$. The initial perturbation from equilibrium, say, as represented by a function $W_{\mathbf{N}}$ of eq. (6g.7), for a macroscopic displacement of such quantities as number density $\rho_\alpha(\mathbf{r})$ of species α of molecules or of energy density $\epsilon(\mathbf{r})$ from uniformity in the system is, in some vaguely defined sense, one of meaningful lack of smoothness. The particular function $W_{\mathbf{N}}(\mathbf{q}^{(\Gamma)}, \mathbf{p}^{(\Gamma)})$ for specific set $\mathbf{N}$ of monatomic molecules given by eq. (6g.7) has the corresponding $\lambda\Psi$ of eq. (6h.17) given by

$$\begin{aligned}
&\lambda\Psi_{\mathbf{N}}(\mathbf{q}^{(\Gamma)}, \mathbf{p}^{(\Gamma)}) \\
&\quad = \frac{W_{\mathbf{N}}}{W_{\mathbf{N}}^{\mathrm{eq}}} - 1 \\
&\quad = \left[\exp - \left((\bar{\phi} - \phi_0)V + \sum_{i=1}^{i=N} \{[\nu_i(\mathbf{r}) - \nu_0] + [\beta(\mathbf{r}_i) - \beta_0 H_i(\mathbf{r}_i, \mathbf{r}^{(N-1)}, \mathbf{p}_i]\} \right)\right] - 1.
\end{aligned} \tag{6j.1}$$

In first order for small gradients, eqs. (6g.10) and (6g.11), it corresponds to a locally smooth function at each position **r** in the system. The equation for the time change of $W_{\mathbf{N}}$ [eq. (4b.19)] is

$$\frac{\partial W_{\mathbf{N}}}{\partial t} = -\mathscr{L} W_{\mathbf{N}} = -\mathscr{L} W_{\mathbf{N}}^{\mathrm{eq}}(1+\lambda \Psi_{\mathbf{N}}). \tag{6j.2}$$

Since $\mathscr{L}$ is a first-order differential operator $\mathscr{L} W^{\mathrm{eq}}\Psi = \Psi \mathscr{L} W^{\mathrm{eq}} + W^{\mathrm{eq}} \mathscr{L}\Psi$, and since the phase space dependence of W^{eq} is only through the Hamiltonian $\mathscr{L} W^{\mathrm{eq}} = 0$ [see eq. (4b.21)], we have

$$\frac{\partial W_{\mathbf{N}}}{\partial t} = \lambda W^{\mathrm{eq}}(-\mathscr{L}\Psi) = \lambda W^{\mathrm{eq}} \frac{\partial \Psi}{\partial t}. \tag{6j.3}$$

The series

$$\begin{aligned} \Psi(t) &= \Psi(t=0) + \sum_{n \geq 1} \frac{t^n}{n!} \frac{\partial^n \Psi}{\partial t^n}\, t=0 \\ &= \sum_{n \geq 0} \frac{t^n}{n!} (-\mathscr{L})^n \Psi(t=0) \end{aligned} \tag{6j.4}$$

can be written formally as

$$\Psi(t) = [\exp(-\mathscr{L} t)] \Psi(t=0). \tag{6j.4′}$$

Now, as we discuss in some mathematical detail later, the functions $(-\mathscr{L})^n \Psi(t=0)$ become extremely intricate functions of the coordinate of the molecules, involving correlations between enormous numbers of molecules. These detailed correlations have no effect on the macroscopically measurable properties of the system. They represent a lack of smoothness in W, but one that is not macroscopically meaningful. The measure $\frac{1}{2}\lambda^2 \mathscr{I}_N W^{\mathrm{eq}} \Psi^2$ of the *total* lack of smoothness remains constant in time in a model isolated system. The terms that have macroscopic significance representing gradients in density of molecules or energy decrease. Those of no physical significance increase.

Gibbs likened the situation to a large drop of carbon-black ink added to a glass of clear water and then stirred. Before stirring there is a volume occupied by pure water, and one of pure ink. After stirring, at every microscopic position there is either an ink particle, or water, but the mixture looks uniform and gray. Somehow we must smooth $W \ln W$ to remove the nonsignificant irregularities before integration. Many suggestions for smoothing have been made. If we simply average $W \ln W$ over volume elements in phase space and then integrate, no change results, since the operator $\mathscr{I}$ is just an averaging operator. If somehow $W(\mathbf{q}^{(\Gamma)}, \mathbf{p}^{(\Gamma)})$ is smoothed to $\bar{W}(\mathbf{q}^{(\Gamma)}, \mathbf{p}^{(\Gamma)})$ by averaging the value of W in the neighborhood of $\mathbf{q}^{(\Gamma)}$, $\mathbf{p}^{(\Gamma)}$, and we then operate by $\mathscr{I}_{\mathbf{N}}$ on the quantity $\langle W\rangle$ ln $\langle W\rangle$, we change $-S/k$ so computed, and one can prove that its value is reduced. The difficulty is that $W^{\text{eq}}(\mathbf{q}^{(\Gamma)}, \mathbf{p}^{(\Gamma)})$ has extremely large gradients due to the term $\exp[-\beta H(\mathbf{q}^{(\Gamma)}, \mathbf{p}^{(\Gamma)})]$. The smoothing must be done in such a way as not to alter the predicted macroscopic values of E and $\mathbf{N}$.

We show a mathematical description of an allowed smoothing and suggest that this smoothing actually occurs in natural laboratory systems as a result of time-dependent fluctuations at the walls.

We define first what we mean by normalization and mutual orthogonality, weighted by a given set of $W_{\mathbf{N}}^{(\text{eq})}(\mathbf{q}^{(\Gamma)}, \mathbf{p}^{(\Gamma)})$, for real functions $\Psi_{\mathbf{N}}(\mathbf{q}^{(\Gamma)}, \mathbf{p}^{(\Gamma)})$, $\Phi_{\mathbf{N}}(\mathbf{q}^{(\Gamma)}, \mathbf{p}^{(\Gamma)})$, and so on. If $N=\sum_\alpha\langle N\rangle_\alpha$ and the $\langle N_\alpha\rangle$'s are the average numbers of molecules of species α, we call Ψ normalized to unity if[(1)]

$$N^{-1}\mathscr{I} W_{\mathbf{N}}{}^{\text{eq}}\Psi_{\mathbf{N}}^2=1, \tag{6j.5}$$

and Ψ and Φ orthogonal if

$$N^{-1}\mathscr{I} W_{\mathbf{N}}^{\text{eq}}\Psi_{\mathbf{N}}\Phi_{\mathbf{N}}=0. \tag{6j.6}$$

A set $\Psi_{\mathbf{k}}$, $\Psi_{\mathbf{m}}, \ldots$ of functions is said to form an orthonormal set if

$$N^{-1}\mathscr{I}_{\mathbf{N}} W_{\mathbf{N}}^{\text{eq}}\Psi_{\mathbf{k}}\Psi_{\mathbf{m}}=\delta(\mathbf{k}-\mathbf{m}), \tag{6j.7}$$

where $d(\mathbf{k}-\mathbf{m})$, the Kronecker delta symbol, is unity if $\mathbf{k}\equiv\mathbf{m}$ and zero otherwise. The set is said to be complete if it consists of an infinite number of members, $\mathbf{1}\le\mathbf{k}\le\infty$, such that any function $F_{\mathbf{N}}(\mathbf{q}^{(\Gamma)}, \mathbf{p}^{(\Gamma)})$ having the same boundary conditions and symmetries of the members of the set can be written as a linear combination of its members,

$$F_{\mathbf{N}}(\mathbf{q}^{(\Gamma)}, \mathbf{p}^{(\Gamma)})=\sum_{\mathbf{k}} a_{\mathbf{k}}\Psi_{\mathbf{k}}(\mathbf{q}^{(\Gamma)}, \mathbf{p}^{(\Gamma)}). \tag{6j.8}$$

We implicitly assume here that we can meaningfully assign one index $\mathbf{k}$ (or $\mathbf{m}$) for all number sets $\mathbf{N}$ of molecules. In the desire to not overburden the symbols with subscripts we omit the $\mathbf{N}$ in the more explicit notation $\Psi_{\mathbf{N},\mathbf{k}}$. If the set is complete, we have

$$N^{-1}\mathscr{I} W_{\mathbf{N}}^{\text{eq}}F_{\mathbf{N}}^2=N^{-1}\mathscr{I} W_{\mathbf{N}}^{\text{eq}}\sum_{km}\sum a_{\mathbf{k}}a_{\mathbf{m}}\Psi_{\mathbf{k}}\Psi_{\mathbf{m}}=\sum_k a_{\mathbf{k}}^2 \tag{6j.9}$$

from eqs. (8) and (7). Similarly, from eqs. (8) and (7),

$$a_{\mathbf{k}}=\mathscr{I} W_{\mathbf{N}}^{\text{eq}}\Psi_{\mathbf{k}}F_{\mathbf{N}}. \tag{6j.10}$$

(1) For the type of functions we wish to use, the factor N^{-1} makes the normalization independent of system size.

Thus, if one has a complete set of phase space functions, any initial perturbation $\lambda\Psi(t=0)$ can be written as a linear combination of the members of the set with coefficients $a_{\mathbf{k}}^{(0)}$, and the function $\lambda\Psi(t)$ as a linear combination with coefficients $a_k(t)$.

The operator $\mathscr{L}$ is real and antisymmetric. This means that, if we define $l_{\mathbf{k},\mathbf{m}}$ by

$$l_{\mathbf{k},\mathbf{m}} = \mathscr{I} W_{\mathbf{N}}^{\mathrm{eq}} \Psi_{\mathbf{k}} \mathscr{L} \Psi_{\mathbf{m}} = \mathscr{I} \Psi_{\mathbf{k}} \mathscr{L} W_{\mathbf{N}}^{\mathrm{eq}} \Psi_{\mathbf{m}} \tag{6j.11}$$

(where the extreme right-hand expression follows from $\mathscr{L} W^{\mathrm{eq}} = 0$), for the real functions $\Psi_{\mathbf{k}}$, $\Psi_{\mathbf{m}}$, $l_{\mathbf{km}}$ is real and

$$l_{\mathbf{k},\mathbf{m}} = -l_{\mathbf{m},\mathbf{k}}. \tag{6j.12}$$

(For complex functions $\Phi_{\mathbf{k}}$, $\Phi_{\mathbf{m}}$ the real parts of $l_{\mathbf{k},\mathbf{m}}$ change sign on exchange of indices, but the imaginary parts are equal.) This follows by partial integration of the right-hand form of eq. (11) for each single-phase space derivative of $\mathscr{L}$, remembering that $W^{\mathrm{eq}}\Psi$ is zero at the boundaries of the phase space variables. We may regard the number $l_{\mathbf{k},\mathbf{m}}$ of dimensions reciprocal time as elements of a matrix $\mathbf{L}$ in the representation of the particular set of functions $\Psi_{\mathbf{k}}$. The matrix $\mathbf{L}$ is real and antisymmetric.

The partial time derivative of

$$\Psi(t) = \sum_{\mathbf{k}} a_{\mathbf{k}}(t) \Psi_{\mathbf{k}} \tag{6j.13}$$

is, from eqs. (10) to (12),

$$\frac{\partial \Psi(t)}{\partial t} = -\mathscr{L}\Psi(t) = -\sum_{\mathbf{k}} \Psi_{\mathbf{k}} \sum_{\mathbf{m}} l_{\mathbf{k},\mathbf{m}} a_{\mathbf{m}}(t) = \sum_{\mathbf{m}} \sum_{\mathbf{k}} a_{\mathbf{k}}(t) l_{\mathbf{k},\mathbf{m}} \Psi_{\mathbf{m}}, \tag{6j.14}$$

and the nth derivative is

$$\begin{aligned} \frac{\partial^n \Psi(t)}{\partial t^n} &= \sum_{\mathbf{m}} \sum_{\mathbf{k}} a_{\mathbf{k}}(t) \sum_{\boldsymbol{\nu}_1} \sum_{\boldsymbol{\nu}_2} \cdots \sum_{\boldsymbol{\nu}_{n-1}} l_{\mathbf{k},\boldsymbol{\nu}_1} l_{\boldsymbol{\nu},\boldsymbol{\nu}_2} \cdots l_{\boldsymbol{\nu}_{n-1},\mathbf{m}} \Psi_{\mathbf{m}} \\ &= \sum_{\mathbf{m}} \sum_{\mathbf{k}} a_{\mathbf{k}}(t) [\mathbf{L}^n]_{\mathbf{k},\mathbf{m}} \Psi_{\mathbf{m}}, \end{aligned} \tag{6j.15}$$

where $[\mathbf{L}^n]_{\mathbf{k},\mathbf{m}}$ is the $\mathbf{k}$, $\mathbf{m}$ element of the matrix $\mathbf{L}^n$: the matrix $\mathbf{L}$ multiplied by itself n times. If we make a Taylor series expansion,

$$\Psi(t) = \sum_{n \geq 0} \frac{t^n}{n!} \left[\frac{\partial^n \Psi(t)}{\partial t^n} \right]_{t=0}, \tag{6j.16}$$

we have

$$\Psi(t)=\sum_{\mathbf{m}}\sum_{\mathbf{k}} a_{\mathbf{k}}(t=0)\left[\sum_{n\geq 0}\frac{(\mathbf{L}t)^n}{n!}\right]_{\mathbf{k,m}}\Psi_{\mathbf{m}}, \qquad (6j.16')$$

which can be written symbolically as

$$\Psi(t)=\sum_{\mathbf{m}}\sum_{\mathbf{k}} a_{\mathbf{k}}(t=0)(\exp \mathbf{L}t)_{\mathbf{k,m}}\Psi_{\mathbf{m}}. \qquad (6j.16'')$$

Now our present purpose in developing this exercise is to discuss the nature of the disappearance of negative entropy with time. The function $W_N(t=0)$ corresponding to an ensemble of systems prepared in some prescribed and reproducible macroscopic nonequilibrium state is a function like that given in eq. (6g.7), for which the corresponding function $\Psi(t=0)$ is given in eq. (1). To first order in λ this Ψ consists of a sum of terms [see eq. (6g.5) for H_i] of coordinate and momenta of single molecules, $\sum_{i=1}^{i=N}[\nu_i(\mathbf{r})=\beta(\mathbf{r}_i)(\mathbf{p}_i\mathbf{p}_i/2m_i)]$, and another sum $\sum\sum_{i>j}[\frac{1}{2}\beta(\mathbf{r}_i)+\frac{1}{2}\beta(\mathbf{r}_j)]u_{ij}(r_{ij})$ of functions of pairs of molecules, which are nonzero only when the molecules of the pair are close together in coordinate space. In second order of λ there are more complicated terms, including some involving the simultaneous positions of sets of four molecules. No macroscopically significant functions $\chi_i(\mathbf{q}^{(\Gamma)}, \mathbf{p}^{(\Gamma)})$ contain terms involving correlations between the coordinates or momenta of any very large subsets of molecules. The maximum entropy ensemble always lacks such terms at $t=0$, since additional terms only add to the negative entropy.

This gives us a motivation for making a particular type of orthonormal set of functions. Choose first a complete set of functions which are sums of functions of the μ-space variables of single molecules, symmetric in the exchange of identical molecules,

$$\Psi_{1,\mathbf{k}}(\mathbf{q}^{(\Gamma)}, \mathbf{p}^{(\Gamma)})=\sum_{\alpha}\sum_{i\alpha=1_\alpha}^{i\alpha=N_\alpha}\psi_{\mathbf{k}\alpha}(\mathbf{q}_{i\alpha}^{(\Gamma_\alpha)}, \mathbf{p}_{i\alpha}^{(\Gamma_\alpha)}). \qquad (6j.17)$$

For instance, for monatomic molecules in a rectangular box of sides L_x, L_y, L_z, the functions $\psi(\mathbf{r}_{i\alpha}, \mathbf{p}_{i\alpha})$ might be products of sine-cosine functions of the coordinates, $(2/L_x)^{1/2}\sin(\pi k_x x_{i\alpha}/L_x)$, and so on, times three normalized Hermite polynomials[(1)] of the momenta components, $p_{xi\alpha}(2m_\alpha kT)^{-1/2}$, which latter are orthonormal under the weighting function $(2\pi m_\alpha kT)^{-1/2}\exp[-(p_x^2/2m_\alpha kT)]$.

[(1)] The normalization is correct to first order in $\alpha\kappa_x/L^{1/3}$, unless all three Hermite polynomials are $H_0=1$. In this case an additional factor is required, $(\rho\kappa kT)^{-1/2}$ with $\kappa=-V^{-1}(\partial V/\partial P)_T$, the isothermal compressibility.

Add to these a second mutually orthonormal set, all orthogonal to all members of the first and symmetric in the exchange of identical molecules, which are sums over all pairs of molecules of functions of pairs of phase space variables, identical for all pairs $i\alpha$, $j\gamma$ of the same types α, γ of molecules,

$$\Psi_{2,\mathbf{k}}(\mathbf{q}^{(\Gamma)},\mathbf{p}^{(\Gamma)}) = \sum_{\alpha}\sum_{\gamma}\sum_{i\alpha}\sum_{j\gamma=i\alpha} \psi_{\mathbf{k}\alpha\gamma}(q_{i\alpha}, P_{i\alpha}, q_{j\gamma}, p_{j\gamma}). \tag{6j.18}$$

The set must be complete in the phase space of two molecules of type α and γ with an appropriate positive weighting factor which is the reduced equilibrium probability density. The important members of the set are those for which $\psi_{\mathbf{k},\alpha\gamma}$ is nonzero only when the two molecules $i\alpha$ and $j\gamma$ are spatially close, that is, which approach zero values if $r_{ij}\rho^{-1/3} \gg 1$. To be complete of course the set must also include other functions.

We then introduce a third complete orthonormal set $\Psi_{3,\mathbf{k}}$ orthogonal to the members of the first two sets, consisting of a sum of functions of triples of molecules, each depending only on the molecular species in functional form, and so on to sets $\Psi_{n,\mathbf{k}}$ of sums of functions depending on the coordinates and momenta of n molecules each, for all $n \leq N$.

Now all macroscopic thermodynamic extensive coordinates X_i, except the volume V, are averages of functions $\chi_i(\mathbf{q}^{(\Gamma)}, \mathbf{p}^{(\Gamma)})$ in the γ-space,

$$X_i = \langle\chi_i\rangle = \mathcal{I} W_{\mathbf{N}}(\mathbf{q}^{(\Gamma)}, \mathbf{p}^{(\Gamma)})\chi_i(\mathbf{q}^{(\Gamma)}, \mathbf{p}^{(\Gamma)}). \tag{6j.19}$$

The functions x_i are sums of functions of subsets of molecular phase space variables for small subsets only, usually single molecules or pairs only. Let us say they never involve subsets of greater than n_i molecules, so that we can write them as a linear combination of the orthonormal set functions $\Psi_{n,\mathbf{k}}$, with $n \leq n_i$,

$$\chi_i(\mathbf{q}^{(\Gamma)}, \mathbf{p}^{(\Gamma)}) = \sum_{n\geq 1}^{n\leq n_i}\sum_{k} a_{n,\mathbf{k}}^{(i)}\Psi_{n,\mathbf{k}}(\mathbf{q}^{(\Gamma)}, \mathbf{p}^{(\Gamma)}), \tag{6j.20}$$

and consequently for any $n > n_i$,

$$\mathcal{I}_{\mathbf{N}} W^{(\mathrm{eq})}\chi_i(\mathbf{q}^{(\Gamma)}, \mathbf{p}^{(\Gamma)})\Psi_{n,}{}^{\mathbf{k}}(\mathbf{q}^{(\Gamma)}, \mathbf{p}^{(\Gamma)}) = 0, \qquad n > n_i. \tag{6j.21}$$

It follows that, if in $\Psi(t)$ there is any nonzero amplitude, $a_{n,\mathbf{m}}(t) \neq 0$, of a function $\Psi_{n,\mathbf{m}}$ for any $n > n_i$, for all i, then all thermodynamic extensive variables will be unaltered by the existence or amplitude of this function $\Psi_{n,\mathbf{m}}$. Similarly, no meaningful macroscopic function χ_i involves high Hermite polynomials of the momenta, so that all functions $\psi_{n,\mathbf{k}}$ even for $n = 1$ or 2 having high Hermite polynomials do not alter thermodynamic variables.

The operator $\mathcal{L}$ of eq. (6i.1), if used with cartesian coordinates, will contain additive terms $(p_{xi}/m_i)\partial/\partial x_i$ and $-(\partial U/\partial x_i)\partial/\partial p_{xi}$ for all x_i, y_i, z_i, and i. Unless the potential $U(\mathbf{r}^{(N)})$ is strictly zero, which of course is never the case, the term $-(\partial U/\partial x_i)\partial/\partial p_{xi}$ operating on any function of the coordinates and momenta of $n<N$ molecules, containing p_{xi} to a nonzero power, introduces new terms multiplying the coordinate part by a function $-\partial u_{ij}(\mathbf{r}_i, \mathbf{r}_j)/\partial x_i$, in which j is a molecule *not* in the original subset of n molecules. The term $p_{xi}/m_i(\partial/\partial x_i)$, however, always introduces a higher power of p_{xi} without increasing the number of molecules involved in the function. It follows that the matrix $\mathbf{L}$ contains nonzero elements $l_{m\mathbf{k},n\mathbf{m}}$, with $\Psi_{n,\mathbf{m}}$ having powers of the momenta higher by one unit than in $\Psi_{n,\mathbf{k}}$, and also nonzero elements $l_{n\mathbf{k},(n+1)\mathbf{m}}$, with $\Psi_{(n+1)\mathbf{m}}$ having one lower power of momenta but with one more number of molecules in its constituent functions. We hasten to add that it does *not* follow that $-(\partial u_{ij}/\partial x_i)\partial/\partial p_{xi}\Psi_{n,\mathbf{k}}$ is necessarily orthogonal to all functions $\Psi_{n,\mathbf{m}}$ and $\Psi_{n-\nu,\mathbf{m}}$, but cannot be represented completely by a linear combination of functions $\Psi_{\nu,m}$ with $\nu \leq n$ only.

Now from this it follows that the matrix $\mathbf{L}^{2n}$ must contain nonzero elements $[\mathbf{L}^{2n}]_{\nu\mathbf{k},(\nu+n)\mathbf{m}}$, as well as terms $[\mathbf{L}^{2n}]_{\nu\mathbf{k},\nu\mathbf{m}}$ in which $\Psi_{\nu,\mathbf{m}}$ involves Hermite polynomials H_{2n}. The quantity $t_0=(m/\rho^{2/3}kT)^{1/2}$ has the dimension of time and is approximately the time between collisions for a single molecule. For a normal liquid at room temperature t_0 is of order 10^{-12} sec. One can expect that, at $t=1$ sec, $t/t_0 \sim 10^{12}$, the elements of $(\mathbf{L}t)^{10^{12}}$ are not negligible. The coefficients

$$a_{n,\mathbf{m}}(t)=\sum_{\nu}\sum_{\mathbf{k}} a_{\nu,\mathbf{k}}(t=0)(\exp \mathbf{L}t)_{\nu\mathbf{k},n\mathbf{m}}, \tag{6j.22}$$

for n of the order of 10^{12} at times as low as a second may well be nonzero.

From eqs. (9) and (6h.27), the excess of $-S/k$ above the equilibrium value is

$$\frac{-S}{k}=\tfrac{1}{2}\mathcal{J}W_N^{\text{eq}}(\lambda\Psi_{\mathbf{N}})^2=\tfrac{1}{2}N\sum_{n}\sum_{\mathbf{m}} a_{n,\mathbf{m}}^2(t). \tag{6j.23}$$

The sum on the right, if extended over *all* n and $\mathbf{m}$, remains constant. If we smooth $W_{\mathbf{N}}$ by removing the terms of high n, and also those for which $\Psi_{n,m}$ has very high Hermite polynomials, which represent extremely fine detail in the momentum distribution, we reduce the sum and succeed in finding a $-\Delta S/k$ that decreases in time.

The example of the Hahn spin echo serves as a warning against arbitrary truncation of the series $\sum a_{n,\mathbf{k}}(t)\Psi_{n,\mathbf{k}}$ to exclude the functions $\Psi_{n,\mathbf{k}}$ that are orthogonal to all known macroscopically meaningful functions χ_i. It is possible that the elements $[\mathbf{L}]_{n,\mathbf{k};n'\mathbf{k}'}$ connecting "meaningless" functions $\Psi_{n,\mathbf{k}}$ with significant functions $\Psi_{n',\mathbf{k}'}$ not orthogonal to some χ_i are such as to cause a positive feedback from $a_{n,\mathbf{k}}$ to $|a_{n',\mathbf{k}'}|$. This evidently occurs in the spin echo system. We propose, however, that in real experimental systems random temporally fluctuating forces at the walls decrease the amplitude square $a_{n,k}^2$ for very large n values, for which the functions $\Psi_{n,\mathbf{k}}$ involve correlations in the phase space of n molecules not representable as a sum of correlations of smaller subsets, and also those for which the index $\mathbf{k}$ indicates very high Hermite polynomials of the momenta components. This reduction in the complete sum $\frac{1}{2}\sum a_{n,\mathbf{k}}^2$ is due to a mechanism not included in the Liouville operator $\mathscr{L}$ for the molecules of the system.

A purist might correctly insist that we must then include the surroundings in our mathematical model and, eventually, since nothing is completely isolated, the whole cosmos. The Liouville operator $\mathscr{L}$ then includes very weak coupling terms between system and surroundings, which are never totally absent. We must also introduce functions $\Psi_{n+N,\mathbf{k}}$ having correlations between the phase space of molecules n within the system and N in the surroundings; the amplitudes $a_{n+N,k}(t)$ of these grow in time. The total sum $\frac{1}{2}\sum_{n,N,\mathbf{k}} a_{n+N,\mathbf{k}}^2(t)$ stays constant in time, but the limited sum $\frac{1}{2}\sum_{n,\mathbf{k}} a_{n,\mathbf{k}}^2(t)$ decreases. The entropy of system and surroundings is no longer strictly additive, but negative entropy terms due to their interaction pile up. The negative entropy due only to the system decreases.

This solution to the paradox appears to have an unpleasant feature. We still retain in the sum that measures the negative entropy all amplitude-squared terms $a_{n,\mathbf{k}}^2(t)$, even for those functions with $n \sim 10^{19}$ whose actual amplitudes can never be measured. The statistically defined entropy $S = -k\mathscr{I} W_{\mathbf{N}} \ln W_{\mathbf{N}}$ is then more negative in a system slowly approaching equilibrium than would be inferred from a measurement at time t of its macroscopic variables $T(\mathbf{r}, t)$, $\rho\alpha(\mathbf{r}, t), \ldots$. Furthermore, the extent to which this is true depends on the degree to which the system can be isolated from its surroundings.

However, the example of the Hahn spin echo demonstrates that systems exist for which this is true. There is at least one other example of a computer experiment done on a linear chain of interacting point masses that shows a similar echo. In the case of the nuclear spins the surroundings are the other degrees of freedom of the material in which the spins exist, for instance, the lattice vibrations of a crystal. The coupling of the spin system to the surroundings is extremely weak but increases with increasing lattice vibration temperature. The true negative entropy loss rate of the spin system, measured across the echo peaks, indeed increases with crystal temperature.

Consider a more classical system approaching equilibrium, say, a vertical cylinder of liquid or gas at room temperature slightly heated at the top and then isolated. Depending on dimensions the time required for heat conduction to reduce the temperature difference between top and bottom by a factor $1/e$ may be $15\ \text{min} \cong 10^3$ sec. Single molecules collide about 10^{12} times per second. Suppose that a miraculously clever experimenter actually succeeds in exactly reversing every momentum at a time t_0. The equivalent quantum-mechanical feat would be to change the state function $\Psi(t_0)$ into its conjugate complex $\Psi^*(t_0)$. How long after t_0 will the molecules exactly reverse their previous path? Any billiard player aware of the sensitivity to its initial cued direction of the angle with which a cue ball bounces off another at its second collision would be very skeptical if exact time reversal persisted for many collisions before a minute perturbation, propagating with sound velocity from some random event at the walls, destroyed the programmed sequence of events. Suppose that with very careful isolation one could prolong the exact retracing of collisions for a million collisions. This time is of the order of 10^{-6} sec or 10^{-9} of the characteristic decay time of the system to equilibrium. If we refer back to Fig. 6f.2, this means that the echo peak would go up by about the 10^{-9}-th part of the initial negative entropy displacement. This is, then, the measure of the amount by which in that experimental system the sum $\frac{1}{2}\sum a_{n,\mathbf{k}}^2$ is contributed to by the squared amplitudes of the macroscopically meaningless functions $\Psi_{n,\mathbf{k}}$.

6k. APOLOGIA

We have entirely neglected reference to a very large, erudite, and intrinsically interesting body of mathematical literature stimulated by the problems of statistical mechanics. This literature extends back over many decades but is being added to very significantly with new, certainly elegant, and probably important mathematical techniques which can be used for problems other than the special cases they were designed to prove. As we have emphasized, we find the real physical systems of the laboratory significantly different than the idealized models that form the basis of most of this work.

We have chosen to treat the ensemble average of Gibbs as a more realistic representation of experimental observation than the time average of an isolated system. An observation of any macroscopic property of a real system completely destroys its path through phase space, and a later observation is not made on a continuation of the path of the original isolated system. Naturally, one guesses that the time average and ensemble average are identical, but the proof was far from being trivial. A discussion was given in 1932 by Birkhoff, Birkhoff, and Koopman, and independently by von Neumann.[1]

The problem of ergodicity has a long history. Boltzmann originally postulated that the phase path of an isolated system passes through every point on the energy surface. Rosenthal[2] and also Plancherel[3] showed that this was untenable, and the Ehrenfests[4] postulated the so-called quasi-ergodic hypothesis that the phase path comes infinitesimally close to every point in phase space. Both Birkhoff and von Neumann, in previously quoted articles, in 1932 proved this theorem under restricted circumstances. More recently, Sinai[5] has presented a proof for hard spheres. An article by A. Wightman[6] has a good discussion of the problems.

(1) G. D. Birkhoff, *Proc. Nat. Acad. Sci.*, **17,** 656 (1931); G. D. Birkhoff and O. Koopman, *Proc. Nat. Acad. Sci.*, **18,** 279 (1932); J. von Neumann, *Proc. Nat. Acad. Sci. USA*, **18,** 70 (1932).

(2) A. Rosenthal, *Ann. Phys.*, **42,** 796 (1913).

(3) M. Plancherel, *Ann. Phys.*, **42,** 1016 (1913).

(4) Ehrenfest and T. Ehrenfest, *Encyclopädi der Mathematischen Wissenschaften* Vol. **IV**, Pt. 32, B. G. Teubner, Leipzig, 1911.

(5) Ya. G. Sinai, *Russ. Math. Surv.*, **25,** 137 (1970).

(6) A. Wightman, in *Statistical Mechanics at the Turn of the Decade*, E. G. D. Cohen, Ed., Marcel Dekker, New York, 1971.

PART THREE

In Chapters 3 to 6 which constitute Part 2 of this book we derived the fundamental equations for the probability $W_{\mathbf{K}}$ that a member of an ensemble of equilibrium thermodynamic systems will be found in a quantum state of the total system designated by a multicomponent quantum number **K**. Some completely general results were discussed in Chapter 5, and a critique and defense of the derivation were discussed in Chapter 6. Those who are unhappy with long expositions of abstract principles without reference to the applicability of the derived equations to explicit situations in the real laboratory world are encouraged to study this third part of the book, which deals with applications to specific examples before undertaking the study of Part 2. Many or most probably prefer a compromise sequence. Those less thoroughly acquainted with the methods and terminology of classical and quantum mechanics may find Chapter 3 obligatory reading before undertaking Part 3 of the book, even if only to understand our own use of symbols and terms. Sec. 3j of that chapter presents the general ensemble probability equation without derivation. In this following part of the book we use eq. (3j.1), as a master equation applicable to any equilibrium thermodynamic ensemble of systems.

CHAPTER 7

PERFECT GASES

7a. THE GENERAL EQUATIONS

We use the equation for a grand canonical ensemble consisting of an infinite number of systems each of volume V surrounded by walls through which both energy and molecules can pass to and from an infinite reservoir of temperature T having only one species of molecules at chemical potential μ. The probability $W_{N,\mathbf{K}}$ that a randomly selected member of the ensemble will contain exactly N molecules in V and be in a quantum state $\mathbf{K}$ is [eq. (3i.7)]

$$W_{N,\mathbf{K}} = \exp\{-\beta[PV - N_{\mu} + E(\mathbf{K})]\} \tag{7a.1}$$

where

$$\beta = \frac{1}{kT}, \tag{7a.2}$$

where k is Boltzmann's constant, the gas constant per molecule, and $E(\mathbf{K})$ is the energy of the whole system in the completely specified quantum state $\mathbf{K}$. The pressure P as a function of T and μ is determined by the condition that the sum of the probabilities $W_{N,\mathbf{K}}$ must be unity when summed over all quantum states $\mathbf{K}$ consistent with each N, and then over all values of N from zero to infinity,

$$\sum_{N} \sum_{\mathbf{K}(N)} W_{N,\mathbf{K}} = 1. \tag{7a.3}$$

With this and eq. (1) we have

$$PV = kT \ln \sum_N \sum_{\mathbf{K}(N)} \exp \beta[N\mu - E(\mathbf{K})], \tag{7a.4}$$

where the sum under the logarithm is called the partition function for the grand canonical ensemble (Sec. 3i).

The condition that the system be a perfect gas is ensured by assuming that there is no mutual potential energy and therefore no forces between the molecules. This approximation will be asymptotically valid if the chemical potential μ is so low that the average number density,

$$\rho = \frac{\langle N \rangle}{V}, \tag{7a.5}$$

of molecules is sufficiently low. The Hamiltonian is then separable as the sum of those of the individual molecules (Sec. 3d) and, since the molecules are identical, the molecular Hamiltonians are identical in form. The *molecules* are then in quantum states for which we use the index **k**, having energy $\epsilon_{\mathbf{k}}$. The quantum state **K** of the *system* is given by the numbers $n_{\mathbf{k}}$ of molecules in each quantum state **k** of the molecules,

$$\mathbf{K} \equiv n_0, n_1, \ldots, n_{\mathbf{k}}, \ldots, \tag{7a.6}$$

and the energy $E(\mathbf{K})$ of the system is that of the sum of all molecular energies,

$$E(\mathbf{K}) = \sum_{\mathbf{k}} n_{\mathbf{k}} \epsilon_k. \tag{7a.7}$$

For given values N of the total number of molecules in V one must have

$$\sum_{\mathbf{k}} n_{\mathbf{k}} = N, \tag{7a.8}$$

since every molecule must be in some state **k**. One may therefore write

$$N\mu = \sum_{\mathbf{k}} n_{\mathbf{k}} \mu, \tag{7a.9}$$

and

$$N\mu - E(\mathbf{K}) = \sum_{\mathbf{k}} n_{\mathbf{k}}(\mu - \epsilon_{\mathbf{k}}). \tag{7a.10}$$

The exponent $\exp \beta[N\mu - E(\mathbf{K})]$ in eq. (4) is therefore the exponent of a sum which is a product of exponents. The product must be summed over all **K** for each N and over all N, which means that we first sum over all values of $n_{\mathbf{k}}$ for each **k** subject to the restriction of eq. (8), and then over all N-values, which simply removes the restriction. We write

$$kT \ln \sum_{N} \sum_{\mathbf{K}(N)} \prod_{\mathbf{k}} [\exp n_{\mathbf{k}}(\beta\mu - \beta\epsilon_{\mathbf{k}})] = kT \ln \sum_{n_{\mathbf{k}} \geq 0} \prod_{\mathbf{k}} [\exp n_{\mathbf{k}}(\beta\mu - \beta\epsilon_{k})], \tag{7a.11}$$

and then permute the order of taking the product and sum to obtain, with eq. (4), that

$$PV = kT \ln \prod_{\mathbf{k}} \sum_{n_{\mathbf{k}} \geq 0} [\exp n_{\mathbf{k}}(\beta\mu - \beta\epsilon_{\mathbf{k}})]$$

$$= kT \sum_{\mathbf{k}} \ln \sum_{n_k \geq 0} [\exp n_k(\beta\mu - \beta\epsilon_{\mathbf{k}})], \tag{7a.12}$$

since the logarithm of a product is the sum of the logarithms. We have thus reduced the problem from summation over quantum states of the whole system to the much more tractable problem of summing over quantum states of the individual molecules.

At this stage we have to distinguish two cases depending on the nature of the molecules. If the molecules contain an even number of individual fermion constituents, protons, neutrons, and electrons, they form a Bose-Einstein system (Sec. 3e); if they contain an odd number, the system is Fermi-Dirac. Since, for neutral molecules the number of protons and electrons is equal, the criterion for uncharged molecules is whether the number of neutrons is even or odd. For Bose-Einstein systems any number of molecules may be in the same quantum state; the sum over $n_{\mathbf{k}}$ goes from zero to infinity. For Fermi-Dirac systems only one molecule can occupy a quantum state $\mathbf{k}$; the sum over $n_{\mathbf{k}}$ goes over zero and unity only. Now, since

$$\sum_{n \geq 0}^{\infty} x^n = 1 + x + x^2 + \cdots = (1-x)^{-1}$$

with $x = \exp \beta(\mu - \epsilon_{\mathbf{k}})$, we find from eq. (11) that the two cases yield, with $\beta = 1/kT$ [eq. (2)] that

$$\beta PV = \sum_{\mathbf{k}} \mp \ln(1 \mp e^{\beta(\mu - \epsilon_{\mathbf{k}})}), \qquad \begin{array}{l} -\text{, Bose–Einstein} \\ +\text{, Fermi–Dirac} \end{array} \tag{7a.13}$$

as the rigorous expression for any perfect gas.

We must emphasize here that the sum over $\mathbf{k}$ involves summation over *all* quantum numbers describing the state of the molecule, including all nuclear spin states of the constituent atoms.

One other remark is pertinent here. The chemical potential μ for this one-component system, is the Gibbs free energy G per molecule,

$$\mu = \frac{G}{N} \qquad \text{(one chemical component).} \tag{7a.14}$$

The value of $G = E + PV - TS$ depends on the arbitrary zero from which the energy is measured. Equation (1) is valid only if $N\mu$ and $E(\mathbf{K})$ are measured from the same zero, which means that in eq. (12) the energy zero of μ is that of the lowest energy quantum state, $\mathbf{k} \equiv 0$, of the molecules, which usually, in statistical mechanical treatments, is assigned zero energy, $\epsilon_0 = 0$. The conversion to other zeros is discussed in Sec. 5b in some detail, but the simple rule is that the energy zero of μ is that of the lowest quantum state $\mathbf{k}$ of the molecules.

7b. BOLTZMANN COUNTING

An approximation to eq. (7a.13) called the Boltzmann approximation is numerically adequate for all real molecular or monatomic gases under the conditions that the perfect gas approximation of zero mutual potential energy is applicable. The condition for the validity of the Boltzmann approximation is that, for all $\mathbf{k}$, $\beta(\epsilon_\mathbf{k} - \mu)$ be positive and considerably greater than unity, so that $\exp \beta(\mu - \epsilon_\mathbf{k})$ is very much smaller than unity. The approximation is completely erroneous for blackbody radiation which can be treated as a Bose-Einstein perfect gas of photons with $\mu = 0$ (Sec. 11.d), and also for electrons at the number density of the valence electrons in a metal, for which one can make the rather crude approximation of treating them as a perfect Fermi-Dirac gas trapped in a uniform effective potential due to the positive ions. For helium near its boiling point, 4.2 K, and at its vapor pressure, the error of the Boltzmann approximation is not negligible but is of the same order as that due to the assumption of a perfect gas.

If $\exp \beta(\mu - \epsilon_\mathbf{k})$ is sufficiently small, we use the approximation

$$\mp \ln(1 \mp x) = x$$

in eq. (7a.13) to write

$$\beta PV = \sum_\mathbf{k} e^{\beta(\mu - \epsilon_\mathbf{k})} = e^{\beta\mu} \sum_\mathbf{k} e^{-\beta\epsilon_\mathbf{k}}. \tag{7b.1}$$

Since normally the thermodynamic state of the gas is given by P and T (or $\beta = 1/kT$) rather than by P and μ, we find the more useful explicit expression for $\mu(P, T)$ by solving eq. (1) for μ,

$$\mu = kT \ln\left[\frac{PV}{kT}\left(\sum_\mathbf{k} e^{-\beta\epsilon_\mathbf{k}}\right)^{-1}\right]. \tag{7b.2}$$

Now make use of the fact that the Hamiltonian (Sec. 3a) of the molecule is necessarily separable (Sec. 3d) into a sum of a translational term depending only on the coordinates and momenta of the center of mass, and an internal term depending only on the internal coordinates and momenta, if these latter have to be considered at all. If cartesian coordinates x, y, z are used for the center of mass, the translational Hamiltonian is itself separable into three terms, $H_x = (1/2m)p_x^2$, and so on. The quantum states of fixed energy for the molecule are then (in a rectangular volume of sides parallel to x, y, z) represented by a product of three functions x, y, z, respectively, and a function of the internal coordinates $\mathbf{q}_i$. The quantum number $\mathbf{k}$ of the molecule is given by

$$\mathbf{k} = k_x, k_y, k_z, \mathbf{n}, \tag{7b.3}$$

with $\mathbf{n}$ an internal quantum number having as many components as there are internal degrees of freedom. The energy is a sum,

$$\epsilon_{\mathbf{k}} = \epsilon_{kx} + \epsilon_{ky} + \epsilon_{kz} + \epsilon_{\mathbf{n}}, \tag{7b.4}$$

so that $\exp(-\beta\epsilon_{\mathbf{k}})$ is a product. The summation over $\mathbf{k}$ is an independent summation over k_x, k_y, k_z, and $\mathbf{n}$, so that the sum of the product is the product of the four sums.

We define what we call an internal molecular partition function Q_i, dependent on temperature only,

$$Q_i = \sum_{\mathbf{n}} e^{-\beta\epsilon_{\mathbf{n}}} \tag{7b.5}$$

the value of which depends completely on the nature and structure of the molecule, as discussed in later sections. We only remark here that for the noble gases the ground level of energy, $\mathbf{n} = 0$, to which we can assign the energy $\varepsilon_0 = 0$ is a single quantum state, and the first excited state is an electronic excitation of several electronvolts. Since for 1 eV, $\beta \times 1\text{ eV} \cong 1.1 \times 10^4\, T^{-1}$, the excited levels contribute nothing to Q_i at laboratory temperatures, and Q_i is unity with this choice of the zero of energy.

Now return to the translational terms. In a cubic box of edge L, volume $V = L^3$, the stationary states along the x-axis are the solutions of the time-independent Schrödinger equation,

$$\mathcal{H}\psi_k(x) = \frac{1}{2m}\left(\frac{h}{2\pi i}\frac{\partial}{\partial x}\right)^2 \psi_k(x) = \epsilon_k \psi_k(x), \tag{7b.6}$$

which obey the boundary conditions of being zero at $x = 0$ and $x = L$. The normalized solutions are, with k_x a positive integer,

$$\psi_{kx}(x) = \left(\frac{2}{L}\right)^{1/2} \sin\frac{\pi k_x x}{L} \tag{7b.6$'$}$$

and

$$\epsilon_{kx} = \frac{1}{2m}\left(\frac{h}{2L}\right)^2 k_x^2, \tag{7b.7}$$

with two other similar terms for y and z.

The function $\psi_{kx}(x)$ of eq. (6) is an eigenfunction of the operator $[(h/2\pi i)\partial/\partial x]^2$ which is the quantum-mechanical operator for the square of the x-component of momentum p_x^2. This means that the value of p_x^2 in the state k_x has the precise[(1)] value

$$p_x^2 = \left(\frac{h}{2L}\right)^2 k_x \tag{7b.8}$$

and that the energy ε_{kx} is just the kinetic energy of a molecule reflected back and forth between the walls at $x = 0$ and $x = L$ with the momentum value of eq. (8),

$$\epsilon_{kx} = \frac{1}{2m} p_x^2. \tag{7b.9}$$

With

$$p^2 = p_x^2 + p_y^2 + p_z^2, \tag{7b.10}$$

the total translational energy is

$$\epsilon(k_x, k_y, k_z) = \frac{1}{2m} p^2(k_x, k_y, k_z), \tag{7B.11}$$

where the p_x, p_y, p_z are related to k_x, k_y, k_z by eq. (8).

The translational factor in $\sum_{\mathbf{k}} \exp(-\beta\epsilon_{\mathbf{k}})$ is then

$$\sum\sum\sum_{k_x, k_y, k_z \geq 0} \exp\left[-\frac{\beta}{2m} p^2(k_x, k_y, k_z)\right]. \tag{7b.12}$$

The factor $(1/2m)(h/2L)^2$ of dimension energy in eq. (7) is extremely small compared to kT if L is the edge of a box of macroscopic size, 1 mm or larger. An enormous number of states with positive integer values of k_x, k_y, k_z lie in a phase volume $\omega = (2L/h)^3 \, \Delta p_x \, \Delta p_y \, \Delta p_z$, with all Δp's small enough that the energy multiplied by β is effectively constant, and the number of these is equal to ω in the asymptotic limit. We can replace summation over the positive integers k_x, k_y, k_z by integration over positive p_x, p_y, p_z values of $(2L/h)^3 \, dp_x \, dp_y \, dp_z$ or, for an even function of the momenta, by integration from minus to plus infinity of $(L/h)^3 \, dp_x \, dp_y$

(1) Strictly, p_x^2 is precise only in an infinite box, $L \to \infty$; for finite L there is an uncertainty $\Delta p_x^2 = (h/L)^2$.

dp_z, which is integration from zero to infinity of $Vh^{-3}\, 4\pi p^2\, dp$.

$$\sum\sum\sum_{k_x,k_y,k_z\geq 0} e^{-(\beta/2m)p^2} \to \int_0^\infty\int_0^\infty\int_0^\infty \frac{8V}{h^3}\, dp_x\, dp_y\, dp_z\, e^{-(\beta/2m)p^2}$$

$$= Vh^{-3}\int_0^\infty 4\pi p^2\, dp\, e^{-(\beta/2m)p^2}. \tag{7b.13}$$

With eq. (5) for Q_i, we have

$$\sum_{\mathbf{k}} e^{-\beta\epsilon_{\mathbf{k}}} = Q_i V h^{-3}\int_0^\infty 4\pi p^2\, dp\, e^{-(\beta/2m)p^2}$$

$$= Q_i V\left(\frac{2\pi mkT}{h^2}\right)^{3/2}, \tag{7b.14}$$

and with this, in eq. (2) for μ,

$$\mu = kT \ln\left[\frac{P}{kT}\left(\frac{h^2}{2\pi mkT}\right)^{3/2} Q_i^{-1}\right]. \tag{7b.15}$$

The quantity $(h^2/2\pi mkT)^{1/2}$ has the dimension of a length and is frequently called the de Broglie wavelength or, more properly, the de Broglie thermal wavelength; it is usually designated by the symbol λ,

$$\lambda = \left(\frac{h^2}{2\pi mkT}\right)^{1/2}. \tag{7b.16}$$

It is discussed in more detail in Sec. 7d. With its use eq. (15) becomes

$$\mu = kT \ln \frac{P}{kT}\lambda^3 Q_i^{-1}. \tag{7b.17}$$

The whole discussion from the paragraph starting after eq. (5) and continuing up to eq. (13) could have been eliminated by making use of the assertion made in Sec. 3f. The assertion is that, when the quantum states for any separable degrees of freedom lie sufficiently close together in energy compared to kT, the sum over quantum states of any operator in the phase space can be replaced by the integral of the classical function in phase space over the phase space variables divided by h raised to the power of the number of degrees of freedom. The function in this case is $\exp(-\beta H)$, with H the Hamiltonian which is $(1/2m)p^2$ in the absence of any potential energy. Integration over the coordinate space gives the volume V. The integral over the momentum space is $\int_0^\infty 4\pi p^2\, dp \exp[-(\beta/2m)p^2]$. The number of degrees of freedom is 3, so we divide by h^3. The discussion before eq. (6) and up to eq. (13) serves only to prove in detail for this particular case that the assertion of Sec. 3f is indeed correct.

Actually, the Boltzmann counting can be derived directly from eq. (7a.1) without going through the procedure developed in this section. We use first the pseudo-quantum states $\mathbf{K}^{\dagger}$ described in Sec. 3e, giving all product solutions of the time-independent Schrödinger equation for each *numbered* molecule i, $1 \leq i \leq N$. Since the Hamiltonian is separable, the energy $E(\mathbf{K}^{\dagger})$ is the sum of the energies of the individual molecules i and, since the quantum states of translation lie so close together, we use the classical Hamiltonian $(1/2m)p_i^2$ for these degrees of freedom. One has

$$E(\mathbf{K}^{\dagger}) = \sum_{i \geq 1}^{i=N} \left[\left(\frac{1}{2m}\right) p_i^2 + \epsilon_{\mathbf{n}i} \right], \tag{7b.18}$$

where $\epsilon_{\mathbf{n}i}$ is the internal energy of the molecule numbered i. The Boltzmann approximation now consists of assuming that there are $N!$ different states $\mathbf{K}^{\dagger}$ required to produce one state $\mathbf{K}$ of correct symmetry, either completely antisymmetric in all pair exchanges or completely symmetric, an approximation which is numerically good if the average number $\langle n_{\mathbf{k}} \rangle$ of molecules per quantum state $\mathbf{k}$ of the molecules is small compared to unity.

Since $\sum_{\mathbf{N}} \sum_{\mathbf{K}} W_{N,\mathbf{K}} = 1$, we multiply both sides of this relation by the constant factor $\exp + \beta PV$ to write

$$e^{\beta PV} = \sum_{N \geq 0} \frac{1}{N!} \sum_{K\dagger} \exp \sum_{i=1}^{i=N} \left[\beta\mu - \beta \frac{p_i^2}{2m} - \beta\epsilon_{\mathbf{n}i} \right]. \tag{7b.19}$$

Summation over $\mathbf{K}^{\dagger}$ is now made by integration over $3N$ coordinates x_i, y_i, z_i in V, and of the $3N$ momenta p_{xi}, p_{yi}, p_{zi} of the centers of mass, division by h^{3N}, and summation for each i over $\mathbf{n}_i$. The exponent of the sum in eq. (17) is the product of the single exponentials, and the integration and summation go over each of the products separately. For each N-value one finds the Nth power of a single term,

$$e^{\beta\mu} V \left(\int_0^{\infty} 4\pi p^2 \, dp \, e^{-(\beta p^2/2m)} \right) h^{-3} Q_i = e^{\beta\mu} V Q_i \left(\frac{2\pi mkT}{h^2} \right)^{3/2}, \tag{7b.20}$$

with Q_i given by eq. (5). The sum over N is now

$$e^{\beta PV} = \sum_{N \geq 0} \frac{1}{N!} \left[e^{\beta\mu} V Q_i \left(\frac{2\pi mkT}{h^2} \right)^{3/2} \right]^N$$

$$= \exp(e^{\beta\mu} V \lambda^{-3} Q_i). \tag{7b.21}$$

Solving for μ we find eq. (17).

We note here that the exponent in eq. (19) is proportional to the probability density of the N molecules in the phase space of the centers of mass multiplied by the probability of the set of internal quantum numbers $\mathbf{n}_i$. Since the probabilities are all independent, that is, they are products of probabilities for single molecules, the probabilities for a single molecule are proportional to the single terms in the exponent.

7c. MOMENTUM PROBABILITIES AND MOMENTS

The probability density $W(\mathbf{p})$ in the three-dimensional momentum space, $\mathbf{p} = p_x, p_y, p_z$, is defined so that (Sec. 1c) $W(\mathbf{p})\, dp_x\, dp_y\, dp_z$ is the probability that the momentum vector is $\mathbf{p}$ in the volume element between p_x and $p_x + dp_x$, p_y and $p_y + dp_y$, p_z and $p_z + dp_z$. Its value depends only on the magnitude $p = \sqrt{p_x^2 + p_y^2 + p_z^2}$ and is

$$W(\mathbf{p}) = \left(\frac{\beta}{2\pi m}\right)^{3/2} e^{-\beta p^2/2m}, \tag{7c.1}$$

normalized so that

$$\iiint_{-\infty}^{+\infty} dp_x\, dp_y\, dp_z\, W(\mathbf{p}) = \int_0^\infty 4\pi p^2\, dp\, W(\mathbf{p}) = 4\pi^{-1/2} \int_0^\infty dz\, z^2 e^{-z^2} = 1, \tag{7c.2}$$

where $z = p(2mkT)^{-1/2}$.

The average value $\langle |\mathrm{p}|^n \rangle$ of the nth power of the magnitude of p is called the nth moment of the magnitude of momentum. It is given by

$$\langle |\mathbf{p}|^n \rangle = \int_0^\infty 4\pi p^2\, dp\, p^n W(\mathbf{p}) = 2\pi^{-1/2}(2mkT)^{n/2} \int_0^\infty dz\, 2z^{2+n} e^{-z^2}. \tag{7c.3}$$

The values are given in Table 7c.1.

Table 7c.1 Moments $\langle |\mathbf{p}|^n \rangle$ of the Magnitude of Momentum

n	$\int_0^\infty dz\, 2z^{2+n} e^{-z^2}$	$\langle [\lvert\mathbf{p}\rvert (2mkT)^{-1/2}]^n \rangle$	$\langle [\lvert\mathbf{p}\rvert (2mkT)^{-1/2}]^n \rangle^{1/\lvert n\rvert}$
−2	$\pi^{1/2}$	$2 = 2.00000$	1.41421
−1	1	$2\pi^{-1/2} = 1.12838$	1.12838
0	$\frac{1}{2}\pi^{1/2}$	$1 = 1.00000$	1.00000
1	1	$2\pi^{-1/2} = 1.12838$	1.12838
2	$\frac{3}{4}\pi^{1/2}$	$\frac{3}{2} = 1.50000$	1.22474
3	2	$4\pi^{-1/2} = 2.25676$	1.31166
4	$\frac{15}{8}\pi^{1/2}$	$\frac{15}{4} = 3.75000$	1.39103
5	6	$12\pi^{-1/2} = 6.77027$	1.46504
6	$\frac{105}{16}\pi^{1/2}$	$\frac{105}{8} = 13.12500$	1.53623

7d. THE DE BROGLIE WAVELENGTH

The sine wave function $\psi_{kx}(x)$ of eq. (7b.6) has a wavelength λ, given by $\sin(2\pi x/\lambda) = \sin(\pi k_x x/L)$ or $\lambda = 2L/k_x$. The x-component of momentum squared is, from eq. (7b.8), $p_x^2 = (h/2L)^2 k_x^2 = (h/\lambda)^2$. The state is one in which the momentum is h/λ or $-(h/\lambda)$, strictly so only in the limit $L \to \infty$, since the uncertainty Δp_x of the momentum component is of order h/L. The general statement is that the wavelength of a plane wave corresponding to the motion of a free particle moving in a direction x is related to the particle momentum p_x by

$$\lambda_x = \frac{h}{p_x}, \tag{7d.1}$$

and is called the de Broglie wavelength. For a progressive wave in three dimensions the wavelength λ, which is the distance between crests, is given by

$$\lambda^{-2} = \lambda_x^{-2} + \lambda_y^{-2} + \lambda_z^{-2} = h^{-2}(p_x^2 + p_y^2 + p_z^2) = \left(\frac{p}{h}\right)^2. \tag{7d.2}$$

The average value $\langle\lambda\rangle$ of the wavelengths defined by eq. (2) for molecules in thermal equilibrium of a perfect gas is then $h\langle p^{-1}\rangle$. From Table 7c.1, $\langle p^{-1}\rangle = 2(2\pi mkT)^{-1}$. If this is compared with the length $\lambda = (h^2/2\pi mkT)^{1/2}$ of eq. (7b.16), we see that it would be somewhat more proper to name it the half-de Broglie thermal wavelength. The distinction is too pedantic to be useful, and we continue to use the conventional designation of de Broglie wavelength or de Broglie thermal wavelength for the quantity $\lambda = (h^2/2\pi mkT)^{1/2}$.

Numerically, with M the molecular weight, one finds that

$$\begin{aligned} \lambda &= A \times 10^{-7}\, M^{-1/2} T^{-1/2}, \\ A &= 1.746 \text{ cm deg}^{1/2}, \\ A^2 &= 3.049 \text{ cm}^2 \text{ deg}, \\ A^3 &= 5.324 \text{ cm}^3 \text{ deg}^{3/2}. \end{aligned} \tag{7d.3}$$

A convenient method of remembering is that

$$\lambda \cong \left(\frac{300}{MT}\right)^{1/2} \times 10^{-8} \quad \text{cm}, \tag{7d.3'}$$

to better than 2%, so that at room temperature $T = 300$ K,

$$\lambda(T = 300\text{ K}) = M^{-1/2} \times 10^{-8} \quad \text{cm}. \tag{7d.3''}$$

7e. PERFECT GAS THERMODYNAMIC FUNCTIONS

The expression of eq. (7b.17) for the chemical potential μ of a perfect gas,

$$\mu = kT\left(\ln\left\{\frac{P}{kT}\,\lambda(T)]^3\right\} - \ln Q_i(T)\right), \tag{7e.1}$$

with eq. (7b.16) for λ,

$$\lambda = \left(\frac{h^2}{2\pi mkT}\right)^{1/2} = \left(\frac{h^2\beta}{2\pi m}\right)^{1/2}, \tag{7e.2}$$

and the definition [eq. (7b.5)], of the internal molecular partition function Q_i as

$$Q_i(T) = \sum_{\mathbf{n}\text{ (all internal states)}} e^{-\beta\epsilon_{\mathbf{n}}} \tag{7e.3}$$

permits us to derive equations readily for all thermodynamic functions of any perfect gas in terms of its molecular weight and internal partition function.

The thermodynamic relation that $(\partial\mu/\partial P)_T = V/N$, with eq. (1), immediately gives us the familiar perfect gas equation of state,

$$\left(\frac{\partial\mu}{\partial P}\right)_T = \frac{V}{N} = \frac{kT}{P}, \qquad P = \frac{NkT}{V} = \frac{RT}{V_0}, \tag{7e.4}$$

where R is the normally used gas constant per mole, and V_0 is the volume per mole. This, in eq. (1), gives us an expression for μ in terms of the volume and temperature,

$$\mu = kT\left(\ln\frac{N\lambda^3}{V} - \ln Q_i\right). \tag{7e.5}$$

The temperature derivative of μ at constant pressure is the negative entropy per molecule. From eq. (1), with eq. (2) for λ, we find

$$s = \frac{S}{N} = -\left(\frac{\partial\mu}{\partial T}\right)_P = k\left(\frac{5}{2} - \ln\frac{P\lambda^3}{kT} + \frac{d}{dT}\,T\ln Q_i\right)$$

$$= k\left(\ln\frac{kT}{P\lambda^3}\,e^{5/2} + \ln Q_i + T\frac{d\ln Q_i}{dT}\right), \tag{7e.6}$$

$$s = k\left(\ln\frac{V}{N\lambda^3}\,e^{5/2} + \ln Q_i + T\frac{d\ln Q_i}{dT}\right). \tag{7e.6′}$$

The enthalpy, $H = G + TS = N\mu + TS$, per molecule is

$$\eta = \frac{H}{N} = \mu + \frac{TS}{N} = kT\left(\frac{5}{2} + T\frac{d \ln Q_i}{dT}\right), \tag{7e.7}$$

and the energy per molecule is this minus $PV/N = kT$,

$$\epsilon = \frac{H - PV}{N} = kT\left(\frac{3}{2} + T\frac{d \ln Q_i}{dT}\right). \tag{7e.8}$$

The two heat capacities per molecule are, since η and ϵ are independent of P and V,

$$c_p = \frac{C_P}{N} = \left(\frac{\partial}{\partial T}\right)_P \frac{H}{N} = k\left(\frac{5}{2} + \frac{d}{dT} T^2 \frac{d \ln Q_i}{dT}\right), \tag{7e.9}$$

and

$$c_V = \frac{C_V}{N} = \left(\frac{\partial}{\partial T}\right)_V \frac{E}{N} = k\left(\frac{3}{2} + \frac{d}{dT} T^2 \frac{d \ln Q_i}{dT}\right), \tag{7e.10}$$

at constant pressure and constant volume, respectively.

In Sec. 3j the fact was discussed that for all equilibrium ensembles the entropy is given by the sum over all quantum states $\mathbf{K}$ of the system of the quantity $-kW_{\mathbf{K}} \ln W_{\mathbf{K}}$.

The probability $W_{\mathbf{n}}$ that a molecule is in the internal quantum state $\mathbf{n}$ is proportional to $e^{-\beta\epsilon_{\mathbf{n}}}$ and, since the sum over all n of $W_{\mathbf{n}}$ must be unity, the coefficient is Q_i^{-1},

$$W_{\mathbf{n}} = Q_i^{-1} e^{-\beta\epsilon_{\mathbf{n}}}, \tag{7e.11}$$

$$\ln W_{\mathbf{n}} = -\ln Q_i - \beta\epsilon_n. \tag{7e.11$'$}$$

The quantity $Td \ln Q_i/dT$ is given by

$$T\frac{d \ln Q_i}{dT} = -\beta\frac{d \ln Q_i}{d\beta} = -\beta\sum_{\mathbf{n}} \frac{d}{d\beta} e^{-\beta\epsilon_n}$$

$$= \sum_{\mathbf{n}} \beta\epsilon_n W_n, \tag{7e.12}$$

and from eq. (11′),

$$\ln Q_i + T\frac{d \ln Q_i}{dT} = -\sum_{\mathbf{n}} W_{\mathbf{n}} \ln W_{\mathbf{n}}. \tag{7e.13}$$

The contribution to $-S/k$ of eq. (6) due to the internal degrees of freedom is just the sum of $W_{\mathbf{n}} \ln W_{\mathbf{n}}$ for a single molecule.

The other additive term, in $-S/k$, $\ln(P\lambda^3/kTe^{5/2})$, is N^{-1} times the sum of the probability times the logarithm of the probability over the quantum states of the system due to the translational degrees of freedom of the centers of mass of the molecules. However, because of the identity of the molecules, it is not simply the sum of $W_{\mathbf{k}} \ln W_{\mathbf{k}}$ for the translational states of a single molecule. It differs from this by an additive term, $N^{-1} \ln N! = (N^{-1} \ln N) - 1$. The additive unity to S/k in this is sometimes referred to as the communal entropy and has occasioned a certain amount of mysticism in some of the past literature. For this reason it may be worth a detailed analysis.

The probability density of a single molecule in the six dimensional cartesian phase space x, y, z, p_x, p_y, p_z is proportional to $\exp[-(\beta/2m)p^2]$, where $p = (p_x^2 + p_y^2 + p_z^2)^{1/2}$. The factor by which this must be multiplied to give the classical probability density $W_{\text{cl}}(x, y, z, p_x, p_y, p_z)$ is found by requiring that, within the volume V,

$$\int \cdots \int dx\, dy\, dz\, dp_x\, dp_y\, dp_z\, W_{\text{cl}} = 1, \tag{7e.14}$$

which leads us to

$$W_{\text{cl}} = V^{-1}(2\pi mkT)^{-3/2} e^{-(\beta/2m)p^2}, \tag{7e.15}$$

$$\ln W_{\text{cl}} = -\left[\ln V + \tfrac{3}{2} \ln(2\pi mkT) + \frac{\beta}{2m} p^2\right]. \tag{7e.15$'$}$$

We now have

$$-\iint \cdots \int d\mathbf{r}\, d\mathbf{p}\, W_{\text{cl}} \ln W_{\text{cl}} = \ln V + \tfrac{3}{2} \ln(2\pi mkT) + \frac{\beta}{2m} \langle p^2 \rangle$$

$$= \ln V + \tfrac{3}{2}[\ln(2\pi mkT) + 1], \tag{7e.16}$$

where the last line uses the value of $\langle p^2 \rangle$ from Table 2c.1. To evaluate the entropy, however, we should *sum* $W_{\mathbf{K}} \ln W_{\mathbf{K}}$ over the translational quantum states $\mathbf{K}$, where $W_{\mathbf{K}}$ is normalized by the condition that its sum over $\mathbf{K}$ is unity rather than the integral of eq. (14). Since there is one translational quantum state of a single molecule per volume h^3 in the six-dimensional classical phase space, the new normalization just adds a term $-3 \ln h$ to $\ln W_{\text{cl}}$. With eq. (2) that $\lambda = (h^2/2\pi mkT)^{1/2}$, we now have

$$-\sum_K W_K \ln W_K = \ln \frac{V}{\lambda^3} e^{3/2}. \tag{7e.17}$$

This begins to resemble the correct term $\ln(V/N\lambda^3)e^{3/2}$ for the translational entropy contribution divided by k in eq. (6$'$) and is indeed correct for the total S/k of a system consisting of a single isolated molecule in a total volume V, insofar as it has meaning to speak of the entropy of a single isolated molecule at all. However, it is obviously not satisfactory as the expression for thermodynamic entropy, since now for a system of $2N$ molecules in $2V$ at the same temperature the entropy *per molecule* has an additive term $k \ln 2$ instead of being independent of the system size.

Now one can solve this last difficulty in a sloppy and erroneous manner by saying that obviously one should have used the volume per molecule V/N in eq. (17) instead of V. One still needs to add $\ln e = 1$ to S/k in order to obtain the correct expression. This has sometimes been done by assigning to it the term communal entropy and ascribing its origin to the freedom of the molecules to exchange positions. It would be more correct to ascribe its origins to the fluctuations in local density, since in the gas volume elements of size V/N may contain zero, one, two, or any number of molecules.

The correct procedure is to recognize that the total entropy per *system* in the ensemble is given by the sum over the *system* quantum states $\mathbf{K}$ of $-kW_{\mathbf{K}} \ln W_{\mathbf{K}}$. It is this sum divided by N that gives the entropy per molecule. The states $\mathbf{K}$ of the system are described by giving the number $n_{\mathbf{k}}$ of (identical) molecules in each molecular quantum state $\mathbf{k}$ rather than $\mathbf{k}_i$, by giving the state $\mathbf{k}_i$ for each of the N artificially numbered molecules i. The true number of states of the system is then less (approximately) by a factor $N!$ than that we have implicitly assumed in deriving eq. (17) by assigning its own quantum state $\mathbf{k}_i$ to each molecule i. Since the number of states is smaller the probability of the individual states increases by the factor $N!$; the logarithm has an added term, $\ln N! = N[\ln N - 1]$, and N^{-1} times the negative of this added to the expression on the right of eq. (17) gives the correct translational contribution to S/k.

We turn now to the term $kT^2\, d \ln Q_i/dT$, which is the additive contribution of the internal degrees of freedom to the enthalpy per molecule η [eq. (7)] and the energy per molecule [eq. (8)]. Since, with $\beta = (kT)^{-1}$, we have that $T\, d \ln Q_i/dT = -\beta d \ln Q_i/d\beta$ and Q_i is the sum of $\exp(-\beta\epsilon_{\mathbf{n}})$ [eq. (3)], we have from eq. (11) for W_n that

$$kT^2 \frac{d \ln Q_i}{dT} = -Q_i^{-1} \sum_{\mathbf{n}} \frac{d}{d\beta} e^{-\beta\epsilon_{\mathbf{n}}}$$

$$= \sum_{\mathbf{n}} \epsilon_{\mathbf{n}} Q_i^{-1} e^{-\beta\epsilon_{\mathbf{n}}} = \sum_{\mathbf{n}} \epsilon_{\mathbf{n}} W_{\mathbf{n}} = \langle \epsilon_i \rangle, \tag{7e.18}$$

the average internal energy per molecule, as it must be.

The contribution $k\, d/dT\, (T^2\, d \ln Q_i/dT)$ of the internal degrees of freedom to the heat capacities per molecule [eq. (9) and (10)], is also simply related to the average of the internal energy squared minus the square of the average internal energy. Use

$$\sum_{\mathbf{n}} \frac{d}{dT} \epsilon_{\mathbf{n}} Q_i^{-1} e^{-\beta\epsilon_{\mathbf{n}}} = -k\beta^2 \sum_{\mathbf{n}} \frac{d}{d\beta} \epsilon_{\mathbf{n}} Q_i^{-1} e^{-\beta\epsilon_{\mathbf{n}}}$$

$$= \sum_{\mathbf{n}} k\beta^2 \left(\epsilon_n^2 Q_i^{-1} e^{-\beta\epsilon_n} - \epsilon_n Q_i^{-1} e^{-\beta\epsilon_n} \sum_{\mathbf{n}} \epsilon_n Q_i^{-1} e^{-\beta\epsilon_n} \right)$$

$$= k\beta^2 \left(\sum_{\mathbf{n}} \epsilon_{\mathbf{n}}^2 W_{\mathbf{n}} - \langle \epsilon_i \rangle \sum_{\mathbf{n}} \epsilon_{\mathbf{n}} W_{\mathbf{n}} \right) \tag{7e.19}$$

in eq. (18) for $kT^2\, d \ln Q_i/dT$. The two sums are $\langle \epsilon_i^2 \rangle$ and $\langle \epsilon_i \rangle^2$, respectively, or, with eq. (18),

$$k \frac{d}{dT} T^2 \frac{d \ln Q_i}{dT} = k\beta^2 (\langle \epsilon_i^2 \rangle - \langle \epsilon_i \rangle^2), \tag{7e.20}$$

for the contribution of the internal degrees of freedom to the heat capacity, eqs. (9) and (10).

This form of relationship between heat capacity and the difference, average square of the energy minus square of the average energy, is completely general for any classical or quantum-mechanical system. The relation

$$C_V = k^2 (\langle E^2 \rangle - \langle E \rangle^2) \tag{7e.21}$$

for the total heat capacity at constant volume of any system is derived in Sec. 5a [eq. (5a.10)]. In Sec. 5d we derived that for a separable Hamiltonian the additive contributions of the different separable parts of the Hamiltonian to C_V each individually obeys eq. (20). From Table 7c.1 for the moments of $|\mathbf{p}|$ one finds for the kinetic energy,

$$\langle \epsilon_{\mathrm{kin}} \rangle = (2m)^{-1} \langle |p| \rangle = \tfrac{3}{2} \beta^{-1} = \tfrac{3}{2} kT, \tag{7e.22}$$

$$\langle \epsilon_{\mathrm{kin}}^2 \rangle = (2m)^{-2} \langle |p| \rangle^4 = \tfrac{15}{4} \beta^{-2} = \tfrac{15}{4} (kT)^2, \tag{7e.22$'$}$$

that

$$c_{V,\mathrm{kin}} = \tfrac{3}{2} k = k\beta^2 (\langle \epsilon_{\mathrm{kin}}^2 \rangle - \langle \epsilon_{\mathrm{kin}} \rangle^2), \tag{7e.23}$$

in accord with eq. (10).

We should, however, draw attention to one enormous numerical difference between eqs. (20) and (21). Of course, since $\langle E \rangle = N \langle \epsilon \rangle$, with $N = 10^{20}$, the values of $\langle E \rangle^2$ and $\langle \epsilon^2 \rangle$ as well as $\langle E \rangle^2$ and $\langle \epsilon \rangle^2$ differ by factors of order 10^{40}. This is trivial. What is not so trivial is that $\langle E \rangle^{-2} (\langle E^2 \rangle - \langle E \rangle^2)$ is of order $10^{-20} = N^{-1}$, whereas $\langle \epsilon^2 \rangle^{-2} (\langle \epsilon^2 \rangle - \langle \epsilon \rangle^2)$ is of order unity; for instance, for the kinetic energy part it is $\frac{2}{3}$. This, of course, only reflects the fact that, whereas the energy of the whole system fluctuates negligibly (see Sec. 5c), that of single molecules fluctuates greatly.

7f. THE VALIDITY OF BOLTZMANN COUNTING

The expression in eq. (7e.5) for the chemical potential μ gives us a quick and easy way of estimating a posteriori the error make in using the Boltzmann counting instead of the exact Bose-Einstein or Fermi-Dirac equations. In deriving eq. (7b.1) we used the first term only in the approximation $\mp\ln(1\mp x) = x \pm \frac{1}{2}x^2 + 0(x^3)$ for the sum over states of $\mp\ln[1\mp e^{\beta(-\epsilon_k)}]$. The worst error in the terms is at $\epsilon_{\mathbf{k}} = 0$, so that one can expect the fractional error in pressure at given chemical potential to be somewhat less than $\frac{1}{2}e^{\beta\mu}$. From eq. (7e.5) we find

$$\tfrac{1}{2}e^{\beta\mu} = \frac{N\lambda^3}{2V} Q_i^{-1}, \tag{7f.1}$$

and since Q_i can never be less than unity,

$$\tfrac{1}{2}e^{\beta\mu} \le \frac{N\lambda^3}{2V}. \tag{7f.2}$$

At given P and T, since λ^3 varies as the molecular weight M to the power $-\frac{3}{2}$, eq. (7d.3′), the greatest error is for low molecular weights and for noble gases for which $Q_i = 1$. At room temperature, $T = 300$ K, and 1 atm pressure, the value of the volume per molecule V/N is about $4\times10^4\times(10^{-8}\text{ cm})^3$, whereas from eq. (7d.3″), $\lambda^3 \cong M^{-3/2}\times(10^{-8}\text{ cm})^3$. For helium, $M = 4$, we have

$$\left(\frac{N\lambda^3}{2V}\right)_{\text{He, 300 K, 1 atm}} \cong \frac{1}{6.4}\times10^{-5},$$

the condition for the validity of Boltzmann counting is amply satisfied. For higher-molecular-weight noble gases, and even for H_2 at this temperature and pressure, since Q_i is greater than 1, the condition is even better satisfied.

However, the boiling point of helium is about 4.5 K, and $N\lambda^3/V$ for fixed pressure and molecular weight varies as $T^{-5/2}$. We have that at the boiling point $\frac{1}{2}e^{\beta\mu}$, from eq. (5), is greater by the factor $(1.5)^{-5/2}\times10^5 \cong 0.36\times10^{+5}$. The error in using Boltzmann counting can be expected to be of the order of 6% in the pressure at constant chemical potential. Actually, it is somewhat less.

7g. THE GENERAL INTERNAL PARTITION FUNCTION

The thermodynamic functions for a perfect gas listed in eqs. (7e.1) to (7e.10) are all additively composed of two terms, one due to the translational degrees of freedom and one due to the internal quantum states. The translational contribution to the energy, heat content or enthalpy, and to the two heat capacities, is the same for all gases. Only the entropy contribution, hence also the free-energy term, depends on the species of molecule, and then only through the molecular weight M in λ [eq. (7d.3)].

The essential difference in the thermodynamic properties of gases composed of different molecules is in the internal partition function,

$$Q_i(T) = \sum_{\mathbf{n}} e^{-\beta\epsilon_{\mathbf{n}}}, \quad \mathbf{n} = \text{internal quantum states} \tag{7g.1}$$

This function can be highly complicated for a sufficiently complicated molecule but, for most molecules that actually exist experimentally as gases, the Hamiltonian of the internal degrees of freedom is, to a reasonably accurate approximation, separable. The sum in eq. (1) is then a product of simpler sums; Q_i is a product of terms due to different degrees of freedom, and $\ln Q_i$ is a sum. Each of the separable degrees of freedom contributes additively to the thermodynamic functions.

The actual number of the electronic degrees of freedom is normally very large: three coordinates of position and one spin degree for each of the electrons in the atoms. Fortunately, the energy of excitation of the first electronic level above that of the ground level is usually of the order of several electronvolts. For the energy of 1 eV the value of $\beta\epsilon$ is $1.16 \times 10^4 T^{-1}$, so that up to several thousand degrees all excited electronic levels can be ignored, since $e^{-\beta\epsilon}$ is so small. In those few cases in which electronic excitation must be considered, so few levels are involved that the summation, using the experimental values of energy of excitation, is not tedious. However, the ground level may be degenerate, that is, consist of more than one state. In a monatomic gas having total electronic angular momentum J in units of $h/2\pi$, in the lowest level there are $2J+1$ states, differing in a quantum number m_J which can take values between $-J$ and $+J$ differing by integers. These correspond to different orientations of the total electronic angular momentum vector having projection $(h/2\pi)m_J$ on some unique axis in space. In this case, and in the absence of appreciable excitation, Q_i has the constant value $2J+1$, and its temperature derivative is zero. The contribution to entropy is $k \ln(2J+1)$ per molecule, but there is no contribution to energy or heat capacity.

Even this minor complication is relatively rare. The noble gases have zero electronic spin and a nondegenerate lowest electronic level. Most, but not all, of the elements that have nonzero electronic angular momentum in their lowest level either condense at laboratory temperatures to metal or nonmetallic crystals, or they form small molecules with zero electronic angular momentum in the ground state. In the relatively few exceptions, and O_2 is one of them, the electronic angular momentum interacts with that due to the rotation of the molecule as a whole and slightly complicates the rotational spectrum.

One other source of a degenerate lowest energy level exists. The nuclei of many atoms have a nonzero nuclear spin S_n. For isotopes with even atomic weights S_n is an integer, and very often zero in the nuclear ground level, but for odd atomic weight it is a half-integer, and never zero. There are $2S_n + 1$ different orientations possible and, although these often have in principle different energies due to interaction with other angular momenta, such as the electronic or molecular, the energy differences are completely negligible compared to kT. There is then altogether a factor $2S_{n\alpha} + 1$ in Q_i from each atom α in the molecule, and an additive term $\ln(2S_{n\alpha} + 1)$ from each atom in $\ln Q_i$. We hasten to add a caveat regarding the statement of equal energies. The total quantum state must be antisymmetric in exchange of pairs of identical nuclei of odd atomic weight, and symmetric in pair exchange of identical nuclei of even atomic weight. The relative orientation of the spins of a pair of nuclei in the same molecule affects the exchange symmetry character of the nuclear spin state, and this in turn places a requirement on the symmetry of the quantum state of nuclear motion. For instance, in H_2 one relative orientation of the half-integer nuclear spin requires an even quantum number of rotation of the molecule, whereas the other requires an odd value (see Sec. 7k). Since the energy differences of rotational states of H_2 are large, the effect is marked in the thermodynamic properties of the gas at low temperatures. For all other molecules, except possibly methane, the assumption of a constant additive term $\ln(S_{n\alpha} + 1)$ in $\ln Q_i$ for each atom α is adequate, at least in the gaseous phase and at other than the lowest temperatures in other thermodynamic phases.

The additive nuclear spin entropy $k \ln(2S_{n\alpha} + 1)$ for each atom of species α is present in all thermodynamic phases, except in extreme cryogenic experiments. It does not affect the equilibrium between different thermodynamic phases, nor does it affect equilibrium in chemical reactions to any appreciable extent, since the number of atoms of each species in the reaction is conserved. Especially since until recent decades the values of most nuclear spins were imperfectly known, the nuclear spin entropy is frequently ignored. Along with the similarly unimportant entropy of isotope mixing discussed in Sec. 7o it is very often omitted from tabulations of entropy values.

In nonmonatomic molecules consisting of n atoms the temperature-dependent contribution to the internal partition function usually comes solely from the $3n-3$ internal degrees of freedom due to the coordinates necessary to specify the relative positions of the n atoms with respect to their center of mass. The Hamiltonian is usually separable, to a reasonable approximation, into rotational and vibrational coordinates and momenta. If the minimum potential energy configuration is linear, as it necessarily is in all diatomic molecules and also is in some polyatomic molecules such as CO_2, two angular coordinates are necessary to specify the direction of the axis, and there remain $3n-5$ nearly separable modes of vibration. In nonlinear molecules, for which $n \geq 3$ necessarily, three angles are necessary and the number of separable vibrational coordinates is $3n-6$.

7h. DIATOMIC MOLECULES, APPROXIMATE EQUATIONS

If we neglect nuclear spin and electronic degrees of freedom, the diatomic molecule has 6 degrees of freedom, 3 each for the two atoms composing the molecule. The coordinates of choice are those that make the potential energy a function of only one of them, and which make the Hamiltonian as nearly separable as possible. These are the three cartesian coordinates X, Y, Z of the center of mass, the distance r between the atoms, and two eulerian angles, θ and ϕ, which specify the direction of the vector from one of the two atoms to the other. The equations relating these to the cartesian coordinates x_1, y_1, z_1, x_2, y_2, z_2 of the two atoms are, with

$$M = m_1 + m_2, \tag{7h.1}$$

the total molecular mass, and

$$\mu = m_1 m_2 / M, \tag{7h.2}$$

the reduced mass,

$$\begin{aligned}
X &= M^{-1}(m_1 x_1 + m_2 x_2),\\
Y &= M^{-1}(m_1 y_1 + m_2 y_2),\\
Z &= M^{-1}(m_1 z_1 + m_2 z_2),\\
r &= [(x_2 - x_1)^2 + (y_2 - y_1)^2 + (z_2 - z_1)^2]^{1/2},\\
\theta &= \cos^{-1}(z_2 - z_1)[(x_2 - x_1)^2 + (y_2 - y_1)^2 + (z_2 - z_1)^2]^{-1/2},\\
\phi &= \sin^{-1}(y_2 - y_1)[(x_2 - x_1)^2 + (y_2 - y_1)^2]^{-1/2},
\end{aligned} \tag{7h.3}$$

$$
\begin{aligned}
x_2 &= X + \frac{m_1}{M} r \sin\theta \sin\phi, \\
x_1 &= X - \frac{m_2}{M} r \sin\theta \cos\phi, \\
y_2 &= Y + \frac{m_1}{M} r \sin\theta \sin\phi, \\
y_1 &= Y - \frac{m_2}{M} r \sin\theta \sin\phi, \\
z_2 &= Z + \frac{m_1}{M} r \cos\theta, \\
z_1 &= Z - \frac{m_2}{M} r \cos\theta,
\end{aligned}
\tag{7h.3$'$}
$$

so that the kinetic energy K is

$$
\begin{aligned}
K &= \tfrac{1}{2} m_1(\dot{x}_1^2 + \dot{y}_1^2 + \dot{z}_1^2) + \tfrac{1}{2} m_2(\dot{x}_2^2 + \dot{y}_2^2 + \dot{y}_3^2) \\
&= \tfrac{1}{2} M(\dot{X}^2 + \dot{Y}^2 + \dot{Z}^2) + \tfrac{1}{2}\mu\dot{r}^2 + \tfrac{1}{2}\mu\dot{r}^2\dot{\theta}^2 + \tfrac{1}{2}\mu r^2 \sin^2\theta\dot{\phi}^2.
\end{aligned}
\tag{7h.4}
$$

The conjugate momentum p_i to the coordinate q_i is, by definition (see Sec. 3a),

$$
p_i = \frac{\partial K}{\partial \dot{q}_i}, \tag{7h.5}
$$

so that

$$
\begin{aligned}
p_X &= M\dot{X}, & \dot{X} &= M^{-1} p_X \\
p_Y &= M\dot{Y}, & \dot{Y} &= M^{-1} p_Y \\
p_Z &= M\dot{Z}, & \dot{Z} &= M^{-1} p_Z \\
p_r &= \mu\dot{r}, & \dot{r} &= \mu^{-1} p_r \\
p_\theta &= \mu\dot{r}^2\dot{\theta}, & \dot{\theta} &= (\mu r^2)^{\ 1} p_\theta \\
p_\phi &= \mu r^2 \sin^2\theta\dot{\phi}, & \dot{\phi} &= (\mu r^2 \sin^2\theta)^{-1} p_\phi.
\end{aligned}
$$

The quantity

$$
I(r) = \mu r^2 = m_1 m_2 (m_1 + m_2)^{-1} r^2 \tag{7h.7}
$$

is known as the moment of inertia. The potential energy u depends only on the distance r between the two atoms in the absence of external fields,

$$
u = u(r), \tag{7h.8}
$$

so that, if the velocities $\dot{q}_i$ in expression (4) for K are replaced by the values in eq. (6) in terms of the momenta p_i, one finds for the classical

Hamiltonian $H(\mathbf{p}, \mathbf{q})$ that

$$H=(2M)^{-1}(p_X^2+p_Y^2+p_Z^2)+[(2\mu)^{-1}p_r^2+u(r)] \\ +[2I(r)]^{-1}(p_\theta^2+\sin^{-2}\theta p_\phi^2). \quad (7h.9)$$

This is strictly separated into three translational terms, each dependent on only one of the three cartesian momenta of the center of mass, and the internal term

$$H_{\text{int}}=[(2\mu)^{-1}p_r^2+u(r)]+[2I(r)]^{-1}[p_\theta^2+\sin^{-2}\theta\, p_\phi^2]. \quad (7h.10)$$

One now makes two independent approximations in eq. (10) for H_i. The first of these is to eliminate the dependence of the term involving the angular momenta, p_θ and p_ϕ, on r, so that H_i is separated into a term dependent on r, p_r and one dependent on the angle θ and the momenta p_θ, p_ϕ alone. This approximation assumes that the potential $u(r)$ has such a sharp minimum at one equilibrium distance of separation r_0 that no appreciable excursions in r from this value occur, so that we can replace $I(r)$ by a constant,

$$I(r)=I(r_0)=I_0=\mu r_0^2. \quad (7h.11)$$

The second approximation is to assume that the potential energy function $u(r)$ has a very simple form, namely, that with

$$\Delta r=r-r_0, \quad (7h.12)$$

we can approximate $u(r)$ by a quadratic expression,

$$u(r)=u_0+\tfrac{1}{2}\kappa\,\Delta r^2. \quad (7h.12')$$

With these,

$$H_{\text{int}}=H_{\text{vib}}+H_{\text{rot}}, \quad (7h.13)$$

$$H_{\text{vib}}=u_0+(2\mu)^{-1}p_r^2+\tfrac{1}{2}\kappa\,\Delta r^2, \quad (7h.14)$$

$$H_{\text{rot}}=(2I_0)^{-1}\left[p_\theta^2+\left(\frac{p_\phi}{\sin\theta}\right)^2\right]. \quad (7h.15)$$

The numerical value of u_0 is arbitrary; its value gives the arbitrary energy from which the chemical potential μ is to be measured. The Hamiltonian H_{vib} of eq. (14) is that of classical harmonic oscillation along the displacement Δr with frequency ν and arbitrary amplitude a,

$$\nu=\frac{1}{2\pi}\left(\frac{\kappa}{\mu}\right)^{1/2}, \quad (7h.16)$$

$$\Delta r=a\sin(2\pi\nu t+\delta). \quad (7h.16')$$

With the internal Hamiltonian separable the internal partition function

Q_i becomes a product of two partition functions,

$$Q_i = Q_{\text{vib}} Q_{\text{rot}}, \tag{7h.17}$$

each of which is given semiclassically by h raised to the negative power of the number of coordinates times the integral over coordinates and momenta of the exponent of $-\beta$ times the corresponding Hamiltonian.

$$Q_{\text{vib}} = h^{-1} \iint_{-\infty}^{+\infty} d\,\Delta r\, dp_r \exp\{-\beta[u_0 + (2\mu)^{-1} p_r^2 + \tfrac{1}{2}\kappa\,\Delta r^2]\}$$

$$= \frac{1}{h\beta} \frac{2\pi\sqrt{\mu}}{\kappa} e^{-\beta u_0}$$

$$= \frac{kT}{h\nu} e^{-\beta u_0}, \tag{7h.18}$$

$$Q_{\text{rot}} = \sigma^{-1} h^{-2} \int_0^{2\pi} d\phi \int_0^{\pi} d\theta \iint_{\infty}^{\infty} dp_\theta\, dp_\phi \exp\left\{-\frac{\beta}{2I_0}\left[p_\theta^2 + \left(\frac{p_\phi}{\sin\theta}\right)^2\right]\right\}$$

$$= \sigma^{-1} \frac{8\pi^2 I_0 kT}{h^2}. \tag{7h.19}$$

The symmetry number σ in eq. (19) is unity unless the two atoms of the molecule are identical isotopes of the same element, in which latter case $\sigma = 2$. It arises from the requirement for division of the classical integral over coordinates and momenta by $n!$, when there are n identical atoms. The symmetry number is discussed fully in Sec. 7m in connection with polyatomic molecules, where it is seen to be less trivial than division by $\prod_\alpha n_\alpha!$ for all identical atoms of type α.

The term $-\beta u_0$ in $\ln Q_{\text{vib}}$ contributes an additive constant u_0 to the energy per molecule, $\epsilon = kT\,d \ln Q_i/d \ln T$ [eq. (7e.8)]. Apart from this term both Q_{vib} and Q_{rot} are linear in temperature, $d \ln Q/d \ln T$ is unity, and both add kT to ϵ. The contribution to c, the heat capacity per molecule, for each is k. That due to rotation is exactly the $\frac{1}{2}k$ from the kinetic energy contribution for each of the 2 degrees of freedom, p_θ and p_ϕ. This is general for all classical degrees of freedom, as discussed in Sec. 5d. The contribution of k to c due to vibration is made up of two terms of $\frac{1}{2}k$ each; one of these is always required from the kinetic energy, and the other equal term from the special approximation of eq. (12′) that the potential energy $u(r)$ is quadratic in the displacement Δr from the minimum.

The classical expression [eq. (19)] for Q_{rot} is numerically fairly adequate for all gases at temperatures for which the vapor pressure is of order of an atmosphere, except for H_2 and D_2, but the classical Q_{vib} of eq. (18) is in complete disagreement with experiment. The contribution of vibration to the heat capacity is very nearly zero for the lighter truly stable diatomic molecules such as H_2, O_2, N_2, CO, even well above room temperature. The quantum-mechanical summation over allowed stationary energy levels must be used for the vibrational degree of freedom and, for H_2 and D_2, also for rotation at room temperature and below.

The harmonic oscillator characterized by the Hamiltonian H_{vib} of eq. (14) has allowed quantum states characterized by a single integer quantum number n of energy,

$$\epsilon_n = u_0 + (n + \tfrac{1}{2})h\nu, \tag{7h.20}$$

with ν the classical frequency of oscillation [eq. (16)]. The partition function is then

$$Q_{\text{vib}} = e^{-\beta[u_0+1/2h\nu]} \sum_{n\geq 0} e^{-\beta nh\nu}. \tag{7h.21}$$

The sum $1+x+x^2+x^3+\cdots$, with $x = e^{-\beta h\nu} < 1$, is $(1-x)^{-1}$, so that

$$Q_{\text{vib}} = e^{-\beta[u_0+1/2h\nu]}(1-e^{-\beta h\nu})^{-1}. \tag{7h.22}$$

We have, for the contribution to the energy per molecule,

$$\begin{aligned}\epsilon_{\text{vib}} &= kT\frac{d \ln Q_{\text{vib}}}{d \ln T} = \frac{-d \ln Q_{\text{vib}}}{d\beta} \\ &= u_0 + \tfrac{1}{2}h\nu + h\nu(e^{\beta h\nu} - 1)^{-1}.\end{aligned} \tag{7h.23}$$

With the dimensionless quantity

$$\theta = \theta(T) = \beta h\nu = \frac{h\nu}{kT} \tag{7h.24}$$

this can be written as

$$\epsilon_{\text{vib}} = u_0 + \tfrac{1}{2}h\nu + kT\frac{\theta}{e^{\theta}-1}. \tag{7h.25}$$

For H_2, $\theta = 5958\ T^{-1}$, so that even at $T = 1000$ K the last term is almost zero, since for $\theta = 6$ the function $\theta/(e^{\theta}-1)$ is 0.0149. The energy above $u_0 + \frac{1}{2}h\nu$, which is that of the ground level, is only 1.5% of the classical value kT. Typical values of θT for low-molecular-weight stable molecules are 3336 for N_2, 2228 for O_2, and 4131 for HCl, but for higher-molecular-weight molecules, or those of lower binding energy, the values are much lower, for instance, 305 for I_2, 460 for Br_2, and 132 for K_2.

For small values of θ we can develop

$$e^{\theta}-1=\theta+\tfrac{1}{2}\theta^2+\tfrac{1}{6}\theta^3+\tfrac{1}{24}\theta^4+\cdots, \tag{7h.26}$$

and

$$\theta(e^{\theta}-1)^{-1}=1-\tfrac{1}{2}\theta+\tfrac{1}{12}\theta^2-\tfrac{1}{720}\theta^4+\cdots. \tag{7h.27}$$

Thus at high temperatures eq. (23) becomes

$$\epsilon_{\text{vib}}=u_0+kT\left[1+\frac{1}{12}\left(\frac{h\nu}{kT}\right)^2+\frac{1}{720}\left(\frac{h\nu}{kT}\right)^4+\cdots\right]. \tag{7h.28}$$

The arbitrary energy u_0 is the minimum of the potential $u(r)$ [eq. (12′)], and the classical equations give this as the energy at $T=0$, with a linear increase kT as T increases. The quantum system starts with an energy $\frac{1}{2}h\nu$ above u_0 at $T=0$ but increases in energy more slowly than the classical equations, approaching the classical energy from above.

The Hamiltonian of eq. (15) is that of a rigid two-dimensional rotator of moment of inertia I_0. The stationary quantum-mechanical states of such an object are characterized by two integer quantum numbers, j and m. The former of these determines the total angular momentum square p_ω^2,

$$p_\omega^2=\left(p_\theta^2+\frac{p_\phi^2}{\sin^2\theta}\right)=j(j+1)\left(\frac{h}{2\pi}\right)^2, \tag{7h.29}$$

and the second gives the projection on the z-axis of the angular momentum vector $\mathbf{p}_\omega$ as $m(h/2\pi)$. The values that m can take are between $+j$ and $-j$,

$$-j\leq m\leq j, \qquad 2j+1 \text{ values}. \tag{7h.30}$$

The energy depends only on the square of the angular momentum and not on the orientation of the vector; it does not depend on m. The energy is

$$\begin{aligned}\epsilon(j,m)=\epsilon_j&=(2I_0)^{-1}p_\omega^2\\&=j(j+1)\frac{h^2}{8\pi^2I_0}.\end{aligned} \tag{7h.31}$$

The partition function Q_{rot} is then

$$\begin{aligned}Q_{\text{rot}}&=\sum_{j\geq 0}^{\infty}\sum_{m\geq -j}^{m\leq +j}e^{-(\beta h^2/8\pi^2I_0)j(j+1)}\\&=\sum_{j\geq 0}(2j+1)e^{-\zeta j(j+1)},\end{aligned} \tag{7h.32}$$

where

$$\zeta = \zeta(T) = \frac{\beta h^2}{8\pi^2 I_0} = \frac{h^2}{8\pi^2 I_0 kT}. \tag{7h.33}$$

If ζ is sufficiently small, which it will be if $I_0 = m_1 m_2 (m_1 + m_2)^{-1} r_0^2$ is large and the temperature not too low, the sum of eq. (32) can be approximated by treating j as a continuous variable and integrating the function under the summation sign over dj from zero to infinity. However, an extra factor of σ^{-1} is needed where σ, the symmetry number, is normally 1 but is 2 if the two atoms of the molecule are identical.

The origin of the factor σ^{-1} lies in the requirement that the state function have a definite prescribed symmetry in the exchange of identical particles. If the two nuclei 1 and 2 exchange positions, the vector going from atom 1 to atom 2 reverses its direction. The angle θ goes to $\pi - \theta$, and ϕ goes to $\pi + \phi$. If one examines the state function $\psi_{j,m}(\theta, \phi)$, one finds that for even j values $\psi_{j,m}(\theta, \phi) = \psi_{j,m}(\pi - \theta, \pi + \phi)$, whereas for odd j values the change in angle values makes $\psi_{j,m}$ go to its negative. For even j's the rotational function of the nuclei is symmetric in exchange, whereas for odd j values it is antisymmetric. Now consider the example of two identical nuclei of even atomic weight and zero nuclear spin, which is a very common case. The total wave function of the molecule is the product of those for the center-of-mass coordinates, that for the vibrational function of the distance r between the nuclei, the rotational function, and the electronic function. The center-of-mass coordinates and the distance r between the nuclei are unaltered by the exchange of the identical nuclei, so that the functions do not change and are therefore symmetric in this exchange. The electronic function in this case is also symmetric in nuclear exchange. The nuclei have an even number of fermions, protons plus neutrons, and therefore only a symmetric total function is allowed. Only rotation functions of even j are allowed; odd j values are forbidden. The rotational spectrum shows this experimentally.[(1)]

Thus for a diatomic molecule of identical atoms only every second value of j is allowed for a given nuclear spin configuration, whether it is symmetric or antisymmetric, either only even values of j or only odd values. If the states of successive j values lie sufficiently close in energy, the sum over only even j's, or only odd j's, is just half of the sum over all

[(1)] It is of historical interest that the nucleus of ^{14}N was assumed, before the discovery of the neutron, to consist of 14 protons and 7 electrons, an odd number of fermions. Spectroscopists found it existed in states of even j only, namely, in states symmetric in nuclear exchange. This was one of the strongest reasons for suspecting the existence of neutrons before their discovery in 1932 by Chadwick.

of them. We then have

$$Q_{\text{rot}} \cong \sigma^{-1} \int_0^\infty (2j+1)\, dj\, e^{-\zeta j(j+1)}$$

$$= (\sigma\zeta)^{-1} \int_0^\infty d[\zeta j(j+1)]\, e^{-\zeta j(j+1)}$$

$$= (\sigma\zeta)^{-1} = \sigma^{-1} \frac{8\pi^2 I_0 kT}{h^2}, \tag{7h.34}$$

which is the semiclassical expression (19) showing again the correspondence asserted in Sec. 3f between number of quantum states and phase space.

The Euler-Maclaurin summation formula (Appendix II) can be used to obtain a somewhat better approximation when ζ is not sufficiently small. The result is

$$Q_{\text{rot}} = (\sigma\zeta)^{-1}(1 + \tfrac{1}{3}\zeta + \tfrac{1}{15}\zeta^2 + \tfrac{4}{315}\zeta^3 + \cdots). \tag{7h.35}$$

For the rotational energy per molecule this gives

$$\epsilon_{\text{rot}} = kT(1 - \tfrac{1}{3}\zeta - \tfrac{1}{45}\zeta^2 - \tfrac{8}{745}\zeta^3 - \quad , \tag{7h.36}$$

analogous to the high-temperature expression (27) for the energy of an oscillator.

Numerical values of ζT are 84.971 K for H_2, and quite closely half of this for D_2. All other molecules have considerably lower values: 2.847 K for N_2, 2.059 for O_2, 0.3477 for Cl_2, and as low as 0.0534 for I_2. However, the hydrides of even heavy elements have low reduced masses; ζT for HCl is 14.746 and for HI it is 9.191. Thus the purely classical approximation of eq. (34) for H_2 at $T = 300$ K is in error by nearly 10%, but for N_2 and O_2 at the same temperature it is more like 0.3 and 0.2%, respectively. The classical result for the heat capacity per molecule due to rotation is much better. The approximation equivalent to eq. (36) is, by differentiation, found to be

$$c = k(1 + \tfrac{1}{45}\zeta^2 + \tfrac{16}{945}\sigma^3 + \cdots), \tag{7h.37}$$

and the first correction term of $\frac{1}{45}\zeta^2$ is less than 0.2% at $T = 300$ K even for H_2.

There is close analogy between the partition function Q_{rot} of eq. (34) and that for translation of a single molecule $V\lambda^{-3}$. The quantity $(h^2/2\pi\mu kT)^{1/2}$ of dimension length is strictly analogous to the de Broglie wavelength $\lambda = (h^2/2\pi mkT)^{1/2}$ (Sec. 7d), with μ the reduced mass replacing m the total mass. The three-dimensional volume V of translation is replaced by the two-dimensional area $4\pi r_0^2$ swept out by any one of the atoms moving with respect to the other as center. The factor $\sigma^{-1} = (n!)$, with $n = 2$ for identical atoms corresponds to the division of the total translational contribution of all molecules by N!

7i DIATOMIC MOLECULES, THERMODYNAMIC FUNCTIONS

With rather high numerical precision the equations of the preceding section give simple analytical equations for the thermodynamic functions of most of the more stable diatomic molecules in the temperature range above their normal boiling points up to 1000 or 2000 K. The chief exceptions are H_2 and D_2 at and below room temperature, for which the rotational contribution must be treated by summation of the quantum-mechanical partition function.

Although in principle we can imagine a completely a priori calculation using the equations of quantum mechanics to evaluate the potential energy function $u(v)$ and from it the distance r_0 of the potential minimum and the force constant κ[eq. (7h.12′)], such calculations are not only extremely difficult but also, at the present time at least, quite imprecise. The necessary molecular constants are almost always available to very high precision from spectroscopic data and are tabulated as the frequency ν or wave number, $\omega = c^{-1}\nu$, of vibration [eqs. (7h.16) and (7h.20)] and a constant B,

$$B = \frac{h}{8\pi^2 I_0 c}, \tag{7i.1}$$

of dimensions l^{-1} related to the moment of inertia I_0 at $r = r_0$ [eq. (7h.7)]. In terms of these the internal energies of the quantum levels of the molecule are

$$\begin{aligned} \epsilon_{\text{int}} &= u_0 + \tfrac{1}{2}h\nu + nh\nu + Bhcj(j+1) \\ &= u_0 + hc[(\tfrac{1}{2} + n)\omega + Bj(j+1)] \end{aligned} \tag{7i.2}$$

for the integer quantum numbers n of vibration and j of rotation.

The thermodynamic functions G, E, H, A, having the dimensions of energy, always contain the temperature-independent additive term $N(u_0 + \frac{1}{2}h\nu)$. The lowest energy state, $n = 0$, $j = 0$, of the molecule has energy $\epsilon_0 = u_0 + \frac{1}{2}h\nu$. It is convenient to choose the arbitrary energy constant u_0 to be

$$u_0 = -\tfrac{1}{2}h\nu, \tag{7i.3}$$

so that

$$(n = 0,\ j = 0) = \epsilon_0 = 0. \tag{7i.3′}$$

The thermodynamic energy functions are then zero for the hypothetical perfect gas state at the absolute zero of temperature.

This convention is not universally used. For molecules containing different isotopes of the same element the potential energy difference between the isolated atoms and the stable molecules is quite precisely the same at the minimum. Since the frequency ν [eq. (7h.16)] has the square root of the reduced mass in the denominator, the frequencies are different for different isotopic compositions. The choice of eq. (3) for u_0 then means that in chemical reactions there is a difference in the energy of the reaction at absolute zero for different isotopic compositions, and this of course enters into the expressions for the chemical equilibria. Especially since at sufficiently high temperatures, $h\nu/kT \ll 1$, the energy of vibration above u_0 approaches the classical value NkT [eq. (7h.28)], it is usually more convenient in computing isotopic separation factors at equilibrium to use a convention assigning the same u_0 to all isotopic compositions and to leave the additive $\frac{1}{2}h\nu$ per molecule in the thermodynamic expressions; see Sec. 7*m* on isotopic equilibria.

By using the convention of eq. (3) that $u_0 = -\frac{1}{2}h\nu$, with

$$\lambda = \left(\frac{h^2}{2\pi mkT}\right)^{1/2} \tag{7i.4}$$

for the de Broglie wavelength, and with σ the symmetry number unity unless the two atoms are identical, in which case $\sigma = 2$, the thermodynamic functions for 1 mol of a diatomic gas are

$$G_0 = -RT\left\{\ln\frac{kT}{P\lambda^3} + \ln\frac{kT\sigma}{Bhc} + \ln\left[1 - \exp\left(\frac{-hc\omega}{kT}\right)\right]^{-1}\right\}, \tag{7i.5}$$

$$A_0 = -RT\left\{\ln\frac{kT}{P\lambda^3}e + \ln\frac{kT\sigma}{Bhc} + \ln\left[1 - \exp\left(\frac{-hc\omega}{kT}\right)\right]^{-1}\right\}, \tag{7i.6}$$

$$H_0 = RT\left[\frac{7}{2} + \frac{hc\omega}{kT}\left(\exp\frac{hc\omega}{kT} - 1\right)^{-1}\right], \tag{7i.7}$$

$$E_0 = RT\left[\frac{5}{2} + \frac{hc\omega}{kT}\left(\exp\frac{hc\omega}{kT} - 1\right)^{-1}\right], \tag{7i.8}$$

$$S_0 = R\left\{\ln\frac{kT}{P\lambda^3}e^{7/2} + \frac{hc\omega}{kT}\left[\exp\frac{hc\omega}{kT} - 1\right]^{-1} + \ln\left[1 - \exp\left(\frac{-hc\omega}{kT}\right)\right]^{-1}\right\}, \tag{7i.9}$$

$$C_P = R\left[\frac{7}{2}+\left(\frac{hc\omega}{kT}\right)^2\left(\exp\frac{hc\omega}{kT}\right)\left(\exp\frac{hc\omega}{kT}-1\right)^{-2}\right], \tag{7i.10}$$

$$C_V = R\left[\frac{5}{2}+\left(\frac{hc\omega}{kT}\right)^2\left(\exp\frac{hc\omega}{kT}\right)\left(\exp\frac{hc\omega}{kT}-1\right)^{-2}\right]. \tag{7i.11}$$

Convenient aids to numerical evaluation are given in Appendix IX.

7j. DIATOMIC MOLECULES, MORE EXACT EQUATIONS

Actually, the potential $u(r)$ is of course never a true parabola, $u(r) = u_0+\frac{1}{2}\kappa\,\Delta r^2$ [eq. (7h.12′)], but rises more steeply for negative Δr and for infinite Δr takes the constant value of the two separated atoms. In addition, the moment of inertia $I(r)$ is not the constant $I_0=\mu r_0^2$. Its average value increases for high vibrational levels because of the anharmonicity of the potential function and for high rotational levels because of centrifugal force stretching. An empirical expression for the energy levels,

$$\epsilon(n, j) = u_0 + hc[(n+\tfrac{1}{2})\omega_e + j(j+1)B_e - (n+\tfrac{1}{2})^2 x_e\omega_e - nj(j+1)\alpha - j^2(j+1)^2 D_e], \tag{7j.1}$$

is commonly used, and the values of x_e, α, and D_e listed along with ω_e and B_e from the spectroscopic data. The centrifugal-force constant D_e is related to the force constant κ of $u(r)$, hence to the frequency. The relation is $D_e=4B_e/\omega_e^2$. By a comparatively long but moderately straightforward perturbation procedure, given on pages 160–166 of the first edition of this book, one can evaluate an approximate additive correction $\ln Q_c$ to $\ln Q_{\mathrm{int}}$. The correction is appreciable only at relatively high temperatures, namely, when $\theta=\beta h\nu$ is fairly small. It is

$$\ln Q_c = \theta^{-1}\left[8\frac{B_e}{\omega_e}+\frac{\alpha}{B_e}\frac{\theta}{e^\theta-1}+2x_e\frac{\theta^2}{(e^\theta-1)^2}\right]. \tag{7j.2}$$

The dimensionless temperature-independent quantities B_e/ω_e, α/B_e, and x_e are of the order of 10^{-2} for all molecules, having the largest values 0.0143, 0.0517, and 0.0273, respectively, for H_2 but considerably smaller ones (down to one-tenth of these values) for most of the other diatomic gases. Since even at 1000 K the value of θ for H_2 is nearly 6, and nearly 20 at room temperature, it is seen that the additive $\ln Q_c$ is really rather small except at very elevated temperatures, or for molecules that have very low vibrational frequencies.

The evaluation of $\ln Q_c$ given in eq. (2) was made before the advent of high-speed computers. It is convenient to use for estimation of the temperature up to which the approximate expressions of Sec. 7i are adequate. Probably, with present-day computers most workers prefer, when the approximate equations are judged inadequate, to carry out numerical summation of the partition function using eq. (1) directly for the energy levels, or even experimental values, rather than the semiempirical equation (1). The same comments apply to evaluation of the rotational contribution, Q_{rot}, as a factor to the partition function. The series developments of eqs. (7h.35) to (7h.37) serve to indicate the range of validity of the simple classical expression $Q_{\mathrm{rot}}=(\sigma\zeta)^{-1}=\sigma^{-1}(8\pi^2 I_0 kT/h^2)$ [eqs. (7h.19) and (7h.34)]. They are also certainly not valueless in giving better values than the classical when ζ is fairly large. However, direct summation of the quantum-mechanical partition function (7h.32), even for comparitively low values of ζ, with a modern computer is trivial.

7k. ORTHO AND PARA HYDROGEN

In Sec. 7g we briefly discussed the nuclear spin entropy, stating that the difference in energy of the different spin orientations was negligible, but called attention to an effect on the allowed rotational states. Since this effect is important in hydrogen, we discuss it here in some detail. It exists, but is far less numerically important, in other molecules.

The proton spin is $\frac{1}{2}$, so that there are two states, say, state α with the nuclear magnetic moment in the direction of any ambient magnetic field, and state β oppositely directed. A molecule with two protons numbered 1 and 2 has four states, $\alpha(1)\alpha(2)$, $\alpha(1)\beta(2)$, $\beta(1)\alpha(2)$, and $\beta(1)\beta(2)$. The first and fourth of these are symmetric in exchange of the two nuclei. The second and third are degenerate in energy both with zero net magnetic moment, but have no symmetry. Two linear combinations, $2^{-1/2}[\alpha(1)\beta(2)+\beta(1)\alpha(2)]$ and $2^{-1/2}[\alpha(1)\beta(2)-\beta(1)\alpha(2)]$, however, are symmetric and antisymmetric, respectively. The molecule, then, has one antisymmetric nuclear state of zero spin and three symmetric states of total spin unity with projections of $+1$, 0, and -1 on any magnetic field.

The total state of the molecule must be antisymmetric in proton exchange and can be written as the product of the translational, electronic, vibrational, rotational, and nuclear spin states. As discussed in Sec. 7h, the first three of these are symmetric, since their coordinates are unchanged by permutation of the nuclei, and the rotational state is antisymmetric for odd j values and symmetric for even j values. The total product then changes sign on exchange of protons only if the spin state is antisymmetric and j is even, or if the spin state is symmetric and j is odd.

The molecules in the three symmetric nuclear spin states of parallel spin are called ortho hydrogen, and those in the one antisymmetric state of zero spin are called para hydrogen. The conversion of one to the other is slow, especially at low temperatures, not so much because the nuclei rigidly resist reorientation but because only an inhomogeneous magnetic field exerts any force tending to reorient the two nuclei with respect to each other. Collisions with strongly paramagnetic molecules such as O_2, or with a wall containing magnetic atoms catalyze the conversion.

At high temperatures the equilibrium is an ortho/para ratio 3 : 1 due to the ratio of the number of nuclear spin states. If the gas is cooled to its 1-atm pressure boiling point of 20.38 K without exposing it to a catalyst, the ratio remains 3 : 1, and the ortho hydrogen is frozen in the first excited rotational level of $j = 1$ and energy $\epsilon_1 = Bhc\, j(j+1)$, having energy of kT at 169.9 K. At 20.38 K the value of $\beta\epsilon_1$ is 8.35, and at equilibrium at this temperature the ratio of ortho to para hydrogen is less than 10^{-3}. The rapidly cooled gas is then regarded as a mixture of two distinct chemical species, with the ortho hydrogen slowly converting to the stable para hydrogen with the evolution of a considerable amount of heat.

7*l*. POLYATOMIC MOLECULES

The same approximate separability of the internal Hamiltonian exists in polyatomic molecules as that used in Sec. 7h for diatomic molecules. For some moderately complicated rigid molecules the approximation is numerically excellent; for others which have a less rigidly fixed minimum potential configuration it is far less satisfactory.

A molecule of n atoms has $3n$ degrees of freedom, of which 3 are translational, leaving $3n - 3$ internal degrees of freedom. Of the $3n - 3$ internal coordinates either two or three angles can be chosen which determine the orientation of the rigid frame of the atoms with their mutual distances fixed in the position of minimum potential energy. The potential energy does not depend on these angles. Two angles will be sufficient if this minimum corresponds to a linear molecule with all the atoms in a straight line. Three angles will be needed if the potential minimum corresponds to a nonlinear figure.

If the molecule is linear, one can assume as a first approximation that the single moment of inertia is constant, and the variables in the Hamiltonian will be separable. One obtains an additive contribution from the rotational quantum number j to the energy of each quantum state, and corresponding additive contributions to the thermodynamic functions. These are the same as the rotational contributions to a gas of diatomic molecules, with the same value of the moment of inertia I_0. The moment of inertia of the polyatomic linear molecule is given by

$$I_0 = \sum_{i=1}^{i=n} m_i x_i^2, \tag{7l.1}$$

where x_i is the distance, in the equilibrium configuration, of the ith atom of mass m_i from the center of mass, so that $\sum_i m_i x_i = 0$.

If the molecule is nonlinear, one can assume, as a first approximation at least, that the moments of inertia will be fixed as those corresponding to the position of the minimum of potential energy. The Hamiltonian will again be separable, and the three angles will make an additive contribution to the energies of the quantum levels of the molecule and to thermodynamic properties of the gas. These calculations are carried out later in this section.

The potential energy of the molecule depends only on the remaining coordinates of the molecule, which are called the vibrational coordinates. There remain $\gamma_{\text{vib}} = 3n-6$ or $3n-5$ degrees of freedom in nonlinear and linear molecules, respectively. That is, there are γ_{vib} coordinates q_i, and the potential energy can be expected to depend, at least to some extent, on all of them. These coordinates can be so chosen that zero value for all of them corresponds to the position of minimum potential, which is assigned the energy u_0. These coordinates then represent displacements of the figure from the position of equilibrium.

If the total potential energy u of the molecule is developed as a power series in the coordinates q_i, the condition that this energy be a minimum, that is, $\partial u/\partial q_i = 0$ for all q_i's when all q_i's are zero, requires that the power series begin with the constant u_0 followed by the quadratic terms; all linear terms are zero. The cubic and higher-order terms are smaller than the quadratic terms at sufficiently small displacements from the equilibrium position, at small values of the q_i's. As a first approximation they may be neglected, so that the potential can be written as u_0 plus a sum of terms, each of which is quadratic in q_i's, but among which, in general, there occur cross-product terms of the type $q_i q_j$, $i \neq j$.

It is a mathematical theorem that, whatever the values of the force constants (the coefficients of the various terms) and whatever the values of the reduced masses (one-half the inverse of the coefficients of the terms p_i^2 in the kinetic energy), one can always transform to new normal coordinates q_λ such that, with their conjugated momenta p_λ, the Hamiltonian has the form

$$H_{\text{vib}}(q_\lambda, p_\lambda) = u + \sum_{\lambda=1}^{\lambda=\gamma_{\text{vib}}} \tfrac{1}{2}(\kappa_\lambda q_\lambda^2 + \mu_\lambda^{-1} p_\lambda^2). \tag{7l.2}$$

In this equation the essential simplification that has been reached is the elimination of second-order cross-product terms of the type $q_i q_j$ in the potential energy (without the introduction of cross products $p_i p_j$ in the kinetic energy). The κ's are the generalized force constants, and the μ_λ's are the generalized reduced masses. (By a linear change in scale the coordinates are often so determined that the reduced masses are all unity.)

The third-order terms in the potential, if included in the above equation, usually contain cross products. Their neglect is justified only if they are negligible compared with the quadratic terms for such displacements q_λ that the quadratic terms $(\kappa_\lambda/2)q_\lambda^2$ are of the order of magnitude of kT.

If the third-order terms are neglected, the Hamiltonian is separable, that is, it consists of a sum of terms each depending on one only of the γ_{vib} coordinates with its conjugated momentum. The wave function is a product, and the energies of the quantum states are sums of terms, each depending on the one quantum number n_λ associated with the normal coordinate q_λ. The additive part of the Hamiltonian for each coordinate with its conjugate momentum is exactly the Hamiltonian of the harmonic oscillator with the force constant κ_λ and the mass μ_λ. The solution in classical or quantum mechanics consists of independent harmonic vibrations of the system along all the normal coordinates, each with its own frequency $\nu_\lambda = (1/2\pi)(\kappa_\lambda/\mu_\lambda)^{1/2}$. The contribution of each coordinate to the energy of a quantum state of the molecule is, as for the vibration of a diatomic molecule, $\epsilon_\lambda = (n_\lambda + \frac{1}{2})h\nu_\lambda$.

The partition function Q is then a product, and ln Q a sum, of γ_{vib} terms, each of the same type as the Q_{vib} and ln Q_{vib} calculated for the diatomic molecule in Sec. 7h, in which the frequency ν_λ must be used in the term Q_λ or ln Q_λ.

The analysis of the motion of the molecule as harmonic oscillation along the normal coordinates is essentially a formal one. It has no more, but also no less, physical significance than the arbitrary analysis of a wave as a Fourier sum or integral of sine and cosine waves. In the case of white light the spectrograph makes a physical analysis of the wave, which is exactly that of the formal mathematical analysis into a Fourier integral. So also, certain experiments analyze the motion of a large-scale model of a molecule, which obeys the classical laws, into harmonic vibration along the normal coordinates.

If such a model is distorted from its equilibrium shape and then released, the rather complicated motion that ensues may be analyzed in the formal manner described as harmonic vibration along the normal coordinates. If the original displacement were such that only one, $q_{\lambda'}$, of the normal coordinates q_λ was different from zero, *and if the cubic*

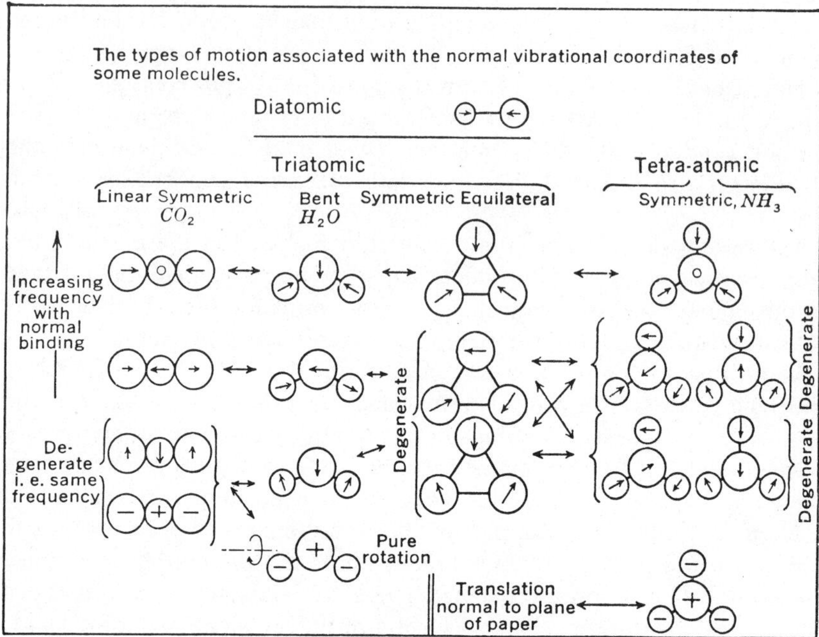

Figure 7i.1 The frequencies and their order depend on the masses and the force constants. That of the upper two, for instance, is reversed in CO_2.

interaction terms were really zero, the subsequent motion would actually be true harmonic vibration along this coordinate $q_{\lambda'}$ with the frequency $\nu_{\lambda'}$, all the other coordinates remaining zero during the motion.

If such a model is shaken with a variable frequency, say, by an electric motor, the amplitude of the motion of the molecule will be small at most frequencies. If, however, the frequency of the shaking becomes that of one of the normal coordinates, the amplitude will increase markedly and the molecule will vibrate along that coordinate.[(1)]

In Fig. 7*l*.1 the directions of the motion given by displacements along the normal coordinates are shown for some of the simpler types of molecules. The exact angles of the displacements depend on the numerical values of the forces and of the masses, but certain of the characteristics of the motion can be deduced from considerations of the symmetry of the molecule alone. These considerations are of prime importance in the case of such a relatively complicated but extremely symmetric molecule as

[(1)] C. F. Kettering, L. W. Shutts, and D. H. Andrews, *Phys. Rev.*, **36,** 531 (1930); D. E. Teets and D. H. Andrews, *J. Chem. Phys.*, **3,** 175 (1935).

benzene, C_6H_6.[1] The most useful tool for these considerations is the mathematical theory of groups.

In general, if the molecule has some degree of symmetry, not all the γ_{vib} frequencies ν_λ will have different numerical values, but there may be several normal coordinates for which the frequencies will be necessarily identical. In this case it is also true that the choice of the normal coordinates will not be unique, since any linear combination of two coordinates with the same frequency will also be a normal coordinate of the system. However, the number of normal coordinates, and therefore the number of terms $\ln Q_\lambda$ that enter into the thermodynamic functions, is uniquely fixed.

In such a case the coordinates of identical frequencies are said to be degenerate, and two or more $\ln Q_\lambda$ terms enter into the thermodynamic expressions with the same frequencies.

The numerical values of the frequencies are always obtained from an analysis of the spectrum of the molecule. They may be observed in the infrared, the Raman, or the visible or ultraviolet spectrum of the gas.

If the molecule is completely unsymmetric all γ_{vib} frequencies will be essentially different, although of course one cannot completely rule out the possibility that by pure chance two of them may be very close to each other in value. In this case of a completely unsymmetric molecule all γ_{vib} frequencies are, in principle at least, observable in any one of the above spectra.

If all γ_{vib} different frequencies are actually observed, no mechanical analysis of the motion of the molecule will be necessary. The contribution of the γ_{vib} degrees of freedom to the thermodynamic properties of the gas is given by γ_{vib} terms of the same type as the contribution due to vibration in a diatomic molecule, each depending on one frequency alone.

In a symmetric molecule, however, certain difficulties are encountered. The frequencies due to certain of the normal coordinates are absent in one, or even conceivably in all three, of the above types of spectra. For instance, the first type of motion for the symmetric linear triatomic molecule CO_2 shown in Fig. 7l.1 is one in which the oxygens always move in opposite directions and the carbon remains fixed. For this motion there is no dipole displacement of the electric charge, and the frequency associated with this normal coordinate is absent, or at least very weak, in the infrared spectrum of the gas.

Even if all the different frequencies are actually observed, their total number is often less than γ_{vib}, owing to the essential degeneracies present in a molecule of the given symmetry. One must then ascertain which of

[1] E. B. Wilson, Jr., *Phys. Rev.*, **45**, 706 (1934).

the observed frequencies are to be used twice or more often in the thermodynamic terms. In order to do this an analysis of the mechanical motions of the molecule is necessary. For some of the simpler molecules this may be done almost intuitively; for others it is more complicated.

For CO_2 a few qualitative considerations are sufficient to enable one to place the observed frequencies uniquely. The four modes of motion of the four normal coordinates can be seen intuitively to be those sketched in Fig. 7*l*.1. The first motion shown does not occur in the infrared but is present in the Raman spectrum of the molecule. If the polarizability of the molecule at positive and negative values of q_λ is the same, the frequency is Raman-forbidden, if the polarizability is different, the spectral line is Raman-active. This means essentially that, if the change $q_\lambda \rightarrow -q_\lambda$ can be accomplished by rotation alone, it is Raman-forbidden. This type of motion, in which the oxygen atoms move oppositely, has a frequency between the other two (about 7.5 μm). The second type of motion, in which the oxygens move together, should be observed in the infrared, since in this type of motion the center of gravity of the negative charge, associated with the oxygen atoms, moves with respect to the center of gravity of the positive charge. This frequency is missing in the Raman spectrum. This type of motion has the highest frequency (about 4.7 μm). The third type of motion is degenerate, and it is this frequency that must be used twice in the thermodynamic terms. This type of motion is a bending of the molecule and is infrared-active but Raman-inactive. However, the forces resisting bending in a molecule are much weaker than those resisting changes in distance between the atoms, so that this frequency is expected to be decidedly lower in value than either of the others (about 15 μm).

One expects, then, for CO_2, to find two strong infrared frequencies reported, one of which should be considerably lower than the other. In the Raman spectrum one expects only one strong line, the frequency of which should lie between those of the two infrared-active vibrations. The lower of the two infrared-active frequencies is the degenerate one.

The actual situation in CO_2 is complicated by the occurrence of what is called an accidental degeneracy. The frequency of the bending motion is almost exactly half of that of the frequency of the Raman-active vibration, so that the two quantum levels, one in which there are two quanta in the bending degrees of freedom and one in which there is one quantum in the stretching degree of freedom, have the same energy. These two levels combine, that is, they form two new levels, one of lower and one of higher energy, each of which has some of the mechanical properties of both of the original levels.[(1)] Consequently, two Raman lines are observed

[(1)] Enrico Fermi, *Z. Phys.*, **71,** 250 (1931).

instead of the single line we have been led to expect. The regular equal spacing of the vibrational levels in the energy scale is also distorted, and so for CO_2 one cannot expect that the thermodynamic properties can be correctly calculated by the simple equations derived here. The magnitude of the energy splitting that arises when two levels of the molecule approach each other in energy, due to such an accidental numerical relationship between the values of the different frequencies, is dependent on the magnitude of the cubic and higher-order terms containing cross products between the Q_λ's. If the coefficients of these terms are identically zero, the splitting will be zero.

For CO_2 molecules, as for most molecules consisting of only three or four atoms, simple qualitative considerations are sufficient to enable one to predict the type of spectral frequencies that will be found and to interpret any anomalies that occur. For more complicated molecules, such as benzene, a careful mathematical analysis of the mechanical problem is necessary before the observed frequencies can be utilized for statistical calculations.

The usual method of attack involves first a group theoretical analysis of the normal coordinates, making use of the symmetry properties of the molecule. The number of *different* frequencies is determined in this manner, the degeneracy of each type of motion is found, and for each frequency it is determined if it is infrared- of Raman-active. By qualitative considerations, an attempt is made to order the different modes of motion in order of the numerical values of their frequencies, and thus to associate the observed infrared and Raman lines with the different normal coordinates.

If, as in benzene, some of the frequencies are completely absent in both the Raman and the infrared spectra, and also are not known from the electronic transition spectrum, a further numerical analysis will be necessary. The force constants for the different normal coordinates may be expressed as functions of the forces between the individual atoms. By neglecting the forces between distant atoms and making ample use of the symmetry of the molecule, the number of different force functions between the atoms can be reduced to equal or less than the number of observed frequencies. One has then, in principle, enough observed data to estimate the unknown forces and so to calculate numerical values for the unobserved frequencies. Needless to say, the operations are rather difficult[(1)] and not very precise.

In order to deduce that the contribution of the γ_{vib} oscillatory coordinates to the thermodynamic function X of the gas was to add γ_{vib} terms of the type X_v, calculated in Sec. 7i, it was necessary to assume that the

[(1)] R. C. Lord, Jr., *J. Phys. Chem.*, **41,** 149 (1937).

cubic terms properly present in eq. (2) are negligible. It is obvious that for sufficiently small displacements q_λ this is legitimate. However, the actual displacements at any temperature are approximately given by the relation that the quadratic term has the value kT, so that the temperature range of validity can be estimated by these conditions. It is not always true that the approximation is justified even at room temperature.

Two rather different types of deviations may occur.

One of these is that the cubic and higher-order terms for one of the normal coordinates alone must be considered, but the cross-product terms in which the coordinate is multiplied by others may be neglected. In this type of deviation the Hamiltonian is still separable, but the energy of a quantum level due to the coordinate q_λ now is not given by the simple equation $(n_\lambda + \frac{1}{2})h\nu_\lambda$, even for energies of about kT or less, but by some more complicated expression. The additive contribution of this coordinate to $\ln Q$ may still be calculated as one term, independently of the others, but is not given by the form of $\ln Q_{\text{vib}}$ due to simple harmonic vibration.

The problem is again essentially mechanical and not statistical. If the actual quantum levels due to this coordinate can be found in the spectra, then numerical summation of $e^{-\epsilon_\lambda/kT}$ will yield the desired value of Q_λ. If, instead, the actual form of the potential were known, or could be guessed with reasonable certainty, it would always be possible to solve for the quantum levels, if necessary, by numerical integration of the one-dimensional wave equation, and to find Q_λ by direct summation.

A problem of this sort arises in ethane, $H_3C{-}CH_3$, and, for instance, $Cl_3C{-}CCl_3$. One of the normal coordinates of these molecules corresponds to equal and opposite rotation of the two CX_3 groups about the axis of the C—C bond. Now two extreme cases are conceivable, and the true state of affairs lies between them. One might assume that the three symmetric positions of minimum potential energy for this coordinate are separated by such low potential hills that it would be possible to treat the potential energy as though it were independent of the value of this coordinate. The quantum-mechanical solution is then easy[1] and leads to the energy levels

$$\epsilon_n = \frac{n^2h^2}{16\pi^2 C}, \tag{7l.3}$$

where n is an integral quantum number, and C is the moment of inertia of *one* of the CH_3 groups about the C—C axis of the molecule.

The other extreme is to assume that the minima along this coordinate are so steep that one can use the quadratic term in the potential

[1] J. E. Mayer, S. Brunauer, and M. Geoppert Mayer, *J. Am. Chem. Soc.*, **55,** 37 (1933).

expression alone, and assume that the rotational vibrational amplitude is never great enough to leave the region near one of the minima where this is allowable. This is of course justifiable only if the potential hills separating the minima are much higher than the value of kT. If this is assumed, $\epsilon_n = (n+\frac{1}{2})h\nu_\lambda$.

A decent approximation of the potential for all values of the coordinate q_r is probably obtained by the equation

$$u(q_r) = \tfrac{1}{2}A(1+\cos 3q_r). \tag{7l.4}$$

At q_r equal to zero, $2\pi/3$, and $4\pi/3$, the cosine has the value unity, and the potential has the maximum value of A, the top of the hill between the minima. At $q_r = \pi/3$, π, and $5\pi/3$, the potential is zero; these are the positions of the three minima.

The quantum levels can be calculated for this type of potential, and their values are obtained as a function of the unknown A.[1] Motion along this coordinate is neither infrared- nor Raman-active, so that the energy levels are not directly observed in the low-pressure gas. However, at sufficiently high pressure the selection rules break down, and these internal-rotation transitions can be observed. The value[2] is 2.93 kcal mol^{-1} for the barrier height A in ethane with the staggered configuration having minimum energy, so that βA at $T = 300$ K is almost 5.

Another difficulty that sometimes reduces the accuracy of the values of ln Q obtained by the method outlined here is the stretching of the molecule in the higher vibrational and rotational levels. This results in a dependence of the moment of inertia on the angular momentum and on the quantum numbers of vibration. This effect was recorded for diatomic gases in Sec. 7j. It is impossible to make as general a calculation for all types of polyatomic molecules and, since the rotational spectrum is not usually analyzed, the constants for such an empirical formula as eq. (7j.1) are unknown. For any particular molecule it is always possible to ascertain the extent of the stretching if all the frequencies are known,[3] but this is usually a moderately involved mechanical calculation.

It is to be expected that for such rigid molecules as CH_4, or C_6H_6, this effect of lack of constancy of the moments of inertia is not very important. For propane, $CH_3CH_2CH_3$, in which the carbons are not in a straight line, the effect may be very appreciable at room temperature.

It is seen that one is usually restricted to making much less accurate calculations for polyatomic than for diatomic molecules. Nevertheless,

(1) E. Teller and K. Weigert, *Nachr. Ges. Wiss. Göttingen*, 218 (1933).
(2) S. Weiss and G. E. Leroi, *J. Chem. Phys.*, **48**, 962 (1968).
(3) E. B. Wilson, Jr., *J. Chem. Phys.*, **4**, 526 (1936).

the methods outlined in this chapter are capable of giving rather good results for many molecules.

The rotational contribution to the partition function of a linear polyatomic molecule is the same as that for a diatomic molecule with the same moment of inertia. The moment of inertia of the polyatomic molecule can be calculated by eq. (1) if the positions of the atoms are known from X-ray or electron-diffraction data or even estimated from empirical bond length information. If the rotational spectrum has been analyzed, it of course furnishes the most precise information.

For nonlinear polyatomic molecules we must calculate the partition function due to the three degrees of freedom of rotation. The general quantum-mechanical solution for the rotational coordinates of a rigid body cannot be made but, since the moments of inertia are almost invariably large, the quantum levels are closely spaced compared with the value of kT at the boiling point of the gas, and the classical approximation can be safely used.

The moment of inertia A of a body composed of n mass points of masses $m_1, \ldots, m_i, \ldots, m_n$ about any axis in space is given by the equation

$$\sum_{i=1}^{i=n} m_i r_i^2 = A, \tag{71.5}$$

where r_i is the perpendicular distance of the mass point from the axis. If the axis passes through the center of mass of the molecule, it follows that

$$\sum_{i=1}^{i=n} m_i \mathbf{r}_i = 0, \tag{71.6}$$

in which $\mathbf{r}_i$ is a vector.

If the magnitudes of the moments of inertia of any rigid body about the various axes passing through the center of mass are plotted along the directions of the axes, they fall on the surface of an ellipsoid with its center at the origin of the plot. This means that three perpendicular axes can be found such that the moment of inertia about one of them is a maximum (is larger or equal to the moment about any other axis), the moment about the second is a minimum, and the moment about the third is at a saddle point, so that it is smaller than the moment of any other axis in the plane common to it and the first axis, and greater than that of any other axis in the plane common to it and the second axis.

These three moments of inertia about the center of mass of a molecule are called its three principal monents of inertia A, B, C, and the three axes are referred to as the principal axes of the molecule.

If the three principal moments are all equal, as in methane, the molecule is called a spherical top, and of course the moments about *all* axes through the center of mass then have the same value. If two moments are equal, but the third has a different value, the molecule is said to be a symmetric top, and then all axes in the plane of the two axes with equal moments have the same moments of inertia. Benzene, ethane, and chloroform molecules are symmetric tops.

If the position of the ith atom in cartesian coordinates is x_i, y_i, z_i, then

$$\sum_{i=1}^{i=n} m_i x_i = \sum_{i=1}^{i=n} m_i y_i = \sum_{i=1}^{i=n} m_i z_i = 0, \tag{7l.7}$$

if the center of mass is taken as the origin. The moments of inertia about the x-, y-, and z-axis are, respectively,

$$I_{xx} = \sum_{i=1}^{i=n} m_i(z_i^2 + y_i^2), \qquad I_{yy} = \sum_{i=1}^{i=n} m_i(z_i^2 + x_i^2), \qquad I_{zz} = \sum_{i=1}^{i=n} m_i(x_i^2 + y_i^2), \tag{7l.8}$$

and products of inertia I_{yz}, I_{xz}, I_{xy} may be defined as

$$I_{yz} = \sum_{i=1}^{i=n} m_i y_i z_i, \ldots. \tag{7l.9}$$

The three equations

$$\begin{aligned} \alpha(I_{xx} - \eta) - \beta I_{xy} - \gamma I_{xz} &= 0, \\ -\alpha I_{xy} + \beta(I_{yy} - \eta) - \gamma I_{yz} &= 0, \\ -\alpha I_{xz} - \beta I_{yz} + \gamma(I_{zz} - \eta) &= 0, \end{aligned} \tag{7l.10}$$

with

$$\alpha^2 + \beta^2 + \gamma^2 = 1,$$

can be solved for three different values of η, which are the three principal moments of inertia. The corresponding values of α, β, γ are the direction cosines of the three principal axes. It is seen that, if all the products of inertia are zero, the x-, y-, and z-axis are the principal axes and their moments are the principal moments of inertia of the molecule.

The three eulerian angles θ, ϕ, and ψ are used to describe the orientation of a rigid body in space.

If x, y, and z are taken as the three cartesian coordinates fixed in space, and ξ, η, ζ as the three principal (perpendicular) axes of the body, then θ is the angle between the body axis ζ and the space axis z. A line in the xy-plane perpendicular to the plane common to ζ and z is called a nodal line. The angle between a nodal line and the x-axis is ϕ. Thus θ and ϕ completely determine the direction of the ζ-axis in space. The angle between the nodal line and the ζ-axis in the body is ψ. This, then, completely defines the orientation of the whole body with a fixed center of gravity.

The angle θ may vary between 0 and π; the two angles ϕ and ψ take all values from 0 to 2π.

The Hamiltonian of the rigid body with a fixed center of gravity is just the kinetic energy, written as a function of these angles and their conjugated momenta, namely,

$$H=\frac{\sin^2\psi}{2A}\left[p_\theta-\frac{\cos\psi}{\sin\theta\sin\psi}(p_\phi-\cos\theta\,p_\psi)\right]^2+\frac{\cos^2\psi}{2B}\left[p_\theta+\frac{\sin\psi}{\sin\theta\cos\psi}(p_\phi-\cos\theta\,p_\psi)\right]^2+\frac{1}{2C}p_\psi^2. \tag{7l.11}$$

This can be transformed into an expression that will be found more convenient for future operations,

$$\frac{H}{kT}=\frac{1}{2kT}\left(\frac{\sin^2\psi}{A}+\frac{\cos^2\psi}{B}\right)\times\left[p_\theta+\left(\frac{1}{B}-\frac{1}{A}\right)\frac{\sin\psi\cos\psi}{\sin\theta\left(\dfrac{\sin^2\psi}{A}+\dfrac{\cos^2\psi}{B}\right)}(p_\phi-\cos\theta p_\psi)\right]^2+\frac{1}{2kTAB\sin^2\theta}\frac{1}{\left(\dfrac{\sin^2\psi}{A}+\dfrac{\cos^2\psi}{B}\right)}(p_\phi-\cos\theta p_\psi)^2+\frac{1}{2kTC}p_\psi^2. \tag{7l.11'}$$

The partition function due to the rotation of this body is

$$Q_{\text{rot}}=\int_{-\infty}^{+\infty}\int_{-\infty}^{+\infty}\int_{-\infty}^{+\infty}\int_0^{\pi}\int_0^{2\pi}\int_0^{2\pi}\frac{1}{h^3}e^{-H(\mathbf{p},\mathbf{q})/kT}\,d\psi\,d\phi\,d\theta\,dp_\psi\,dp_\phi\,dp_\theta. \tag{7l.12}$$

The substitution of eq. (11′) in eq. (12) appears to lead to a rather formidable-looking integral, but direct integration in the order p_θ, p_ϕ, p_ψ actually offers no difficulties. It is necessary to remember that

$$\int_{-\infty}^{+\infty}e^{-a(x+b)^2}\,dx=\int_{-\infty}^{+\infty}e^{-ax^2}\,dx=\left(\frac{\pi}{a}\right)^{1/2}.$$

Integration over p_θ leads to the factor

$$(2\pi kT)^{1/2}\left(\frac{\sin^2\psi}{A}+\frac{\cos^2\psi}{B}\right)^{-1/2}$$

Subsequent integration over p_ϕ yields

$$(2\pi kTAB)^{1/2}\sin\theta\left(\frac{\sin^2\psi}{A}+\frac{\cos^2\psi}{B}\right)^{1/2}$$

as a factor, which cancels part of that obtained in the first integration. Integration over p_ψ yields the factor

$$(2\pi kTC)^{1/2}.$$

Integration of $\sin\theta\,d\theta$ from 0 to π gives 2, and the other angles each give a factor 2π, so that with the symmetry number σ we have

$$Q_R=\sigma^{-1}\pi^{1/2}\left(\frac{8\pi^2AkT}{h^2}\right)^{1/2}\left(\frac{8\pi^2BkT}{h^2}\right)^{1/2}\left(\frac{8\pi^2CkT}{h^2}\right)^{1/2} \qquad (7l.13)$$

If, in conformity with the notation adopted in Sec. 7h for the diatomic molecule, we define

$$\zeta_A=\frac{h^2}{8\pi^2AkT},\qquad \zeta_B=\frac{h^2}{8\pi^2BkT},\qquad \ldots, \qquad (7l.14)$$

then

$$\ln Q_R=\tfrac{1}{2}\ln\frac{\pi}{\zeta_A\zeta_B\zeta_C\sigma^2}. \qquad (7l.15)$$

Using eqs. (6.26′) to (6.29′) for the contribution of these three degrees of rotational freedom to the thermodynamic properties of the gas, one obtains

$$G_R=A_R=-RT\ln Q_R=RT\,\tfrac{1}{2}\ln\frac{\zeta_A\zeta_B\zeta_C\,\sigma^2}{\pi}, \qquad (7l.16)$$

$$H_R=E_R=\tfrac{3}{2}RT, \qquad (7l.17)$$

$$S_R=R\frac{d}{dT}(T\ln Q_R)=R\left(\tfrac{3}{2}+\tfrac{1}{2}\ln\frac{\pi}{\zeta_A\zeta_B\zeta_C\,\sigma^2}\right), \qquad (7l.18)$$

$$C_R=\tfrac{3}{2}R. \qquad (7l.19)$$

7m. THE SYMMETRY NUMBER

The term symmetry number was introduced by Ehrenfest and Trkal[(1)] who first demonstrated its requirement in the molecular partition function. We give here a similar demonstration, starting with the semiclassical requirement of division of the classical phase space integral by the

(1) P. Ehrenfest and V. Trkal, *Proc. Amst. Acad.*, **23,** 162–183 (1920).

product $\prod N_\alpha!$ for a system containing N_α identical entities of type α. Consider a perfect gas consisting of N_m molecules each containing n_α atoms of type α, $1 \le \alpha \le \nu$. There are in all $N_\alpha = n_\alpha N_m$ atoms of type α in the system. We know the prescription for the semiclassical phase space integration. Use the $3\sum_\alpha N_\alpha = 3N_a$ numbered atomic coordinates with their conjugate momenta to form a Hamiltonian H and integrate $\exp(-\beta H)$ over the total phase space; then divide by $h^{3N_a}\prod_\alpha(N_\alpha!)$. Now this coordinate choice and integration would be much too painful. Instead, we limit the integration to the much smaller configurational volume of phase space for which the Hamiltonian H is small in value, namely, where the atoms are arranged close to the equilibrium positions in N_m molecules.

To accomplish this we make a transformation to $3N_m$ coordinates of the centers of mass of N_m numbered molecules and $3N_m(\sum_\alpha n_\alpha - 1)$ internal coordinates and their conjugate momenta, using an approximate potential energy function that is numerically satisfactory near the equilibrium configuration but which goes to infinity if any of the atoms assigned to one molecule departs too far from the molecular center of mass. We still divide by h^{3N_α}, for each α, but what do we do with the division by $\prod_\alpha(N_\alpha!)$? The original prescription called for integration over the *total* atomic phase space, including all possible assignments of the N_α atoms of species α to the numbered N_m molecules. The number of ways we can put $N_\alpha = n_\alpha N_m$ numbered objects in N_m numbered receptacles with n_α in each receptacle is $N_\alpha!\,(n_\alpha!)^{-N_m}$. We must therefore multiply $(\prod_\alpha N_\alpha!)^{-1}$ by $N_\alpha!\,(n_\alpha!)^{-N_m}$ for each species α, leaving a factor $(\prod_\alpha n_\alpha!)^{-1}$ for each molecule if we integrate only over the phase space of the N_m molecules with a fixed assignment of atoms. But when we integrate over all configurations of the numbered molecules, we include $N_m!$ different positions of the numbered coordinates and momenta which correspond only to permutations of the molecular positions in the phase space. When we computed the factor $N_\alpha(n_\alpha!)^{-N_m}$ for atoms of type α, we assumed the molecules to be numbered, all of which would be proper if they were at different fixed positions in phase space, but integration counts too large a volume of phase space by a factor of $N_m!$. We are then left with the conclusion that, to be consistent with our prescription of initial division by $\prod_\alpha N_\alpha!$ for the atoms, we must now divide the integral over the total system phase space by $N_m!\,(\prod_\alpha n_\alpha!)^{N_m}$, or by $\prod_\alpha n_\alpha!$ for each molecular partition function, where n_α is the number of identical atoms of type α in one molecule.

Now, however, when we evaluate the partition function for the molecule, we must integrate $\exp(-\beta H_{\text{int}})$ over all possible positions of the atoms in the molecule. Consider a completely unsymmetric molecule such as formic acid, HCOOH. The normal coordinates of vibration we use represent displacements of the atoms from the equilibrium positions and use a simple potential energy function which takes very large positive values for large displacements. Exchange of the positions of the two hydrogen atoms, or of the two oxygens, from one equilibrium potential minimum brings the molecule into another minimum, but with the usual choice of approximate potential function is represented by very large normal coordinate displacements and an exorbitantly high potential energy. We must therefore multiply the integral by $2!\,2! = 4\prod_\alpha n_\alpha!$ to correctly cover all permutations of the identical atoms. This cancels the factor we had previously deduced. *For a completely unsymmetric molecule the factor multiplying the classical integral over phase space of the internal partition function is just h raised to the negative power of the number of internal degrees of freedom.*

In a diatomic molecule of identical atoms, however, the normal integration over the rotational angles θ, ϕ covers the portion of phase space in which the vector between the two atoms goes to its negative; the two identical atoms are permuted. In general, in polyatomic molecules with high symmetry the rotational coordinates cover an integral number σ of permutations of identical atoms. The total number of permutations is $\prod_\alpha (n_\alpha!)$. The number $\sigma^{-1}\prod_\alpha n_\alpha!$ is the number of different molecules that would exist if the n_α identical atoms of type α were replaced by atoms of different elements, for all α. The factor σ^{-1} is then the factor that multiplies the semiclassical integral for the internal partition functions.

For instance, the tetrahedral molecule CH_4 has a symmetry number $\sigma = 12$, since 12 positions corresponding to different permutations of the four hydrogens are covered by rotation, three values differing by 120° of the angle around the vertical axis for each of the four hydrogens "up." The value $\prod n_\alpha!$ is $4! = 24$. There are $2 = \prod n_\alpha!/\sigma = \frac{24}{12}$ optical isomers in the corresponding tetrahedral substituted methane with four different atoms CHXYZ. The symmetry number of benzene, C_6H_6, is also 12, since 6 positions differing by 60° rotation about the axis perpendicular to the plane of the ring, and with each of these two, positions differing by 180° rotation about the axis passing through (say) C-1 and C-4 correspond to differing permutations of identical atoms. Were the ring of benzene puckered, with the even numbered CH groups "up" and the odd ones "down," the symmetry number would be 6.

Actually, a more careful statement of the value of the symmetry number than that it is the number of different permutations of identical atoms covered by rotations is evident from the derivation. *The symmetry number σ is the number of permutations of identical atoms included in the integration with the correct potential energy minimum.*

Consider, for example, molecules of type AX_3 in which the minimum potential is when the three X-atoms are at the vertices of an equilateral triangle and the A atom lies on a line perpendicular to the plane of the X's passing through the center of the triangle. One normal mode of vibration is with the X-atom plane and the A-atom moving oppositely in the direction normal to the plane (Fig. 7*l*.1, lower right), but presumably also with the equilateral triangle of the X's expanding and contracting at least slightly. If the minimum of potential is with the A atom in the plane of the X's, the symmetry number is $\sigma = 6$; the potential can presumably be well approximated by a quadratic rise in the amplitude of the displacement. There is a single vibrational frequency with very nearly equispaced quantum states of energy $\epsilon(n) = (n + \frac{1}{2})h\nu$ above the minimum potential. A different situation is that the potential minimum has the A atom relatively far off the plane of the X's. There are two symmetrically placed equal minima, one with the A atom above, and the other below, the plane, the potential having a maximum when the A atom is in the plane. If the potential maximum is sufficiently high, we can treat a single one of the two minima as harmonic vibration within this minimum and use the appropriate symmetry number $\sigma = 3$ for a trigonal pyramid.

However, unless the barrier is infinite, the true quantum-mechanical solution for the stationary states of this molecule consists of two different linear combinations, $2^{-1/2}(\psi_+ + \psi_-) = \psi_S$ and $\psi_A = 2^{-1/2}(\psi_+ - \psi_-)$, of the functions ψ_+ and ψ_- representing two states of the same energy in the two minima with A above and below the plane of X's, respectively. The two states ψ_S and ψ_A do not have identical energies but have $\Delta\epsilon = \epsilon_A - \epsilon_S = h\nu'$, with ν' the frequency with which the molecule crosses the potential maximum. The values of ν' and $\Delta\epsilon$ then depend strongly (actually exponentially) on the height of the barrier. We remark parenthetically that, if ν' is of the order of reciprocal months or more, then molecules of three different species replacing X, namely, AXYZ, will have two stable optical isomers. If the barrier is low, $\Delta\epsilon$ may be appreciable compared to kT. The spectrum (Fig. 7m.1) of stationary-state energy levels is complicated, consisting of pairs of energy levels close together at low energies but more widely spaced at higher ones. The partition function for this degree of freedom should then be calculated by summing $\exp(-\beta\epsilon_n)$ numerically.

The main reason for this rather detailed discussion of one example is to emphasize that no discontinuity in the values of thermodynamic functions occurs when one considers the treatment of a continuous series of molecules changing gradually from a planar AX_3 to an extreme trigonal pyramid. For the planar molecule we use $\sigma = 6$, and at the other extreme we use $\sigma = 3$. In going through the series planar to trigonal, at some stage

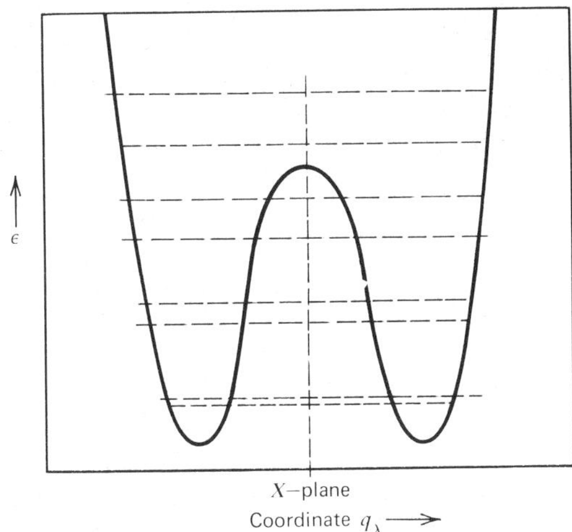

Figure 7m.1 Schematic representation of the potential energy and allowed energy levels for a trigonal pyramidal molecule, such as NH_3, for vibration of the N atom normal to the plane of the H atoms.

one decides that the counting of states in one of the potential minima is adequate, halving the number of states counted but changing the denominator σ from the value 6 to the value 3. Obviously, many other examples occur in which the inclusion of the number of equal potential minima that permute identical atoms is a matter of choice in the treatment of the vibrational coordinates. In all these the correct choice is dictated by the numerical values of the potential barrier between the minima, the precision of the calculation demanded, and the temperature range for which the calculation is expected to be applicable.

In discussing the factor $\sigma = 2$ for symmetric diatomic molecules (Sec. 7h) and in Sec. 7k on ortho and para hydrogen, we pointed out that the condition of symmetry in permutations required that, for given other quantum states such as the nuclear spin states, only half of the rotational levels are allowed. The equivalent relation holds in symmetric polyatomic molecules but is appreciably more complicated to derive in detail.[(1)] In a sequence of rotational states of increasing energy only the fraction σ^{-1} has the correct symmetry character to be allowed with all other quantum numbers fixed.

The arguments of this section apply equally to any case in which we treat as a fundamental unit objects consisting of smaller entities that keep their identity when combined in a larger unit, for example, nuclei which consist of protons and neutrons, and atoms which contain a nucleus and

(1) J. E. Mayer, S. Brunauer, and M. G. Mayer, *J. Am. Chem. Soc.*, **55**, 37, (1933).

electrons, as well as the example chosen of molecules made up of constituent atoms, and even of crystals composed of molecules.

The rule is always the same. If we actually count the observed existing quantum states of the composite unit, no extraneous factor such as symmetry number is needed. If, however, we use an approximate theory of counting all solutions of the time-independent Schrödinger equation without regard to symmetry, or use the semiclassical integration, then a factor such as the reciprocal of the symmetry number or of $N!$ enters.

7n. GAS MIXTURES AND ENTROPY OF MIXING

The law of partial pressures is well known and old: The total pressure P of a mixture of gases is the sum of the partial pressures P_a of each species a of molecule, where the partial pressures P_a are those that would be exerted by the same mass of component a in the same total volume at the same temperature. The only need to amplify this in a treatment of statistical mechanics is to point out that it follows, almost trivially, from the general equation equivalent to eq. (7a.1) for the probability of a quantum state $\mathbf{K}$ of N_a molecules of type a, N_b of type b, and so on, in a grand canonical ensemble of open systems of fixed V, T and chemical potential set $\mu_a, \mu_b, \ldots$, and the assumption of no interaction between molecules.

If we use a boldface $\mathbf{N}$ to represent the set of numbers

$$\mathbf{N} \equiv N_a, N_b, \ldots, \tag{7n.1}$$

and $\boldsymbol{\mu}$ for the set of chemical potentials

$$\boldsymbol{\mu} \equiv \mu_a, \mu_b, \ldots, \tag{7n.2}$$

with the scalar product notation

$$\mathbf{N} \cdot \boldsymbol{\mu} \equiv \sum_a N_a \mu_a, \tag{7n.3}$$

then eq. (7a.1) for the probability is altered only by the replacement of the number N of the single species and μ its chemical potential by $\mathbf{N}$ and $\boldsymbol{\mu}$:

$$W_{\mathbf{n},\mathbf{K}} = \exp\{-\beta[PV - \mathbf{N} \cdot \boldsymbol{\mu} + E(K)]\}. \tag{7n.4}$$

The condition that the sum over all possible states of their probabilities $W_{\mathbf{N},\mathbf{K}}$ be unity [eq. (7a.3)] is now altered only in that there is an independent summation from zero to infinity of N_a for each species a, and from this, with the same change in the meaning of summation over $\mathbf{N}$

[eq. (7a.4)] for PV follows. Thus far the equations are general for any system. The assumption of the perfect gas makes $E(\mathbf{K})$ the sum of energies of the individual molecules, hence the sum over $\mathbf{N}$ becomes a product of sums for each species a over N_a of exactly the same function as that for the single species in eq. 7a.4. The logarithm of the product is the sum of the logarithms, so that our total PV is the sum $\sum_a P_a V$, with P_a the partial pressure of species a a function of V, T, and μ_a only.

The thermodynamic relation [eq. (3h.20)]

$$\left[\frac{\partial(PV)}{\partial\mu_a}\right]_{V,T,\mu_b,\ldots} = N_a, \tag{7n.5}$$

with P the sum of partial pressures, and with each P_a dependent only on its *own* chemical potential μ_a, leads to the same relation between $\rho_a = N_a/V$ and μ_a, T that exists in a pure gas of species a. All this is derived without going over to the approximation of Boltzmann counting. The statistical repulsion in the Fermi-Dirac gas, and attraction in the Bose-Einstein gas, exists only between identical entities, and there is no effect on the behavior of species a when another (sufficiently dilute) species b is present.

Consider now a mixture of gases of mole fractions x_a, x_b, ... of species a, b, ..., respectively, and use the approximation of Boltzmann counting. The enthalpy η per molecule and energy ε per molecule, as well as the two heat capacities C_P and C_V [eqs. (7e.7) to (7e.10)], are independent of the concentration $\rho = N/V$. For 1 mol of gas these quantities are the sum over all species present of the mole fractions times the values for 1 mol of each of the pure gases, for instance,

$$H^{(0)}(T) = \sum_a x_z H_a^{(0)}(T), \ldots. \tag{7n.6}$$

The entropy s per molecule has an additive term, $-k \ln N/V = -k \ln \rho$ [eq. (7e.6′)]. For a given total number $N = \sum_a N_a$ of molecules, $N_a = x_a N$, so that, for 1 mol of the mixture,

$$S^{(0)}(P, T) = \sum_a x_a S_a^{(0)}(P, T) - R \sum_a x_a \ln x_a. \tag{7n.7}$$

The extra term

$$S_{\text{mix}}^{(0)} = -R \sum_a x_a \ln x_a \tag{7n.8}$$

in the entropy per mole is known as the entropy of mixing.

This term is intimately connected with the division by $\prod_a N_a!$ in the counting of solutions of the Schrödinger equation or in the semiclassical integral over phase space. Entropy of mixing is inherently present in condensed thermodynamic phases as well as in perfect gases, although it is often masked by other effects due to the interaction of molecules of the different species.

7o. ISOTOPIC EQUILIBRIUM

Molecules that differ only in having different isotopes of the same element in the same configuration have almost the same mechanical properties. Except in the case of H_2 and D_2, for which the mass difference is a factor 2, at equilibrium in chemical reactions the isotopic compositions of reactants and products are very nearly equal. It is difficult to separate isotopes by chemical reaction at equilibrium. Indeed, we prove in this section that in the classical limit no separation at all occurs. The only essential difference in the thermodynamic properties of a mixture of isotopes and of a pure single isotope is in the entropy of mixing terms which of course also appears in the two free energies G and A. The same term appears in products and reactants, so that the change ΔG in the reaction is unaltered.

The quantum-mechanical effect that makes separation possible lies in the zero-point energy $\sum_\lambda \frac{1}{2}h\nu_\lambda$ summed over all vibrational coordinates. The potential energy u_0 per molecule in the equilibrium configuration is independent of the isotopic composition to a very high approximation, since it is determined by the electronic state, and the electrons are affected only by the nuclear charge. The frequency ν_λ is inversely proportional to the square root of a reduced mass μ_λ [eqs. (7h.2) and (7h.16) or eq. (7i.2)]. If the reduced mass μ_λ of frequency mode λ is smaller for isotopic composition a than for b by a small fraction α,

$$\mu_{\lambda a} = \mu_{\lambda b}(1-\alpha) \tag{7o.1}$$

then, since $(1-\alpha)^{-1/2} = 1 + \frac{1}{2}\alpha + O(\alpha^2)$,

$$\tfrac{1}{2}h\nu_{\lambda a} \cong \tfrac{1}{2}h\nu_{\lambda b}(1+\tfrac{1}{2}\alpha) = \tfrac{1}{2}h\nu_{\lambda b} + \frac{\alpha}{4}\,h\nu_{\lambda b}. \tag{7o.2}$$

The zero-point energy is higher for the lighter isotope, and higher by an amount proportional to the frequency. The lighter isotopes tend to concentrate in the molecules for which the binding is less strong and the frequencies lower.

We proceed to the proof that in the classical limit no separation of isotopes occurs in a chemical equilibrium reaction and prove this without any assumption that we deal only with gaseous systems. We use eq. (3i.5) for the petite canonical ensemble, which gives an expression for the Helmholtz free energy A for fixed volume V, temperature T, and number set $N_a, N_b, \ldots$ of constituents,

$$-\beta A = \ln \sum_{\mathbf{K}} \exp[-\beta E(K)], \tag{7o.3}$$

and use classical integration divided by h^Γ and $\prod_a N_a!$ for the partition function,

$$A = -kT \ln \iint \cdots \int \frac{dp^{(\Gamma)}\, dq^{(\Gamma)}}{h^\Gamma \prod_a N_a!} \exp[-\beta H(p^{(\Gamma)} q^{(\Gamma)})]. \tag{7o.4}$$

We now treat a system consisting of

$$N = \sum_a N_a$$

atoms which are all isotopes of the same element, the isotope of species a having mass m_a, and define the atomic fraction x_a of isotope a as

$$x_a = \frac{N_a}{N}, \qquad N_a = x_a N, \tag{7o.5}$$

and a geometric mean mass,

$$m = \prod_a m_a^{x_a}, \tag{7o.6}$$

so that

$$\prod_a m_a^{3N_a} = m^{3N}. \tag{7o.6$'$}$$

Use cartesian coordinates of the atoms so that $\Gamma = 3N$ and the Hamiltonian is a sum of $3N$ kinetic energy terms each of the form $(1/2m_a)p_{ia\alpha}^2$, $\alpha = x, y, z$, and a potential energy term $U_N(\mathbf{q}^{(\Gamma)})$ depending only on the coordinates. Include the denominator h in each of the momentum integrals. Since

$$\int_{-\infty}^{+\infty} h^{-1}\, dp \exp\left[-\frac{1}{2m_a kT} p^2\right] = \left(\frac{2\pi m_a kT}{h^2}\right)^{1/2} = \lambda_a^{-1}, \tag{7o.7}$$

and we have $3N_a$ such terms for each a, we obtain, as a factor from the integral over the momenta, using eq. (6′),

$$\prod_a \left(\frac{2\pi m_a kT}{h^2}\right)^{3N_a/2} = \left(\frac{2\pi m kT}{h^2}\right)^{3N/2} = \lambda^{-3N}, \tag{7o.8}$$

where λ is the de Broglie wavelength for mass m of eq. (6).

The potential energy function $U_N(\mathbf{q}^{(3N)})$ is, for isotopes, symmetric in exchange of the coordinates of all the N atoms, whether or not they are identical isotopes. That is, it is the same function of the atoms numbered *seriata* from 1 to N, independent of isotopes and of the composition. We define the configuration integral, independent of composition, to be

$$Q_\tau(N, V, T) = V^{-N}\int\int_V \cdots \int dx_1 \cdots dz_N \exp(-\beta U_N). \qquad (7o.9)$$

Now look at the logarithm of the factor $(\prod_a N_a!)^{-1}$ and use the Stirling approximation (Appendix III)

$$\ln N_a! = N_a(\ln N_a - 1), \qquad (7o.10)$$

which, with eq. (5) for x_a is,

$$\ln N_a! = N[x_a(\ln N - 1) + x_a \ln x_a], \qquad (7o.11)$$

or, since $\sum_{x_a} = 1$ and $\ln e = 1$,

$$-\sum_a \ln N_a! = -N\left[\ln \frac{N}{e} + \sum_a x_a \ln x_a\right]. \qquad (7o.11')$$

With these in eq. (4) for A we have that

$$A(V, T, N_a, N_b, \ldots) = -NkT\left(\ln \frac{Ve}{N\lambda^3}\right) - kT \ln Q_\tau + NkT \sum_a x_a \ln x_a$$

$$= A(V, T, N) + NkT\left(\sum_a x_a \ln x_a\right),$$

since the Helmholtz free energy $A(V, T, N)$ of a single-component system of the mass m given by eq. (6) is $-NkT \ln(Ve/N\lambda^3)Q_\tau$. The pressure P is $-(\partial A/\partial V)_{T,N_a}, \ldots$ and, since the sum of $x_a \ln x_a$ is independent of V, the pressure is that of the pure one-component system, and $G = A + PV$ is also that of the pure system plus the term $NkT \sum_a x_a \ln x_a$. The entropy is $S = -(\partial A/\partial T)_{V,N_a,\ldots}$ and is that of the single-component system minus $Nk \sum_a x_a \ln x_a$. The extra term is absent in H and E, but in the chemical potential μ_a of the isotope we find

$$\mu_a = \left(\frac{\partial A}{\partial N}\right)_{V,T,N_b,\ldots} = \mu_0 + kT \ln x_a, \qquad (7o.12)$$

where μ_0 is the chemical potential of the hypothetical one-component system.

The condition of isotopic equilibrium between two thermodynamic phases at equilibrium is that the chemical potentials μ_a of each isotope a is the same in both phases. From eq. (12) we see that this is satisfied if and only if μ_0 and x_a for all a's are equal in both phases. At equilibrium there is no fractionation of isotopes between liquid and vapor or between either of these and the crystal under conditions when all degrees of freedom can be treated classically.

At the expense of a somewhat more cumbersome notation eq. (4) can be treated for a system of N_{1a}, $N_{1b}, \ldots$ isotopes of the first element number 1, N_{2a}, $N_{2b}, \ldots$ isotopes of element 2, and so on, which may be combined in molecules. Using $N_1 = \sum_a N_{1a}, \ldots$ for the numbers of atoms of element 1, and so on, one finds, with $x_{ia} = N_{ia}/N_i$, that the entropy of mixing is

$$S_{\text{mix}} = \sum_i N_i k \left(\sum x_{ia} \ln x_{ia} \right), \tag{7o.13}$$

and with μ_{i0} the chemical potential of a hypothetical system containing only one kind of atoms of element i,

$$\mu_{ia} = \mu_{i0} + kT \ln x_{ia}. \tag{7o.14}$$

In a chemical reaction the total number of atoms of element i in reactants and products is the same. It follows again that equality of the chemical potentials μ_{ia} in reactants and products at equilibrium will be satisfied if and only if the equilibrium is the same as that of molecules composed of identical atoms of each species i of masses m_i given by eq. (6), and if the mole fractions x_{ia} of the isotopes are the same in reactants and products. Isotopic fractionation at equilibrium is purely a quantum effect, and, since it is a minor effect except in H_2 and D_2, the entropy of isotope mixing is often completely omitted from entropy tabulations.

A direct but tedious method of calculating the isotopic fractionation occurring in some chemical reactions is to use eq. (7b.17) for the chemical potential μ of each species of molecule in the reaction, and then to use the condition that $\Delta\mu$ for the reaction should be zero at equilibrium. Since the chemical potentials μ for differing isotopic compositions are very nearly the same, high precision of calculation is necessary to obtain the small differences correctly. A much simpler method is to use the theorem we have proved above, namely, that in a chemical reaction at equilibrium no fractionation occurs if the partition functions have their semiclassical values Q_{cl} and then calculate the *differences* for different isotopic compositions directly.[1] Fractionation is then governed by the ratio Ψ of the true quantum-mechanical partition function Q_{qm} to the

[1] J. Bigeleisen and Maria G. Mayer, *J. Chem. Phys.*, **15**, 261, (1947).

semiclassical,

$$\Psi = \frac{Q_{\rm qm}}{Q_{\rm cl}}, \tag{7o.15}$$

which differs slightly for molecules of different isotopic composition. Since the partition functions are products of terms from the separable degrees of freedom, only those need be considered for which the contribution differs considerably from the semiclassical value.

Suppose that for an element X there are two isotopes X_a and X_b. For a molecule RX consider the hypothetical perfect gas equilibrium of the dissociation reaction

$$\mathrm{RX} \rightleftharpoons \mathrm{R} + \mathrm{X}. \tag{7o.16}$$

With $Q_{\rm qm} = \Psi Q_{\rm cl}$ and $P/kT = N/V$, we use eq. (7b.17) for the chemical potential,

$$\mu = kT \ln \frac{P}{kT} \lambda^3 Q_i^{-1} = kT \ln \frac{N}{V} \lambda^3 (Q_{\rm cl}\Psi)^{-1}, \tag{7o.17}$$

so that for any species of molecule the number $N(V, T, \mu)$ of molecules present in volume V at temperature T and chemical potential μ is

$$N(V, T, \mu) = \Psi N^{\rm cl}(V, T, \mu), \tag{7o.18}$$

where $N^{\rm cl}$ is the number that would be calculated from the semiclassical equations. In the reaction in eq. (16) at equilibrium let $N_{\mathrm{R}a}$ and $N_{\mathrm{R}b}$ be the numbers of molecules RX_a and RX_b, respectively, and N_a and N_b the number of dissociated atoms, respectively. For the monatomic species X_a and X_b the internal partition function Q_i is normally unity, but if not, it is equal for both isotopic species, since it arises from the electronic states. We then have $N_a/N_b = N_a^{\rm cl}/N_b^{\rm cl}$, but we have proved that $N_{\rm Ra}^{\rm cl}/N_{\rm Rb}^{\rm cl} = N_a^{\rm cl}/N_b^{\rm cl}$, so that

$$\frac{N_{\mathrm{R}a}}{N_{\mathrm{R}b}} = \left(\frac{N_a}{N_b}\right)\frac{\Psi_{\mathrm{R}a}}{\Psi_{\mathrm{R}b}}. \tag{7o.19}$$

Now we would hardly expect to establish equilibrium between a molecule RX_a and its dissociated products R and monatomic X_a, but consider the exchange reaction

$$\mathrm{RX}_a + AX_b \rightleftharpoons \mathrm{RX}_b + \mathrm{AX}_a, \tag{7o.20}$$

and use the dissociated $\mathrm{R} + \mathrm{A} + \mathrm{X}_a + \mathrm{X}_b$ as hypothetical intermediates simply to establish a zero of chemical potential. We find

$$\frac{N_{\mathrm{R}a}}{N_{\mathrm{R}b}} = \left(\frac{N_{\mathrm{A}a}}{N_{\mathrm{A}b}}\right)\left(\frac{\Psi_{\mathrm{R}a}\Psi_{\mathrm{A}b}}{\Psi_{\mathrm{R}b}\Psi_{\mathrm{A}a}}\right). \tag{7o.21}$$

As mentioned previously, the Ψ's are products of contributions from the separable degrees of freedom, and normally these single terms differ from unity only for vibrational degrees of freedom. For a normal coordinate q_λ of harmonic oscillation of frequency ν_λ the classical partition function is given by eq. (7h.18) as

$$Q_{cl} = \frac{kT}{h\nu_\lambda} \exp\left(\frac{-u_0}{kT}\right), \tag{7o.22}$$

and the quantum-mechanical partition function by eq. (7h.22) as

$$Q_{qm} = \left[\exp\left(-\tfrac{1}{2}\frac{h\nu_\lambda}{kT}\right)\right]\left[1-\exp\left(-\frac{h\nu_\lambda}{kT}\right)\right]^{-1} \exp\left(\frac{-u_0}{kT}\right). \tag{7o.23}$$

The factor ψ_λ in Ψ due to this degree of freedom, with

$$\xi = \frac{h\nu_\lambda}{kT},$$

is then

$$\psi_\lambda = \xi e^{(-1/2)\xi}(1-e^{-\xi})^{-1} = \xi(e^{(+1/2)\xi} - e^{(-1/2)\xi})^{-1}. \tag{7o.24}$$

For $\xi \leq 1$ we can expand the function $\psi_\lambda^{-1}(\xi)$, which is even in ξ, very easily as

$$\psi_\lambda^{-1} = 1 + \tfrac{1}{24}\xi^2 + \tfrac{1}{1920}\xi^4 + (\sim 3\times 10^{-6})\xi^6 + \cdots, \tag{7o.25}$$

with reciprocal

$$\psi_\lambda = 1 - \tfrac{1}{24}\xi^2 + \tfrac{7}{5760}\xi^4 + \cdots. \tag{7o.25'}$$

For the chemical exchange reaction in eq. (20) we seek the ratio $\psi_\lambda(\xi_1)\psi_\lambda^{-1}(\xi_2)$ for the normal coordinate λ in either RX or AX with two different isotopes X_1 and X_2. Let

$$\xi_0 = \tfrac{1}{2}(\xi_1+\xi_2) = \frac{h}{kT}\tfrac{1}{2}(\nu_1+\nu_2) = \frac{h\nu_0}{kT},$$

$$\Delta = \xi_2 - \xi_1 = \frac{h}{kT}(\nu_2-\nu_1) = \frac{h\,\Delta\nu}{kT}, \tag{7o.26}$$

so that

$$\xi_1 = \xi_0 - \tfrac{1}{2}\Delta, \qquad \xi_2 = \xi_0 + \tfrac{1}{2}\Delta, \tag{7o.26'}$$

and with these in eqs. (25) and (25′) we find

$$\begin{aligned}\psi(\nu_1)[\psi(\nu_2)]^{-1} &= 1 + \tfrac{1}{12}\xi_0\,\Delta(1-\tfrac{1}{60}\xi_0^2) + \tfrac{1}{288}(\xi_0\Delta)^2 + 0(\Delta^3)\\ &= 1 + \frac{1}{12}\frac{\Delta\nu}{\nu_0}\left(\frac{h\nu_o}{kT}\right)^2\left[1-\frac{1}{60}\left(\frac{h\nu_0}{kT}\right)^2\right]\\ &\qquad + \frac{1}{288}\left(\frac{\Delta\nu}{\nu_o}\right)^2\left(\frac{h\nu_0}{kT}\right)^4.\end{aligned} \tag{7o.27}$$

For $h\nu_0/kT > 1$ we develop differently. Write

$$\frac{\psi(\nu_1)}{\psi(\nu_2)} = \frac{\xi_1}{e^{(1/2)\xi_1} - e^{(-1/2)\xi_1}} \frac{e^{(1/2)\xi_2} - e^{(-1/2)\xi_2}}{\xi_2}$$

$$= \frac{1-(\Delta/2\xi_0)}{1+(\Delta/2\xi_0)} \frac{e^{(1/2)\xi_0}e^{(1/4)\Delta} - e^{(-1/2)\xi_0}e^{(-1/4)\Delta}}{e^{(1/2)\xi_0}e^{(-1/4)\Delta} - e^{(-1/2)\xi_0}e^{(1/4)\Delta}}$$

$$= \left[1 - \frac{\Delta}{\xi_0} + \frac{1}{2}\left(\frac{\Delta}{\xi_0}\right)^2 - \cdots\right] \frac{1+(e^{(1/2)\Delta}-1)(1-e^{-\xi_0})^{-1}}{1-(e^{(1/2)\Delta}-1)e^{-\xi_0}(1-e^{-\xi_0})^{-1}}, \tag{7o.28}$$

and develop $e^{(1/2)\Delta}-1$ up to quadratic terms in Δ, $e^{(1/2)\Delta}-1 = \frac{1}{2}\Delta + \frac{1}{8}\Delta^2 + \cdots$. Expanding the denominator of eq. (28) and introducing

$$B = \frac{\xi_0(1+e^{-\xi_0})}{2(1-e^{-\xi_0})} - 1 = \xi_0(1-e^{-\xi_0})^{-1} - \tfrac{1}{2}\xi_0 - 1$$

$$= \frac{h\nu_0}{kT}(1-e^{-h\nu_0/kT})^{-1} - \frac{h\nu_0}{2kT} - 1, \tag{7o.29}$$

one obtains, with a bit of tedious algebra,

$$\frac{\psi(\nu_1)}{\psi(\nu_2)} = 1 + \frac{\Delta\nu}{\nu_0} B + \frac{1}{2}\left(\frac{\Delta\nu}{\nu_0}\right)^2 B^2 + \cdots, \tag{7o.30}$$

up to quadratic terms in $\Delta\nu/\nu$. Since for $h\nu_0/kT < 1$ we find that

$$B = \tfrac{1}{12}\xi_0^2 - \tfrac{1}{720}\xi_0^4 = \frac{1}{12}\left(\frac{h\nu_0}{kT}\right)^2\left[1 - \frac{1}{60}\left(\frac{h\nu_0}{kT}\right)^2\right], \tag{7o.31}$$

we recover eq. (27) up to terms in $(h\nu_0/kT)^4$ for the coefficients of $\Delta\nu/\nu_0$ and $(\Delta\nu/\nu_0)^2$.

For two isotopes of nitrogen differing by one mass unit the mass differences are about 0.07 times the mass. If the normal coordinate q_λ is that of pure vibration of the N atom this will be the value of $-\Delta\mu_\lambda/\mu_\lambda$ [eq. (7h. 2)]; if it is a more complicated motion, $\Delta\mu_\lambda/\mu_\lambda$ will be less. The frequency ratio $\Delta\nu/\nu$ will be half of this, since the frequency is proportional to $\mu^{-1/2}$ [eq. (7h.16)]. At room temperature $h\nu_0/kT$ may be as high as 10 for a very strong bond, in which case $h\,\Delta\nu/2kT$ could be as high as 0.15, but one would not expect the chemical exchange reaction in eq. (20) to equilibrate in reasonable time for such a strong binding. More usually, $h\,\Delta\nu/2kT$ is of the order of 10^{-2}, so that eq. (27) or (30) is adequately approximated by the linear terms only.

The ratio of eq. (30) is that due to any *one* vibrational degree of freedom for which the frequencies with isotopes X_1 and X_2 are ν_1 and ν_2, respectively. For the molecule the ratio Ψ_1/Ψ_2 is the product for all modes of vibration, so that, with the linear approximation,

$$\frac{\Psi(RX_1)}{\Psi(RX_2)} = 1 + \sum_\lambda \frac{\Delta\nu_\lambda}{\nu_{0\lambda}} B_\lambda, \tag{7o.32}$$

and for the chemical exchange reaction of eq. (20) we find the ratio of numbers of the two isotopes in products and reactants to be

$$\frac{N(RX_1)}{N(RX_2)} = \frac{N(AX_1)}{N(AX_2)}\left[1 + \sum_{\lambda C(RX)} \frac{\Delta\nu_\lambda}{\nu_{0\lambda}} B_\lambda - \sum_{\kappa C(AX)} \frac{\Delta\nu_\kappa}{\nu_{0\kappa}} B_\kappa\right], \tag{7o.33}$$

where

$$\Delta\nu_\lambda = \nu_\lambda \text{ (isotope } X_2) - \nu_\lambda \text{ (isotope } X_1). \tag{7o.34}$$

Fractionation in a single equilibrium is seldom much more than a percent difference in reactant and product and usually very much less. Of course, if the principle of a fractionation of many plates can be employed, reasonable separation can be obtained.

7p. CHEMICAL EQUILIBRIUM IN PERFECT GASES

The Helmholtz free energy A in a perfect gas containing N_ζ molecules of species ζ in volume V at temperature T is, from eq. (7e.5) for $\mu_\zeta(V, T, N)$,

$$A = G - PV = \sum_\zeta N_\zeta(\mu_\zeta - kT)$$

$$= \sum_\zeta N_\zeta kT \ln \frac{N_\zeta \lambda_\zeta^3}{Ve} Q_{\zeta i}^{-1} \tag{7p.1}$$

where $\lambda_\zeta = (h^2/2\pi m_\zeta kT)^{1/2}$ is the de Broglie wavelength for molecules of mass m_ζ, e is the Naperian base, $\ln e = 1$, and $Q_{\zeta i}$ is the internal partition function per molecule of species ζ.

Suppose that a chemical reaction can occur in which n_ζ molecules of type ζ are produced for unit change dN_r in the number of molecules that have reacted,

$$dN_\zeta = n_\zeta \, dN_r. \tag{7p.2}$$

Thus n_ζ is a small integer *negative* number if ζ is one of the reactants, and *positive* if ζ is one of the products. Let

$$\Delta n = \sum_\zeta n_\zeta \tag{7p.3}$$

be the change in total number of molecules in the reaction as written. The condition of equilibrium at constant V, T and amount of matter is that A be a minimum or, for this reaction, that [since $(\partial A/\partial N_\zeta)_{V,T,N_\xi} = \mu_\zeta$]

$$\left(\frac{dA}{dN_r}\right)_{V,T} = 0 = \sum_\zeta n_\zeta \mu_\zeta = kT \sum_\zeta n_\zeta \ln \frac{N_\zeta \lambda^3}{VQ_{\zeta i}}$$

$$= kT \ln V^{-\Delta n} \prod_\zeta \left(\frac{N_\zeta \lambda_\zeta^3}{Q_{\zeta i}}\right)^{n_\zeta}, \tag{7p.4}$$

which is recognized as the familiar condition at equilibrium that the Gibb free energy of reactants and products be equal.

The number mass action constant,

$$K_N = \prod_\zeta N_\zeta^{n_\zeta}, \tag{7p.5}$$

is then, from eq. (4),

$$K_N = V^{\Delta n} \prod_\zeta \left(\frac{Q_{\zeta i}}{\lambda_\zeta^3}\right)^{n_\zeta}. \tag{7p.6}$$

Alternatively, we can write μ_ζ in terms of P_ζ and T [eq. (7e.1)], which simply replaces N_ζ/V by P_ζ/kT in eq. (4).

$$\ln \prod_\zeta P_\zeta^{n_\zeta} \left(\frac{\lambda_\zeta^3}{Q_{\zeta i} kT}\right)^{n_\zeta} = 0, \tag{7p.7}$$

or,

$$K_P = \prod_\xi P_\xi^{n_\xi} = \prod_\zeta \left(\frac{Q_{\zeta i} kT}{\lambda_\zeta^3}\right)^{n_\zeta}$$

$$= -\sum_\zeta n_\zeta \left(\frac{\mu_\zeta^{(0)}}{kT}\right) = \frac{-G^{(0)}}{N_0 kT}, \tag{7p.8}$$

where $\Delta G^{(0)}$ is the difference in Gibbs free energies at unit pressure of products minus reactants for n_ζ moles of each, if N_0 is Avogadro's number.

One obvious caution must be taken in using equations like (6) or (8). The partition functions Q_i have a factor $\exp(-\beta u_0)$ [eq. (7h.21)] in which u_0 is an arbitrary zero of energy. This is often chosen so that the lowest quantum state of the molecule has zero energy or, as in the case of isotopic exchange reactions, so that the minimum of potential energy of the molecule has zero energy. In the use of the partition function to express chemical equilibrium it is essential that *the same zero of energy be used fo products and reactants.* This offers no fundamental problem, since there are necessarily the same number of atoms of each element in reactants and products. If $Q_{\zeta i}$ is computed assuming the lowest quantum state to have zero energy, then $\prod_\zeta Q_{\zeta i}^{n_\zeta}$ must be multiplied by $\exp(-\beta\Delta\epsilon_0)$, where $\Delta\epsilon_0$ is the energy of reaction per molecule at 0 K.

In all of the preceding there is nothing different than the classical thermodynamic argument equating the mass-action constant to the change in Gibbs free energy. The condition that the entropy S be a maximum for a system of fixed volume and energy requires that $A = E - TS$ be a minimum when V and T are fixed, and $G = E + PV - TS$ be a minimum for fixed P and T. The logic and semantics are identical in statistical mechanics and classical thermodynamics. The only addition statistical mechanics makes to the classical equations is in the use of the equations relating the thermodynamic functions to λ_ξ and to the internal partition function $Q_{\zeta i}$ for each molecular species ζ.

However, one illuminating relation is obtained for the mass-action constant if the semiclassical integral is used for the internal partition function, and this relation is usually sufficiently good numerically to serve as a useful method of estimating simple equilibrium constants in gaseous systems.

Assume that the molecule of species ζ contains ν_ζ atoms and designate its lowest potential energy measured from some zero defined in terms of a zero for the elements composing the molecule by the symbol $u_{0\zeta}$. The energy

$$\Delta u_0 = \sum_\zeta n_\zeta u_{0\zeta}, \tag{7p.9}$$

which differs from $\Delta\epsilon_0$ by the sum of the $\frac{1}{2}h\nu_\lambda$'s of zero-point energy change in products and reactants, is the difference in potential energy in the equilibrium configuration of products minus reactants.

The dimensionless quantity

$$Q_\zeta = \frac{e^{\beta u_{0\zeta}} V Q_{\zeta i}}{\lambda_\zeta^3} \tag{7p.10}$$

is given semiclassically by the integral over the phase space of $d\mathbf{q}^{(3\nu_\zeta)}\, d\mathbf{p}^{(3\nu_\zeta)}$ of the ν_ζ atoms of $h^{-3\nu_\zeta} \exp(-\beta H_\zeta)(\mathbf{q}^{(3\nu_\zeta)}, \mathbf{p}^{(3\nu_\zeta)})$ divided by the symmetry number σ_ζ. The zero of energy in the Hamiltonian H_ζ is now that of the equilibrium configuration of the molecule. We may, if we choose, use cartesian coordinates of the atoms. Integration over each of the $3\nu_\zeta$ momenta gives us the familiar factor

$$\int_{-\infty}^{+\infty} h^{-1}\, dp_{\alpha i} \exp[-(\beta/2m_i)P_{\alpha i}^2] = \left(\frac{2\pi m_i kT}{h^2}\right)^{1/2} = \lambda_i^{-1} \tag{7p.11}$$

for each of the three momenta of an atom i of mass m_i. Since the total number of atoms of each element is the same in products and reactants, these terms cancel in the product $\prod_\zeta Q_{\zeta i}^{n_\zeta}$ that appears in the expression (6) for K_N. With

$$N = \sum_\zeta N_\zeta, \tag{7p.12}$$

and

$$Q_{\tau\zeta}^{\text{cl}} = N^{-1}\sigma^{-1} \iint_V \cdots \int dx_1 \cdots dz_{\nu\zeta} \exp(-\beta u_\zeta), \tag{7p.13}$$

we can now rewrite eq. (6) as

$$K_N = e^{-\beta \Delta u_0} \prod_\zeta (N Q_{\tau\zeta}^{\text{cl}})^{n_\zeta}. \tag{7p.14}$$

With the mole fraction x_ζ,

$$x_\zeta = \frac{N_\zeta}{N}, \tag{7p.15}$$

we have for the mole fraction mass-action constant

$$K_x = \prod_\zeta x_\zeta^{n_\zeta} = N^{-\Delta n_{K_N}} = e^{-\beta \Delta u_0} \prod_\zeta (Q_{\tau\zeta}^{\text{cl}})^{n_\zeta}. \tag{7p.16}$$

The quantity $Q_{\tau\zeta}^{\text{cl}}$ has the dimensions of volume raised to the power of the number of atoms ν_ζ in the molecule ζ. For instance, in a diatomic molecule hold one atom fixed and integrate over the coordinates of the second atom. With r_0 the equilibrium distance between atoms and Δr an average amplitude of vibration, with κ the force constant, $u = \frac{1}{2}\kappa(r - r_0)^2$,

$$\Delta r = \int_{-\infty}^{+\infty} dr \exp(-\tfrac{1}{2}\beta\kappa r^2) = \left(\frac{2\pi kT}{\kappa}\right)^{1/2}, \tag{7p.17}$$

one obtains $4\pi r_0^2 \Delta r$ from the integration over the second atom. Integration over the coordinates of the remaining first atom gives the total volume V of the container, which in eq. (13) must be divided by N, the total number of molecules. With

$$\frac{V}{N} = v_0, \tag{7p.18}$$

the volume per molecule, in the gas we have

$$Q_\tau^{\text{cl}}\,(\text{diatomic}) = \sigma^{-1} v_0\, 4\pi r_0^2\, \Delta r. \tag{7p.19}$$

Since Q_τ for an atomic species is v_0, the mass-action constant K_x for the dissociation of the diatomic molecule AB,

$$AB \rightleftharpoons A+B, \tag{7p.20}$$

is in the semiclassical approximation

$$K_x = \frac{x_A x_B}{x_{AB}} = e^{-\Delta u_0/kT} \frac{v_0}{4\pi r_0^2 \, \Delta r}, \tag{7p.21}$$

with a factor $(2v_0/4\pi r_0^2\,\Delta r)$ for the dissociation of element A_2 for which $\sigma = 2$.

Since one is interested only in the dissociation of a diatomic molecule at temperatures for which some dissociation occurs, one usually finds that the semiclassical approximation is quite good; many vibrational states are excited. For the dissociation of a compound more complicated, such as CH_3I to CH_3+I, no great error is introduced by assuming that the internal partition function contributions due to the CH_3 group are the same in products and reactants even if these contributions are strictly quantum mechanical. However, if this is done, the volume allowed the I atom in CH_3 is the product of three Δr's of vibration rather than $4\pi r_0^2\,\Delta r$, since for fixed position and orientation of the CH_3 group the I atom has one stretching and two equal bending modes of vibration.

The volumes that enter eq. (21) for the simple dissociation reaction in eq. (20) can be accurately evaluated from spectroscopic information. It is convenient to use angströms (1 Å $=10^{-8}$ cm) as a length unit and 10^{-24} cm^3 as a volume unit. The volume per molecule v_0 in the gas is quite nearly

$$v_0 = 4\times 10^4\, P_{\text{atm}}^{-1} \frac{T}{300} \quad \text{Å}^3. \tag{7p.22}$$

From $B_0 = h/8\pi^2\mu r_0^2 c$ [eq. (7i.1)], with $I_0 = \mu r_0^2$ [eq. (7h.7)], one can relate r_0^2 to B_0 and the reduced mass, $\mu = m_1 m_2/(m_1+m_2)$. Using atomic weight units, $M_i = N_0 m_i$, and

$$M_\mu = N_0 \mu = \frac{M_1 M_2}{M_1 + M_2}, \tag{7p.23}$$

one finds numerically that, with B_0 in cm^{-1},

$$4\pi r_0^2 = 21 \frac{10}{M_\mu B_0} \quad \text{Å}^2. \tag{7p.24}$$

The values of B_0 vary from 59 cm^{-1} for H_2, 1.99 cm^{-1} for N_2 and 1.48 cm^{-1} for O_2 down to 0.0373 for I_2, but $M_\mu B_0/10$ is nearer unity and a little more constant being 2.85 for H_2, 1.4 for H_2, 1.2 for 0_2 but down to 0.23 for I_2. The frequency $\nu = \omega c$ of vibration is related to the force constant κ in eq. (17) for Δr by $\nu = (\frac{1}{2}\pi)(\kappa/\mu)^{1/2}$ so that Δr can be related to the wave number ω, μ, and T. The numerical relation is

$$\Delta r = 0.066 \times \frac{1000}{\omega}\left(\frac{10}{M_\mu}\right)^{1/2}\left(\frac{T}{300}\right)^{1/2} \quad \text{Å}. \tag{7p.25}$$

The values of $(1000/\omega)(10/M_\nu)^{1/2}$ are reasonably close to constant and near unity, being 1.1, 0.5, 0.7, and 1.9 for H_2, N_2, O_2, and I_2, respectively. A quite good empirical relation between $\omega^2 M_\nu$ and the energy D of dissociation per mole leads to the very simple relation

$$\Delta r = \left(\frac{RT}{D}\right)^{1/2} \quad \text{Å}, \tag{7p.26}$$

which is a fairly good alternative to eq. (25).

The quantity $Q_{\tau\zeta}^{\text{cl}}$ of eq. (13) having the dimensions of volume to the power ν_ζ is the classical approximation to

$$Q_{\tau\zeta} = e^{\beta u_{0\zeta}} \frac{V}{N\lambda_\zeta^3} Q_{\zeta i} \prod_{i<\zeta} \lambda_i^3, \tag{7p.27}$$

in which $Q_{\zeta i}$ is the (dimensionless) internal partition function of the molecule ζ, λ_ζ is the de Broglie wavelength of the molecule, and $\prod \lambda_i^3$ is the product over the ν_ζ atoms in the molecule ζ of their individual de Broglie wavelengths cubed. It is only in the classical approximation of eq. (13) that it is obvious to interpret this quantity as the product of the volumes within which the atoms can move in the free molecule confined to the total volume per molecule, $v_0 = V/N$,

$$Q_{\tau\zeta} = \left\langle \prod_i v_i \right\rangle. \tag{7p.28}$$

If the potential energy U in eq. (13) is approximated by having the value zero in the neighborhood of the equilibrium configuration, and infinity elsewhere, then the integral of eq. (13) gives precisely the ω_ζ-dimensional volume for which $U = 0$. For a real potential the configurations for which $U > 0$ are diminished by the probability factor $e^{-\beta v}$ that the configuration be reached, and $Q_{\tau\zeta}^{\text{cl}}$ is an averaged product of volume.

In a rather sophisticated treatment of the quantum-mechanical equations involving the Wigner representation of the density matrix (see Chapter 13), one can draw a similar interpretation of $Q_{\tau\zeta}$ from eq. (27). The mass-action constant in mole fractions [eq. (16)] is then, with eq. (28),

$$K_x = \prod_\zeta x_\zeta^{n_\zeta} = e^{-\beta \Delta u_0} \frac{\langle \prod_i v \text{ (products)} \rangle}{\langle \prod_i v_i \text{ (reactants)} \rangle}. \tag{7p.29}$$

In the limit $T \to 0$, $\beta \to \infty$, the classical integral of eq. (13) goes to zero, except for monatomic species, since the potential energy must be greater than zero except at one minimum position along at least one coordinate. Since $Q_{\zeta i}$ can never be less than unity, $Q_{\tau\zeta}$ of eq. (27) remains finite. The quantum-mechanical lengths have a minimum value due to the uncertainty principle that $\Delta p \, \Delta q \geq h$. We can estimate this by equating the potential energy $\frac{1}{2}\kappa \, \Delta r^2$ at the amplitude Δr to the zero-point energy $\frac{1}{2}h\nu$ and using the relation for the classical frequency, $\nu = (1/2\pi)(\kappa/\mu)^{1/2}$, to relate Δr to $\omega = \nu/c$ and $M_\mu = N_0 \mu$. One finds

$$\Delta r \simeq 0.6 \times 10^{-7} M_\mu^{-1/2} \omega^{-1/2} \quad \text{cm}, \tag{7p.30}$$

or with

$$\left[\left(\frac{10}{M_\mu}\right)^{1/2} \frac{1000}{\omega}\right]^{1/2} = 1,$$

$$\Delta r = 0.06(10/M_\mu)^{1/4} \quad \text{Å}. \tag{7p.30'}$$

For H_2 this is about the value the classical equation (25) gives at $T = 1200$ K. The quantum-mechanical lengths are larger than those computed from the classical equation (25), approaching them asymptotically from above at high temperatures. A rough rule is that for $\Delta r \geq \lambda$ the classical equation (25) is adequate. Using eq. (7d.3′), where $\lambda = (300/M_\mu T)^{1/2}$, with eq. (25) for Δr and assuming that $(1000/\omega)$ $10/M_\mu)^{-1/2}$ is unity, one finds that this requires

$$T \geq 4500 \, M_\mu^{-1/2} \quad \text{K}, \tag{7p.31}$$

which is 6350 K for H_2 with $M_\mu = \frac{1}{2}$, but only about 560 K for I_2.

CHAPTER 8

DENSE GASES

8a. INTRODUCTION, THE POTENTIAL ENERGY

A simple order-of-magnitude calculation leads us to expect that at T_{bp}, the 1-atm boiling point of a liquid of low molecular weight, the equilibrium vapor will deviate from perfect gas behavior by possibly as much as 10% if T_{bp} is not too far from room temperature. The gas volume per mole is of the order of 25×10^3 cm^3 at $T = 300$ K, $P = 1$ atm, whereas that of the liquid is more nearly 25 cm^3; the liquid is of the order of 1000-fold denser. In the liquid each molecule is surrounded by about 10 others at close to the distance r_0, for which the pair potential has a minimum value, but in the 1000-fold less dense gas we expect a 1000-fold fewer neighbors; approximately 1 molecule in 100 have a random chance of having a neighbor at nearly the distance r_0. Actually, if the pair potential is $-u_0$ at $r = r_0$, the probability of a neighbor is increased over the random value by a factor of the order of $e^{u_0/kT}$. Now with 10 neighbors per molecule the energy of vaporization per mole ΔE_0 is a little more than $5N_0u_0$, with N_0 Avogadro's number. From Trouton's rule $\Delta H_0 = \Delta E_0 + RT = 11\,RT = 11\,N_0kT$, so that u_0/kT may be nearly as great as 2; $e^{u_0/kT_{\mathrm{bp}}} \cong e^2 \cong 7.5$, and the number of molecules with a close neighbor may be expected to be as great as 7%. For the same gas at $T = 2T_{\mathrm{bp}}$ and $P = 1$ atm the gas density is half as great and $e^{u_0/2kT_{\mathrm{bp}}} \cong e \cong 2.7$, so that more nearly 1% of the molecules are expected to be paired.

The empirical experimental values at the melting point T_m are indeed not far from $(u_0/kT_m) \cong 2$, but at the 1-atm boiling point T_{bp} one has more nearly $(u_0/kT_{\mathrm{bp}}) \cong 1.15$. These values appear to be fairly characteristic not only for the noble gases without internal degrees of freedom, but also for many molecules that have no large dipole moments. The critical temperature T_c is approximately three-halves of T_{bp}, so that $u_0/kT_c \cong 0.77$ and $\exp(-u_0/kT_c) \cong 2.16$. At the critical volume V_c the critical pressure P_c is very far from the perfect gas value; we have $P_cV_c/NkT_0 \cong 0.3$ instead of unity.

We have very little reliable and precise information about the form of the total potential energy $U_N(\mathbf{r}_1, \mathbf{r}_2, \ldots, \mathbf{r}_N)$ in a system of N molecules. For molecules with internal degrees of freedom U_N must depend, to some extent at least, on the internal quantum states of the molecules and not only on the positions $\mathbf{r}_1, \mathbf{r}_2, \ldots, \mathbf{r}_N$ of their centers of mass. However, empirically the law of corresponding states, according to which P/P_c is a function of V/V_c which is almost the same for many molecular gases including the noble gases with no internal degrees of freedom, indicates that $u_0^{-1}U_N(r_0^{-1}\mathbf{r}_1, r_0^{-1}\mathbf{r}_2, \ldots, r_0^{-1}\mathbf{r}_N)$, with u_0, r_0 depending on the molecular species, must be close to a universal function for these molecules. It is known[1] theoretically that for neutral spherical or nearly spherical molecules without dipole moments the potential $u(r)$ between a pair as a function of their distance r must fall off to zero for large r-values as r^{-6}, and that for all distances r_{ij} between pairs of molecules large enough the potential of many molecules is the sum of that between the pairs.[2] We have that, for identical molecules,

$$\left.\begin{aligned} U_N &\to \sum_{N\geq i>j\geq 1}\sum u_2(r_{ij}), \\ u_2(r_{ij}) &\to -u_0\left(\frac{r_{ij}}{r_0}\right)^{-6}. \end{aligned}\right\} \quad \text{all } r_{ij}/r_0 \gg 1. \tag{8a.1}$$

There is no assumption involved in developing U_N as

$$\begin{aligned} U_N = {} & \sum_{N\geq i>j\geq 1}\sum u_2(r_{ij}) + \sum\sum_{N\geq i>j>k\geq 1}\sum u_3(r_{jk}, r_{ki}, r_{ij}) \\ & + \sum_{N\geq i>j>k>l\geq 1}\cdots\sum u_4(\mathbf{r}_i, \mathbf{r}_j, \mathbf{r}_k, \mathbf{r}_l) + \cdots + \sum\cdots\sum u_N, \end{aligned} \tag{8a.2}$$

since for each U_{N+1} we can add a new term u_{N+1}. However, we need an implicit assumption that in some sense the terms u_n decrease in importance for $n \gg 2$. As we show later in this chapter, the first density correction to the perfect gas law, $PV/NkT = 1 + \alpha(T)(N/V) + \cdots$, depends only on the pair potential $u_2(r)$ in eq. (2), and in a known way as an

[1] F. London, *Z. Phys. Chem.* (B) **11**, p. 222 (1930).

[2] Actually, for very large distances where $u(r)$ is very small the falloff is more rapid, going as r^{-7}.

integral over $\exp[-u(r)/kT]-1$. If $\alpha(T)$ were known with infinite precision for the complete range $0 \leq T \leq \infty$, an inversion of the integral would give $u_2(r)$.[1] One of the first attempts to use experimental information of the second virial coefficient [essentially $\alpha(T)$ above] was made by Lennard-Jones,[2] although obviously the values of $\alpha(T)$ were not infinitely precise nor did they cover a large range of temperature. Lennard-Jones tried $u(r) = Ar^{-m} - Br^{-n}$, A, $B > 0$, $m > n$, and finally suggested what is now known as the Lennard-Jones 6–12 potential which can be written in the form

$$u(r) = 4u_0\left[\left(\frac{\sigma}{r}\right)^{12} - \left(\frac{\sigma}{r}\right)^{6}\right]. \tag{8a.3}$$

This equation gives $u(r)=0$ at $r=\sigma$. Differentiate eq. (3) to find that $r\,du/dt = -4u_0[12(\sigma/r)^{12} - 6(\sigma/r)^6]$ which is zero at $r = 2^{1/6}\sigma$, where $u(r)$ has the value $-u_0$. This is the minimum value of $u(r)$. Set

$$r_0 = 2^{1/6}\sigma, \tag{8a.3$'$}$$

and we can write eq. (3) as

$$u(r) = u_0\left[\left(\frac{r_0}{r}\right)^{12} - 2\left(\frac{r_0}{r}\right)^{6}\right]. \tag{8a.3$''$}$$

Both eqs. (3) and (3″) are used in the literature. It is convenient in interpreting volumes per molecule to remember that, in close packing at the equilibrium distance r_0 between neighboring molecular centers, the volume per molecule is $(r_0/2^{1/6})^3 = \sigma^3$. Since most spherically symmetric molecules have close-packed crystals, the volume per molecule of the crystal is almost σ^3, and that of the liquid at its boiling point is about 17% greater; the critical volume per molecule is about $3\sigma^3 = 2.1r_0^3$.

Although the Lennard-Jones 6–12 potential of eq. (3) or (3′) is certainly not a very good representation of that of any real molecular species, and although the next triple term u_3 in the development of eq. (2) for U_N is surely not zero and may be quite considerable at certain configurations of a triple of molecules, it has become quite standard to assume that

$$U_N(\mathbf{r}_1, \ldots, \mathbf{r}_N) = \sum_{N \geq i > j \geq 1}\sum u(r_{ij}), \tag{8a.4}$$

with $u(r)$ given by eq. (3) or (3′). The greatest competitor of the Lennard-Jones 6–12 form is the less realistic but very convenient hard

(1) See, for instance, Bruce W. Davis, *J. Chem. Phys.* **57**, 5098 (1972).
(2) J. E. Lennard-Jones, *Proc. R. Soc. London, Ser. A*, **106**, 463 (1924).

sphere potential,

$$\begin{aligned} u(r) &= +\infty \qquad r < r_0 = \sigma, \\ &= 0 \qquad r \geq r_0. \end{aligned} \tag{8a.5}$$

One also assumes in applying the equations to other than the noble gases that the potential is independent of the internal quantum states which are assumed to have the same spectrum as in a perfect gas.

One should mention here that there exists a Stockmayer potential[(1)] which corrects eq. (3) for the case of molecules with a permanent dipole moment, and also that for charged ions, as in a plasma, the electrostatic potential $e_i e_j / r_{ij}$ should be added. This latter long-range potential poses an extra problem in the treatment and requires a considerable change in approach (Sec. 8n). We also remark here that in speaking of dense gases we are usually considering temperatures higher than the critical temperature of the gas where u_0/kT is less than 0.77, so that even at r_0, the equilibrium distance, $\exp[-u(r)/kT]$ seldom exceeds 2.

Actually, there is one empirical justification for using the sum [eq. (4)] of pair potentials only. One may compute u_0 from virial coefficient data for the gas using eq. (3) or (3′) for $u(r)$. The potential energy compared to that of the perfect gas of a close-packed crystal, with 12 nearest neighbors at a distance r_0 from each molecule, is $-6N_0u_0$ plus a calculable addition due to more distant neighbors of $-1.38\ Nu_0$ or $-7.38 Nu_0$. After making a minor correction for the zero-point energy $\sum \frac{1}{2} h\nu_\lambda$, the gas u_0 value can be compared with the energy of vaporization extrapolated to 0 K. The agreement is reasonably satisfactory, which indicates that, although the higher terms of $u_3, u_4, \ldots$ in eq. (2) certainly are nonzero, their effect is either small individually or tends to cancel.

The formal development of the next sections does not depend on the use of the Lennard-Jones 6–12 potential of eq. (3), nor even on the assumption of eq. (4) that the total potential is a sum of pair terms. It depends on the implicit assumption that in the extended development of $u_2, u_3, u_4, \ldots$ given by eq. (2) the higher terms u_n, $n \gg 2$, decrease in importance. This appears to be the case for gases composed of saturated molecules. It is not the case, for instance, in a gas composed of hydrogen atoms. In such a gas, when two hydrogen atoms approach closely, a strong saturated valence bond is formed and other atoms are repelled. The energy of three atoms with all three pair distances equal to that of the equilibrium distance of one pair is not nearly as negative as that of three independent pairs at this distance.

(1) W. H. Stockmayer, *J. Chem. Phys.*, **9**, 398 (1941).

8b. THE VAN DER WAALS EQUATION

The first really successful treatment of the equation of state at high pressures was made over 100 years ago by van der Waals.[1] Since the molecules repel each other at short distances, as evidenced by the nonzero volume of liquids and solids, each molecule must exclude a volume b in the system not available to other molecules. The volume V of the system in the perfect gas equation should be replaced by $V-Nb$. But there is also a net attraction between molecules due to longer range forces, since otherwise matter would not condense. This decreases the observed pressure that the molecules exert on the walls. We must add a term to the experimental pressure P in the perfect gas equation. The term to be added to the effect of each molecule on P is proportional to the number density N/V of other molecules, but the total pressure is N/V times an effect for each molecule. We replace P by $P+a(N/V)^2$ and write the van der Waals equation,

$$\left(P+\frac{N^2a}{V^2}\right)(V-Nb)=NkT. \tag{8b.1}$$

If we introduce the volume per molecule,

$$v=\frac{V}{N}, \tag{8b.2}$$

and solve for the pressure, we find

$$P=kT(v-b)^{-1}-av^{-2}. \tag{8b.3}$$

Multiply both sides of this by $v^2(v-b)P^{-1}$ and collect all terms on one side. One finds a cubic expression for v,

$$v^3-[(kTP^{-1})+b]v^2+aP^{-1}v-abP^{-1}=0. \tag{8b.3$'$}$$

If P is plotted as a function of v, it goes from zero at $v=\infty$ to infinity at $v=b$. At high temperatures the pressure decreases monotonically as v increases, but for low values of temperature there is a loop (Fig. 8b.1), so that there are three real roots for v of eq. (3′) for fixed P and a low value of T. As T increases, the pressure for which there are three roots increases and the roots become closer, coalescing to one threefold real value at one temperature T_c and one pressure P_c, the temperature and pressure of the critical point. For $T>T_c$ there is only one real root (and two in the complex plane) for all pressures.

If v_c is the one triply degenerate root of eq. (3′) for $P=P_c$, $T=T_c$, we

[1] J. D. van der Waals, Sr., Doctoral Dissertation, University of Leiden, 1873.

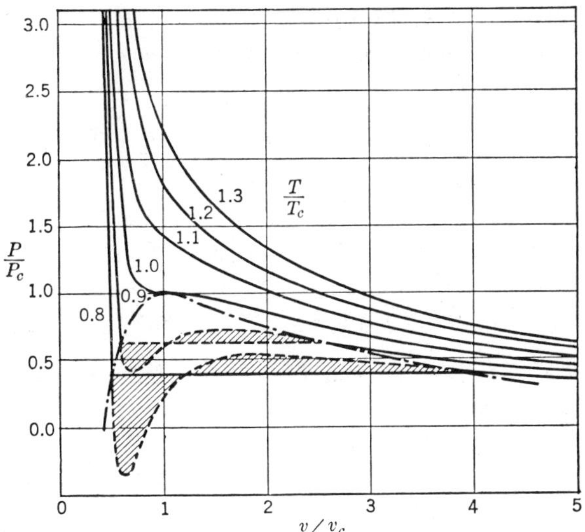

Figure 8b.1

must have

$$(v-v_c)^3 = v^3 - 3v_cv^2 + 3v_c^2v - v_c^3 = 0 \tag{8b.3$''$}$$

as the cubic equation. Equating coefficients in eq. (3′) at P_c, T_c with eq. (3″), we have

$$\begin{aligned} 3v_c &= \frac{kT_c}{P_c} + b, \\ 3v_c^2 &= \frac{a}{P_c}, \\ v_c^3 &= \frac{ab}{P_c}. \end{aligned} \tag{8b.4}$$

If these are solved for a and b,

$$\begin{aligned} a &= 3v_c^2 P_c, \\ b &= \tfrac{1}{3} v_c, \end{aligned} \tag{8b.5}$$

one finds $kT_c = \frac{8}{3}P_cv_c$, or

$$\frac{P_cv_c}{kT_c} = \frac{3}{8} = 0.375. \tag{8b.6}$$

which is higher than the experimental values, which are close to 0.3, but not far off for such a simple equation.

Using the relations of eq. (5) for a and b in van der Waals equation (1), we can express the equation of state for all gases in one relation with reduced dimensionless variables,

$$T^* = \frac{T}{T_c}, \qquad P^* = \frac{P}{P_c}, \qquad v^* = \frac{v}{v_c} = \frac{V}{V_c}, \tag{8b.7}$$

namely, as

$$(v^* - \tfrac{1}{3})\left[P^* + \frac{3}{(v^*)^2}\right] = \frac{8}{3}T^*, \tag{8b.8}$$

or

$$P^* = 8T^*(3v^* - 1)^{-1} - 3(v^*)^{-2}, \tag{8b.8$'$}$$

which of course gives $P^* = 1$ when $T^* = v^* = 1$. Differentiation of eq. (8′) with respect to v^* gives

$$\left(\frac{\partial P^*}{\partial v^*}\right)_{T^*} = -24T^*(3v^* - 1)^{-2} + 6(v^*)^{-3}, \tag{8b.9}$$

which can be written in the form

$$\begin{aligned}\left(\frac{\partial P^*}{\partial v^*}\right)_{T^*} &= -\frac{24(T^* - 1)}{(3v^* - 1)^2} - \frac{6[4(v^*)^3 - (3v^* - 1)^2]}{(v^*)^3(3v^* - 1)^2} \\ &= -\frac{24(T^* - 1)}{(3v^* - 1)^2} - \frac{6(v^* - 1)^2(4v^* - 1)}{(v^*)^3(3v^* - 1)^2}.\end{aligned} \tag{8b.9$'$}$$

Further differentiation gives, from eq. (9),

$$\left[\frac{\partial^2 P^*}{(\partial v^*)^2}\right]_{T^*} = 144T^*(3v^* - 1)^{-3} - 18(v^*)^{-4}. \tag{8b.10}$$

At the critical point, $P^* = 1$, $T^* = 1$, and $v^* = 1$, we see from eq. (9) or (9′) that $\partial P^*/\partial v^* = 0$; the compressibility is infinite. From eq. (10) we find the second derivative $\partial^2 P^*/(\partial v^*)^2$ to be zero at the critical point. The expression of eq. (9′) has two additive terms in $-(\partial P^*/\partial v^*)_{T^*}$, both of which are always positive or zero for $T^* \geq 1$ and allowed values, $v^* > \frac{1}{3}$, of the reduced volume. If $T^* < 1$, the first of the two terms is positive and, since the second term, independent of T^*, is zero at $v^* = 1$, it follows that the equation predicts $(\partial P/\partial v)_T$ to be positive at and near $V = V_c$.

The negative value of the compressibility $-(\partial v/\partial P)_T v^{-1}$ is of course nonphysical. In the region of the looped curve below T_c (Fig. 8b.1) the Helmholtz free energy of the material of uniform density predicted by van der Waals equation is greater than that of a two-phase mixture of the same mass and total volume. The two-phase system is thermodynamically stable. For each temperature draw a horizontal line between the smallest, v_l, and the largest, v_g, of the three roots of eq. (3) for v, at such a pressure

P that the area between the line and the van der Waals curve above and below the line are equal, as in Fig. 8b.1. Since $(dA)_T = -P\,dV$, the difference $A_l(v_l) - A_g(v_g) = \Delta A = PN(v_g - v_l)$ for both the two-phase system and that following the van der Waals curve, and ΔG is zero in both cases. That both $G_l = G_g$ *and* $P_l = P_g$ is the condition of equilibrium. The Helmholtz free energy A is higher on the van der Waals loop (see the discussion in Sec. 5b) than for the two thermodynamic phases.

Van der Waals equation is remarkable in that it gives a very simple two-parameter equation of state which has many of the features of experimental behavior for the gas-liquid region. The critical region is not correctly portrayed. The P-V curve at $T = T_c$ near $V = V_c$ is actually much flatter than the van der Waals cubic, and the envelope of the two-phase region is more nearly an inverted cubic parabola rather than the quadratic parabola that follows from the van der Waals equation.

The Lennard-Jones 6–12 potential of eq. (8a.3) has two characteristic parameters: u_0, the minimum absolute value of the pair potential and σ, the distance between a pair when the potential crosses zero. If it is assumed that all of some simple class of molecules, for instance, all nearly spherical molecules, including diatomic molecules without dipole moments, have the same shape of pair potential, $\beta u(r) = f(\beta u_0, \sigma/r)$, whether or not this is the Lennard-Jones 6–12 potential, it follows readily from the equations of the next section that the P-V-T diagram will be the same for all these gases if plotted as functions P^*, V^*, T^*. This holds remarkably well for all the heavier simple nonpolar gases. It is known as the law of corresponding states.

8c. THE CLASSICAL EQUATION

The natural approach to a theory for a dense gas is to use the equations in such a manner that the first term is that for a perfect gas and succeeding terms measure the effect of the nonzero mutual potential energy between the molecules in terms of ascending complexity. We limit ourselves first to the simplest possible case: a one-component system composed of identical monatomic molecules without internal degrees of freedom. The thermodynamic state of the system is given by the volume V, the temperature T, and the chemical potential μ per molecule. We assume that the classical treatment is adequate and use the grand canonical equation for an ensemble of equilibrium open systems with the same values of V, T, and μ (see Secs. 3i and 3j),

$$e^{\beta PV} = \sum_{N \geq 0}^{\infty} e^{N\beta\mu} \iint_V \cdots \int \frac{d\mathbf{r}^{(N)}\, d\mathbf{p}^{(N)}}{h^{3N} N!} \exp(-\beta H_N), \tag{8c.1}$$

where $\beta = 1/kT$, and H_N is the Hamiltonian for N molecules. The potential energy $U_N(\mathbf{r}_1, \ldots, \mathbf{r}_N)$ in the Hamiltonian is assumed to be a sum of pair terms,

$$u_N = \sum_{N\geq i>j\geq 1}\sum u(r_{ij}), \tag{8c.2}$$

which are functions of the distance, $r_{ij} = |\mathbf{r}_i - \mathbf{r}_j|$, between molecules i and j, and which are assumed to have at least the qualitative characteristics of the Lennard-Jones 6–12 potential in approaching infinity as $r_{ij} \to 0$ and in approaching zero more rapidly than r_{ij}^{-3} as $r_{ij} \to \infty$.

The Hamiltonian is then

$$H_N = \sum_{i=1}^{i=N} \sum_{\alpha=x,y,z} \frac{p_{\alpha i}^2}{2m} + U_N. \tag{8c.3}$$

The integrals in the sum of eq. (1), then, each have $3N$ identical factors into each of which we enter the denominator h,

$$\int_{-\infty}^{+\infty} h^{-1}\, dp_{\alpha i} \exp\left(-\frac{p_{\alpha i}^2}{2mkT}\right) = \left(\frac{2\pi mkT}{h^2}\right)^{1/2} = \lambda^{-1} \tag{8c.4}$$

where λ is the de Broglie wavelength (Sec. 7d). By defining

$$z = \lambda^{-3} e^{\beta\mu}, \tag{8c.5}$$

and what we call a dimensionless configuration integral,

$$Q_\tau(N, V, T) = V^{-N} \int\int \underset{V}{\cdots} \int d\mathbf{r}^{(N)} e^{-\beta U_N} \tag{8c.6}$$

we bring eq. (1) into the form

$$e^{\beta PV} = \sum_{N\geq 0}^{\infty} \frac{(zV)^N}{N!} Q_\tau(N, V, T). \tag{8c.7}$$

In the limit $V \to \infty$ we can neglect, in every term $Q_\tau(N, V, T)$ of finite N, the relatively small part of the configuration space for which U_N is not zero.

$$\lim_{V\to\infty} [Q_\tau(N, V, T)] = 1, \tag{8c.8}$$

so that for small values of z,

$$\lim_{z\to 0} [e^{\beta PV}] = \sum_{N\geq 0} \frac{(zV)^N}{N!} = e^{zV}, \tag{8c.9}$$

which leads to the perfect gas relation

$$\lim_{z\to 0} [z] = \beta P = \frac{N}{V} = \rho, \tag{8c.10}$$

where ρ is the number density,

$$\rho = \frac{N}{V}. \tag{8c.11}$$

The quantity z of dimension l^{-3}, namely, of a concentration, is often called the absolute activity or absolute fugacity and can be defined as

$$z(\rho, T) = \lim_{\rho_0 \to 0} \{\rho_0 \exp \beta[\mu(\rho, T) - \mu(\rho = \rho_0, T)]\}. \tag{8c.12}$$

The word activity was introduced early in this century by G. N. Lewis as a convenient quantity of the dimension of a concentration c. The concentration dependence of the chemical potential in a perfect gas or a perfect solution is given by

$$\mu(c, T) = \mu(c = c_0, T) + kT \ln \frac{c}{c_0}, \tag{8c.13}$$

but this relation fails at high concentrations as the gas or solution becomes imperfect. We can make the equation exact by *defining* the activity a so that

$$\mu(c, T) = \mu(c = c_0, T) + kT \ln \frac{a(c, T)}{c_0}, \tag{8c.13$'$}$$

$$a(c, T) = \lim_{c_0 \to 0} \{c_0 \exp \beta[\mu(c, T) - \mu(c = c_0, T)]\}, \tag{8c.14}$$

where the limit $c_0 \to 0$ is a formal way of indicating that we must find $\mu(c_0, T)$ at a concentration so low that the behavior is perfect. The quantity a, called the activity, can then be used to replace the concentration c in the very useful colligative relations such as the mass-action constant, and the equations now become exact even in the range of c for which the behavior is no longer perfect. Since $a(c) = c$ for the low-concentration perfect behavior region, we can hope to find useful and simple relations by writing $a = c_0[1 + f(\mathbf{c}, T)]$, where f is a small dimensionless function of all the concentrations $c_1, c_2, .1..$ in the gas or solution, such as a power series. The z of eq. (12) is a special case where the concentration c is the number density; $c = \rho = N/V$.

In the case of a gas having internal degrees of freedom for which the internal partition function Q_i [eq. (7b.5)] is not unity, if we assume that there is no dependence of the potential energy U_N on the internal degrees of freedom, then the Hamiltonian H_N of eq. (3) will have N separable independent terms for the internal energy of each molecule. In addition to integrating over all $3N$ momenta, obtaining the factor λ^{-3} for each molecule, we now sum over the internal quantum states for each molecule, obtaining a factor Q_i, and define the absolute activity by

$$z = \lambda^{-3} Q_i e^{\beta\mu}, \tag{8c.15}$$

instead of eq. (5). With eq. (6) for $Q_\tau(N, V, T)$ we again find eq. (7). Since, for the perfect gas μ has the additive term $-kT \ln Q_i$ [eq. (7b.15)], eq. (15) is equivalent to the definition of eq. (12) for z.

8d. THE CLUSTER DEVELOPMENT

We now turn to the evaluation of the partition function integral Q_τ of eq. (8c.6), using the sum of pair term approximation (8c.2) for U_N. With this assumption $e^{-\beta U_N}$ is a product of $\frac{1}{2}N(N-1)$ terms, one for each of the $\frac{1}{2}N(N-1)$ pairs of molecules,

$$e^{-\beta U_N} = \prod_{N \ge i > j \ge 1}\prod e^{-\beta u(r_{ij})}. \tag{8d.1}$$

Define functions of the pair distances,

$$f_{ij} = f(r_{ij}) = e^{-\beta u(r_{ij})} - 1, \tag{8d.2}$$

so that eq. (1) becomes

$$e^{-\beta U_N} = \prod_{N \ge i > j \ge 1}\prod (1 + f_{ij}) = 1 + \sum_{N \ge i > j \ge 1}\sum f_{ij} + \sum \cdots \sum f_{ij} f_{kl} + \sum \cdots . \tag{8d.3}$$

The potential $u(r_{ij})$ is $+\infty$ at $r_{ij} = 0$, so that $f(r_{ij} = 0) = -1$ and remains close to -1 for some distance, rising very rapidly to zero at $r_{ij} = \sigma$, the distance at which $u(r_{ij}) = 0$, continuing steeply to a maximum value $e^{\beta u_0} - 1$ at the minimum value, $-u_0$, of $u(r_{ij})$ at $r_{ij} = r_0$, are then descending to zero for large r_{ij} (Fig. 8d.1). For the Lennard-Jones 6–12 potential of eq. (8a.3) or (8a.3′) the distance, $r_{ij} = r_0$, of the maximum is $r_0 = 2^{1/6}\sigma = 1.2\sigma$ and the descent to zero of $f(r_{ij})$ is as r_{ij}^{-6}. At the critical temperature the maximum of $f(r_{ij})$ is about 1.2, and lower at higher temperatures.

We use the development of the sum of eq. (3) for $e^{-\beta U_N}$ to evaluate the integral over the N-dimensional 3-space of $\mathbf{r}_i$, $1 \le i \le N$, in V. The integral of the first term, unity, is just V^N contributing unity to Q_τ. For one of the $\frac{1}{2}N(N-1)$ terms f_{ij} we integrate first over all the $N-2$ molecular coordinates $\mathbf{r}_k$, $k \ne i, j$, to find the factor V^{N-2}. Then we integrate over $d\mathbf{r}_j$, holding $\mathbf{r}_i$ fixed, far from a wall, obtaining a value we call $2b_2$ of the order of the volume of a molecule. Integration over the last coordinate $\mathbf{r}_i$ gives a factor V if we neglect the small effect due to vessel walls. The total contribution per pair to the integral is $2V^{N-1}b_2$, or to Q_τ is $2b_2/V$, which is of the order of N^{-1} times the ratio of the volume of the condensed phase to that of the gas, in all, very, very small. But there are

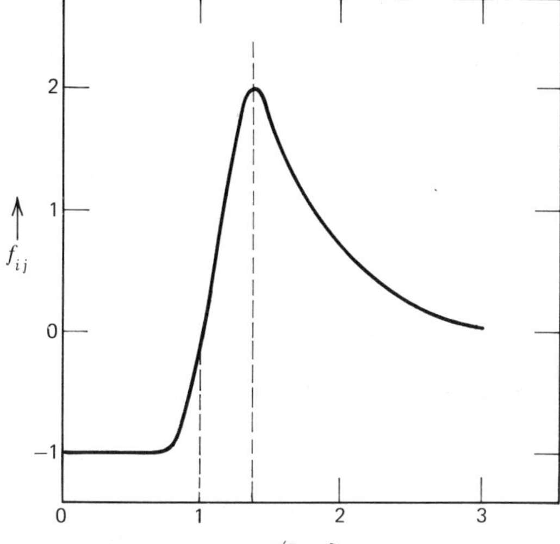

Figure 8d.1 Plot of $f_{ij}(r/\sigma) = e^{-\beta u(r/\sigma)} - 1$.

$\frac{1}{2}N(N-1)$ such terms, and the total contribution of these terms is enormous compared to the first term. Similarly, since the sum of the next terms $f_{ij}f_{kl}$ has the order of N^4 members, their contribution is still larger. We need to be more sophisticated than to consider only the first few terms in the development of eq. (3) and to examine the general term in the series.

For most people, visualization of such a term is aided by consideration of a simple diagrammatic representation of a general term in the sum. The sum includes all possible products of λ *different* f_{ij}'s for all values of λ from zero to $\frac{1}{2}N(N-1)$. We can draw a diagram (Fig. 8d.2) with N numbered circles corresponding to molecules and with a line drawn

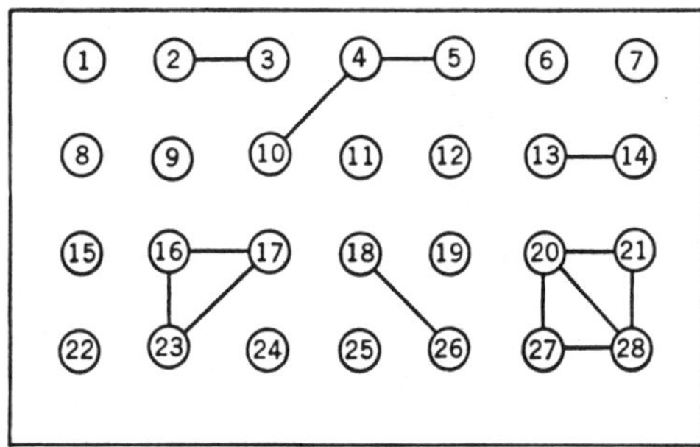

Figure 8d.2 Diagram for the product of 28 molecules $f_{2,3}(f_{4,10}f_{4,5})f_{13,14}(f_{16,17}f_{16,23}f_{17,23}) \times f_{18,26}(f_{20,21}f_{20,27}f_{20,28}f_{21,28}f_{27,28})$; part of b_1^{12}, b_2^3, b_3^2, b_4.

between circles i and j for every f_{ij} in the product. Every term of the sum in eq. (3) can be represented by one such figure, and every figure corresponds to exactly one term in the sum.

The first term, unity, in the sum of eq. (3) corresponds to the figure that has no line. The $\frac{1}{2}N(N-1)$ figures that can be drawn with only one line connecting any two of the numbered circles each correspond to one of the $\frac{1}{2}N(N-1)$ different terms containing only one f_{ij}.

The functions f approach zero for large values of their arguments r_{ij} (large compared to molecular distances of 10^{-8} cm). The contribution to the configuration integral Q_τ of one term arises therefore only from that part of the space for which all the distances represented by a line in the figure are small. We may speak of the molecules connected by lines in the figure, or functions f_{ij} in the term, as being bound to each other in that term.

In any specified figure, that is, any specified product of f_{ij}'s, such as that sketched in Fig. 8d.2, there are groups or clusters of molecules that are all bound to each other directly or indirectly by lines and not bound to any molecules that are not members of the cluster. Such molecules are said to be part of a cluster, and by this criterion every molecule in any figure may be said to be one of a cluster of a certain number of molecules.

The simplest cluster is that consisting of a single molecule not bound to any other, that is, its index does not occur as a subscript to any f in the term. The number of these single clusters of one molecule each, in any term, is designated by ν_1.

A cluster of two consists of a bound pair of molecules neither of which is bound to any other molecule. In the term the two indices i and j of the molecules in one cluster of two occur as subscripts to the same f, but to no other f. The number of such clusters of two is ν_2.

A cluster of three specified molecules, i, j, and k, may be formed in any of four ways.

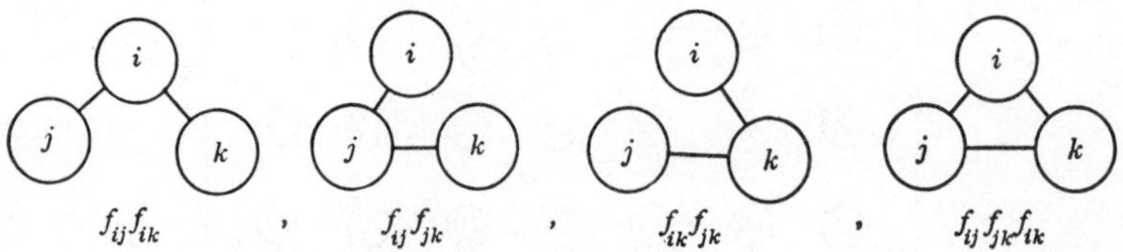

The terms in which the same molecules are bound to each other in clusters have in common the property that they differ from zero only in that part of the configuration space for which the molecules in the same cluster are close to each other. Since the larger clusters may be formed from the same molecules in several ways, there is a considerable number of such terms. We now propose to collect these terms. The sum of all such terms we call a cluster function, symmetric in exchange of the coordinates of the different numbered molecules.

In any term the number of clusters of n molecules is designated ν_n. The total number N of molecules is the sum of the number per cluster n, times the number of clusters of this size ν_n, or

$$N = \sum_{n \geq 1}^{N} n\nu_n \tag{8d.4}$$

The integrals over the molecules in different clusters of one term are independent of each other, since the clusters are so defined that the integrand contains no functions that depend on the coordinates of two molecules in different clusters. The integral of the term is a product of the integrals over the molecules in the same cluster. We sum the integrals of all the products that occur when the same n molecules are in one cluster and designate this the cluster integral b_n after multiplication by a normalization factor $1/n!V$. That is, the cluster integral b_n is defined as

$$b_n = \frac{1}{n!\,V} \int\int \cdots \int \underbrace{\sum \prod_{n \geq i > j \geq 1}}_{\substack{\text{sum over all}\\ \text{products consistent}\\ \text{with single}\\ \text{cluster}}} f_{ij}\, d\mathbf{r}_1 \cdots d\mathbf{r}_n. \tag{8d.5}$$

The dimension of b_n is volume to the power $n-1$. There are at least $n-1$ f's in the product, and at most $\frac{1}{2}n(n-1)$ f's in any term of the integrand of the cluster integral.

The first three cluster integrals are

$$b_1 = \frac{1}{V} \int d\mathbf{r}_1 = 1, \tag{8d.6}$$

$$b_2 = \frac{1}{2V} \int\int f(r_{12})\, d\mathbf{r}_2\, d\mathbf{r}_1 = \frac{1}{2} \int_0^\infty 4\pi r^2 f(r)\, dr, \tag{8d.7}$$

$$b_3 = \frac{1}{6V} \int\int\int (f_{31}f_{21} + f_{32}f_{31} + f_{32}f_{21} + f_{32}f_{31}f_{21})\, d\mathbf{r}_1\, d\mathbf{r}_2\, d\mathbf{r}_3. \tag{8d.8}$$

The first integral b_1 is identically unity. The second integral b_2 was used earlier in this section. The first three terms of the integral b_3 have the same numerical value, since the products differ only in the numbering on the molecules. The value of each is actually $V(2b_2)^2$, a fact that is discussed in greater detail in Sec. 8i.

The value of the integral over $n-1$ of the molecules is independent of the position of the nth, since the integrand drops rapidly to zero for large distances between the molecules. This is true,[1] however, only if n has reasonable values and if the total volume of the system is of macroscopic size, that is, if the ratio V/n is considerably larger than the volume of a single molecule. The integral over the nth particle, then, leads to a factor V which cancels the volume in the denominator of the normalization factor. The cluster integral is consequently independent of the volume of the system, at least as long as n is not too large or V too small.

With this definition of the cluster integrals, the contribution to the configuration integral of all the terms for which the same numbered molecules occur together in clusters is, for ν_n clusters of size n,

$$V^{-N}\prod_{n\geq 1}(n!\,Vb_n)^{\nu_n},$$

and comes only from that part of the configuration space for which the specified numbered molecules are close to each other. We now collect all the identical contributions of this sort that have the same number ν_n of clusters of n molecules each. The total number of these terms is the number of ways in which N numbered objects can be distributed in ν_n unnumbered piles of n objects each, or (Appendix VI)

$$\frac{N!}{\prod_{n\geq 1}\nu_n!\,(n!)^{\nu_n}}.$$

With the product of these two terms, remembering that $N=\sum n\nu_n$ [eq. (4)], we have that the contribution of the terms of fixed value ν_n to $(zV)^N(N!)^{-1}Q_\tau(N, V, T)$ is $\prod_n (Vz^nb_n)^{\nu_n}/\nu_n!$. We must sum this over all values of ν_n consistent with $\sum n\nu_n = N$ to find

$$\frac{(zV)^N}{N!}Q_\tau(N, V, T)=\sum_{\nu_1}\sum_{\nu_2}\cdots\sum_{\nu_n(\sum n\nu_n=N)}\cdots\prod_n\frac{(Vz^nb_n)^{\nu_n}}{\nu_n!}. \tag{8d.9}$$

We now use eq. (9) in eq. (8c.7) for $e^{\beta PV}$ as the sum of eq. (9) over all N from zero to infinity. This summation removes the restriction $\sum n\nu_n = N$ on the values of ν_n in the summation of eq. (9). Since now the summations over the ν_n's for different n-values are independent, we can invert the order of summation and product, writing eq. (8c.7) as

$$\begin{aligned} e^{\beta PV} &= \prod_n\left[\sum_{\nu_n\geq 0}\frac{(Vz^nb_n)^{\nu_n}}{\nu_n}\right] = \prod_n \exp Vb_nz^n \\ &= \exp V\sum_{n\geq 1} b_nz^n \end{aligned} \tag{8d.10}$$

[1] We are interested only in the bulk properties proportional to N, N/V constant. Surface terms proportional to $N^{2/3}$ are always discarded without further apology.

Taking the logarithm of both sides and eliminating the volume, we find an equation for βP,

$$\beta P = \frac{P}{kT} = \sum_{n\geq 1} b_n z^n \tag{8d.10$'$}$$

as a power series in the absolute activity z defined by eq. (8c.5) for monatomic gases, and in general by eq. (8c.12) or (8c.15) for a gas of molecules with a nonunit internal partition function Q_i. In the limit of small z-values we have $z = \rho = N/V$ [eq. (8c.10)] so that, since $b_1 = 1$ [eq. (6)],

$$\lim_{z\to 0} (\beta P) = z = \rho,$$

$$P = \frac{NkT}{V}, \tag{8d.11}$$

the perfect gas relation.

For changes at constant numbers of molecules and constant temperature the change $(dG)_{N,T}$ of the Gibbs free energy G is

$$(dG)_{N,T} = V\,dP, \tag{8d.12}$$

or, for a one-component system $\mu = G/N$, $(d\mu)_T = \rho^{-1}\,dP$, where $\rho = N/V$, so that, with $\beta = 1/kT$,

$$\left(\frac{\partial P}{\partial \mu}\right)_T = \rho = \left(\frac{\partial P}{\partial \ln z}\right)_T \left(\frac{\partial \ln z}{\partial \mu}\right)_T = \beta\left(\frac{\partial P}{\partial \ln z}\right)_T. \tag{8d.13}$$

Using this in eq. (10$'$) for βP we find that

$$\left[\frac{\partial(\beta P)}{\partial \ln z}\right]_\beta = \rho = \sum_{n\geq 1} n b_n z^n, \tag{8d.14}$$

and again, since $b_1 = 1$,

$$\lim_{z\to 0}\left(\frac{z}{\rho}\right) = 1. \tag{8d.15}$$

8e. INTERPRETATION OF THE PRESSURE SERIES

A simple interpretation of the pressure given by eq. (8d.10$'$) is obtained if we ask for the most probable number $N_n^{\max}$ of clusters of size n in the member systems of the ensemble. We find that βPV is just the total most probable number $N_{\text{cl}}^{\max} = \sum_n N_n^{\max}$ of clusters

The probability of a given configuration $\mathbf{r}_1, \mathbf{r}_2, \ldots, \mathbf{r}_N$ in the configuration space for fixed N and T is proportional to $\exp(-\beta U_N)$, and for systems of the ensemble of open systems of fixed μ and T is the sum over

N of those weighted by $z^N/N!$. We have broken the integrand $\exp(-\beta U_N)$ into a sum of terms each characterized by a number ν_n of clusters of size n of molecules, for each n, and integrated over the configuration space. We later examine with more care the physical significance of these clusters, but let us first assume that we can assign a reality to them and examine the expression for the most probable number $N_n^{\max}$ of clusters of size n in the resulting sum. In the limit of large systems, $V\to\infty$, we tacitly assume without proof that, as is usual in macroscopic systems, the most probable numbers $N_n^{\max}$ and the average numbers $\langle N_n\rangle$ become identical.

The probability of the term with the set $\nu_1, \nu_2, \ldots, \nu_n, \ldots$ of clusters is proportional to $T(\nu_1, \ldots, \nu_n, ..) = \prod_n \sum_{\nu_n} (Vb_n z^n)^{\nu_n}/\nu_n!$, and the logarithm of $T(\omega_1, \ldots, \nu_n, \ldots)$ is a sum of independent terms for each value of the cluster size n,

$$\ln T(\nu_1, \ldots, \nu_n, \ldots) = \sum_n \ln \sum_{\nu_n \geq 0} \frac{(Vb_n z^n)^{\nu_n}}{\nu_n!}. \tag{8e.1}$$

The sum for given n under the logarithm depends only on $Vb_n z^n$ and is independent of ν_k for $k \neq n$. The maximum term, $\nu_n = N_n^{\max}$, is that for which

$$\left(\frac{\partial}{\partial \nu_n}\right)\frac{(Vb_n z^n)^{\nu_n}}{\nu_n!} = 0, \qquad \nu_n = N^{\max}, \tag{8e.2}$$

where, for large enough V, we can treat ν_n as a continuous variable. Use the Stirling approximation (Appendix III), $\ln \nu_n! = \nu_n(\ln \nu_n - 1)$, to write

$$\frac{\partial}{\partial \nu_n}\exp\{\nu_n[\ln(Vb_n z^n) - \ln \nu_n + 1]\}$$

$$= \frac{[\ln(Vb_n z^n) - \ln \nu_n](Vb_n z^n)^{\nu_n}}{\nu_n!} = 0, \qquad \text{at } \nu_n = N_n^{\max}, \tag{8e.2$'$}$$

or

$$N_n^{\max} = Vb_n z^n. \tag{8e.2$''$}$$

The most probable *total* number of clusters $N_{\text{cl}}^{\max}$ in the system is

$$N_{\text{cl}}^{\max} = \sum_{n\geq 1} N_n^{\max} = \sum_{n \geq 1} Vb_n z^n. \tag{8e.3}$$

Refer now to eq. (8d.10$'$) for βP and we find that

$$\beta P = \frac{P}{kT} = \frac{N_{\text{cl}}^{\max}}{V}, \tag{8e.4}$$

which is the perfect gas equation with the number N of molecules replaced by $N_{\text{cl}}^{\max}$, the most probable number of clusters in the equilibrium system. The clusters appear to behave like independent entities without interaction between them in their contribution to the pressure.

From eq. (8d.14) for ρ, with $V\rho = \langle N\rangle$, we have with eq. (8e.2″) for $N_n^{\max}$ that

$$\langle N\rangle = V\sum_{n\geq 1} nb_n z^n = \sum_{n\geq 1} nN_n^{\max}, \tag{8e.5}$$

that is, that the total average number of molecules is the sum over all cluster sizes of the number of molecules n in each cluster times the most probable number of clusters of that size. We note that the integrands of b_n [eqs. (8d.5) to (8d.8)] contain the functions $f = [\exp(-\beta u)] - 1$ which are negative for small values of r, so that b_n can be negative, hence also $N_n^{\max}$. As we discuss in the following, this means simply that the number of molecules clustered together in groups of n molecules is *smaller* than a random grouping.

Now obviously since the integrals b_n of eq. (8d.5) come only from that part of the configuration space $\mathbf{r}_1, \mathbf{r}_2, \ldots, \mathbf{r}_n$ in which each of the molecules is close to at least one other and all are linked together by short bonds f_{ij}, there is some relation between the clusters and the intuitive notion of molecules polymerized in bonded n-mers. But the contribution to terms in which molecules i and j are in *different* clusters includes the part of the configuration space for which the distance $r_{ij} = |\mathbf{r}_i - \mathbf{r}_j|$ is small and, indeed, even the position $r_{ij} = 0$, $\mathbf{r}_i = r_j$, contributes. It is because the integrand is independent of the relative positions of the clusters that the simple relation of eq. (4) is found.

The physical significance of the clusters is made clearer if we consider the following problem. Suppose we have given a static array of many mass points in a volume V by a set $\mathbf{r}_1, \mathbf{r}_2, \ldots, \mathbf{r}_N$ of positions and are asked whether the distribution is random, whether they tend to cluster in pairs, or whether they tend to avoid each other. If the number density $\rho = N/V$ is small and there is a considerable number of pairs at a distance apart of approximately $r_0 \ll \rho^{-1/3}$, we can count these and call them paired points, but we will have an ambiguity in knowing at what distance to cut off the counting.

Now count around each molecule the total number $Nn(r)\,dr$ of other molecules at a distance between r and $r+dr$ so that $n(r)\,dr$ is the average number in this distance range around any one. Define $g(r)$ by setting $n(r) = \rho 4\pi r^2 g(r)$. If the distribution is purely random, the value of $g(r)$ will be unity, since the random number in the volume $4\pi r^2\,dr$ will be $\rho 4\pi r^2\,dr$ [more strictly, $V^{-1}\int \rho' 4\pi r^2\,dr = \rho' = (N-1)/V = \rho(1-N^{-1}) \cong \rho$]. The number in *excess* of random between r and $r+dr$ is then

$\rho 4\pi r^2\, dr[g(r)-1]$, and the total average number of pairs in excess of random around each molecule will be the integral $\rho\int_0^\infty 4\pi r^2\, dr[g(r)-1] = 2b_2'$ if $g(r)$ goes sufficiently rapidly to unity for large r. The total number of paired points is N times b_2', since we have counted each pair twice, once with one of the pair as center and once with the other as center.

The procedure above is not precisely the one that defines our clusters at high densities, because there is presumably an excess of random number of triples close to each other, and any set of three or more influences the number of close pairs. If, however, we replace $g(r)$ by the influence in an equilibrium ensemble of the extra weighting $e^{-\beta u_2(r)}$ due *only* to the one neighbor (and appropriately manipulate ρ), we find that the probability of there being pairs *in excess* of random is proportional to $\frac{1}{2}\int 4\pi r^2\, dr[e^{-\beta u(r)}-1] = b_2$ [eqs. (8d.2) and (8d.7)], where we have *not* included the effect due to triples and higher clusters. Similarly, for the triples we use $\exp[-\beta U_3(\mathbf{r}_1, \mathbf{r}_2, \mathbf{r}_3)] = (f_{23}+1)(f_{31}+1)(f_{12}+1)$, but subtract from this the excess probability of the three pairs $f_{23}+f_{31}+f_{12}$ with one random point and unity for the random probability for three independent points. With the factor $1/3!$ for multiple counting the integral is b_3 [eq. (8d.8)].

Thus the clusters in the equations represent the numbers of molecules *in excess of the random number*, due only to the mutual potential energy of the molecules in the cluster, and subtracting the excess already accounted for in clusters of smaller size.

8f. GENERALITY OF THE CLUSTER DEVELOPMENT

The treatment in Sec. 8d was limited to the simplest possible realistic system. The assumptions made can be listed as follows.

1. Classical mechanics is adequate, and there is no need to use quantum-mechanical summations over states.

2. The potential energy $U_N(\mathbf{r}_1, \ldots, \mathbf{r}_N)$ for N molecules is a sum of pair terms $u(r_{ij})$ dependent only on the distance r_{ij} between the molecules i and j.

3. If there are excited internal degrees of freedom, U_N will be a function of the positions $\mathbf{r}_1, \ldots, \mathbf{r}_N$ only, independent of the internal excitations; in short, the Hamiltonian is separable in the internal degrees of freedom and the translational ones.

4. The system is a one-component system consisting of only one kind of identical molecules.

5. At large distances $u(r)$ goes to zero more rapidly than r^{-3}.

The first three of these restrictions can readily be removed without any alteration in the formal part of the development, but at the cost of far greater difficulty in the numerical evaluation of the coefficients b_n for $n \geq 2$. This difficulty is partly due to ignorance in cases 2 and 3 of what to assume in any real system for the Hamiltonian. The fourth restriction to a one-component system can be readily resolved by a more sophisticated notation to result in almost identical-looking equations. The fifth restriction requires more change in the formalism and becomes necessary only in the treatment of a plasma consisting of electrically charged ions. In this case, if the total plasma is neutral, there must be at least two kinds of ions, positive and negative, with mutual pair potentials $z_a z_b e^2/r$. The individual cluster integrals for pairs b_{aa}, b_{ab}, and b_{bb} all go to infinity but cancel each other, since b_{aa} and b_{bb} are positive and $b_{ab} = b_{ba}$ are negative. By reordering the sum one finally obtains a series in the square root of the number density, of which the first term is the Debye–Hückel correction. We examine the restrictions one by one, the fourth restriction in Sec. 8g and the fifth in Sec. 8n.

1. In the classical treatment the probability density $W_N(\mathbf{r}_1, \mathbf{r}_2, \ldots, \mathbf{r}_N)$ in the $3N$-dimensional coordinate space $\mathbf{r}_1, \mathbf{r}_2, \ldots, \mathbf{r}_N$ is proportional to $\exp[-\beta U_N(\mathbf{r}_1, \mathbf{r}_2, \ldots, \mathbf{r}_N)]$. In Chapter 13 we define and discuss a density matrix whose diagonal elements in the coordinate representation are equivalent to the quantum-mechanical probability density $W_N(\mathbf{r}_1, \ldots, \mathbf{r}_N)$. With a known factor A, AW_N goes over to $\exp(-\beta U_N)$ in the classical limit that $h \to 0$. If we know AW_N for $N = 2, 3, \ldots$, we can write its logarithm as a sum,

$$\ln AW_N = \sum_{N \geq i > j \geq 1}\sum \omega_2(r_{ij}) + \sum_{N \geq i > j > k \geq 1}\sum \omega_3(r_i, r_j, r_k) + \cdots, \tag{8f.1}$$

as we did for W_N in eq. (8a.2). As we show in Chapter 13, if the classical potential energy $U_N(\mathbf{r}_1, \ldots, \mathbf{r}_N)$ is a sum of pair terms, then $\ln AW_N$ of eq. (1) is also a sum of pair terms only; $\omega_3(\mathbf{r}_i, \mathbf{r}_j, \mathbf{r}_k)$ and all higher terms vanish. If the classical potential U_N includes triple terms but none higher the sum in eq. (1) terminates at ω_3. The ω_ν functions are sums of increasing derivatives of the corresponding classical u_ν functions below with coefficients proportional to increasing powers of Planck's constant h (actually βh^2), and retain the same qualitative features of convergence for large distances that the classical potential functions possess. The interpretation of Sec. 8e is still valid.

We note that at low temperatures, where ω_n differs appreciably from $-u_n/kT$, the temperature dependence is *not* as T^{-1}. It is known empirically that helium, and even neon to a lesser extent, show a deviation in behavior from the more classical heavier noble gases, but that nothing very drastic is different in the gaseous portion of the equation of state (see also Secs. 7a and 7b on Bose–Einstein versus Boltzmann counting).

2. The potential energy U_N is almost certainly not given exactly as a sum of pair terms, but we can always formally write

$$U_n = \sum_{N\geq i>j\geq 1}\sum u_2(r_{ij}) + \sum_{N\geq i>j>k\geq 1}\sum u(\mathbf{r}_i, \mathbf{r}_j, \mathbf{r}_k) + \sum_{N\geq i>j>k>l\geq 1}\sum\sum\sum u_4(\mathbf{r}_i, \mathbf{r}_j, \mathbf{r}_k, \mathbf{r}_l) + \cdots, \tag{8f.2}$$

where

$$u_2(r_{ij}) = U_2(r_{ij}), \tag{8f.2$'$}$$

$$u_3(\mathbf{r}_i, \mathbf{r}_j, \mathbf{r}_k) = U_3(\mathbf{r}_i, \mathbf{r}_j, \mathbf{r}_k) - U_2(r_{jk}) - U_2(r_{ki}) - U_2(r_{ij}), \tag{8f.2$''$}$$

$$\begin{aligned} u_4(\mathbf{r}_i, \mathbf{r}_j, \mathbf{r}_k, \mathbf{r}_l) = {} & U_4(\mathbf{r}_i, \mathbf{r}_j, \mathbf{r}_k, \mathbf{r}_l) - U_3(\mathbf{r}_j, \mathbf{r}_k, \mathbf{r}_l) - U_3(\mathbf{r}_i, \mathbf{r}_k, \mathbf{r}_l) \\ & - U_3(\mathbf{r}_i, \mathbf{r}_j, \mathbf{r}_l) - U_3(\mathbf{r}_i, \mathbf{r}_j, \mathbf{r}_k) + U_2(r_{ij}) + U_2(r_{ik}) \\ & + U_2(r_{il}) + U_2(r_{jk}) + U_2(r_{jl}) + U_2(r_{kl}). \end{aligned} \tag{8f.2$'''$}$$

The general formula for u_n is proved in Sec. 9e to be that which one can guess from the above, namely, that u_n is a sum of $(-1)^{n-\nu}U_\nu$ for all possible subsets of ν molecules out of the set of n, with $U_1 \equiv 0$.

The pair potential $U_2(r)$ goes to zero for large r-values. Similarly, the potential U_3 for a triple becomes $U_2(r_{ij})$ if the molecule k is distant from the other two, and of course goes to zero if all three are far apart. From eq. (2$''$) we see that $u_3(\mathbf{r}_i, \mathbf{r}_j, \mathbf{r}_k)$ is zero unless all three molecules, i, j, k, are reasonably close. Similarly, if r_{ij} is small and r_{kl} is small but the distances r_{ik}, r_{il}, r_{jk}, and r_{jl} are all large, $U_4(\mathbf{r}_i, \mathbf{r}_j, \mathbf{r}_k, \mathbf{r}_l) \to U_2(r_{ij}) + U_2(r_{kl})$, and if i, j, k are close but if l is far from them, $U_4 \to U_3(\mathbf{r}_i, \mathbf{r}_j, \mathbf{r}_k)$. From eq. (2$'''$) we see that u_4 goes to zero if the positions are such that *any* distance becomes excessively large. In Sec. 9e this relation is proved in general for all u_n, namely, that if the coordinate positions $\mathbf{r}_1, \mathbf{r}_2, \ldots, \mathbf{r}_n$ break into two or more subsets of distant groups, u_n approaches zero value.

Define

$$f_n(\mathbf{r}_1, \ldots, \mathbf{r}_n) = \{\exp[-\beta u_n(\mathbf{r}_1, \ldots, \mathbf{r}_n)]\} - 1 \tag{8f.3}$$

so that the f_n's have the property of going to zero unless all the n molecules are clustered in space. We now have

$$e^{-\beta U_N} = \left[\prod_{N\geq i>j\geq 1}\prod (1+f_{ij})\right]\left[\prod_{N\geq i>j>k\geq 1}\prod\prod (1+f_{ijk})\right]\left[\prod\cdots\prod(1+f_{ijkl})\right]\cdots \tag{8f.4}$$

and can expand the product as a sum of products of f's, collecting as before all products for n specified molecules which are nonzero only with the members all close together. The change is such as to add $f_{ijk}\exp[-\beta u_3(\mathbf{r}_1, \mathbf{r}_2, \mathbf{r}_3)]$ to the integrand in eq. (8d.8) for b_3, and somewhat more complicated additions for b_n, $n>3$. The rest of the procedure is unchanged

In the first three assumptions discussed the fact that they are unnecessary and can be eliminated is more abstract than real, at least at the present time. Our knowledge of the nature of the triplet term $u_3(\mathbf{r}_i, \mathbf{r}_j, \mathbf{r}_k)$ is so primitive that its inclusion would be of no value. In practice, for numerical evaluation one uses the sum of pair potentials with classical mechanics and usually the Lennard-Jones 6–12 potential of eq. (8a.3) or (8a.3″). The influence of internal quantum levels is neglected, except insofar as such attributes of the molecules as dipole moments are approximated by a different potential. Empirically, the close similarity of most gaseous equations of state when scaled with V/V_c and T/T_c offers some justification.

8g. MULTICOMPONENT SYSTEMS

Assume that the open systems of the grand canonical ensemble contain more than one species $a, b, \ldots$ of molecules and that the internal degrees of freedom are uncoupled to the mutual potential of the centers of mass of the molecules. The pressure equation is

$$e^{\beta PV} = \sum_{N_a \geq 0, N_b \geq 0}\sum \cdots [\exp(N_a\beta\mu_a + N_b\beta\mu_b + \cdots)](Q_{\text{int}}^{(a)})^{N_a}(Q_{\text{int}}^{(b)})^{N_b}\cdots$$

$$\times \iint_V \cdots \int \frac{d\mathbf{r}^{(N)}\,d\mathbf{p}^{(N)}}{h^{3N}N_a!\,N_b!\cdots}\exp(-\beta H_{N_a,N_b,\ldots}), \tag{8g.1}$$

where $N = N_a + N_b + \cdots$ is the total number of molecules in each term. Integrate over all the $3N$ momentum components of the molecular centers of mass including the factor h^{-1} in each term, as in Sec. 8c. With the de Broglie wavelength λ_a for species a of molecule,

$$\lambda_a = \left(\frac{h^2}{2\pi m_a kT}\right)^{1/2}, \tag{8g.2}$$

a factor of $\lambda_a^{-3N_a}\lambda_b^{-3N_b}\cdots$ is introduced in each term of given N_a, N_b, $\ldots$.

Define the absolute activity for species a, as we do in eq. (8c.15), to be

$$z_a = \lambda_a^{-3} Q_{\text{int}}^{(a)} e^{\beta\mu_a} \tag{8g.3}$$

and also define, as in eq. (8c.6), the dimensionless configuration integral for each set N_a, N_b, $\ldots$ of molecules,

$$Q_\tau(N_a, N_b, \ldots, V, T) = V^{-N}\iint_V \cdots \int d\mathbf{r}^{(N)}\exp(-\beta U_{N_a,N_b,\ldots}). \tag{8g.4}$$

Instead of eq. (8c.7) we now have

$$e^{\beta PV} = \sum_{N_a \geq 0, N_b \geq 0}\sum \cdots \frac{(z_a V)^{N_a}}{N_a!}\frac{(z_b V)^{N_b}}{N_b!} \cdots Q_\tau(N_a, N_b, \ldots, V, T). \quad (8g.5)$$

The notation $N_a, N_b, \ldots$ is tiresome. We can shorten it by using boldface letters for sets of numbers,

$$\begin{aligned} &\mathbf{N} \equiv N_a, N_b, \ldots, &&\mathbf{N}! \equiv N_a!\ N_b! \ldots, \\ &\sum_{\mathbf{N}\geq 0} \equiv \sum_{N_a\geq 0, N_b \geq 0}\sum \cdots, &&\mathbf{N}\cdot\boldsymbol{\mu} \equiv N_a\mu_a + N_b\mu_b + \cdots, \\ &\mathbf{z} \equiv z_a, z_b, \ldots, &&\mathbf{z}^{\mathbf{N}} \equiv z_a^{N_a} z_b^{N_b} \cdots = \exp \mathbf{N}\cdot \ln \mathbf{z}, \end{aligned} \quad (8g.6)$$

so that eq. (8g.5) becomes

$$e^{\beta PV} = \sum_{\mathbf{N}\geq 0} \frac{(\mathbf{z}V)^{\mathbf{N}}}{\mathbf{N}!} Q_\tau(\mathbf{N}, V, T), \quad (8g.5')$$

identical in appearance to the one-component eq. (8c.7) except for the boldface letters.

We can now proceed to the steps in Sec. 8d. Assume the potential energy $U_{\mathbf{N}}$ is a sum of pair terms as in eq. (8c.2), but now the function $u(r_{ij})$ is different, $u_{ij}(r_{ij})$ depending on the molecular species of the two molecules.

For two species there are three different analytic forms of the pair potential function, $u_{aa}(r_{ij})$, $u_{ab}(r_{ij}) = u_{ba}(r_{ij})$, and $u_{bb}(r_{ij})$, depending on whether i and j are both species a, one is a and the other b, or both b, and for n species $\frac{1}{2}n(n-1)$ forms of f_{ij} of eq. (8d.21),

$$f_{ij} = \{\exp[-u_{ij}(r_{ij})]\} - 1 = f_{ji}, \quad (8g.7)$$

but in counting the sum of $\exp(-\beta U_{\mathbf{N}})$, since either i or j can be a or b, the terms ab have weight 2 for two components and in general $n!/\mathbf{n}!$.

As in eq. (8d.3), we express $\exp(-\beta U_{\mathbf{N}})$ as a product of the $\frac{1}{2}N(N-1)$ different f_{ij}'s and develop the product as a sum of terms, collecting all terms in which the same numbered molecules are linked together in clusters. The resulting cluster functions are now characterized by a set, $\mathbf{n} = n_a, n_b, \ldots$, of the numbers $n_a, n_b, \ldots$ of molecules of each species a, $b, \ldots$ and are symmetric in exchange of the coordinates of all molecules of the same species. As in eq. (8d.5), we define a cluster integral for each such set,

$$b_{\mathbf{n}} = \frac{1}{\mathbf{n}!V}\int\int \underset{V}{\cdots} \int \Big[\sum_{\substack{\text{all cluster} \\ \text{products of} \\ \text{a set } \mathbf{n} \text{ of} \\ \text{molecules}}} \prod f_{ij}\Big] d\mathbf{r}_1 \cdots d\mathbf{r}_n, \quad (8g.8)$$

where, as in eq. (6), $\mathbf{n}! \equiv n_a!\, n_b! \cdots$. The contribution to the configuration integral for the product of cluster functions with the same numbered molecules in each cluster and having $\nu_{\mathbf{n}}$ clusters with each set $\mathbf{n}$ of molecules is now

$$V^{-N} \prod_{\mathbf{n} \geq 1} (\mathbf{n}!\, V b_{\mathbf{n}})^{\nu_{\mathbf{n}}},$$

and the number of ways the set $\mathbf{N}$ of molecules can be distributed in these clusters is

$$\frac{N!}{\prod_{\mathbf{n} \geq 1} \nu_{\mathbf{n}}!\, (\mathbf{n}!)^{\nu_{\mathbf{n}}}},$$

so that the contribution of all these terms to $V^N \mathbf{z}^{\mathbf{N}} (\mathbf{N}!)^{-1} Q_\tau(N, V, T)$ is the product, or

$$\prod_{\mathbf{n} \geq 1} \frac{(V \mathbf{z}^{\mathbf{n}} \mathbf{b}_n)^{\nu_{\mathbf{n}}}}{\nu_n!}.$$

We now sum independently over $\nu_{\mathbf{n}}$ from zero to infinity as in eq. (8d.10) and take the logarithm of both sides of eq. (5) to obtain, as in eq. (8d.10′),

$$\beta P = \frac{P}{kT} = \sum_{\mathbf{n} \geq 1} b_{\mathbf{n}} z^{\mathbf{n}}$$
$$= z_a + z_b + \cdots + b_{aa} z_a^2 + b_{ab} z_a z_b + b_{bb} z_b^2 + \cdots. \qquad (8g.9)$$

For a multicomponent system the thermodynamic relation

$$\left(\frac{\partial P}{\partial \mu_a}\right)_{T, \mu_b, \ldots} = \left(\frac{\partial P}{\partial \ln z_a}\right)_{T, z_b, \ldots} \left(\frac{\partial \ln z_a}{\partial \mu_a}\right)_T = \beta \left(\frac{\partial P}{\partial \ln z_a}\right)_{T, z_b, \ldots} = \rho_a \qquad (8g.10)$$

is equivalent to eq. (8d.13) and yields, from eq. (9),

$$\rho_a = \sum_{\mathbf{n} \geq 1} n_a b_{\mathbf{n}} \mathbf{z}^{\mathbf{n}} = z_a + 2 b_{aa} z_a^2 + b_{ab} z_a z_b + \cdots. \qquad (8g.11)$$

If the various species $a, b, \ldots$ of molecules in the multicomponent system are different isotopic compositions of the same chemical formula, the potential energy function U_N is independent of the isotopic composition to a very high precision (see the discussion in Sec. 7o). The functions f_{ij} will then be the same, independent of the molecular species, and the cluster functions for all sets n of the same number, $n = n_a + n_b + \cdots$, of molecules will be the same. The $b_{\mathbf{n}}$'s then differ only in the factor $(\mathbf{n}!)^{-1}$ [eq. (8)] and can be written as

$$b_{\mathbf{n}} = \left(\frac{n!}{\mathbf{n}!}\right) b_n \qquad \text{(isotopes)}. \qquad (8g.12)$$

With a single activity that is the sum

$$z = z_a + z_b + \cdots, \tag{8g.13}$$

using the multinomial expansion for z^n,

$$z^n = (z_a + z_b + \cdots)^n = \sum_{n_a, n_b, \geq 0, \sum_a n_a = n} \sum \cdots \left(\frac{n!}{\mathbf{n}!}\right)\mathbf{z}^{\mathbf{n}}, \tag{8g.14}$$

with eq. (12) in eq. (9), we find that the equation for the pressure is just that for a single chemical component [eq. (8d.10′)] with an activity which is the sum of that for all isotopic compositions. Similarly, from eq. (11) for ρ_a, we find for the total number density,

$$\rho = \rho_a + \rho_b + \cdots = \sum_n (n_a + n_b + \cdots) b_n z^n = \sum_n n b_n z^n, \tag{8g.15}$$

as for a single component [eq. (8d.14)].

8h. SERIES CONVERGENCE AND CONDENSATION

In the foregoing we tacitly assumed that we are dealing with activities z which are sufficiently low so that the pressure series $\sum b_n z^n$ converges rapidly at relatively small values of n and, indeed, in view of the difficulty in numerical evaluation, we might expect that the series would be of little value unless this were the case. In the next section, however, we transform the series into a power series in the number density, and this at least at higher temperatures, permits numerical evaluation where the $b_n z^n$ series would require evaluation for very high n-values.

The cluster integrals b_n of eqs. (8d.5) to (8d.8) were defined for finite volumes V, and it was noted that for small n-values and macroscopic volumes they are expected to be numerically independent of V.

Let us examine the limits of this assertion. The functions $f(r)$ drop rapidly to zero at large values of r, and no great numerical error will be made if we assume $f(r)$ to be exactly zero for r greater than some r_m. For small molecules r_m is of the order of 10^{-7} cm. The cluster function in the integrand of b_n contains $\frac{1}{2}n!$ terms which differ only in the order of the indices in the product $f_{12} f_{23} f_{34} \cdots f_{n-1,n}$ and contribute the same integral to b_n. This function is zero unless the distance r_{1n} between the two end molecules is less than $(n-1)r_m$. For all other products in the sum of products composing the integrand the maximum distance between any pair is less. For a given position $\mathbf{r}$ of the last molecule over whose coordinates the integration is extended, the integration over the positions of the other $n-1$ is independent of V unless $\mathbf{r}$ is within $(n-1)r_m$ of one of the walls, that is, for a cubic box, in the volume $6V^{2/3} n r_m$ close to the six faces of the cube. It follows that, for

$$n \ll \frac{V^{1/3}}{6r_m}, \tag{8h.1}$$

the integrand of the function of **r** which is now to be integrated in the volume V is $n!\, b_n^{(0)}$, independent of V except in the small fraction $6nr_m/V^{1/3}$ of the total volume, and

$$b_n(V) = b_n^{(0)}\left(1 + O\frac{6nr_m}{V^{1/3}}\right) = \lim_{V\to\infty}[b_n(V)] + b_n^{(0)} O\frac{6nr_m}{V^{1/3}}. \tag{8h.1$'$}$$

For $r_m = 10^{-7}$ cm, $V = 1$ cm^3, and $n \leq 100$, the independence of V should hold to about 1 part in 10^4. However, at fairly high temperatures for which the linear chains $f_{12}\cdots f_{n-1,n}$ are numerically important, the value of n for which $b_n(V) = b_n^{(0)}$ increases only with the one-third power of the system size.

If the pressure P is plotted against $\ln z$, the slope $(\partial P/\partial \ln z)_T$ will be $kT\rho$ [eq. (8d.13)] and necessarily positive. With κ for the isothermal compressibility,

$$\kappa = -V^{-1}\left(\frac{\partial V}{\partial P}\right)_T = \rho^{-1}\left(\frac{\partial \rho}{\partial P}\right)_T, \tag{8h.2}$$

the second derivative,

$$\left[\frac{\partial^2 P}{(\partial \ln z)^2}\right]_T = kT\left(\frac{\partial \rho}{\partial \ln z}\right)_T = kT\left(\frac{\partial \rho}{\partial P}\right)_T\left(\frac{\partial P}{\partial \ln z}\right)_T$$

$$= (\rho kT)^2 \rho^{-1}\left(\frac{\partial \rho}{\partial P}\right)_T = (\rho kT)^2 \kappa, \tag{8h.2$'$}$$

is also necessarily positive. The curve has a positive slope and is concave upward. For temperatures below the critical temperature, $T < T_c$, condensation occurs at an activity $z = z_0(T)$; the slope changes from $\rho_g kT$ to $\rho_l kT$, where ρ_g and ρ_l are the number densities of the equilibrium vapor and liquid phase, respectively. Classical thermodynamic theory assumes this change to be discontinuous at $z = z_0(T)$; that is, $\rho(z)$ represented by the power series $\sum nb_n z^n$ of eq. (8d.14) has a mathematical discontinuity on the positive real axis of z at $z_0(T)$ for $T < T_c$. The sum of $b_n z^n$ must diverge at $z = z_0(T)$, which would be the case if

$$\lim_{n\to\infty} |b_n(T)|^{1/n} = [z_0(T)]^{-1}, \qquad T < T_c. \tag{8h.3}$$

Now one can prove, as follows, that for finite volumes the function $\beta P(z)$ can have no singularity on the positive real axis, which contradicts the above assumption of classical thermodynamics.

The expression of eq. (8c.7) for $e^{\beta PV}$ is a power series in activity z with the Nth coefficient $V^N Q_\tau(V, \beta, N)/N!$ real and positive. The series, in this form, is impractical for numerical use, since its largest term has N equal to the most probable number of molecules in the macroscopic volume V, but nevertheless the series must converge for a realistic potential energy between molecules. Indeed, for a molecular pair potential having a hard core, $u(r_{ij})=\infty$, $r_{ij}<r_{\min}$, the series must terminate at close packing, $Q_\tau(V, \beta, N)=0$, $N>N_{\max}=2^{1/2}V(r_{\min})^{-3}$, and replacing the true $u(r_{ij})$ by infinity when $u(r_{ij})>50\ kT$ makes no numerical difference for finite laboratory volumes. But then the series is a polynomial, and a polynomial of real positive coefficients can have no singularity on the real positive axis.

No essential difficulty exists in the explanation, and the analysis is instructive. We can make use of our thermodynamic knowledge of condensation to deduce certain features of the statistical mechanical partition functions, and from this also to show that the systems undergoing condensation should consist of two parts of different densities. The equation for the Helmholtz free energy [eq. (3i.11″)] is

$$\beta A(N, V, T) = -\ln Q_c(V, \beta, N), \tag{8h.4}$$

where Q_c is the petite canonical partition function. For a classical system with the Hamiltonian separable in the internal molecular degrees of freedom and that due to the coordinates and momenta of the molecular centers of mass, we found in Sec. 8c that

$$\begin{aligned} Q_c(V, \beta, N) &= \int\cdots\int d\mathbf{p}^{(N)}\, d\mathbf{r}^{(N)} (h^\Gamma N!)^{-1} \exp(-\beta H_N) \\ &= \left(\frac{Q_{\text{int}}}{\lambda^3}\right)^N \frac{V^N Q_\tau(V, \beta, N)}{N!}, \end{aligned} \tag{8h.5}$$

where Q_τ [eq. (8c.6)] is the dimensionless configuration integral.

The grand canonical ensemble relation [eqs. (3i.12′) and (3i.12″)], which we have used, is

$$\beta PV = \ln Q_g(V, \beta, (-\beta\mu)) = \ln \sum_{N\geq 0} e^{N\beta\mu} Q_c(V, \beta, N). \tag{8h.6}$$

With eq. (8c.15) for the activity $z=\lambda^{-3}Q_i e^{\beta\mu}$ and eq. (5) we have written Q_g as a power series in the activity with coefficients

$$\begin{aligned} J(V, \beta, N) &= e^{N\beta\mu} Q_c(V, \beta, N) z^{-N} \\ &= \left(\frac{\lambda^3}{Q_i}\right)^N Q_c(V, \beta, N) \\ &= V^N Q_\tau(V, \beta, N)/N!, \end{aligned} \tag{8h.7}$$

namely,

$$Q_g(V, \beta, z) = \sum_{N\geq 0} J(V, \beta, N) z^N. \tag{8h.8}$$

Now with eq. (4) and the second expression on the right of eq. (7), using the thermodynamic relation

$$\beta A = N\beta\mu - \beta PV, \tag{8h.9}$$

we can express $J(V, \beta, N)$ in terms of *thermodynamic* functions of V, β, N instead of statistical mechanical integrals as in eq. (7). Using the symbol

$$v = \rho^{-1} = \frac{V}{N}, \tag{8h.10}$$

which now depends on N in the summation over N with fixed V, for the volume per molecule, and $\zeta(v, \beta)$ for the activity[1] of the system with fixed V, β, N, which differs for consecutive N-values,

$$\begin{aligned}\ln J(V, \beta, N) &= N \ln \frac{\lambda^3}{Q_i} + \ln Q_c(V, \beta, N)\\ &= N\left\{-\left[\beta\mu(v) - \ln \frac{\lambda^3}{Q_i}\right] + \beta v P(v)\right\}\\ &= N\{-\ln \zeta(v, \beta) + \beta v P(v, \beta)\}.\end{aligned} \tag{8h.11}$$

With this in eq. (6) we obtain a rather odd-looking relation involving solely thermodynamic functions on both sides,

$$\beta VP(z, \beta) = \ln \sum_{N\geq 0} \left[\frac{z}{\zeta(v, \beta)}\right]^N \exp V\beta P(v, \beta), \tag{8h.12}$$

where we may, if we wish, write P on the right-hand side as a function of β and $\ln \zeta(v, \beta)$.

The logarithm of the ratio of the $(N+1)$th term to the Nth term in the sum of eq. (8), since V is constant and $V[\partial(\beta P)/\partial \ln \zeta] = N$, is

$$\begin{aligned}\ln \frac{J(V, \beta, N+1) z^{N+1}}{J(V, \beta, N) z^N} &= \ln z - \ln \zeta(v, \beta) - \left(\frac{\partial \ln \zeta}{\partial N}\right)\left[N - \beta V \left(\frac{\partial P}{\partial \ln \zeta}\right)_\beta\right]\\ &= \ln z - \ln \zeta(v, T).\end{aligned} \tag{8h.13}$$

[1] The activities z and ζ defined by the two equivalent relations eqs. (8c.12) and (8c.15) have the dimensions of a number density ρ or a reciprocal volume. Since $J(V, \beta, N)z^N$ is dimensionless, J_N has the dimension l^{3N}. In the equations that follow we commit a sin of serious poor taste by writing, on both sides of the equations, logarithms of quantities with physical dimensions. No error will be introduced if we are careful to use the same units on both sides. It would be more elegant to use the dimensionless activity coefficient $\alpha = v_0 z$ numerically equal to z if v_0 is chosen as unity, and J_N/v_0^N, but the elegance appears not to justify the more complicated nomenclature.

Now $\zeta(V/N, \beta)$ increases with N so that the maximum term is at that N for which $\zeta = z$ and $P = P(z, \beta)$. Thus eq. (12) for the grand canonical ensemble gives $P(z, \beta)$ in terms of the logarithm of a sum, and if, as discussed in Sec. 3i [eqs. (3i.14) to (3i.16)], we substitute the logarithm of the largest term for that of the sum, we find that $P(z, \beta)$ is, to high numerical precision, equal to that for a canonical ensemble of fixed number N_m of molecules, where N_m is the most probable number at V, β, and $-\beta\mu$.

The real reason for displaying the form (12) of the grand canonical expression for βPV was, however, to discuss condensation. Experimentally, we know that the activity ζ and pressure P both remain constant at $\zeta = z_0$ from the volume per molecule $v_g = \rho_g^{-1}$ to $v_l = \rho_l^{-1}$, so that between $N = N_g = \rho_g V$ and $N = N_l = \rho_l V$ the value of $\ln J(V, \beta, N)[z_0(T)]^N$ is constant. If, however, $z = z_0(1 - 10^{-12})$, the maximum (and most probable) term will occur at a value of v negligibly greater than v_g, but $\ln Jz^N$ at $N = N_l$ will be less than that of the maximum by an additive negative term $10^{-12}(N_g - N_l)$, and the probability of observing N_l molecules will be proportional to the exponent of this. For the normal boiling point, N_l/N_g is of the order of 10^3, so that $N_l - N_g$ is of the order of the total number of molecules, say 10^{20}. Similarly, for z greater than 1 part in 10^{-12} the probability of observing other than liquid density is vanishingly small. Condensation in a finite system is not infinitely sharp, but in any macroscopic system it is experimentally indistinguishable from being discontinuous.

We have used experimental knowledge of the behavior of systems during condensation to show [eq. (11), last expression on the right] that the logarithm of the coefficient, $\ln J(V, \beta, N) = \ln V^N Q_\tau(V, \beta, N)/N!$, is proportional to N at constant v, β in the condensation range, but have not deduced this from any property of the definition [eq. (8c.6)] of Q_τ as an integral. We now wish to show that this is reasonable, and that it occurs exactly under such circumstances that the molecules occupy separate contiguous macroscopic volumes at differing number densities.

With eq. (11) for $\ln J$, and eq. (7) where $\ln J$ is $\ln[V^N Q_\tau(V, \beta, N)/N!]$, we write

$$\ln\left[\frac{V^N Q_\tau(V, \beta, N)}{N!}\right] = N \ln q(v, \beta), \tag{8h.14}$$

which defines $q(v, \beta)$, with $v = V/N$, the volume per molecule.

The integral of eq. (8c.6), which defines Q_τ, is an integral over all possible positions of the molecules, including those in which an appreciable fraction N_a/N of the N molecules is in some contiguous macroscopic fraction V_a/V of the total volume V, at a volume per molecule v_a greater than V/N, say, at $v_a = v(1+\alpha)$, $\alpha > 0$, and the remaining molecules occupy the remaining volume $V_b = V - V_a$ at a volume per molecule $v_b = v(1-\gamma)$, with $\gamma > 0$. The conditions $N_a + N_b = N$, $V_a + V_b = V$ require that

$$\frac{N_a}{N} = \frac{\gamma}{\alpha+\gamma}, \qquad \frac{N_b}{N} = \frac{\alpha}{\alpha+\gamma}$$
$$\frac{V_a}{V} = \left(\frac{\gamma}{\alpha+\gamma}\right)(1+\alpha), \quad \frac{V_b}{V} = \left(\frac{\alpha}{\alpha+\gamma}\right)(1-\gamma). \tag{8h.15}$$

If $q^*(v, \beta)$ is defined by eq. (14) with the integration of Q_τ limited, in some vaguely specified manner, to that part of configuration space for which the volume per molecule is v in all parts of V, for small α and γ we can write

$$\begin{aligned} N_a \ln q^*(v_a, \beta) &= N_a\left[\ln q^*(v, \beta) + (v_a - v)\frac{\partial \ln q^*(v, \beta)}{\partial v} \right. \\ &\quad \left. + \tfrac{1}{2}(v_a - v)^2 \frac{\partial^2 \ln q^*(v, \beta)}{\partial v^2} + \cdots\right] \\ &= N\left(\frac{\lambda}{\alpha+\gamma}\ln q^* + \frac{\gamma\alpha v}{\alpha+\gamma}\frac{\partial \ln q^*}{\partial v} + \tfrac{1}{2}\alpha^2 v^2 \frac{\partial^2 \ln q^*}{\partial v^2} + \cdots\right), \end{aligned} \tag{8h.16}$$

and a corresponding expression for $N_b \ln q^*(v_b, T)$. In summing the two expressions, the first derivative terms cancel, and

$$\begin{aligned} N_a \ln q^*(v_a, \beta) &+ N_b \ln q^*(v_b, \beta) \\ &= N\left[\ln q^*(v, \beta) + \tfrac{1}{2}\alpha\gamma v^2 \left(\frac{\partial^2 \ln q^*}{\partial v^2}\right)_T + \cdots\right]. \end{aligned} \tag{8h.17}$$

If the second derivative $v^2(\partial^2 \ln q^*/\partial v^2)$ is negative and of the order of unity, even for the extremely small density difference $\alpha = \gamma = 10^{-8}$, with $N = 10^{20}$, the contribution $[q^*(v_a)]^{N_a}[q^*(v_b)]^{N_b}$ to Q_τ will be less than that due to the configurations of uniform density by a factor of the order of $\exp(-10^{-16}N)$. With $N = 10^{20}$ the contribution to Q_τ will come solely from configurations of uniform density, and fluctuations in volumes of macroscopic size will be essentially absent. If, however, $\partial^2 \ln q^*/\partial v^2$ is positive, the reverse will be true. The contribution to Q_τ will come entirely from regions of the configuration space for which macroscopic volumes[1] are occupied at different densities.

[1] Our equations and statements, when N-values are macroscopic, are numerically valid to high *percentual* precision for $\ln J_N$ and for q, but not at all for J_N itself. Factors differing from unity are introduced in the Q_τ integral at the coordinate values of molecules close to walls

Suppose that there is a range of v for which the second derivative is positive. We seek, for fixed N and V, the values of N_g, v_g, N_l, v_l that contribute predominantly to $V^N Q_\tau/N!$. These are the two reciprocal number densities $\rho_g^{-1} = v_g$ and $\rho_l^{-1} = v_l$ of the equilibrium gas and liquid, respectively, and the numbers, N_g and N_l, of molecules in each phase. To do this we maximize $N_g \ln q^*(v_g) + N_l \ln q^*(v_l) = N \ln q(v)$, subject to the two conditions $N_g + N_l = N$ constant and $V_g + V_l = N_g v_g + N_l v_l = V$ constant. The first condition is most easily handled by substituting $N - N_g$ for N_l. The second condition is used by the method of undetermined multipliers (Appendix V), namely, by subtracting $\alpha(N_g v_g + N_l v_l)$ from $N \ln q$, which now for some values of α must give zero for arbitrary independent variations δv_g, δv_l, and δN_g, leading to three independent equations,

$$\delta[N_g \ln q^*(v_g) + (N - N_g)\ln q^*(v_l) - \alpha N_g v_g - \alpha(N - N_g)v_l = 0, \quad (8h.18)$$

or

$$\begin{gathered} N_g\left\{\left[\frac{\partial \ln q^*(v)}{\partial v}\right]_{v=v_g} - \alpha\right\}\delta v_g = 0, \\ (N - N_g)\left\{\left[\frac{\partial \ln q^*(v)}{\partial v}\right]_{v=v_l} - \alpha\right\}\delta v_l = 0, \\ [\ln q^*(v_g) - \ln q^*(v_l) - \alpha(v_g - v_l)]\delta N_g = 0. \end{gathered} \quad (8h.18')$$

These three relations for any N_g between zero and N, hence for all $v_l < v < v_g$, give the same requirement for v_g and v_l. The construction for the solution is shown in Fig. 8h.1. Since the second derivatives $\partial^2 \ln q^*(v)/\partial v^2$ are negative at v_a and v_b, we can replace $\ln q^*$ by $\ln q$ at these values. We have that the two first derivatives are equal,

$$\left[\frac{\partial \ln q(v, \beta)}{\partial v}\right]_{v=v_g} = \left[\frac{\partial \ln q(v, \beta)}{\partial v}\right]_{v=v_l} = \alpha, \quad (8h.19)$$

and that

$$\ln q(v_g, \beta) - \ln q(v_l, \beta) = (v_g - v_l)\left[\frac{\partial \ln q(v, \beta)}{\partial v}\right]_{v=v_g \text{ or } v_l} \quad (8h.19')$$

For fixed values of N and variable $v = V/N$,

$$N_g = N\frac{v - v_l}{v_g - v_l}, \qquad N_l = N\frac{v_g - v}{v_g - v_l}, \quad (8h.20)$$

---

or in the surface between regions of different number density. The number of such factors is large and their total product value differs greatly from unity, but their logarithm is negligible compared to $N \ln q$. Since surface free energies are positive, the factors are less than unity. The large contributions to Q_τ come from configurations of minimum surface, so that the most probable configuration is that with only two compact macroscopic volumes of differing densities.

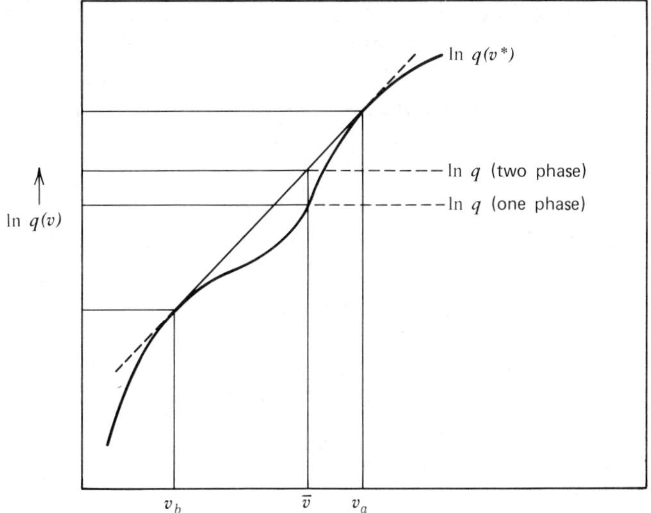

Figure 8h.1 Plot of $\ln q(v)$ versus v in the two-phase region.

$$N \ln q(v,\ \beta) = N\left\{\ln q(v_l,\ \beta) + (v - v_l)\left[\frac{\partial \ln q(v,\ \beta)}{\partial v}\right]_{v=v_l \text{ or } v_g}\right\}, \qquad v_l < v < v_l, \tag{8h.20$'$}$$

and

$$\left[\frac{\partial \ln q(v,\ \beta)}{\partial v}\right]_\beta = -\beta\left(\frac{\partial}{\partial V}\right)_T\left[A - N \ln\left(\frac{\lambda^3}{Q_i}\right)\right]$$
$$= \beta P(v,\ \beta) = \left(\frac{\partial \ln q(v,\ \beta)}{\partial v}\right)_{v=v_l \text{ or } v_g}, \qquad v_l < v < v_g, \tag{8h.21}$$

is constant.

In the grand canonical ensemble $V = Nv$ is constant, and the sum over N is of terms $[q(v,\ \beta)z]^N$. At the activity z of condensation, $z = z_0(\beta)$, the logarithm $N[\ln q(v,\ \beta) + \ln z]$ of the terms for which N lies between Vv_g^{-1} and Vv_l^{-1} is constant. With $dV = v\,dN + N\,dv = 0$ we have $(\partial v/\partial N)_V = -vN^{-1}$,

$$\left(\frac{\partial}{\partial N}\right)_{V\beta}[N \ln q(v,\ \beta)] = \ln q(v,\ \beta) - v\frac{\partial \ln q(v,\ \beta)}{\partial v}, \tag{8h.22}$$

is constant for $v_l < v < v_g$, and the constancy of the terms for different N-values is satisfied if

$$\ln z = \ln z_0(\beta) = -\ln q(v,\ \beta) + v\left[\frac{\partial \ln q(v,\ \beta)}{\partial v}\right]_{v=v_l \text{ or } v_g} \tag{8h.23}$$

A more formal description of the phenomenon of condensation, and indeed of all phase transformations, is due to Yang and Lee,[1] which relates condensation to the distribution of the roots (generally complex), $z = z_k$, of the polynomial $Q_g(V, \beta, -\beta\mu) = 0$, the zeros of the partition function of the grand canonical ensemble.

We start with the general grand canonical equation for a system of finite volume [eq. (8c.7)], making use of the assertion that the sum is finite, all terms for $N > N_m$ being zero. The equation is

$$e^{\beta PV} = Q_g(V, \beta, -\beta\mu) = \sum_{N \geq 0}^{N_m} J_N z^N, \tag{8h.24}$$

with $J_0 = 1$, and all J_N real and positive. The polynomial on the right has N_m roots z_K, $1 \leq k \leq N_m$, none of which lie on the positive real axis and most of which are complex, occurring as complex conjugate pairs, $z_k = z^*_{k'}$, above and below the real axis. We can write $e^{\beta PV}$ as a function of a complex variable z,

$$e^{\beta PV} = \sum_0^{N_m} J_N z^N = B \prod_{k \geq 1}^{N_m} (z - z_k), \tag{8h.25}$$

with $B = J_{Nm}$. At $z = 0$, βPV is zero, $e^{\beta PV} = J_0 = 1$, so that

$$B = \prod_{k \geq 1}^{N_m} (-z_k)^{-1}, \tag{8h.26}$$

and using this eq. (25) becomes

$$e^{\beta PV} = \prod_{k \geq 1}^{N_m} \left(1 - \frac{z}{z_k}\right), \tag{8h.27}$$

$$\beta PV = \sum_{k \geq 1}^{N_m} \ln\left(1 - \frac{z}{z_k}\right). \tag{8h.27$'$}$$

Use the development $\ln(1 - x) = -\sum_{n \geq 1}(x^n/n)$ to write this, for $z < z_k$, as

$$\beta PV = -\sum_{k \geq 1} \sum_{n \geq 1}^{\infty} n^{-1}\left(\frac{z}{z_k}\right)^n = \sum_{n \geq 1}^{\infty} z^n\left(-\sum_{k \geq 1}^{N_m} n^{-1} z_k^{-n}\right), \tag{8h.28}$$

but from eq. (8d.10′) where $\beta P = \sum b_n z^n$ with $b_1 = 1$ we find

$$\begin{aligned} b_1 &= 1 = -V^{-1} \sum_{k \geq 1}^{N_m} (z_k)^{-1}, \\ b_n &= -(nV)^{-1} \sum_{k \geq 1}^{N_m} (z_k)^{-n}. \end{aligned} \tag{8h.29}$$

[1] C. N. Yang and T. D. Lee, *Phys. Rev.*, **87,** 404, 410 (1952).

The sum of the reciprocals of the roots is $-V$, and the b_n's are related to the nth moments of the reciprocals for $n \geq 2$.

The number N_m of roots is very large for any macroscopic volume, considerably larger than the number N_l in the liquid phase at normal pressure, and increases linearly with the volume V. Yang and Lee prove that they lie on a line symmetric in the complex plane above and below the real axis. For odd N_m at least one root must be on the real axis and, since it cannot be on the positive real axis, it must be on the negative real axis. The line of roots crosses the real axis at negative z. From eq. (27′) the pressure is singular for $z - z_k$ at any of the roots, $1 \leq k \leq N_m$, and will be enormously "kinked" if it passes very close to a root. This is exactly what happens for temperatures below the critical, $T < T_c$, for finite V at the real positive activity, $z = z_0(T)$, of condensation. We deduce that for $T < T_c$ the line on which the roots lie crosses the positive real axis at $z = z_0(T)$ with no root *on* the axis but with roots closely spaced above and below. The density of these roots along the line is proportional to the inverse total volume of the system, and presumably the distance of the two nearest complex conjugates from the real axis is also proportional to V^{-1}. In the limit that $V \rightarrow \infty$ the function on the positive real axis becomes truly singular at $z = z_0$.

Yang and Lee have proved several mathematical theorems for the infinite-volume limit one expects from the behavior of experimental systems. These are, first, that the limit $\beta P(V, \beta, -\beta\mu) = \lim V \rightarrow \infty [V^{-1} \ln Q(V, \beta, -\beta\mu)]$ exists and is asymptotically approached for large V-values and, second, that

$$\lim_{V\rightarrow\infty} [\rho(V, \beta, -\beta\mu] = \lim_{V\rightarrow\infty} \left(\frac{N}{V}\right)$$
$$= \lim_{V\rightarrow\infty} \left\{\frac{\partial[V^{-1} \ln Q(V, \beta, -\beta\mu)]}{\partial \ln z}\right\}_{\beta,\mu},$$

where the order of the limit and differentiation can be permuted so that

$$\lim_{V\rightarrow\infty} [\rho(V, \beta, -\beta\mu] = \left(\frac{\partial}{\partial \ln z}\right)_{\beta,\mu} \left\{\lim_{V\rightarrow\infty} [V^{-1} \ln Q(V, \beta, -\beta\mu)]\right\}. \quad (8h.30)$$

Together these simply mean that one can prove that nothing unexpected happens in going from a macroscopic volume V to the limit of infinite volume for which one can speak of a true mathematical discontinuity in the density. One also can deduce that, since the slope of the βP curve versus $\ln z$ is the number density ρ, which cannot exceed N_m/V, the derivative must be finite everywhere; the pressure cannot be discontinuous on the positive real axis of z.

Thus we have a purely formal mathematical theory of the phase change, but one in which there is no direct way of using any knowledge we have of the mutual potential between molecules. Above the critical temperature the line of roots z_k that cross the positive real axis below T_c at the activity $z_0(\beta)$ of condensation to a liquid must pull away from the real axis. However, there is another phase change to a crystal at a higher activity $z_{cr}(\beta)$, and there is reason to believe that this has no critical temperature. If so, as long as the temperatures are not so high that the molecules dissociate or ionize, the line of roots must cross the real axis at this activity.

We return now to the equations for the cluster development for βP [eq. (8d.10′)] and for $\rho = (N/V)$ of eq. (8d.14). For any finite V we must recognize that the cluster integrals depend, in principle at least, on V and write

$$\beta P(V, \beta, z) = \sum_{n \geq 1} b_n(V, \beta) z^n, \tag{8h.31}$$

$$\rho(V, \beta, z) = \sum_{n \geq 1} n b_n(V, \beta) z^n. \tag{8h.32}$$

We can define

$$b_n^{(0)}(\beta) = \lim_{V \to \infty} [b_n(V, \beta)], \ldots. \tag{8h.33}$$

and we know, from the discussion at the beginning of this section, that for volumes of the order of 1 cm^3 or greater $b_n(V, \beta) = b_n^{(0)}(\beta)$ to high numerical precision up to n of the order of 10^2. The function $P(V, \beta, z)$ for finite V has no true singularity on the positive real axis of z, but

$$\lim_{V \to \infty} [\beta P(V, \beta, z)] = \beta P(\beta, z) = \lim_{V \to \infty} \left[\sum_{n \geq 1} b_n(V, \beta) z^n \right] \tag{8h.34}$$

has a singularity at $z = z_0(\beta)$ for $T < T_c$. As long as the series

$$\beta P(\beta, z) = \sum_{n \geq 1} |b_n^{(0)}(\beta)| z^n, \tag{8h.35}$$

where $|b_n^{(0)}(z)|$ is the absolute value of $b_n^{(0)}$, converges, we can commute the order of summation and the limiting process of taking infinite volume and use eq. (35) for the pressure of an infinite-volume system, but we are not in general justified in assuming that the singularity of eq. (35) is the same as that of the singularity in eq. (34). We discuss this further at the end of Sec. 8k.

8i. THE VIRIAL DEVELOPMENT

The extensive dimensionless quantity $-\beta PV$ is a minimum at equilibrium for fixed values of V, β, ln z and has simple derivatives with respect to these variables,

$$-\left[\frac{\partial(\beta PV)}{\partial V}\right]_{\beta,z}=-\beta P, \tag{8i.1}$$

$$-\left[\frac{\partial(\beta PV)}{\partial \beta}\right]_{V,z}=-\left\{E-\lim_{V\to\infty}[E(V,\ T,\ N)]\right\}, \tag{8i.2}$$

$$-\left[\frac{\partial(\beta PV)}{\partial \ln z}\right]_{V,\beta}=-N. \tag{8i.3}$$

For purely thermodynamic manipulations it is convenient to have the pressure given as a function of β and z. However, for the experimentalist it is far preferable to have the activity z and the pressure P as functions of $\rho=N/V$.

As long as the series $\sum nb_n z^n=\rho$ [eq. (8d.15)] converges rapidly, it is quite feasible to write $z=\sum \alpha_n\rho^n$ and with this for z in $b_n z^n$ equate coefficients of the density ρ to find the α_n's in terms of the b_n's. The algebra looks a little less messy if we anticipate the values of α_0 and α_1 from the fact that, in $\rho-\sum nb_n z^n$, $b_0=0$ and $b_1=1$, so that α_0 must be zero and $\alpha_1=1$. For instance, if we carry the algebra up to the fourth power in z and ρ, we use

$$\rho=z+2b_2z^2+3b_3z^3+4b_4z^4+\cdots \tag{8i.4}$$

with

$$\begin{aligned} z&=\rho(1+\alpha_1\rho+\alpha_2\rho^2+\alpha_3\rho^3+\cdots),\\ z^2&=\rho^2[1+2\alpha_1\rho+(\alpha_1^2+2\alpha_2)\rho^2+\cdots],\\ z^3&=\rho^3(1+3\alpha_1\rho+\cdots),\\ z^4&=\rho^4(1+\cdots),\end{aligned} \tag{8i.5}$$

to write

$$\begin{aligned}\rho=\rho&+(\alpha_1+2b_2)\rho^2+(\alpha_2+4b_2\alpha_1+3b_3)\rho^3\\ &+(\alpha_3+2b_2\alpha_1^2+4b_2\alpha_2+9b_3\alpha_1+4b_4)\rho^4+\cdots=0.\end{aligned} \tag{8i.6}$$

If this all converges, the coefficients of all powers of ρ higher than the first must be zero, from which, by successively solving for the α's, we find,

$$\begin{aligned}\alpha_1&=-2b_2,\\ \alpha_2&=-3b_3+8b_2^2,\\ \alpha_3&=-4b_4+30b_2b_3-40b_2^3,\end{aligned} \tag{8i.6$'$}$$

and

$$\begin{aligned}\ln z&=\ln\rho+\ln[1+\rho(\alpha_1+\alpha_2\rho+\alpha_3\rho^2+\cdots)]\\ &=\ln\rho+\rho(\alpha_1+\alpha_2\rho+\alpha_3\rho^2+\cdots)-\tfrac{1}{2}\rho^2(\alpha_1+\alpha_2\rho+\cdots)^2+\tfrac{1}{3}\rho^3(\alpha_1+\cdots)^3\\ &=\ln\rho-2b_2\rho-3(b_3-2b_2^2)\rho^2-4(b_4-6b_2b_3+\tfrac{20}{3}b_2^3)\rho^3+\cdots.\end{aligned} \tag{8i.7}$$

For the pressure we find,

$$\begin{aligned}P &= kT(z+b_2z^2+b_3z^3+b_4z^4+\cdots)\\ &= kT\rho[1-b_2\rho-2(b_3-2b_2^2)\rho^2-3(b_4-6b_2b_3+\tfrac{20}{3}b_2^3)\rho^3-\cdots].\end{aligned} \quad (8i.8)$$

It is no accident that the coefficients β_k in the power series development of $\ln z(\rho)-\ln\rho$,

$$\ln z(\rho)-\ln\rho=-\sum\beta_k\rho^k, \quad (8i.9)$$

with, from eq. (7),

$$\begin{aligned}\beta_1 &= 2b_2,\\ \beta_2 &= 3(b_3-2b_2^2),\\ \beta_3 &= 4(b_4-6b_2b_3+\tfrac{20}{3}b_2^3,\end{aligned} \quad (8i.10)$$

are so simply related to the coefficients c_k in the power series development [eq. (8)] of P/kT,

$$\beta P=\rho\left(1-\sum c_k\rho^k\right). \quad (8i.11)$$

We prove that the relation

$$c_k=\left(\frac{k}{k+1}\right)\beta_k, \quad (8i.12)$$

which holds for the first three terms from eqs. (7) and (8), is general. The proof follows from the thermodynamic relation $[\partial(P/kT)/\partial\ln z]_T=N/V=\rho$ [eq. (3)] and the limiting relationship that as z and ρ approach zero $\ln z(\rho)$ equals $\ln\rho$. We write, from eq. (3),

$$\rho^{-1}\left(d\frac{P}{kT}\right)_T=(d\ln z)_T, \quad (8i.13)$$

$$\lim_{\epsilon\to 0}\left\{\int_\epsilon^\rho d\rho'\frac{1}{\rho'}\left[\frac{\partial(P/kT)}{\partial\rho'}\right]_T-\int_\epsilon^\rho d\ln\rho'\right\}=\ln z(\rho)-\ln\rho. \quad (8i.14)$$

From eq. (11) for P/kT we have

$$\rho^{-1}\left[\frac{\partial(P/kT)}{\partial\rho}\right]_T=\rho^{-1}-\sum_{k\ge 1}(k+1)c_k\rho^{k-1}, \quad (8i.15)$$

and the integrals on the left of eq. (14) then give

$$-\sum_{k\ge 1}\left(\frac{k+1}{k}\right)c_k\rho^k=\ln z-\ln\rho, \quad (8i.16)$$

which with eq. (9) proves eq. (12) as long as the two series converge.

Obviously, the method by which eqs. (7) and (8) were derived could be used to find expressions for β_k far beyond the three relations given in eq.

(10) for the β_k's in terms of the cluster integrals b_n of eq. (8d.5), but the algebra soon becomes tedious. An examination of how the integrals of eq. (8d.5) for b_n could be evaluated shows a startling simplicity for the integrals of the first three β_k's, a simplicity one may suspect to be general and which we prove to be general. First let us solve the relations in eq. (10) for the $n!\,b_n$'s up to $n=4$ in terms of the β_k's. We find

$$\begin{aligned} 2!\,b_2 &= \beta_1, \\ 3!\,b_3 &= 2!\,\beta_2 + 3\beta_1^2, \\ 4!\,b_4 &= 3!\,\beta_3 + 12(2!\,\beta_2)\beta_1 + 16\beta_1^3. \end{aligned} \tag{8i.17}$$

From eq. (8d.5) we see that $n!\,b_n$ is V^{-1} times an integral of a cluster function of n molecules over all n coordinates which, for sufficiently large V, is numerically equal to the integral over any of $n-1$ coordinates in an infinite volume holding the coordinates of one molecule fixed at the origin. The cluster function is a sum of all products of at least $(n-1) f_{ij}$'s, $1 \leq i < j \leq n$, such that every molecule is connected to all others through a path of f_{ij}'s. The four products of the cluster function of three are shown graphically on page 234. The first three products, $f_{ki}f_{ij}$, $f_{ij}f_{jk}$, and $f_{jk}f_{ki}$, lead to the same integral; the third, $f_{ij}f_{jk}f_{ki}$, gives a new integral.

Define $\beta_1 = 2b_2$ by

$$\beta_1 = \int d\mathbf{r}_2 f(|\mathbf{r}_{12}|) = \int_0^\infty 4\pi r^2 f(r)\, dr. \tag{8i.18}$$

In any one of the three first products in the integrand of $3!\,b_3$, say $f_{ki}f_{ij}$, hold $\mathbf{r}_k$ and $\mathbf{r}_i$ fixed and integrate over dr_j. One obtains the factor β_1 times $f(|\mathbf{r}_k - \mathbf{r}_i|)$. Integrate over either $d\mathbf{r}_i$ or $d\mathbf{r}_j$, and the result is β_1^2. There are three such products, so there is an additive term $3\beta_1^2$ in $3!\ b_3$. Define

$$\beta_2 = \frac{1}{3!}\iint d\mathbf{r}_i\, d\mathbf{r}_j\, f_{ij}f_{jk}f_{ki}, \tag{8i.19}$$

and the remaining term is $2\beta_2$. We have the expression for $3!\,b_3$ in eq. (17).

There are 38 different products in the cluster function of 4 molecules. Of these (Fig. 8i.1), there are 12 terms like $f_{12}f_{23}f_{34}$ differing in $4! = 24$ permutations of the order of indices, but dividing by 2 since $f_{ij} \equiv f_{ji}$ and $f_{12}f_{23}f_{34}$ and $f_{43}f_{32}f_{21}$ are the same product. Successively integrating over $d\mathbf{r}_1$, $d\mathbf{r}_2$, and $d\mathbf{r}_3$ gives each time a factor β_1, or for all 12 terms $12\beta_1^3$. But there are four terms like $f_{12}f_{13}f_{14}$ differing only in having any one of the four indices 1, 2, 3, or 4 as common to all three f_{ij}'s. Integration over $d\mathbf{r}_2$, $d\mathbf{r}_3$, and $d\mathbf{r}_4$ in this product gives each a factor β_1. In all, there is an additive term $16\beta_1^3$ in $4!\,b_4$, as in eq. (17). Products of type $f_{12}f_{23}f_{34}f_{42}$

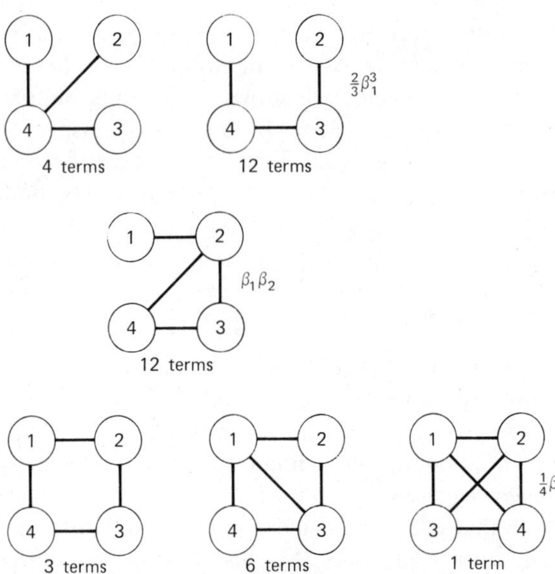

Figure 8i.1 Graphs of products occurring additively in the cluster formulation of four molecules.

occur 12 different times in the cluster functions for 4, since the unique index 1 can be selected in four ways and for each such selection the other unique index 2 in three ways. Integrate over $d\mathbf{r}_1$ to obtain a factor β_1 times the integrand of $2\beta_2$, contributing $12\beta_1(2!\,\beta_2)$. The remaining 10 products, shown graphically in Fig. 8i.1, constitute the integrand of $3!\,\beta_3$,

$$\beta_3 = \frac{1}{3!}\iiint d\mathbf{r}_2\, d\mathbf{r}_3\, d\mathbf{r}_4 (3f_{12}f_{23}f_{34}f_{41} + 6f_{12}f_{23}f_{34}f_{41}f_{13} + f_{12}f_{23}f_{34}f_{41}f_{13}f_{24}). \quad (8i.20)$$

Thus up to $n=4$ we can write b_n as a sum of products of β_k's which are integrals of sums of products of f_{ij}'s, $1 \le i < j \le k+1$, such that there are at least two independent paths of f_{ij}'s (whenever k is greater than unity) connecting any two ij's, the two paths having no common index. In general, a product of f_{ij}'s composing additively a cluster function of n molecules (see Fig. 8i.2) may have one or more nodal vertices or articulation points, such that a nodal vertex i separates the remaining $n-1$ vertices into two or more subsets, and every path of f_{ij}'s connecting two vertices l and j in different subsets passes through i. In this case, holding i fixed, the integral over the remaining $n-1$ vertices gives independent product contributions from each subset. The integrand of a β_k is a product with no nodal vertex.

Define the irreducible integral for $k+1$ molecules by

$$k!\,\beta_k = \lim_{V\to\infty}\left(V^{-1}\iint\cdots\int d\mathbf{r}_1\cdots d\mathbf{r}_{k+1} \underset{\substack{\text{sum over all connected}\\ \text{products, } k+1\ge i>j\ge 1,\\ \text{with no nodal vertices}}}{\sum\prod} f_{ij}\right) \tag{8i.21}$$

We refer to the integrand of eq. (21) as an irreducible function. In the next two sections we prove that

$$b_n(V\to\infty) = n^{-2} \sum_{\substack{\nu_k \\ k\nu_k = n-1}} \prod_k \frac{(n\beta_k)^{\nu_k}}{\nu_k!}, \tag{8i.22}$$

and that eq. (9), namely,

$$\ln z = \ln \rho - \sum_{k\ge 1} \beta_k \rho^k, \tag{8i.23}$$

as well as eq. (11) for βP with eq. (12),

$$\beta P = \rho\left(1 - \sum_{k\ge 1} \frac{k}{k+1} \beta_k \rho^k\right), \tag{8i.24}$$

are both valid as long as two conditions are both satisfied, namely, that with ρ_s the value of ρ at the first singularity of $\sum \beta_k \rho^k$ on the positive real axis,

$$\begin{aligned} &1.\quad \rho < \rho_s, \\ &2.\quad \sum_{k\ge 1} k\beta_k\rho^k < 1. \end{aligned} \tag{8i.25}$$

From eq. (24) the isothermal compressibility κ is given by

$$\kappa = \left(\frac{\partial P}{\partial \ln \rho}\right)_T^{-1} = \beta\left[\left(1 - \sum_{k\ge 1} k\beta_k\rho^k\right)\right]^{-1}. \tag{8i.26}$$

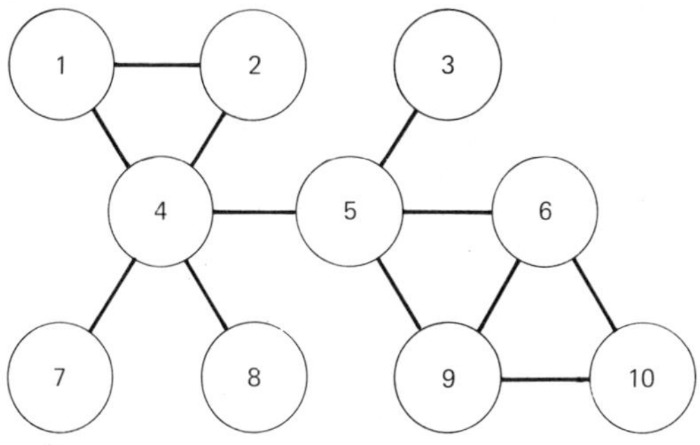

Figure 8i.2 Graph of a term in the cluster function of ten molecules. Vertices 4 and 5 are *nodal* or *articulation* points. The value of λ_4 (see Sec. 8i) is 3, that for 5 is $\lambda_5 = 2$. All other vertices have $\lambda_i = 0$. The integral is proportional to $\beta_1^3\beta_2$ times one term of β_3.

From the conditions in eq. (25) κ can never be negative, an obvious requirement for a stable equilibrium system.

From eq. (8d.15) that $\rho = \sum n b_n z^n$ it follows that

$$\left(\frac{\partial \rho}{\partial \ln z}\right)_T = \sum_{n \geq 1} n^2 b_n z^n. \tag{8i.27}$$

The compressibility κ can be written as

$$\left(\frac{\partial \ln \rho}{\partial P}\right)_T = \kappa = \rho^{-1}\left(\frac{\partial \rho}{\partial \ln z}\right)_T\left(\frac{\partial P}{\partial \ln z}\right)_T^{-1} = \frac{\beta}{\rho^2} \sum n^2 b_n z^n \tag{8i.28}$$

or, from eq. (26),

$$\sum_{n \geq 1} n^2 b_n z^n = \rho\left(\frac{\rho\kappa}{\beta}\right) = \rho\left(1 - \sum_{k \geq 1} k\beta_k \rho^k\right)^{-1}. \tag{8i.28'}$$

We remark here that the analysis of the cluster function constituting the integrand of $n!\, b_n$ as being given as a product of irreducible functions having common coordinate indices at nodal vertices on the corresponding graph does not depend on having the total potential U_N given as a sum of pair terms. If there are explicit triple terms $u_3(\mathbf{r}_i, \mathbf{r}_j, \mathbf{r}_k)$ in the potential, the integrand of $2\beta_2$ will contain an additive term, which is $[\exp - \beta u_3(i, j, k) - 1]\exp\{-\beta[u(r_{jk}) + u(r_{ki}) + u(r_{jk})]\}$, in addition to the irreducible function when only pair terms are nonzero. The irreducible functions for higher k-values become quite complicated, but all the subsequent proofs are unaltered.

8j. PROOF OF EQ. (8i.22) FOR b_n

The proof of the generality of eqs. (8i.22) to (8i.28) involves two completely independent parts, first a comparatively difficult and long combinatorial proof of expression (8i.22) for b_n and, second, from this the proof of eq. (8i.28'). With this, by steps reversing the order of the steps in the last section, we arrive at eqs. (8i.23) and (8i.24) for $\ln z$ and βP. The second proof is in the next section.

The integrands of the irreducible integrals $k!\, \beta_k$ are irreducible functions completely symmetric in permutations of the $k+1$ numbered coordinates on which the functions depend, as are the cluster functions of n molecular coordinates whose integrals are $n!\, b_n$. Including the case that the ν_k of eq. (8i.22) are $\nu_{n-1} = 1$, $\nu_k = 0$ if $k < n-1$ with only one irreducible function, the cluster functions are sums of products of irreducible functions having common coordinates $\mathbf{r}_i$ at nodal vertices i on the corresponding connected graph. With the notation $\boldsymbol{\nu} \equiv \nu_1, \nu_2, \ldots, \nu_{n-1}$

and $C(\boldsymbol{\nu}) \equiv C(\nu_1, \nu_2, \ldots, \nu_{n-1})$ we can write $n!\, b_n$ as a sum of products,

$$n!\, b_n = \sum_{\boldsymbol{\nu}} C(\boldsymbol{\nu}) \prod_k (k!\, \beta_k)^{\nu_k}, \tag{8j.1}$$

in which the coefficients $C(\boldsymbol{\nu})$ are the total number of different products of ν_k irreducible functions, each of $k+1$ molecular coordinates, out of a total of n molecules which appear additively in one cluster function of n. We now prove that

$$C(\boldsymbol{\nu}) = \frac{n!}{\prod_k [(k!)^{\nu_k} \nu_k!]} n^{\sum \nu_k - 2} \tag{8j.2}$$

if $\sum_k k\nu_k = n-1$ and zero otherwise. This, then, leads to eq. (8k.22) for b_n.

In the combinatorial problem that constitutes this proof some may find it helpful to visualize the irreducible functions replaced by unnumbered frames containing $k+1$ symmetrically placed holes, and the numbered molecular indices $1 \le i \le n$ replaced by numbered bolts which, in the completed structure must fill all the $\sum(k+1)\nu_k$ holes and bolt the frames together in one arrangement. The assignment of a molecular index i to a coordinate $\mathbf{r}$ of an irreducible function then corresponds to passing a numbered bolt i through one of the holes in an unnumbered frame.

In a given product of the cluster function we can uniquely assign an order λ_i to each molecule i as one less than the number of irreducible functions in which its coordinates appear as an argument (Fig. 8i.2). A nonodal vertex i on the graph has $\lambda_i = 0$, since it appears in one irreducible function only; a nodal vertex has $\lambda_i > 0$. The coordinates of molecule i appear in $\lambda_i + 1$ irreducible functions. There are n molecular indices i, but $\sum(k+1)\nu_k$ coordinate positions in the irreducible functions to which the indices must be assigned. The sum of λ_i over all i is one less than the total number $\sum_k \nu_k$ of irreducible functions,

$$\sum_k (k+1)\nu_k = n + \sum_{i=1}^{i=n} \lambda_i = n + \sum_k \nu_k - 1,$$

$$\sum_k k\nu_k = n - 1, \tag{8j.3}$$

which proves the limitation stated under eq. (2) on the nonzero values of $C(\boldsymbol{\nu})$.

Now assign $n-1$ of the n numbered molecular indices each to one coordinate position in the irreducible functions, assigning only k indices to each of the ν_k unnumbered irreducible functions of $k+1$ coordinates. This leaves one of the n indices i as yet unplaced, and one of the coordinate positions in each of the symmetric irreducible functions without a molecular index. We refer to this as a dissociated assignment of indices to irreducible functions. The number $N(\mathbf{v})$ of ways that such a dissociated assignment can be made is

$$N(\boldsymbol{\nu}) = \frac{n!}{\prod_k [(k!)^{\nu_k} \nu_k!]}, \qquad \sum k\nu_k = n-1, \tag{8j.4}$$

since the unassigned index can be selected in n ways and the remaining $\sum_k k\nu_k = n-1$ distributed among the ν_k indistinguishable functions of type k in $(n-1)!/\prod_k[(k!)^{\nu_k}\nu_k!]$ ways (see Appendix VI).

However, more than one such dissociated arrangement leads to the same product in the cluster function sum and, indeed, exactly n different dissociated assignments lead to a single cluster function product. This can be seen as follows. Start with a single cluster function with all indices assigned. Select one of the indices and erase it, leaving it as the unique unassigned index in the dissociated assignment of indices to irreducible functions. Obviously, the dissociated assignment depends on which of the n indices has been chosen and, equally obviously, there are n ways of choosing this index. There are therefore *at least n* different dissociated assignments leading to one assignment of indices in the cluster function. Now, as we show, the rest of the dissociation process is uniquely determined and therefore there are exactly n different dissociated assignments that can lead to the same cluster function. That the dissociation process is unique can be seen as follows. Suppose the missing index in the otherwise undissociated function is at a coordinate for which $\lambda_i = 0$. Unless $\nu_{n-1} = 1$, in which case the dissociation is already complete, the irreducible function to which it belonged must be connected at least one place by a coordinate of $\lambda > 0$ to other irreducible functions. Dissociate all other irreducible functions from the one originally having the unique free index but, since this function already has one missing index, and only one missing index per irreducible function is allowed in the dissociated assignment, the index must be removed from each of the partly or completely dissociated products, leaving each of them with an unassigned index in a nonnodal vertex. Subsequent steps in the dissociation process are similarly uniquely determined, completing the proof that there are n different dissociated assignments that can lead to each single connected product. If we define $K(\boldsymbol{\nu})$ as the total number of connected products that can be formed from one dissociated assignment, then

$$C(\boldsymbol{\nu}) = n^{-1} N(\boldsymbol{\nu}) K(\boldsymbol{\nu}) = n^{-1} \frac{n!}{\prod_k [(k!)^{\nu_k} \nu_k!]} K(\boldsymbol{\nu}) \tag{8j.5}$$

We now show that

$$K(\boldsymbol{\nu}) = n^{\sum \nu_k - 1}, \tag{8j.6}$$

which then completes the proof of eq. (2) and thus also of eq. (8i.22) for b_n. The irreducible functions in the dissociated assignment each have at least one indexed coordinate, hence are now distinguishable. Now first assign the orders $\lambda_i > 0$, with $\sum_i \lambda_i = (\sum_k \nu_k) - 1$, to the n numbered indices i, $1 \le i \le n$, including the free index, and count the number of different cluster functions that can be formed with a given assignment of the λ_i's. We then sum over all possible values of λ_i subject to the fixed value on their sum. The number of different cluster functions consistent with a fixed assignment in the dissociated set and a fixed set of λ_i's is shown to be $(\nu-1)! \prod_i \lambda_i!$, with $\nu = \sum_k \nu_k$, as follows.

Out of the $\nu - 1$ distinguishable dissociated functions other than that already containing i and therefore not having that index on any coordinate we can select λ_i of them in $(\nu-1)(\nu-2)\cdots(\nu-\lambda_i)/\lambda_i!$ ways, since the order of selection is immaterial. This combined connected product now has one unindexed coordinate which may not be assigned an index already in the product. For subsequent steps in the combinatorial problem it behaves exactly like a single irreducible function. There remain $\nu - \lambda_i$ products, each with one coordinate unindexed, and for any index j with $\lambda_j > 0$ there are $\nu - 1 - \lambda_i$ unindexed coordinates to which it can be assigned in $(\nu - 1 - \lambda_i)\cdots(\nu - \lambda_i - \lambda_j)/\lambda_j!$ ways. Continue until all indices but the free index whose order is, say λ_f, have been assigned. There will then be $\lambda_f + 1$ independent products, each with one coordinate to which no molecular index has been assigned and which must be given this index. There is only one way to do this, but $\lambda_f!/\lambda_f! = 1$. One then has that for given assignment of order λ_i for every molecule i, $1 \le i \le n$, the number $M_n^{(\nu)}(\lambda_1, \lambda_2, \ldots, \lambda_n)$ of different connected products that can be formed from one dissociated assignment is

$$M_n^{(\nu)}(\lambda_1, \ldots, \lambda_f) = \frac{(\nu-1)!}{\prod_{i=1}^{i=n} \lambda_i!}, \tag{8j.7}$$

where

$$\nu = \sum_k \nu_k = 1 + \sum_{i=1}^{i=n} \lambda_i. \tag{8j.7$'$}$$

Obviously, every different assignment of λ_i's yields different products, so that $K(\boldsymbol{\nu})$ is the sum of $M_n^{(\nu)}$ over all possible values of λ_i subject only to the condition in eq. (7′). However, $M_n^{(\nu)}$ is the multinomial coefficient (Appendix VI),

$$\left(\sum_{i=1}^{i=n} x_i\right)^{\nu-1} = \sum_{\lambda_1}\sum_{\lambda_2}\cdots\sum_{\lambda_n}^{(\Sigma\lambda_i = \nu-1)} M_n^{(\nu)}(\lambda_1, \ldots) \prod_{i=1}^{i=n} x_i^{\lambda_i}, \tag{8j.8}$$

so that the sum over all $M_n^{(\nu)}$'s is the limit when all $x_i = 1$ of $(\sum x_i)^{\nu-1}$, which is $n^{\nu-1}$, so that

$$K(\boldsymbol{\nu}) = \sum_{\lambda_1 \ge 1}^{(\sum \lambda_i = \nu - 1)} \cdots \sum_{\lambda_n \ge 1} M_n^{(\nu)}(\lambda_1, \ldots, \lambda_n) = n^{\nu-1}, \tag{8j.9}$$

which proves eq. (6) and thus eq. (8k.22) for b_n, namely,

$$n^2 b_n = \underset{(\sum k\nu_k = n-1)}{\sum_{\nu_1} \sum_{\nu_2} \cdots \sum_{\nu_{n-1}}} \prod_{k=1}^{k=n-1} \frac{(n\beta_k)^{\nu_k}}{\nu_k!}. \tag{8j.10}$$

8k. DERIVATION OF THE VIRIAL DEVELOPMENT EQUATIONS

In order to derive eqs. (8i.24) to (8i.26) for $\ln z - \ln \rho$, P/kT, and the isothermal compressibility κ as a power series in the number density $\rho = N/V$ we use a theorem from the theory of functions of a complex variable, which follows. If a function $f(\zeta)$ is singular at the origin $\zeta = 0$ but is regular for $0 < |\zeta| < R$, with R a real positive number, the function $f(\zeta)$ can be developed as a power series in ζ including negative powers (Laurent series),

$$f(\zeta) = \sum_{-\infty}^{\infty} a_n \zeta^n, \tag{8k.1}$$

which is valid within the regular region $0 < |\zeta| < R$. The coefficient a_{-1} of ζ^{-1} in the series in eq. (1) is called the residue of the function and is given by the integral

$$a_{-1} = \frac{1}{2\pi i} \oint d\zeta \; f(\zeta) \tag{8k.2}$$

over a closed path surrounding the origin and no other singularity of the function. (If the integration path encloses two points of singularity, the integral gives the sum of the two residues.) It follows that the general coefficient a_n of ζ^n in eq. (1) is given by

$$a_n = \frac{1}{2\pi i} \oint d\zeta \; \zeta^{-(n+1)} f(\zeta). \tag{8k.2'}$$

The function $\exp \sum_{k \ge 1} n\beta_k \zeta^k$ can be developed as a power series in ζ,

$$e^{\sum n\beta_k \zeta^k} = \sum_{\nu_1} \sum_{\nu_2} \cdots \sum_{\nu_k} \cdots \prod_k \frac{(n\beta_k \zeta^k)^{\nu_k}}{\nu_k!}, \tag{8k.3}$$

so that from eq. (8j.10) for $n^2 b_n$ we see that $n^2 b_n$ is the coefficient of ζ^{n-1}

in this, and with eq. (2′) we can express $b^2 b_n z^n$ as

$$n^2 b_n z^n = \frac{1}{2\pi i}\oint d\zeta\left(\frac{z}{\zeta}e^{\Sigma\beta_k\zeta^k}\right)^n. \tag{8k.4}$$

The function, with real and positive z,

$$f(z, \zeta) = z\zeta^{-1}e^{\Sigma\beta_k\zeta^k} \tag{8k.5}$$

has a first-order pole,

$$\lim \zeta = 0[f(z, \zeta)] = \frac{z}{\zeta}, \tag{8k.5′}$$

at the origin $\zeta = 0$, and no other singularity for $|\zeta| < |\zeta_s|$ if $|\zeta_s|$ is the lowest value for which

$$\sum_k \beta_k \zeta_s^k = \text{singular}. \tag{8k.6}$$

Now $f(z, \zeta)$ is unity if $\zeta = \zeta_0$, the lowest real positive value of ζ satisfying the implicit equation

$$z = \zeta_0 e^{-\Sigma\beta_k\zeta_0^k}. \tag{8k.7}$$

For sufficiently small z values we have

$$\zeta_0 < |\zeta_s| \tag{8k.8}$$

and, as we show, can find a closed path of ζ surrounding the pole of $f(z, \zeta)$ at the origin with

$$\zeta_0 < |\zeta| < |\zeta_s|, \tag{8k.9}$$

which is just outside the closed path with $|f(z, \zeta)| = 1$. On this path

$$|f(z, \zeta)| < 1. \tag{8k.10}$$

We want the sum of $n^2 b_n z^n$ for $n \geq 1$, which is given by the sum of the expressions on the right of eq. (4). With eq. (10) satisfied we can permute the order of summation and integration. Using

$$\sum_{n\geq 1} f^n = \frac{f}{1-f}, \tag{8k.11}$$

we have

$$\sum_{n\geq 1} n^2 b_n z^n = \frac{1}{2\pi i}\oint d\zeta\, f(z, \zeta)[1 - f(z, \zeta)]^{-1} = \frac{1}{2\pi i}1\oint d\zeta\, z(\zeta e^{-\Sigma\beta_k\zeta^k} - z)^{-1}, \tag{8k.12}$$

where in the last expression we have multiplied numerator and denominator by $\zeta e^{-\Sigma\beta_k\zeta^k}$.

The summation over n has removed the pole at the origin where $f(1-f)^{-1}=-1$, but the path of integration encloses a new pole where $f=1$ and eq. (7) is satisfied. That this is a simple (first-order) pole follows from the fact that the derivative of $[f(z, \zeta)]^{-1}$ is finite at $\zeta=\zeta_0$, so that the value of the residue is found by replacing the denominator with its derivative. We have

$$\sum_{n\geq 1} n^2 b_n z^n = \left\{\left[\frac{\partial}{\partial \zeta} f^{-1}(z, \zeta)\right]_{\zeta=\zeta_0}\right\}^{-1} = \zeta_0\left(1-\sum_{k\geq 1} k\beta_k \zeta_0^k\right)^{-1}. \quad (8k.12')$$

From eq. (7) for z we have

$$\frac{d \ln z}{d \ln \zeta_0} = 1-\sum_k k\beta_k \zeta_0^k, \quad (8k.13)$$

and from eq. (8d.14), $\sum nb_n z^n = \rho$, we have, with eq. (12),

$$\rho = \sum_{n\geq 1} nb_n z^n = \int_0^z d\ln z' \sum_{n\geq 1} n^2 b_n (z')^n = \int_0^{\zeta_0} \zeta\, d\ln \zeta = \int_0^{\zeta_0} d\zeta = \zeta_0, \quad (8k.14)$$

which identifies ζ_0 with the number density ρ. Further integration gives

$$\begin{aligned}\beta P &= \sum_{n\geq 1} b_n z^n = \int_0^z d\ln z' \sum_{n\geq 1} nb_n (z')^n \\ &= \int_0^\rho d\ln \zeta\left(\zeta - \sum_{k\geq 1} k\beta_k \zeta^{k+1}\right) \\ &= \rho\left[1-\sum_{k\geq 1}\left(\frac{k}{k+1}\right)\beta_k \rho^k\right]. \end{aligned} \quad (8k.15)$$

With $\zeta_0=\rho$, eq. (7) yields

$$\ln z = \ln \rho - \sum_{k\geq 1} \beta_k \rho^k, \quad (8k.16)$$

while from eq. (12), with eq. (8i.28), the compressibility is

$$\kappa = \beta \rho^{-1}\left(1-\sum_{k\geq 1} k\beta_k \rho^k\right)^{-1}. \quad (8k.17)$$

We have thus completed the derivation of the set of equations (8i.23), (8i.24), and (8i.26), which were asserted in Sec. 8i to be general after showing them to hold up to $k=3$. The derivation is valid for sufficiently small z-values. From eq. (12′) we see that $\sum n^2 b_n z^n$ is singular if $\sum k\beta_k \rho^k = 1$. From eq. (5) for $f(z, \zeta)$, we see that

$$\frac{\partial \ln f(z, \zeta)}{\partial \ln \zeta} = -\left(1-\sum_k k\beta_k \zeta^k\right) \quad (8k.18)$$

and is negative if $\sum k\beta_k\zeta^k < 1$, which justifies our assertion that a path *outside* that for which $|f(z, \zeta)| = 1$ has $|f(z, \zeta)| < 1$. As long as eq. (8), $\zeta_0 < |\zeta_s|$, is satisfied and $\sum k\beta_k\rho^k < 1$, the derivation is valid. The sums $\sum b_n z^n$, $\sum nb_n z^n, \ldots$ are seen to be singular from eqs. (14), (15), and (17) if the sums $\sum \beta_k\rho^k$, $\sum k\beta_k\rho^k, \ldots$ are singular. It follows that the two conditions of eq. (8i.25),

$$\begin{aligned}&1.\quad \rho < \rho_s, \qquad \sum \beta_k\rho_s^k \text{ singular,}\\ &2.\quad \sum k\beta_k\rho^k < 1\end{aligned} \tag{8k.19}$$

must be satisfied if the $b_n z^n$ sums are to converge. Condition 2 of eq. (19) is seen from eq. (17) to assure us that the isothermal compressibility κ is positive.

The irreducible function that forms the integrand of $k!\,\beta_k$ for very large k-values, $k = N \cong 10^{20}$, includes products $\prod f_{ij}$ with all but a negligible fraction of the molecules bonded by a function $f(r)$ to 12 others. These products have their maximum value when the coordinate configuration is that of the N molecules in a close-packed lattice at nearest-neighbor distance r_0 of the minimum value, $u(r_0) = -u_0$, of the pair potential. In addition, the function includes all products differing from one of these by inclusion of any or all f_{ij}'s between molecules not already bonded as nearest neighbors, in other words, by multiplying each such product by the product of $1 + f_{ij} = e^{-\beta u(r_{ij})}$ for all non-nearest-neighbor pairs in the original product. It further includes a very large fraction of the products in which the original nearest-neighbor f_{ij}'s are missing and therefore replaced by unity. In all, more than half of the total of 2^N products of

$$\prod_{N \geq i > j \geq 1}\!\!\prod (1 + f_{ij}) = \exp\left(-\frac{U_N}{kT}\right) \tag{8k.20}$$

contribute to the irreducible function, hence to $N!\,\beta_N$, and at sufficiently low temperature for which $u_0/kT \gg 1$ it is just these products that contribute most to the integral of $\exp(-U_N/kT)$.

We conclude that for large values of N, with eq. (8c.6) for Q_τ,

$$\lim_{T \to 0} (\ln \beta_N) = \ln[V^N (N!)^{-1} Q_\tau(N, V(P = 0), T)]. \tag{8k.21}$$

From eqs. (8h.4) and (8h.5) for $\ln V^N(N!)^{-1}Q_\tau$, and eq. (8c.15) for the activity z, remembering that at $P = 0$, $A = G = N\mu$, we have

$$\begin{aligned}\lim_{T \to 0} (\ln \beta_N) &= -\frac{A}{kT} + N \ln \lambda^3 - N \ln Q_i\\ &= -N \ln z(P = 0, T).\end{aligned} \tag{8k.21′}$$

Since at low temperatures the equilibrium vapor pressure is low, the vapor above the condensed phase obeys the perfect gas equation,

$$[z(P=0, T)] = \rho \text{ (eq. vapor)} = \rho_0, \tag{8k.22}$$

we have that, for very large N,

$$\lim_{T\to 0} (\beta_N)^{1/N} = \rho_0^{-1}. \tag{8k.23}$$

If the β_N's are all positive, as they are from eq. (21) at low temperatures, and if the limit as $N\to\infty$ of $\beta_N^{1/N} = \beta_0$ exists, as it does from eq. (21′), then the series $\sum \beta_k \rho^k$ will have its first singularity when $\rho = \rho_s$ on the positive real axis at $\rho_s = \beta_0^{-1}$, $(\beta_0 \rho_s)^N = 1$. It follows that condition 1 of eq. (19) is violated as ρ becomes equal to ρ_0, the number density of the equilibrium vapor at very low temperatures. At low temperatures condition 1 of eq. (19) determines the number density at which the gas becomes thermodynamically unstable.

At low temperatures β_0 is approximately proportional to $\exp 6u_0/kT$, actually with a larger exponential due to the effect of next-nearest-neighbor interaction. For small k-values with fewer average bonds per molecule $(\beta_k)^{1/k}$ is much less than β_0 and the various sums $\sum \beta_k \rho^k$, $\sum k\beta_k \rho^k$, $\sum k^2 \beta_k \rho$ approach small values as the singularity $\rho = \rho_s$ is approached from below on the real axis. The dimensionless quantity $\rho\kappa kT$ is unity for a perfect gas, and below the Boyle temperature increases with pressure as the gas becomes imperfect, becoming infinite at the critical point. From eq. (17) for the compressibility κ,

$$\rho\kappa kT = \left(1 - \sum_{k\geq 1} k\beta_k \rho^k\right)^{-1}, \tag{8k.24}$$

so that the second condition of eq. (19), $\sum k\beta_k \rho^k = 1$, is reached just at the critical point. The $P(\rho)$ above the critical temperature is presumed to be regular on the real axis of ρ, which means that the singularity of the function defined by the series $\sum [k/(k+1)]\beta_k \rho^k$ breaks away from the real axis into the complex plane (symmetrically above and below the real axis) for $T > T_c$. It is not obvious from anything we know why the disappearance of the singularity from the real positive axis occurs when the $k\beta_k \rho^k$ sum just becomes unity. Were $\sum k\beta_k \rho^k < 1$ when the various series $\sum \beta_k \rho^k$ represent functions for which the singularity has just disappeared from the positive real axis, we would have experimentally the disappearance of a two-phase region at a cusp in the $V-T$ diagram with a noninfinite compressibility at the cusp. There would be no thermodynamic inconsistency in such behavior, but it seems never to occur in any known real system.

It seems to us to be likely that this is connected with the problem discussed at the end of Sec. 8h, namely, whether the limit $V\rightarrow\infty$ of the sum $\sum b_n(V, T)z^n$ becomes singular at the same z as $\sum\{\lim V\rightarrow\infty [b_n(V, T)]\}z^n$. The derivation of this section is rigorous for b_n with n finite and V sufficiently large, and for the limit $V\rightarrow\infty$ of $b_n(V, T)$ as $n\rightarrow\infty$, namely, for

$$b_n^{(0)}(T) = \lim_{V\rightarrow\infty} [b_n(V, T)]. \tag{8k.25}$$

Now when the singularity of the $\beta_k\rho^k$ sum determines the singularity in the sum b_nz^n, the value of $b^{(0)}(T)$ for large N is determined primarily by β_N, and this, at least at low temperatures, is the integral of an irreducible function which is large only in a compact volume proportional to N. We expect that the asymptotic value of $[b_N(V, T)]^{1/N}$ will become independent of V if V/N is sufficiently large.

If, however, the singularity in the series $\sum b_nz^n$ is due to the pole $(1-\sum k\beta_k\rho^k)^{-1}$ when condition 2 of eq. (19) is violated, the predominant terms in b_N will be due to the product of high powers of β_k of low k-values. As discussed at the beginning of Sec. 8h [eq. (8h.1′)], the asymptotic independence of $b_N(V)$ on V when the chainlike products $f_{12}f_{23}\cdots f_{n-1,n}$ play a predominant role occurs only if V/N^3 is sufficiently large.

We have seen that in the limit $T\rightarrow 0$ the number density ρ_s of the singularity in $\sum\beta_k\rho^k$ is indeed that of condensation, which means that, empirically at least, the singularity of βP at condensation is the same value of z as the singularity of $\sum b_n^{(0)}z^n$. This offers some presumptive evidence for its validity at higher finite temperatures, but it is not unlikely that precisely as $\sum k\beta_k\rho^k\rightarrow 1$ the relationship fails.

8*l*. MULTICOMPONENT VIRIAL EXPANSION

In Sec. 8g we discussed the development as a power series in the activities z_a, $z_b, \ldots$ of βP for a system composed of several chemical species a, $b, \ldots$ of molecules. By using a notation of boldface symbols,

$$\begin{aligned}
\mathbf{n} &\equiv n_a,\ n_b, \ldots,\\
\mathbf{n}! &\equiv n_a!\, n_b! \ldots,\\
\mathbf{z} &\equiv z_a,\ z_b, \ldots,\\
\mathbf{z}^{\mathbf{n}} &\equiv z_a^n a z_b^n b \cdots,\\
b_{\mathbf{n}} &\equiv b_{n_1,n_2,\ldots,}
\end{aligned} \tag{8l.1}$$

where $b_{\mathbf{n}}$ in $(1/\mathbf{n}!)V^{-1}$ of an integral of the cluster function for a set $\mathbf{n}$ of molecules, we wrote

$$\beta P = \sum_{\mathbf{n}\geq 1} b_{\mathbf{n}} \mathbf{z}^{\mathbf{n}}. \tag{8l.2}$$

The cluster function of the set **n** of molecules is, as for the one-component system, a sum of products of the functions $f(r_{ij})$, but now the individual functions $f_{ij}(r_{ij})$ are no longer identical in their dependence on their arguments, $r_{ij} = |\mathbf{r}_i - \mathbf{r}_j|$, but depend on the species of molecules numbered i and j. As in the one-component case the integrals of the individual products can, for many products, be written as a sum of integrals of a smaller number of molecules, and the cluster integral $\mathbf{n}!\, b_{\mathbf{n}}$ as a sum of products of integrals of irreducible functions. The notation used for the one-component case defined integrals $k!\,\beta_k$ [eq. (8i.21)] numbered by a subscript k which was one less than the number, $k+1$, of molecular coordinates appearing as arguments in the irreducible function. Since $m = k+1$ molecules, the irreducible function is different for different sets

$$\mathbf{m} \equiv m_a, m_b, \ldots, \qquad \sum_a m_a = m, \tag{8l.3}$$

of m_a molecules of species a, m_b of species b, and so on, it is almost obligatory to use **m** as a labeling subscript and define

$$\mathbf{m}!\, B_{\mathbf{m}} = \lim_{V\to\infty} \left(V^{-1} \iint \cdots \int d\mathbf{r}_1 \cdots d\mathbf{r}_{\mathbf{m}} \underset{\substack{\text{sum over all products}\\ \text{of the set } \mathbf{m} = m_a,\, m_b, \ldots\\ \text{of molecules with no}\\ \text{nodal vertices}}}{\sum \prod f_{ij}} \right). \tag{8l.4}$$

For a one-component system, comparing with eq. (8i.21),

$$B_m = \left(\frac{1}{m}\right) \beta_{m-1}, \tag{8l.5}$$

and eqs. (8i.24), (8i.23), and (8i.26) become

$$\beta P = \rho - \sum_{m\geq 2} (m-1) B_m \rho^m, \tag{8l.6}$$

$$\ln z = \ln \rho - \sum_{m\geq 2} m B_m \rho^{m-1}, \tag{8l.7}$$

$$\kappa = \beta \left[\rho - \sum_{m\geq 2} m(m-1) B_m \rho^m \right]^{-1}. \tag{8l.8}$$

Both the combinatorial problem of finding the coefficient of a general product of $B_{\mathbf{m}}$'s in the sum composing $b_{\mathbf{n}}$, handled in Sec. 8i for one component, and that of then evaluating the sums $\sum b_{\mathbf{n}}\mathbf{z}^{\mathbf{n}}$, as in Sec. 8j for one component, are appreciably more difficult in the multicomponent case. For two components the problem has been solved by one of the authors,[(1)] and for the general case by Fuchs.[(2)] We do not present the formal mathematical proofs but state the resulting equations and then produce an a posteriori argument for their validity, which appears to us to be completely compelling. The equations are: for the pressure, with $\mathbf{x} = x_a, x_b$, the mole fraction set,

$$\begin{aligned}\beta P &= \rho - \sum_{\mathbf{m}\geq 2} (m-1)B_{\mathbf{m}}\boldsymbol{\rho}^{\mathbf{m}} \\ &= \rho - \sum_{\mathbf{m}\geq 2} (m-1)B_{\mathbf{m}}\rho^{m}\mathbf{x}^{\mathbf{m}},\end{aligned} \qquad (8l.9)$$

where

$$\rho = \sum \rho_a, \qquad m = \sum_a m_a, \qquad \boldsymbol{\rho}^{\mathbf{m}} = \rho_a^{m_a}\rho_b^{m_b}\cdots; \qquad (8l.10)$$

for the activity z_a of species a,

$$\ln z_a = \ln \rho_a - \rho_a^{-1}\sum_{\mathbf{m}} m_a B_{\mathbf{m}}\boldsymbol{\rho}^{\mathbf{m}}; \qquad (8l.11)$$

and for the compressibility κ

$$\kappa = \beta\left[\rho - \sum_{\mathbf{m}} m(m-1)B_{\mathbf{m}}\boldsymbol{\rho}^{\mathbf{m}}\right]^{-1}. \qquad (8l.12)$$

Equations (9), (11), and (12) are consistent with thermodynamic relationships, namely, from eq. (9), at constant temperature,

$$d(\beta P)_T = \sum_a\left[1 - \sum_{\mathbf{m}\geq 2} \rho_a^{-1} m_a(m-1)B_{\mathbf{m}}\boldsymbol{\rho}^{\mathbf{m}}\right] d\rho_a, \qquad (8l.13)$$

whereas, from eq. (11),

$$\begin{aligned}\sum_a \rho_a(d\ln z_a)_T &= \sum_a\left[1 - \sum_{m\geq 2} \rho_a^{-1} m_a(m_a-1)B_{\mathbf{m}}\boldsymbol{\rho}^{\mathbf{m}}\right] d\rho_a \\ &\quad - \sum_a\left(\sum_{b\neq a} m_a m_b \rho_a^{-1} \sum_{\mathbf{m}\geq 2} B_{\mathbf{m}}\boldsymbol{\rho}^{\mathbf{m}}\right) d\rho_a \\ &= \sum_a\left(1 - \sum_{\mathbf{m}\geq 2} \rho_a^{-1} B_{\mathbf{m}}\boldsymbol{\rho}^{\mathbf{m}}\right) d\rho_a \\ &= d(\beta P)_T,\end{aligned} \qquad (8l.14)$$

(1) Joseph E. Mayer, *J. Phys. Chem.*, **43,** 71 (1939).
(2) K. Fuchs, *Proc. R. Soc., Ser. A*, **179,** 408 (1942).

as required by the thermodynamic relation $(\partial\beta P/\partial \ln z_a)_{T,z_b,\ldots} = \rho_a$. Since at the limit that all $\rho_a = z_a \to 0$ we have $\beta P \to 0$ and $\rho_a \ln z_a = \rho_a \ln \rho_a \to 0$, it follows that, if either eq. (9) or (11) is valid, the other is correct. Similarly, for the isothermal compressibility $\kappa = \rho^{-1}(\partial\rho/\partial P)_{T,x_a,x_b,\ldots} = [\rho(\partial P/\partial\rho)_{T,\mathbf{x}}]^{-1}$, we find that differentiation of eq. (9) leads to eq. (12). Equations (9), (11), and (12) reduce to the one-component cases, eqs. (6), (7), and (8), respectively, if $x_a \to 1$, $x_b = x_c = \cdots \to 0$.

Consider the case in which all species a, b, ... have the same mutual pair potentials $u_{aa}(r) = u_{ab}(r) = u_{bb}(r) = \cdots$, which is very nearly the case for heavy molecules of differing isotopic composition. For this case the functions for different sets $\mathbf{m} = m_a, m_b, \ldots$ depend on the number m only, and the $B_{\mathbf{m}}$'s differ only in the factorials $\mathbf{m}!$ [eq. (4)],

$$B_{\mathbf{m}} = \left(\frac{m!}{\mathbf{m}!}\right) B_m, \qquad \text{(all } u_{ij}\text{'s equal).} \tag{8l.15}$$

Now in this case it is easy to show from the fundamental grand canonical partition function that βP must be completely independent of composition, that is, of the mole fraction set $x_a, x_b, \ldots, \sum x_a = 1$. But if one assumes that the function $P(\rho_a, \rho_b, \ldots)$ is regular at the origin, it can be developed as a power series,

$$\beta P = \sum_{\mathbf{m}} C_{\mathbf{m}} \boldsymbol{\rho}^{\mathbf{m}} = \sum_m \rho^m \sum_{\mathbf{m}, \sum m_a = m} C_{\mathbf{m}} \mathbf{x}^{\mathbf{m}}, \tag{8l.16}$$

and the only way βP can be independent of the set $\mathbf{x}$ is that

$$C_{\mathbf{m}} = \left(\frac{m!}{\mathbf{m}!}\right) C_m \tag{8l.17}$$

for all sets $\mathbf{m}$ of given $\sum_a m_a = m$, since then, for every m,

$$\sum_{\mathbf{m}(\sum m_a = m)} C_{\mathbf{m}} \mathbf{x}^{\mathbf{m}} = C_m \sum_{\mathbf{m}} \frac{m!\, \mathbf{x}^{\mathbf{m}}}{\mathbf{m}!} = C_m \left(\sum_a x_a\right)^m = C_m, \tag{8l.18}$$

since $\sum x_a = 1$. This proves that eq. (9) is valid for identical u_{ij}-potentials, but this could be an artifact valid only when eq. (15) holds.

A more cogent argument follows from a consideration of the limit as $\rho_a \to 0$ of $(\partial^\nu \beta P/\partial \rho_a^\nu)_{T,\rho_b,\ldots}$ which, using eq. (9), is found to be [with $\delta(\nu - 1) = 1$ if $\nu = 1$, and zero otherwise]

$$\lim_{\rho_a \to 0} \left(\frac{\partial^\nu \beta P}{\partial \rho_a^\nu}\right)_{T,\rho_b,\ldots} = \delta(\nu - 1) - \rho_a^{-\nu} \sum_{\mathbf{m}(m_a = \nu)} \nu!\, B_{\mathbf{m}} \boldsymbol{\rho}^{\mathbf{m}}, \tag{8l.19}$$

namely, that it involves integrals of irreducible functions containing exactly ν molecules of species a in its coordinate arguments. It is worthwhile noting here that from the activity series $\beta P = \sum b_{\mathbf{n}} \mathbf{z}^{\mathbf{n}}$ a similar result follows,

$$\lim_{z_a \to 0} \left(\frac{\partial^\nu \beta P}{\partial z_a^\nu} \right)_{T, z_b, \ldots} = z_a^{-1} \sum_{\mathbf{n}(n_a = \nu)} \nu! \, b_{\mathbf{n}} \mathbf{z}^{\mathbf{n}}, \tag{8l.20}$$

namely, that only integrals of functions having ν coordinates of molecules of type a occur.

From eq. (13) for $d(\beta P)_T$ one can obtain an expression for a quantity which is trivial in a one-component system, namely, the partial molecular volume,

$$\bar{v}_a = \left(\frac{\partial V}{\partial N_a} \right)_{T, P, N_b, \ldots}.$$

Set $\rho_a = N_a/V$, $d\rho_a = V^{-1}[dN_a - \rho_a \, dV]$ for $d\rho_a$, in eq. (13) to write

$$d(\beta P)_T = V^{-1} \left\{ \sum_a \left[1 - \sum_{\mathbf{m} \geq 2} \rho_a^{-1} m_a (m-1) B_{\mathbf{m}} \boldsymbol{\rho}^{\mathbf{m}} \right] dN_a - \left[\rho - \sum_{\mathbf{m} \geq 2} m(m-1) B_{\mathbf{m}} \boldsymbol{\rho}^{\mathbf{m}} \right] dV \right\}. \tag{8l.22}$$

With $d(\beta P) = 0$ and $dN_b = 0$ for $b \neq a, \ldots$ we find, from eqs. (21) and (22),

$$\bar{v}_a = \frac{1}{\rho} \frac{\rho - x_a^{-1} \sum_{\mathbf{m} \geq 2} m_a (m-1) B_{\mathbf{m}} \boldsymbol{\rho}^{\mathbf{m}}}{\rho - \sum_{\mathbf{m} \geq 2} m(m-1) B_{\mathbf{m}} \boldsymbol{\rho}^{\mathbf{m}}}. \tag{8l.23}$$

$$\sum_a x_a v_a = v = \frac{V}{N} = \rho^{-1}$$

is obeyed by eq. (23), as it must be. From eq. (12) for the isothermal compressibility κ, we can write

$$\bar{v}_a = \frac{\kappa}{\rho \beta} \left[\rho_a - x_a^{-1} \sum_{\mathbf{m} \geq 2} m_a (m-1) B_{\mathbf{m}} \boldsymbol{\rho}^{\mathbf{m}} \right], \tag{8l.24}$$

or

$$\rho \bar{v}_a = (\rho \kappa k T) \left[1 - \rho_a^{-1} \sum_{\mathbf{m} \geq 2} m_a (m-1) B_{\mathbf{m}} \boldsymbol{\rho}^{\mathbf{m}} \right]. \tag{8l.24'}$$

8m. SIMPLY INTEGRABLE DIAGRAMS

Integration of a general function of many variables, even with modern computers, is impossibly time-consuming, especially if the integrand has both positive and negative domains in the total region of integration, since then the desired integral is a difference, which may be small, of the integrals over the positive and negative domains. Consider the brute force method of integrating a function $f(x)$ of a single variable between two limits a and b, $I = \int_a^b dx\, f(x)$. One can compute the value of $f(x)$ at values of $x_i = a + i(a-b)n^{-1}$ for integer i- and n-values, sum $\sum_{i=0}^{n-1} f(x_i)$, and multiply by the interval $(a-b)n^{-1}$,

$$I \cong (a-b)n^{-1} \sum_{i=0}^{n-1} f(x_i). \tag{8m.1}$$

In the limit $n \to \infty$ this is exact if $f(x)$ is reasonably behaved and finite in value. But suppose we have a function of ν variables $f(x_1, \ldots, x_\nu)$. We now need to evaluate f at n^ν values $f(x_{1i}, \ldots, x_{\nu j})$. If for the single variable we were satisfied with $n = 10^2$, we need $10^{2\nu}$ evaluations of f, and it is clear that the method soon becomes impossible. Of course, there are better methods of computer use, but we are obviously limited to comparatively small values of ν for functions as complicated as the irreducible integrands of the β_k's.

There is, however, an analytical procedure that can be used up to all values of k, $1 \le k \le \infty$, for certain classes of diagrams appearing in the irreducible integrands $\prod_{i>j} f_{ij}$. In going from the cluster functions, which included diagrams having nodal vertices (Sec. 8i), we were able to sum rigorously $\sum b_n z^n$ to $n = \infty$ in terms of $\sum \beta_k \rho^k$, where part of the contribution of very large n-values of $b_n z^n$ is included in $\sum \beta_k \rho^k$ for small values of k. In this section we use an analogous procedure to make feasible a computation for *parts* of the integral $\beta_k \rho^k$ for very large k-values.

The method starts with a procedure known as folding. With x a single real variable, and given x_0 and x_{n+1}, let us define an integral as a function of $|x_{n+1} - x_0|$ by the equation

$$I_n(|x_{n+1} - x_0|) = \iint_{-\infty}^{\infty} \cdots \int dx_1 \cdots dx_n \prod_{i \ge 1}^{n+1} f(|x_i - x_{i-1}|), \tag{8m.2}$$

namely, as a single chain of $n+1$ functions between x_0 and x_{n+1}. This will be finite if $f(x)$ goes to zero more rapidly than x^{-1} as $x \to \infty$. Use the Fourier integral transform (see Appendix VIII), but with $ph^{-1} = t$ and $h^{-1}\, dp = dt$),

$$g(t) = \int_{-\infty}^{+\infty} dx\, f(x)\, \varphi(tx), \tag{8m.3}$$

where

$$\varphi(tx) = e^{2\pi i t x}, \qquad \varphi^*(tx) = e^{-2\pi i t x}, \tag{8m.4}$$

for which

$$\int dx\, \varphi(t'x)\varphi^*(tx) = \delta(t'-t),$$

$$dt\, \varphi(tx')\varphi^*(tx) = \delta(x'-x), \tag{8m.5}$$

with δ the Dirac delta function,

$$\int dx\, \delta(x'-x)f(x)=f(x'). \tag{8m.6}$$

$$\int dt\varphi^*(tx)g(t)=\iint dx'\, dtf(x')\varphi^* tx')\varphi(tx)$$

$$=\int dx' f(x')\delta(x-x')=f(x),$$

$$f(x)=\int dt g(t)\varphi^*(tx). \tag{8m.7}$$

Now use eq. (7) for the functions $f(|x_i-x_{i-1}|)$ in eq. (2), so that, since from eq. (4) $\varphi^*(-xt)=\varphi(xt)$,

$$I_n(|x_n-x_0|)=\iint\cdots\int dx_1\cdots dx_n\, dt_1\cdots dt_n\, dt_{n+1}$$

$$\times\prod_{i=1}^{n+1} g(t_i)\varphi^*(t_i|x_i-x_{i-1}|)$$

$$=\iint\cdots\int dt_1\cdots dt_n\prod_{i=1}^{n+1} g(t_i)\varphi(t_1x_0)\varphi^*(t_{n+1}x_{n+1})$$

$$\times\iint\cdots\int dx_1\, dx_n\prod_{i=1}^{n}\varphi^*(t_ix_i)\varphi(t_{i+1}x_i)$$

$$=\iint\cdots\int dt_1\cdots dt_n\prod_{i=1}^{n+1} g(t_i)\prod_{i=1}^{n}\delta(t_i-t_{i+1})$$

$$=\int dt[g(t)]^{n+1}\varphi(t|x_{n+1}-x_0|). \tag{8m.8}$$

For a coordinate $r=(x^2+y^2+z^2)^{1/2}$ with volume element $dx\,dy\,dz=4\pi r^2\,dr$, the appropriate transform[(1)] for a real dimensionless function $f(r)$ is a function $\chi(t)$ of dimension l^3,

$$\chi(t)=\int 4\pi r^2\, dr f(r)(2\pi tr)^{-1}\sin 2\pi tr, \tag{8m.9}$$

with inverse (where, since tr is dimensionless, $t^2\,dt\,\chi$ is also)

$$f(r)=\int 4\pi t^2\, dt\,\chi(t)(2\pi tr)^{-1}\sin 2\pi tr, \tag{8m.9′}$$

[(1)] See, for instance, Elliott W. Montroll and Joseph E. Mayer, *J. Chem. Phys.*, **2,** 626 (1941).

so that

$$I_k(r_{0,k+1}) = \int\int_0^\infty \cdots \int d\mathbf{r}_1 \cdots d\mathbf{r}_k \prod_{i=1}^{k+1} f(r_{i,i-1})$$

$$= \int 4\pi t^2 \, dt [\chi(t)]^{k+1} (2\pi t r_{0,k+1})^{-1} \sin 2\pi t r_{0,k+1}. \quad (8m.10)$$

The simplest calculation we can now perform is to reduce the computation of the total contribution of the ring graphs to the sum $\sum \beta_k \rho^k$ to two successive single-variable integrations. By ring graph we mean those with $k \geq 2$ of $k+1$ molecules having integrands of the form $f(r_{01})f(r_{12})\cdots f(r_{k-1,k})f(r_{k0})$. The integration for a single specified product over the coordinates of k molecules, holding one fixed at the origin, is $I_k(r_{0,k+1}=0)$. As $r_{0,k+1} \Rightarrow 0$, $(2\pi t r_{0,k+1})^{-1} \sin 2\pi t r_{0,k+1} \Rightarrow 1$. We have

$$\lim_{r_{0,k+1}\to 0} [I_k(r_{0,k+1})] = \int_0^\infty 4\pi t^2 \, dt [\chi(t)]^{k+1}. \quad (8m.11)$$

The number of ways we can permute the order of the numbered molecules in the ring is $\frac{1}{2}k!$, since, holding the zeroth molecule fixed, the others can be arranged in $k!$ different ways clockwise; but for each such permutation the anticlockwise order of the same permutation is the identical product $\prod_{i=1}^n f(r_{i,i-1})$. Since β_k is defined as $(k!)^{-1}$ times the integral, the factorials cancel, leaving the factor $\frac{1}{2}$. We have

$$\sum_{k\geq 2} \beta_k^{\text{ring}} \rho^k = \int 2\pi t^2 \, dt \, \chi(t) \sum_{k\geq 2} [\rho\chi(t)]^k$$

$$= \int 2\pi t^2 \, dt \, \chi(t) [\rho\chi(t)]^2 [1-\rho\chi(t)]^{-1}. \quad (8m.12)$$

Now we can proceed much further. Let $C(r)$ be a dimensionless function of the distance r between two specified numbered molecules defined by the sum

$$C(r) = \sum_{k\geq 1} \rho^k I_k(r)$$

$$= \int 4\pi t^2 \, dt \, \chi(t) \left\{ \sum_{k\geq 1} [\rho\chi(t)]^k \right\} (2\pi r t)^{-1} \sin 2\pi r t.$$

$$= \int 4\pi t^2 \, dt \, \chi(t) \rho\chi(t) [1-\rho\chi(t)]^{-1} (2\pi r t)^{-1} \sin 2\pi r t. \quad (8m.13)$$

This is the sum of $(k!)^{-1}$ times the integrals over $\prod_{i=1}^k \int \rho \, d\mathbf{r}_i$ of all single chains containing one or more intermediate numbered molecules linking the two end molecules. Since the order in the chain has $k!$ permutations, the factorials cancel. The factor $\frac{1}{2}$ in eq. (12) is now missing, since the two ends of the chains are distinguishable by the numbered end molecules. Then go to other dimensionless functions $S'_\lambda(r_{12})$ which are $\rho^n(n!)^{-1}$ times the sum of the integrals for a total of n intermediate molecules of products forming all numbers ν, $\lambda = \nu \leq \infty$, of at least λ parallel independent chains. For $\lambda \geq 2$ we refer to these as a spindle of chains, each containing one or more intermediate molecules. If there are ν_k chains with k molecules each, $\sum_{k\geq 1} \nu_k = \nu \geq \lambda$ and $\sum_{k\geq 1} k\nu_k = n$. The chains are distinguishable only by the number of molecules in them. We can assign the n numbered molecules to the ν chains in $n!(\prod_k (K!)^{\nu_k} \nu_k!)^{-1}$ ways. This is to be divided by $n!$ and multiplied by $\prod (k!)^{\nu_k}$, the number of ways we can permute the order of the molecules indices within each chain. For a given set of numbers $\nu_1, \nu_2, \ldots, \nu_k, \ldots$ the contribution is $\prod [\rho I_k(r_{12})]^{\nu_k} (\nu_k!)^{-1}$. We sum this over all values of $\nu_k \geq 0$ for all $k \geq 1$, but with $\sum \nu_k = \nu \geq \lambda$. The results are

$$S'_\lambda(r) = \left\{\exp\left[\sum_{k\geq 1} \rho I_k(r)\right]\right\} - \sum_{\nu \geq 0}^{\lambda - 1} (\nu!)^{-1} \left[\sum_{k \geq 1} \rho I_k(r)\right]^\nu$$

or, with eq. (13) for $C(r)$,

$$\begin{aligned} S'_1(r) &= e^{C(r)} - 1, \\ S'_2(r) &= e^{C(r)} - 1 - C(r), \\ S'_3(r) &= e^{C(r)} - 1 - C(r) - \tfrac{1}{2}C^2(r). \end{aligned} \tag{8m.14}$$

We wish to use these to sum the contributions to $\sum \beta_k \rho^k$ of as many products in the irreducible functions as desirable, avoiding impossible numerical integrations over many coordinates.

The caution we must exercise is to avoid counting identical products more than once. For instance, we propose that, by using a Fourier transform $\sigma(t)$ of a function involving those in eq. (14) instead of $\chi(t)$ in eq. (12) for the ring graphs, we can sum the contribution of enormously more products to $\sum \beta_k \rho^k$. Included in this contribution is a ring of three molecules numbered 1, 2, 3, all three pairs 2-3, 3-1, and 1-2 being connected by direct links $f(r_{23}), \ldots$ *and* by f with at least one parallel chain, or by two or more chains. Were we to include the single chain *without* the parallel direct link $f(r_{23})$, we would count the corresponding product in the rings of $n > 3$.

We define the function

$$\begin{aligned} S_2(r) &= f(r) + f(r)S'_1(r) + S'_2(r) \\ &= [f(r) + 1]e^{C(r)} - C(r) - 1 \\ &= e^{-\beta u(r) + C(r)} - C(r) - 1, \end{aligned}$$

since $f(r)+1=e^{-\beta u(r)}$. With $\sigma(t)=\int 4\pi r^2\,dr\,S_2(r)(2\pi rt)^{-1}\sin 2\pi rt$ in eq. (12) replacing $\chi(t)$, we find the contribution of all ring graphs of three or more molecules,

$$\sum \beta_k^{\text{spindle rings}}\rho^k=\int 2\pi t^2\,dt\,\delta(t)[\rho\sigma(t)]^2[1-\rho\sigma(t)]^{-1}. \qquad (8\text{m}.15)$$

Adding $\beta_1=\int 4\pi r^2\,dr\,f(r)$ to eq. (15) we have included as many terms in the $\beta_k\rho^k$ sum as seem worthwhile to compute without recourse to multidimensional integrations. It is of course not obvious without careful analysis that retaining to infinity selected members of an infinite series improves the approximation as compared to using only a limited number of the first few members. A trivial counterexample is the case in which even and odd members of the original series have opposite sign; summing to infinity only even or odd members would hardly be wise. It appears that in some cases at least the procedure used here is useful; see, for instance, it use in Secs. 8n and 8o.

Some relations between the functions $f(r)=e^{-\beta u(r)}-1$ and $\chi(t)$ follow from eqs. (9) and (9′) using $(2\pi tr)^{-1}\sin(2\pi tr)=1$ if either $r\to 0$ or $t\to 0$. They are

$$f(r\to 0)=-1=\int 4\pi t^2\,dt\,\gamma(t), \qquad (8\text{m}.16)$$

$$\chi(t\to 0)=\int 4\pi r^2 dr\,f(r)=\beta_1, \qquad (8\text{m}.16')$$

where β_1, the second virial coefficient, is $-[\partial(\beta P\rho^{-1})/\partial\rho]_{\beta,\rho\to 0}$ and is positive except at temperatures far above the critical temperature. From eq. (10) for I_k, with $k=1$ at $r\to 0$,

$$I_1(r=0)=\int 4\pi r^2\,dr\,f^2(r)=\int 4\pi t^2\,dt\,\chi^2(t). \qquad (8\text{m}.17)$$

8n. PLASMA, DEBYE–HÜCKEL APPROXIMATION

A gas of charged molecules, that is, of ions or of ions and electrons, is called a plasma. An equilibrium, or near-equilibrium, gaseous plasma exists only at temperatures of many thousands of degrees. Its properties are of great interest to astrophysicists, especially in the presence of relatively strong magnetic fields, and in the last few decades to scientists working with atomic bombs or controlled fusion. The mathematics of its treatment, however, can be taken over almost unchanged in the discussion of ionic solutions in water or other high dielectric solvents at more normal temperatures; see Sec. 9j, where the early Debye–Hückel theory is given.

The bulk of a plasma at equilibrium is always electrically neutral. If there were an excess of one charge, that excess charge would go to the surface, leaving a zero macroscopic electric field in the interior, although at a different absolute electrical potential than the surroundings. The plasma must then consist of at least two component species of ions, one positively and the other negatively charged, so that the formalism of a multicomponent system is required.

Subscripts $a, b, \ldots$ are used to designate the different ionic species having number densities $\rho_a = N_a/V$, $\rho_b, \ldots$. The condition of electric neutrality requires that, summed over all species a,

$$\sum_a \rho_a z_a \epsilon = 0, \tag{8n.1}$$

where z_a is the integer electrical charge on ions of species a in terms of the absolute electronic charge ϵ.

There are two unique features of plasma that distinguish its statistical theoretical treatment from that of the neutral molecule multicomponent system. One of these complicates, but the other simplifies, the analysis. The first is that the long-range pair potential energy of $z_a z_b \epsilon^2 r^{-1}$ between ions of species a and b leads to divergent integrals of the irreducible functions. The second feature is that this mutual electric pair potential is so strong that deviation from perfect gas behavior is pronounced at low number densities, so low that, in first approximation at least, the ever-present van der Waals–type short-range pair potential can be neglected, and in higher approximation it can be reasonably well approximated by a simple hard sphere repulsion at pair distance $r \leq r_0$. The simplicity of the electrostatic pair interaction is such that all first-order deviation from perfect gas behavior depends on one parameter only, the so-called Debye kappa, κ, of the dimensions of a reciprocal length, defined by[(1)] its square,

$$\kappa^2 = 4\pi\beta\epsilon^2 \sum_a \rho_a z_a^2, \tag{8n.2}$$

with $\beta = 1/kT$ and ϵ the electronic charge.

The model for which we wish to make calculations is that of hard-sphere ions all of the same diameter r_0. The potential energy is a sum of pair terms,

$$\begin{aligned} u_{ab}(r) &= \infty & r &\leq r_0, \\ &= \epsilon^2 z_a z_b r^{-1}, & r &> r_0, \end{aligned} \tag{8n.3}$$

[(1)] For ions in water solution the mutual pair potential is $z_i z_j \epsilon^2/Dr$, with D the water dielectric constant having a value of about 80. At the same temperature the equations of this section apply with κ^2 of eq. (2) divided by D, that is, 80-fold smaller (Sec. 9j).

so that

$$f_{ab}(r) = e^{-\beta u_{ab}(r)} - 1 = -1, \qquad r \leq r_0,$$
$$= [\exp(-\beta\epsilon^2 z_a z_b r^{-1})] - 1, \qquad r > r_0, \tag{8n.4}$$

but, in order to ensure convergence of certain integrals we initially replace the r^{-1} dependence by $r^{-1}e^{-\alpha r}$ and later let α approach zero. We first artificially construct a function that is a sum of an analytic term $r^{-1}e^{-\alpha r}$ for $0 \leq r \leq \infty$ plus a discontinuous term, zero for $r > r_0$, but which gives $f(r)$ of eq. (4) approximately for all small α and exactly for $\alpha \to 0$.

Let $\theta(r)$ be a discontinuous function of r,

$$\theta(r) = 1, \qquad r \leq r_0,$$
$$= 0, \qquad r > r_0, \tag{8n.5}$$

and

$$\psi(\alpha, r_0, r) = r^{-1}e^{-\alpha r}[1 - \theta(r)], \tag{8n.6}$$

so that

$$r\psi(\alpha, r_0, r) = 0, \qquad r \leq 0,$$
$$= e^{-\alpha r}, \qquad r > 0. \tag{8n.7}$$

The function

$$\Psi_{ab}(\alpha, r_0, r) = \{\exp[-\beta\epsilon^2 z_a z_b \psi(\alpha, r_0, r)]\}[1 - \theta(r)] - 1 \tag{8n.8}$$

is then

$$\Psi_{ab}(r) = -1, \qquad r \leq r_0,$$
$$= \{\exp[-\beta\epsilon^2 z_a z_b r^{-1} e^{-\alpha r}]\} - 1, \qquad r > r_0, \tag{8n.8}$$

and from eq. (4),

$$f_{ab}(r) = \lim_{\alpha \to 0} [\Psi_{ab}(\alpha, r_0, r)] \tag{8n.10}$$

We now turn to the problem of computation using the techniques of the last section, but with several essential differences. The most obvious one is that we now have a system of more than one species, at least two ions, one positively and one negatively charged. This offers no essential difficulty, as we have seen previously (Sec. 8g). However, we have one very great simplification over the multicomponent case with Lennard-Jones–type potentials in that by the assumption of using the *same* hard-sphere diameter for all species pairs the only difference between the different pair species pseudo-potentials proportional to $\psi(\alpha, r_0, r)$ of eq. (6) is a factor $\beta\epsilon^2 z_a z_b$, which of course is also true in the Fourier transform $\varphi(s)$ of the potential we use in this development.

The assumption of a single r_0 for all pairs, although certainly incorrect for real ions, is probably a very good approximation for real systems. First, its value does not enter in the lowest-order term, but only in higher-order terms. Second, ions of equal charge at small distances repel each other strongly. Adding an infinite repulsion for $r<r_0$ probably has little effect on the results. Oppositely charged ions attract each other. We should therefore interpret our single r_0 to be the sum of the hard core radii for the positive and negative species.

Another very real simplification over the Lennard-Jones potential system calculation is that the Fourier transform function of the long-range part $r^{-1}e^{-\alpha r}$ of the potential is a simple analytic function, rather than the $\chi(t)$ of eq. (8m.9) which has to be computed numerically. It is this simplification that motivates us to use the otherwise more awkward method of operating directly with $\beta u_{ab}(r)$ rather than with $f_{ab}(r)=e^{-\beta u_{ab}(r)}-1$. We use the equivalent of $f_{ab}(r)=\sum_{\nu\geq 1}(\nu!)^{-1}[-\beta u_{ab}(r)]^{\nu}$ with Ψ_{ab} of eq. (8), replacing f_{ab} by

$$\Psi_{ab}(r)=-\theta(r)+\sum_{\nu\geq 1}(\nu!)^{-1}[-\beta\epsilon^2 z_a z_b \psi(\alpha, r_0, r)]^{\nu} \tag{8n.11}$$

and ψ given by eq. (6), which is the same function for all species pairs aa, ab, bb, and so on.

We first examine the case in which $r_0\to 0$, which gives the lowest-order correction term to the perfect gas. With

$$\lim_{r_0\to 0}[\psi(\alpha, r_0, r)]=\psi_0(\alpha, r)=r^{-1}e^{-\alpha r}, \tag{8n.12}$$

the Fourier transform, normalized as in eq. (8m.9) is

$$\begin{aligned}\chi_0(t)&=\int_0^{\infty}4\pi r^2\,dr\,r^{-1}e^{-\alpha r}(2\pi rt)^{-1}\sin 2\pi rt\\ &=4\pi s^{-1}\int_0^{\infty}dr\,e^{-\alpha r}\sin sr,\end{aligned} \tag{8n.13}$$

where $s=2\pi t$. It is more convenient, and also more in line with conventional notation in this case, to use s as a variable and a function defined in the general case by

$$\varphi(\alpha, r_0, s)=(4\pi)^{-1}\chi(\alpha, r_0, s)=s^{-1}\int r\,dr\,\psi(\alpha, r_0, r)\sin sr, \tag{8n.14}$$

so that, at $r_0=0$,

$$\varphi_0(\alpha, s)=s^{-1}\int r\,dr\,\psi_0(\alpha, r)\sin sr=s^{-1}\int dr\,e^{-\alpha r}\sin sr \tag{8n.14'}$$

for the Fourier transform. The inverse is now

$$\psi_0(\alpha, r)=r^{-1}e^{-\alpha r}=2\pi^{-1}r^{-1}\int s\,ds\,\varphi_0(\alpha, s)\sin sr. \tag{8n.15}$$

The integral $\int dr e^{-\alpha r} \sin sr$ of eq. (14′) is found in most tables of definite integrals and is $s(\alpha^2+s^2)^{-1}$, so that

$$\varphi_0(s)=(\alpha^2+s^2)^{-1}. \tag{8n.14″}$$

An analogue of the chain integral of eq. (8m.10), the integral over the coordinates of k molecules of the chain between two molecules a distance r apart, is

$$\begin{aligned} J_k^{(\alpha,0)}(r) &= \int\!\!\int_0^\infty\!\!\cdots\!\int d\mathbf{r}_1\cdots dr_k \prod_{i=1}^{i=k+1} \psi_0(\alpha,\, r_{i,i-1}) \\ &= \int 4\pi t^2\, dt[4\pi\varphi_0(2\pi t)]^{k+1}(2\pi rt)^{-1}\sin 2\pi rt \\ &= \frac{2}{\pi} r^{-1}\int_0^\infty s\, ds\, \varphi_0(s)[4\pi\varphi_0(s)]^k \sin rs. \end{aligned} \tag{8n.16}$$

Now we want to sum over all species of ions $(\prod k_a!)^{-1}$ times $I_k^{(0)}$, integrate over $\prod_i \rho_i\, d\mathbf{r}_i$, where ρ_i is the number density of the species to which i belongs, and multiply each $\psi_0(r_{i,i-1})$ by the factor $-\beta\epsilon^2 z_i z_j$ (eq. 8). If the two ends of the chain are of species a and b, respectively, the total factor is $(-)^{k+1}\beta(z_a\epsilon)(z_b\epsilon)\prod_{i=1}^k \beta z_i^2\epsilon^2=(-)^{k+1}\beta(z_a\epsilon)(z_b\epsilon)\prod_a(\beta z^2\epsilon^2)^{k_a}$ if there are k_a molecules of species a in the chain between a and b. The k numbered intermediate molecules can be ordered in $k!$ ways. Multiply by $k!\prod_a \rho_a^{k_a}(k_a!)^{-1}$ and sum over all species a with $\sum_a k_a = k$ to find, with eq. (2) for κ^2,

$$\begin{aligned} &\sum_{k_a}\sum_{k_b}\cdots\sum(-)^k k!\prod_a(4\pi\beta\epsilon^2\rho_a z_a^2)^{k_a}(k_a!)^{-1} \\ &\qquad = (-)^k\sum_a\left(4\pi\beta\epsilon^2\sum_a \rho_a z_a^2\right)^k = (-\kappa^2)^k. \end{aligned} \tag{8n.17}$$

The properly weighted chain integral $C_{k(ab)}^{(0)}(r)$ for two molecules of species a and b at a distance r apart, summed over all species of k intermediate molecules, if $r_0=0$, is then

$$\begin{aligned} C_{k(ab)}^{(\alpha,0)}(r) &= -\beta(z_a\epsilon)(z_b\epsilon)r^{-1}2\pi^{-1}\int_0^\infty s\, ds\, \varphi_0(s)[-\kappa^2\varphi_0(s)]^k \sin rs \\ &= -\beta(z_a\epsilon)(z_b\epsilon)r^{-1}2\pi^{-1}\int_0^\infty s\, ds\,(\alpha^2+s^2)^{-1}[-\kappa^2(\alpha^2+s^2)^{-1}]^k \sin rs, \end{aligned} \tag{8n.18}$$

with eq. (16) for φ_0.

The sum for all k values from zero to infinity is

$$\sum_{k\geq 0}^{\infty} C_{k(ab)}^{(\alpha,0)}(r) = -\beta(z_a\epsilon)(z_b\epsilon)r^{-1}2\pi^{-1}\int_0^{\infty} s\,ds(\alpha^2+s^2)^{-1} \times \sum_{k\geq 0}[-\kappa^2(\alpha^2+s^2)^{-1}]^k \sin rs$$

$$= -\beta(z_a\epsilon)(z_b\epsilon)r^{-1}2\pi^{-1}\int_0^{\infty} s\,ds \times\{(\alpha^2+s^2)[1+\kappa^2(\alpha^2+s^2)^{-1}]\}^{-1}\sin rs$$

$$= -\beta(z_a\epsilon)(z_b\epsilon)r^{-1}2\pi^{-1}\int_0^{\infty} s\,ds(\alpha^2+s^2+\kappa^2)^{-1}\sin rs \tag{8n.18$'$}$$

or, when $\alpha\to 0$,

$$\sum_{k\geq 0} C_{k(ab)}^{(0)} = C_{ab}^{(0)}(r) = -\beta(z_a\epsilon)(a_b\epsilon)r^{-1}2\pi^{-1}\int_0^{\infty} s\,ds(\kappa^2+s^2)^{-1}\sin rs$$

$$= -\beta(z_a\epsilon)(z_b\epsilon)r^{-1}e^{-\kappa r}. \tag{8n.19}$$

To any graph with two unique vertices i and j we can add the contribution of all other graphs with i and j by a chain of k intermediate molecules. Actually, we can also add graphs joined by a spindle of parallel chains without changing the numbering on other molecules of the graphs, although this is not pertinent to this section. Doing this replaces the direct $-\beta u(r) = -\beta(z_i\epsilon)(z_j\epsilon)r^{-1}$ by the convergent function of eq. (19). This one difference between the manipulations in this section and those in Sec. 8m is that now, operating with $-\beta u$ rather than $f = e^{-\beta u}-1$, we can have any number of direct bonds $[-\beta u(r)]^{\nu}(\nu!)^{-1}$ between i and j, whereas in Sec. 8m in any spindle of two or more parallel chains only one could include a direct bond.

A neutral plasma necessarily consists of at least two oppositely charged species, but the fact that the analytic form of the potential $z_iz_j\epsilon^2r^{-1}$ is the same for all species greatly facilitates the operations. Summing over all species of the intermediate ions in the chains we have effectively reduced the problem within the chain to that of a one-component system for which $\rho\beta u(r)$ is $(4\pi)^{-1}\kappa^2r^{-1}$. The sum over all chain lengths gives us the welcome converging factor $e^{-\kappa r}$.

With x_a the mole fraction of species a,

$$x_a = \rho_a\rho^{-1}, \tag{8n.20}$$

we sum $x_ax_bC_{ab}^{(0)}(r)$ of eq. (19) over all species,

$$\sum_{ab}\sum x_ax_bC_{ab}^{(0)}(r) = -\beta\sum_{ab}\sum(x_az_a\epsilon)(x_bz_b\epsilon)r^{-1}e^{-\kappa r}$$

$$= -\beta\rho^{-2}\left(\sum_a \rho_az_a\epsilon\right)^2 r^{-1}e^{-\kappa r} = 0, \tag{8n.21}$$

from the electric neutrality condition of eq. (1). The sum of all single chains is zero.

The ring integral contribution to $\beta_k \rho^k$, including the integral $\int 4\pi r^2\, dr \frac{1}{2}[\beta u(r)]^2$ which contributes to $\beta_1 \rho$, gives the lowest-order correction to the thermodynamic functions. We construct rings by multiplying the chain function $C_{ab}^{(0)}(r)$ of eq. (19) by the direct interaction $-\beta u(r_{ab}) = -\beta(z_a \epsilon)(z_b \epsilon) r_{ab}^{-1}$. Multiply the product by $x_a x_b$ and sum over all species a, b, using eq. (2) for κ^2, to construct a dimensionless function $R^{(0)}(r)$ corresponding to a single species of molecule,

$$R^{(0)}(r) = \sum_{ab}\sum x_a x_b [-\beta(z_a\epsilon)(z_b\epsilon) r^{-1}] C_{ab}^{(0)}(r) \left(\rho^{-1}\beta\epsilon^2 \sum_a \rho_a z_a^2\right) \times \left(\rho^{-1}\beta\epsilon^2 \sum_b \rho_b \epsilon^2\right) r^{-2} e^{-\kappa r}$$

$$= \rho^{-1}[(4\pi)^{-1}\kappa^2]^2 r^{-2} e^{-\kappa r}. \tag{8n.22}$$

The dimensionless integral

$$R^{(0)} = \rho \int 4\pi r^2\, dr\, R^{(0)}(r) = \rho^{-1}(4\pi)^{-1}\kappa^3 \int_0^\infty d\kappa r e^{-\kappa r} = (4\pi)^{-1}\kappa^3 \rho^{-1} \tag{8n.23}$$

includes rings of all numbers, $k+1 \geq 2$, of molecules. Since κ^2 is proportional to ρ, we see the correction terms are proportional to $\rho^{1/2}$.

The ring integral term in $\beta_k \rho^k$ is *one-half* of the integral $[\prod_{i=1}^{i=k} \int \rho\, d\mathbf{r}_i f(r_{i,i-1})] f(r_{k,0})$, as a result of the identity of the products of clockwise and anticlockwise ordering (Sec. 8m). We find, from eq. (23),

$$\sum_{k \geq 1} \beta_k^{\text{ring}} \rho^k = (8\pi)^{-1} \kappa^3 \rho^{-1}. \tag{8n.23'}$$

Define κ^* as $\kappa \rho^{-1/2}$, so that

$$(\kappa^*)^2 = 4\pi\beta\epsilon^2 \sum_a x_a z_a^2 \tag{8n.24}$$

is independent of ρ. We then have

$$\sum_{k \geq 1} \beta_k^{\text{ring}} \rho^k = \frac{1}{8\pi} (\kappa^*)^3 \rho^{1/2}. \tag{8n.23''}$$

The equation for the pressure P is given by eq. (8i.24),

$$\beta P \rho^{-1} = 1 - \sum_{k \geq 1} \left[\frac{k}{k+1}\right] \beta_k \rho^k. \tag{8n.25}$$

Now

$$\rho \frac{\partial}{\partial \rho} \rho^{-1} \int_0^\rho d\rho'\, \beta_k (\rho')^k = k(k+1)^{-1} \beta_k \rho^k,$$

and

$$\rho \frac{\partial}{\partial \rho} \rho^{-1} \int_0^P d\rho' (\rho')^{1/2} = \tfrac{1}{2}\tfrac{2}{3}\rho^{1/2} = \tfrac{1}{3}\rho^{1/2}.$$

We find, from eq. (25),

$$\beta P \rho^{-1} = 1 - (24\pi)^{-1} (\kappa^*)^3 \rho^{1/2} = 1 - (24\pi)^{-1} \kappa^3 \rho^{-1}. \tag{8n.26}$$

In Sec. 9j a derivation essentially the same as that of Debye and Hückel for ionic solutions is given.

8o. HIGHER-ORDER PLASMA TERMS

It is doubtful that there is any practical interest in higher-order correction terms for equilibrium gaseous plasma but, since the treatment is directly applicable to ionic solutions at room temperature, we briefly discuss the procedure here. We now no longer assume $r_0 = 0$, and include the term $-\theta(r) r^{-1} e^{-\alpha r}$ in eq. (8n.6) for $\psi(\alpha, r_0, r)$, with $\theta(r) = 1$ for $r < r_0$ and zero for $r > r_0$. Since the term is zero for $r > r_0$, we can with impunity set $\alpha = 0$ initially in this term. The Fourier transform function $\varphi(s)$ is now, from eqs. (8n.6) and (8n.14),

$$\varphi(s) = \varphi_0(s) - \varphi_c(s) = (\alpha^2 + s^2)^{-1} - \varphi_c, \tag{8o.1}$$

with

$$\varphi_c(s) = s^{-1} \int_0^\infty dr\, \theta(r) \sin sr = s^{-2} \int_0^{sr_0} d(sr) \sin(sr)$$

$$= s^{-2}(1 - \cos r_0 s) \tag{8o.1'}$$

which starts with a term $\frac{1}{2} r_0^2$ plus terms of order $r_0^{2\nu} s^{2(\nu-1)}$, $\nu \geq 2$. Neglecting the higher-order terms in the denominator, $[\kappa^2 + s^2 - \kappa^2(1 - \cos r_0 s)]^{-1}$ which replaces $(\kappa^2 + s^2)^{-1}$ of eq. (8n.19), but retaining the trigonometric $\cos r_0 s$ in the numerator, we end up with a correction proportional to r_0^2 for $\beta P \rho^{-1}$. The additional term is $-r_0^2 (16\pi)^{-1} \kappa^5 \rho^{-1}$ proportional to $\rho^{3/2}$. There are more important corrections proportional to ρ and to $\rho \ln \rho$ which arise from $-\theta(r)$ and from the terms with $\nu > 1$ in the expansion of $\Psi_{ab}(r)$ of eq. (8n.11).

One simplification is worthy of emphasis. Having corrected Sec. 8n to compute chains with $\psi(\alpha, r_0, r)$, $r_0 > 0$, we have $r\psi(\alpha, r_0, r) = 0$ if $r < r_0$ [eq. (8n.7)]. With $\nu = 1$ in eq. (8n.11) we added for every graph with a connection $-\beta(z_a \epsilon)(z_b \epsilon)\psi(\alpha, r_0, r)$ between two molecules at distance r the contribution of all possible chains with $k \geq 1$ intermediate molecules, but omitted the additive term $-\theta(r_{ij})$ in eq. (8n.11) for every $\Psi_{ij}(r_{ij})$ in the chain. We should compute

$$\int \cdots \int \prod_{i=1}^{k} \rho_i \, d\mathbf{r}_i \prod_{i=1}^{k+1} [(-\beta z_i z_{i-1} \epsilon^2 \psi(\alpha, r_0, r_{i,i-1}) - \theta(r_{i,i-1})], \tag{8o.2}$$

and then sum over all species. Since for every intermediate ion there are two bonds, we had the factor $-\beta\epsilon^2 p_i z_i^2$ for each, and the sum over all species led to $-(4\pi)^{-1}\kappa^2$ from eq. (8n.2).

Now $\psi(\alpha, r_0, r)$ is zero for $r<r_0$, and $\theta(r)=0$ for $r>0$. If between ions i and $i-1$ we include $\theta(r_{i,i-1})$, but between i and $i+1$ we have the $\psi(\alpha, r_0, r_{i+1,i})$ term, then ion i has only the single factor $\rho_i\beta z_i\epsilon$. Summing over all species the neutrality condition of eq. (8n.1) that $\sum \rho_a z_a \epsilon = 0$ gives a factor zero. Thus $\theta(r)$ contributes only when *all* links between a and b are $\theta(r_{i,i-1})$ links. The leading term in the ring integrals is the contribution to $\beta_1\rho$ of

$$\rho\int_0^\infty 4\pi r^2\,dr(2!)^{-1}[-\beta z_a z_b \epsilon^2 \psi(\alpha, r_0, r) - \theta(r)], \tag{8o.3}$$

the first term of which was included with $\sum \beta_k^{\text{ring}}\rho^k$ in eq. (8n.23′). The $\theta(r)$ term is

$$-\tfrac{1}{2}\rho\int_0^{r_0} 4\pi r^2\,dr = \frac{2\pi}{3}\rho r_0^3. \tag{8o.3′}$$

For rings of $k\geq 1$ molecules the contribution is proportional to $(\rho r_0^3)^{k+1}$.

The summation over all chain lengths and all ionic species leads to the factor $e^{-\kappa r}$ in the effective reduced potential and removes the difficulty of divergent integrals as $r\Rightarrow\infty$. By introducing $\theta(r)$ in eq. (8n.6) for $\psi(\alpha, r_0, r)$, so that $\psi\Rightarrow 0$, $r<r_0$, we have simplified the equations using $\theta(r)$ in the chains and avoided the rightly frowned-upon requirement of setting zero times infinity equal to zero when we use eq. (8n.8) to write $\Psi_{ab}(r)=-1$ for $r\leq r_0$. However, when we attempt to carry through the method for powers of $[-\beta\epsilon^2 z_a z_b \psi(\alpha, r_0, r)]^\nu(\nu!)^{-1}$, we run into very hairy integrals for large values of ν. In principle, having $\psi(\alpha, r_0, r)=0$, $r<r_0$, and using $\theta(r)=1$ for $r\leq r_0$ we should avoid singularities of $r^{-\nu}$ at $r\rightarrow 0$, but the mathematics is not easy.

A fairly extended analysis has been made by Friedman,[1] following essentially the procedure originally used by Mayer[2]

(1) Harold L. Friedman, *Ionic Solution Theory*, Interscience, New York, 1962.
(2) Joseph E. Mayer, *J. Chem. Phys.*, **18**, 1426 (1950).

CHAPTER 9

LIQUIDS

9a. INTRODUCTION

The treatment of real gases starts with approximating the perfect gas, zero mutual potential between molecules with completely random structure, and treating the potential as a perturbation. This was the method of the last chapter.

The crystalline condensed phase also has a relatively simple zeroth-order approximation, that of a completely ordered perfect single crystal with atoms located at each lattice site of a three-dimensional periodic array. The simplest first-order correction to this is the independent vibration of the atoms in an assumed quadratic potential constraining each atom to its own lattice site. A moderately realistic next-order approximation, the Debye model, treats the correlation between these vibrations as normal-mode harmonic vibration of the lattice in standing waves of definite wavelength in the whole crystal and treats the spectrum of these frequencies in a simple approximation (Sec. 10e).

The low-temperature liquid is the thermodynamic phase most difficult to describe in detailed structure and to treat with satisfactory mathematical precision and rigor. Each molecule is in constant direct potential interaction with many nearest neighbors, normally 9 or 10, and with a pair potential of the order of kT or more. In addition, there is a weaker interaction with many more distant ones, and the tightly bound neighbors in turn are strongly affected by their close neighbors. There is, however, no long-range order which lends simplicity as in the crystal treatment.

One precisely defined method of describing liquid structure is conceptually interesting but has apparently been unsuccessful in application. Consider a disordered array of N points representing molecular centers in three-dimensional space of volume V. From the ith point construct a plane perpendicular to, and bisecting, the line joining i to its nearest neighbor j. Make the same construction of a bisecting plane perpendicular to the line joining the next-nearest neighbor and continue, *seriata,* until the planes enclose a polyhedron of n faces. Proceed similarly with all points, $1 \leq i \leq N$. One then has a close-packed array of N irregular polyhedral cells, each containing a molecule. The number, $n \geq 4$, of faces of a cell may be used to define the number of neighbors in the first coordinate sphere around the central molecule, and the distance of the molecular center from each face is just half the distance of the corresponding neighbors. The statistics in an equilibrium ensemble of liquid systems of the number N_n of cells with n faces, $\sum_{n \geq 4} N_n = N$, along with that of the distances, gives an informative picture of the short-range structure. However, no satisfactory method of obtaining or using such data has been devised.

Various lattice models assuming an arbitrary crystal-type lattice with a statistical number of empty and full cells have been used for computational purposes, but all of these have an assumed long-range order and fail by a considerable amount to predict a sufficiently large entropy for the liquid.

The most used rigorously defined conceptual tool for describing a liquid structure in a one-component system of spherical molecules is that of reduced probability densities (see Sec. 1c on probability densities). Consider an ensemble of equilibrium monatomic single-component systems in a fixed thermodynamic state in which only one thermodynamic phase is present. Define $\rho_1(\mathbf{r})$ as being the probability density that a member of the ensemble will have a molecular center of mass at the position $\mathbf{r}$. In a perfect single crystal of fixed orientation and position this is triply periodic in $\mathbf{r}$, with sharp maxima at the lattice sites, but in a fluid liquid, gas, or even in a glass, it is a constant,

$$\rho_1(\mathbf{r}) - \rho = \langle N \rangle V^{-1} \qquad \text{(fluid)}. \tag{9a.1}$$

Define $\rho_2(\mathbf{r}, \mathbf{r}_2)$ as the probability density that in a member of the ensemble there will simultaneously be one molecule at $\mathbf{r}_1$ *and* one at $\mathbf{r}_2$, and in general let $\rho_n(\mathbf{r}_1, \mathbf{r}_2, \ldots, \mathbf{r}_n)$ be the probability density of having, in a member system, simultaneously n molecules at positions $\mathbf{r}_1, \mathbf{r}_2, \ldots, \mathbf{r}_n$.

Now we can make certain generalizations about the nature of these functions. First, sufficiently far from the walls in a fluid and in the absence of an external field (we assume that the earth's gravitational field is too weak to affect this statement), $\rho_2(\mathbf{r}_1, \mathbf{r}_2)$ can depend only on the distance $r_{12} = |\mathbf{r}_1 - \mathbf{r}_2|$, and $\rho_3(\mathbf{r}_1, \mathbf{r}_2, \mathbf{r}_3)$ only on the three distances $r_{23}, r_{31}, r_{12}, \ldots$. Further, since real molecules have strong repulsive forces at short distances, $\rho_2(r)$ must be zero at $r = 0$, and in general $\rho_n(\mathbf{r}_1, \ldots, \mathbf{r}_n)$ is zero if any pair distance r_{ij} is zero. If the distance r_{12} is very large, the probability that two molecules will be simultaneously at $\mathbf{r}_1$ and $\mathbf{r}_2$ in a system of a single thermodynamic phase will be independent of their exact positions; the pair probability density will be the product of the two independent singlet probability densities, $\rho_1(\mathbf{r}_1)=\rho$ and $\rho_1(\mathbf{r}_2)=\rho$,

$$\lim_{r_{12}\to\infty} [\rho_2(r_{12})]=\rho^2. \tag{9a.2}$$

Similarly, for $n+m$ molecules,

$$\lim_{1\le i\le n, n+1\le j\le n+m} \text{all } r_{ij}\to\infty \text{ if } [\rho_{n+m}(\mathbf{r}_1,\ldots,\mathbf{r}_i,\ldots,\mathbf{r}_n,\mathbf{r}_{n+1},\ldots,\mathbf{r}_j,\ldots,\mathbf{r}_{n+m})]$$
$$=\rho_n(\mathbf{r}_1,\ldots,\mathbf{r}_i,\ldots,\mathbf{r}_n)\rho_m(\mathbf{r}_{n+1},\ldots,\mathbf{r}_j,\ldots,\mathbf{r}_{n+m}). \tag{9a.2$'$}$$

These relations are used frequently but are subject to an important caveat. If, in a closed system of exactly N molecules in a volume V, one molecule is known to be at $\mathbf{r}_1$, there will be only $N-1$ *other* molecules which, however, cannot occupy all positions in V but will be excluded by the first molecule from some unknown volume b of the order of V/N or less, so that

$$\begin{aligned}\lim_{r_{12}\to\infty} [\rho_2(V,\beta,N\colon r)] &= \frac{N}{V}\frac{N-1}{V-b}=\left(\frac{N}{V}\right)^2\frac{1-N^{-1}}{1-bV^{-1}}\\ &\cong\left(\frac{N}{V}\right)^2\left(1-\frac{1}{N}+\frac{b}{V}+\cdots\right)\\ &=\rho^2\left[1-\frac{1}{V}(\rho^{-1}-b)\right],\end{aligned} \tag{9a.3}$$

and of course there are similar corrections of the order of V^{-1} to eq. (9a.2′). Now the dimensionless function

$$F_2(r)=\rho^{-2}\rho_2(r) \tag{9a.4}$$

is zero at $r=0$, rises to a value greater than unity at the average nearest-neighbor distance, and then decreases as r increases, oscillating for small r's around the value unity and approaching unity rather rapidly except close to the critical point. From eq. (3) the approach to unity is not perfect in a system of fixed N and V, but for a macroscopic system the difference, of the order of N^{-1},

$$\lim_{r\to\infty}[F_2(V,\beta,N;r)-1]=-V^{-1}(\rho^{-1}-b)=O(N^{-1}), \tag{9a.5}$$

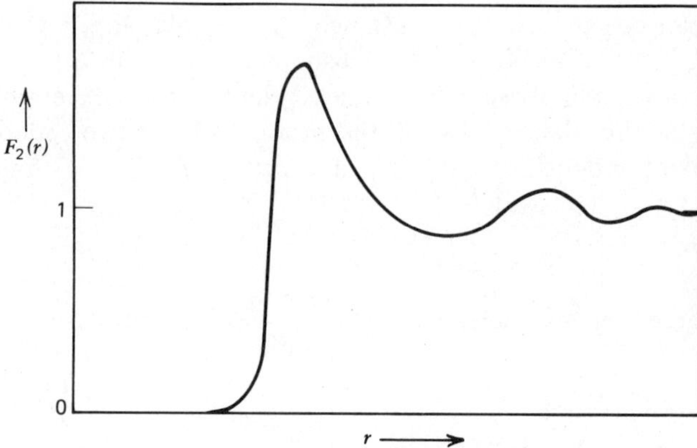

Figure 9a.1 The pair distribution function $F_2(r) = \rho^{-1}\rho_2(r)$.

is normally ignored. However, the dimensionless integral[1]

$$B(V, \beta, N) = \rho \int_V dr[F_2(V, \beta, N; \mathbf{r}) - 1] \tag{9a.6}$$

has what are essentially two independent contributions of the order of unity in a liquid, namely, a short-range contribution $-a$ and a large r-value contribution $-(1-b\rho)$ due to ρ times the correction term $-V^{-1}(\rho^{-1}-b)$ multiplied by V due to integration over the volume V. Since, with fixed N there is a total of $\frac{1}{2}N(N-1)$ pairs of molecules, and one of each pair can be at $\mathbf{r}_1$ or $\mathbf{r}_2$, the double integral is

$$\begin{aligned}\iint_V d\mathbf{r}_1\, d\mathbf{r}_2\, \rho_2(\mathbf{r}_1, \mathbf{r}_2) &= N(N-1) \\ &= V\rho^2 \int_V d\mathbf{r}\, F_2(V, \beta, N; r) \\ &= N\left\{N + \rho \int d\mathbf{r}[F_2(V, \beta, N; r) - 1]\right\} \\ &= N[N + (b\rho - 1 - a)].\end{aligned} \tag{9a.7}$$

The short-range contribution $-a$ to the integral of eq. (6) is equal to ρ times the negative of the average excluded volume b in eq. (3).

[1] We always ignore effects due to the system walls. These would lead to thermodynamic contributions proportional to surface area, not generally proportional to the system size.

Now, however, an ensemble of systems in the thermodynamic state defined by values of V, β, $\beta\mu$ is defined as an ensemble of fixed volume V with walls permeable to the flow of energy and of molecules in contact with an *infinite* reservoir of fixed β and $\beta\mu$. The presence of a molecule at $\mathbf{r}_1$ now does not change the average density $\rho_2(r_{12})$ if $r_{12}=|r_1-r_2|$ is sufficiently large, since molecules can pass through the walls; the correction term of eq. (5) goes strictly to zero, since the appropriate volume in that expression is now that of the system V *plus* the infinite volume of the reservoir. The integral over the *finite* volume V,

$$B(V, \beta, \beta\mu)=\rho\int d\mathbf{r}[F_2(V, \beta, \beta\mu; r)-1], \tag{9a.8}$$

is the integral of an integrand that goes rapidly to a strict zero for r greater than many atomic distances; we may equally well use

$$B(V, \beta, \beta\mu)=\rho\int_0^\infty 4\pi r^2\, d\mathbf{r}[F_2(V, \beta, \beta\mu; r)-1]. \tag{9a.8′}$$

Its value is

$$B(V, \beta, \beta\mu)=-a=-b\rho, \tag{9a.8″}$$

namely, equal to the short-range contribution to eq. (6).

We have dwelt on this distinction between the two ensembles of fixed V, β, N and of V, β, $\beta\mu$ at length, because it is of significance in the later use of expressions for certain thermodynamic quantities. In Sec. 5a [eq. (5a.10)] we derived the relation

$$kT^2C_V=(\langle E^2\rangle-\langle E\rangle^2)_{V,\beta,N}, \tag{9a.9}$$

where $\langle E^2\rangle$ and $\langle E\rangle$ are average square energy and average energy in a system of fixed V, β, N. The relation is general in any ensemble, namely, that the heat capacity C for fixed values of the other natural thermodynamic state variables $\mathbf{x}$ is always

$$kT^2C_\mathbf{x}\equiv kT^2\left(\frac{\partial E}{\partial T}\right)_\mathbf{x}=-\left(\frac{\partial E}{\partial \beta}\right)_\mathbf{x}=(\langle E^2\rangle-\langle E\rangle^2)_\mathbf{x}, \tag{9a.9′}$$

where the averages are those of the ensemble. The difference $\langle E^2\rangle-\langle E\rangle^2$ depends on the constant variables and is quite different if N is held constant than if $\beta\mu$ is constant and N varies as $(\partial N/\partial\beta)_{V,\beta\mu}$. The value of $\langle E^2\rangle$ for a system with the potential a sum of pair terms can be expressed by integrals of pair potentials times reduced probability densities, and among the terms is the quadruple integral over $d\mathbf{r}_1\, d\mathbf{r}_2\, d\mathbf{r}_3\, d\mathbf{r}_4$ of $u(r_{12})u(r_{34})\rho_4(\mathbf{r}_1, \mathbf{r}_2, \mathbf{r}_3, \mathbf{r}_4)$ (Sec. 9i), from which one subtracts that part of the value of $\langle E\rangle^2$ equal to the quadruple integral of

$u(r_{12})\rho_2(r_{12})u(r_{34})\rho_2(r_{34})$, namely, in eq. (9′), the integral of $u(r_{12})u(r_{34})$ $[\rho_4(\mathbf{r}_1, \mathbf{r}_2, \mathbf{r}_3, \mathbf{r}_4) - \rho_2(r_{12})\rho_2(r_{34})]$. From eq. (2′) the term $\rho_4 - \rho_2\rho_2$ is zero at large separations of the two pairs but, since the volume over which the integration is extended is V^2, the small correction proportional to V^{-1} to eq. (2′) for fixed V, β, N gives a term proportional to V, the size of the system. If the grand canonical ensemble of V, β, $\beta\mu$ is used, this correction is absent, but the heat capacity C computed is that at constant V, $\beta\mu$, which by thermodynamic manipulation involving quantities expressible by integrals over other probability densities can be used to find C_V.

In what follows we always assume that the reduced probability densities are defined for the grand canonical ensemble of open systems of fixed V, β, $\beta\mu$, although they could equally well be computed using a V, β, N ensemble and erasing the correction terms of eqs. (2) and (2′) of the order of V^{-1}.

Obviously, the concept of reduced probability densities is equally applicable to systems of $\nu > 1$ components a, b, ... with ν singlet probability densities $\rho_a(r)$, $\rho_b(r), \ldots$, ν^2 pair probability densities, $\rho_{aa}(r_{aa})$, $\rho_{ab}(r_{ab}) = \rho_{ba}(r_{ba})$, $\rho_{bb}(r_{bb}), \ldots$, which have molecules of species a or b at r_i. In general, there are ν^n functions with $\mathbf{n} \equiv n_a, n_b, \ldots$, symbolized by $\rho_{\mathbf{n}}(\mathbf{r}_1, \ldots, \mathbf{r}_n)$, which are numerically independent of the order in which the n_a's are written. The functions are symmetric in exchange of coordinates r_i, r_j occupied by the same species.

Similarly, by introducing internal coordinates, $\mathbf{q}_i \equiv q_{\alpha i}, q_{\gamma i}, \ldots$, we can, in principle at least, define probability density functions $\boldsymbol{\rho}_1(\mathbf{r}_1, \mathbf{q}_1; \mathbf{r}_2, \mathbf{q}_2; \ldots; \mathbf{r}_n, \mathbf{q}_n)$ or, with internal quantum numbers, $\boldsymbol{\kappa}_i \equiv \boldsymbol{\kappa}_{\alpha i}, \boldsymbol{\kappa}_{\gamma i}, \ldots$, functions $\rho_{\mathbf{n}}(\mathbf{r}_1, \boldsymbol{\kappa}_1; \ldots; \mathbf{r}_n, \boldsymbol{\kappa}_n)$. Very little has been done with such esoteric functions, those of assumed spherical molecules presenting sufficient difficulties at present.

9b. REDUCED PROBABILITY DENSITIES, POTENTIALS OF AVERAGE FORCE, THE VIRIAL OF CLAUSIUS

We limit ourselves to systems of spherical or assumed spherical molecules but, at the slight complication of using boldface notation, admit several chemical components. We assume classical mechanics to be adequate. The notation used is, for the numbers of molecules,

$$\mathbf{N} \equiv N_a, N_b, \ldots, \qquad \mathbf{N}! = N_a!\, N_b! \cdots, \qquad N = N_a + N_b + \cdots;$$

for chemical potentials,

$$\boldsymbol{\mu} \equiv \mu_a, \mu_b, \ldots, \qquad \mathbf{N} \cdot \boldsymbol{\mu} = N_a\mu_a + N_b\mu_b + \cdots;$$

for the number densities,

$$\rho_a = \langle N_a \rangle V^{-1}, \qquad \boldsymbol{\rho} = \rho_a, \rho_b, \ldots, \qquad \boldsymbol{\rho}^{\mathbf{n}} \equiv \rho_a^{n_a} \rho_b^{n_b} \cdots, \qquad \rho = \rho_a + \rho_b + \cdots;$$

for the absolute activities, $z_a = A_2 e^{\beta\mu_a}$, with A_a so chosen that $z_a/\rho_a = 1$ in the perfect gas limit for which all $\rho_a \to 0$ [see eq. (8c.12)],

$$\mathbf{z} \equiv z_a, z_b, \ldots, \qquad \mathbf{z}^{\mathbf{n}} \equiv z_a^{n_a} z_b^{n_b} \cdots;$$

and for the coordinates $\mathbf{r}_1, \ldots, \mathbf{r}_N$ of the set $\mathbf{N}$ of molecules, with species a in the first N_a coordinate, species b in the next N_b, and so on,

$$\{\mathbf{N}\} \equiv \mathbf{r}_1, \ldots, \mathbf{r}_N, \qquad d\{\mathbf{N}\} = d\mathbf{r}_1 \, d\mathbf{r}_2 \cdots d\mathbf{r}_N.$$

The classical probability density $W_{\mathbf{N}}\{\mathbf{N}\}$, that there will be exactly the set $\mathbf{N}$ of molecules and no more in the volume V, and that the coordinate positions $\{\mathbf{N}\}$ will be occupied by the set $\mathbf{N}$ of molecules of the appropriate assigned species positions, is for an ensemble of systems of fixed V, β, $\beta\boldsymbol{\mu}$,

$$W_{\mathbf{N}}\{\mathbf{N}\} = \{\exp[-\beta P(\beta, \mathbf{z}) V\} \mathbf{z}^{\mathbf{N}} \exp[-\beta U_{\mathbf{N}}\{\mathbf{N}\}]. \tag{9b.1}$$

(See Sec. 8c for one-component systems, and Sec. 8g for the extension to more chemical components.) Since in integration over the set $\{\mathbf{N}\}$ of numbered coordinates there are $N!$ positions covered with the same species of molecules occupying them, the condition that the sum and integral of the probability densities be unity is that

$$\sum_{\mathbf{N} \geq 0} \frac{1}{\mathbf{N}!} \int\int_V \cdots \int d\{\mathbf{N}\} \, W_{\mathbf{N}}\{\mathbf{N}\} = 1$$

$$= \{\exp[-\beta P(\beta, \mathbf{z}) V\} \sum_{\mathbf{N} \geq 0} \frac{\mathbf{z}^{\mathbf{N}}}{\mathbf{N}!} \int\int \cdots \int d\{\mathbf{N}\} \exp(-\beta U_{\mathbf{N}}\{\mathbf{N}\}), \tag{9b.2}$$

and this relation determines the pressure P as a function of the set of activities $\mathbf{z}$ and of $\beta = 1/kT$.

We want the reduced probability density $\rho_{\mathbf{n}}(\mathbf{r}_1, \ldots, \mathbf{r}_{\mathbf{n}}) \equiv \rho_{\mathbf{n}}\{\mathbf{n}\}$ that, independent of the position of other molecules in the systems, the positions $\{\mathbf{n}\}$ will be occupied by the set $\mathbf{n}$ of molecules. For convenience of precision of definition we assign the first n_a positions to be occupied by molecules of type a, the next n_b by type b, and so on. Since we demand that at least the set $\mathbf{n}$ of molecules be present, we must sum $W_{\mathbf{N}+\mathbf{n}}\{\mathbf{N}+\mathbf{n}\}$ over all $\mathbf{N} \geq 0$ and integrate over $d\{\mathbf{N}\}$, again dividing by $\mathbf{N}!$, since the integration covers this many identical positions occupied by the appropriate species of molecules. We have

$$\rho_{\mathbf{n}}\{\mathbf{n}\} = \sum_{\mathbf{N}} \frac{1}{\mathbf{N}!} \iint \cdots \int d\{\mathbf{N}\}\, W_{\mathbf{N+n}}\{\mathbf{N}+\mathbf{n}\}$$

$$= \{\exp[-\beta P(\beta, \mathbf{z}) V]\} \mathbf{z}^{\mathbf{n}} \sum_{\mathbf{N}} \frac{\mathbf{z}^{\mathbf{N}}}{\mathbf{N}!} \iint \cdots \int d\{\mathbf{N}\} \exp(-\beta U_{\mathbf{N+n}}\{\mathbf{N}+\mathbf{n}\}). \tag{9b.3}$$

We now prove that for this classical system the derivative of ln $\rho_{\mathbf{n}}$ with respect to any coordinate of $\{\mathbf{n}\}$ is β times the average force along the coordinate. For instance, $\partial \ln \rho\{\mathbf{n}\}/\partial x_i$ is $\beta\langle f_{xi}\{\mathbf{n}\}\rangle$, where $\langle f_{xi}\{\mathbf{n}\}\rangle$ is the average force exerted along the positive x-axis on the center of mass of molecule i due to the other molecules at the position $\{\mathbf{n}\}$, *plus* that due to the molecules of the rest of the fluid whose averaged positions are influenced by those occupying the position set $\{\mathbf{n}\} = \mathbf{r}_1, \ldots, \mathbf{r}_n$, This follows directly by differentiation of both sides of eq. (3),

$$\frac{\partial \ln \rho_{\mathbf{n}}\{\mathbf{n}\}}{\partial x_i} = (\rho_{\mathbf{n}}\{\mathbf{n}\})^{-1}[\exp(-\beta PV)]$$

$$\times \mathbf{z}^{\mathbf{n}} \sum_{\mathbf{N}\geq 0} \frac{\mathbf{z}^{\mathbf{N}}}{\mathbf{N}!} \frac{\partial}{\partial x_i} \iint \cdots \int d\{\mathbf{N}\} \exp(-\beta U_{\mathbf{N+n}}\{\mathbf{N}+\mathbf{n}\})$$

$$= (\rho_{\mathbf{n}}\{\mathbf{n}\})^{-1}[\exp(-\beta PV)]\mathbf{z}^{\mathbf{n}} \sum_{\mathbf{N}\geq 0} \frac{\mathbf{z}^{\mathbf{N}}}{\mathbf{N}!} \times \iint \cdots \int d\{\mathbf{N}\}$$

$$\times \left(-\frac{\beta\, \partial U_{N+n}}{\partial x_i}\right) \exp(-\beta U_{\mathbf{N+n}}). \tag{9b.4}$$

The conditional probability density $W_{\mathbf{N}}(\{\mathbf{N}\}|\{\mathbf{n}\})$ [eq. (1c.29)], that if the set $\mathbf{n}$ of molecules is at $\{\mathbf{n}\}$, the disjoint set $\mathbf{N}$ in V will be at the position $\{\mathbf{N}\}$ is just

$$W_{\mathbf{N}}(\{\mathbf{N}\}|\{\mathbf{n}\}) = (\rho_{\mathbf{n}}\{\mathbf{n}\})^{-1}\{\exp[-\beta P(\beta, \mathbf{z}) V]\} \mathbf{z}^{(\mathbf{n}+\mathbf{N})} \exp(-\beta U_{\mathbf{N+n}}\{\mathbf{N}+\mathbf{n}\}), \tag{9b.4$'$}$$

properly normalized so that, from eq. (3),

$$\sum_{\mathbf{N}\geq 0} \frac{1}{\mathbf{N}!} \iint \underset{V}{\cdots} \int d\{\mathbf{N}\}\, W_{\mathbf{N}}(\{\mathbf{N}\}|\{\mathbf{n}\}) = 1. \tag{9b.4$''$}$$

Now

$$-\frac{\partial U_{\mathbf{N+n}}\{\mathbf{N}+\mathbf{n}\}}{\partial x_i} = f_{xi}\{\mathbf{N}+\mathbf{n}\} \tag{9b.4$'''$}$$

is the force along the x-axis on molecule i when the molecules are at the position $\{N+n\}$, so from the definition on an average [eq. (4)] we have

$$\frac{\partial \ln \rho_{\mathbf{n}}\{\mathbf{n}\}}{\partial x_i} = \beta\langle f_{xi}\{\mathbf{n}\}\rangle. \tag{9b.5}$$

In the last section we defined a dimensionless function $F_2(r)$ [eq. (9a.4)] as $F_2(r) = \rho^{-2}\rho_2(r)$. We now define a set of dimensionless functions,

$$F_{\mathbf{n}}\{\mathbf{n}\} = \boldsymbol{\rho}^{-\mathbf{n}}\rho_{\mathbf{n}}\{\mathbf{n}\}, \tag{9b.6}$$

of which $F_2(r)$ is the simplest significant one defined for a single component system. In a fluid the singlet functions $F_a(\mathbf{r})$ are all unity. If now we define a function $\Omega_{\mathbf{n}}(\beta, \mathbf{z}; \{\mathbf{n}\})$ of the dimensions of an energy by

$$\Omega_{\mathbf{n}}(\mathbf{z}, \beta; \{\mathbf{n}\}) = -\beta^{-1} \ln F_{\mathbf{n}}(\mathbf{z}, \beta; \{\mathbf{n}\}), \tag{9b.7}$$

then eq. (5) leads to the relation

$$-\frac{\partial \Omega_{\mathbf{n}}(\mathbf{z}, \beta; \{\mathbf{n}\})}{\partial x_i} = \langle f_{xi}(\mathbf{z}, \beta; \{\mathbf{n}\}\rangle, \tag{9b.8}$$

namely, that the function $\Omega_{\mathbf{n}}(\mathbf{z}, \beta; \{\mathbf{n}\})$ of the coordinate set $\{\mathbf{n}\} \equiv \mathbf{r}_1, \ldots, \mathbf{r}_n$ is the potential of average force for the set $\mathbf{n} = n_a, n_b, \ldots$ of molecules at position $\{\mathbf{n}\}$ in a one-phase system of thermodynamic state defined by the set $\mathbf{z} = z_a, z_b, \ldots$ of activities, and $\beta = 1/kT$. The reduced probability density $\rho_{\mathbf{n}}\{\mathbf{n}\}$ is then

$$\rho_{\mathbf{n}}\{\mathbf{n}\} = \boldsymbol{\rho}^{\mathbf{n}} \exp(-\beta\Omega_{\mathbf{n}}\{\mathbf{n}\}). \tag{9b.9}$$

From the condition that follows from eq. (9a.2),

$$\lim_{1\le i<j\le n} \text{all } r_{ij} \to \infty (F_{\mathbf{n}}\{\mathbf{n}\}) = 1. \tag{9b.10}$$

It follows that the zero of this potential $\Omega_{\mathbf{n}}\{\mathbf{n}\}$ has been taken as infinite separation of all molecules of the set $\mathbf{n}$,

$$\lim_{1\le i<j\le n} \text{all } r_{ij} \to \infty (\Omega_{\mathbf{n}}\{\mathbf{n}\}) = 0. \tag{9b.10′}$$

From eqs. (2) and (3) we see that the integrals of the $\rho_{\mathbf{n}}$'s over $d\{\mathbf{n}\}$ in the volume V are linear combinations of averages of powers of $\mathbf{N}$. If we were to write all possible occupancies of the n coordinates $\mathbf{r}_1, \ldots, \mathbf{r}_n$ by molecules of ν species, there would be ν^n different functions $\rho_{\mathbf{n}}\{\mathbf{n}\}$ for fixed $n = \sum_a n_a$, that is, $2^2 = 4$, which are $\rho_{aa}, \rho_{ab}, \rho_{ba}, \rho_{bb}$ for $n = 2$, $\nu = 2$, and $2^3 = 8$: $\rho_{aaa}, \rho_{aab}, \rho_{aba}, \rho_{baa}, \rho_{abb}, \rho_{bab}, \rho_{bba}, \rho_{bbb}$ for $n = 3$, $\nu = 2$. But $\rho_{ab} = \rho_{ba}$ and the three functions $\rho_{aab}, \rho_{aba}, \rho_{baa}$ have the same dependence on the three distances $r_{aa}, r_{1a,b}, r_{2a,b}$. If we keep the convention of always having the first n_a coordinates $\mathbf{r}_1, \ldots, \mathbf{r}_{na}$ occupied by species a, the next n_b by species b, and so on, we reduce the number of different functions; for instance, ρ_{aa}, ρ_{ab}, ρ_{bb} for $n=2$, $\nu=2$, and ρ_{aaa}, ρ_{aab}, ρ_{abb}, ρ_{bbb} for $n=3$, $\nu=2$, but in general for a given set $\mathbf{n}$ we must weight each function by $n!/\mathbf{n}!$ to count the total number correctly. Annoyances like these, due to the mathematical necessity of numbering the coordinates of functions plague much of statistical mechanics.

If in eq. (3) we renumber $\mathbf{N}+\mathbf{n}$ by $\mathbf{N}$ and compare with eq. (2), we find

$$\iint_V \cdots \int d\{\mathbf{n}\}\frac{n!}{\mathbf{n}!}\rho_{\mathbf{n}}\{\mathbf{n}\} = \sum_{\mathbf{N}\geq\mathbf{n}} \frac{\mathbf{N}!}{(\mathbf{N}-\mathbf{n})} \iint_V \cdots \int d\{\mathbf{N}\}\frac{W_{\mathbf{N}}\{\mathbf{N}\}}{\mathbf{N}!}$$

$$= \left\langle \frac{\mathbf{N}!}{(\mathbf{N}-\mathbf{n})!} \right\rangle = \left\langle \prod_a \frac{N_a!}{(N_a-n_a)!} \right\rangle. \tag{9b.11}$$

Had we used the canonical ensemble of fixed V, β, N, the corresponding integral would have been $\mathbf{N}!/(\mathbf{N}-\mathbf{n})!$, with $\mathbf{N}$ a set of *exact* integer numbers. The difference is of course due to the small but long-range corrections to eqs. (9a.2) and (2′) discussed in the last section.

There is one equation, frequently called the virial equation of Clausius, that stands rather outside the usual statistical mechanical ensemble method of manipulation. It was originally derived by Clausius on the basis of consideration of a time integration of the acceleration of molecules due to the total forces acting on them. We present a proof more closely in line with our ensemble formulation. Consider an ensemble of single-component systems. The grand canonical ensemble probability density $W_N\{N\}$ for N molecules at $\mathbf{r}^{(N)}\equiv\{N\}$ is

$$W_N\{N\} = \exp(-\beta PV)z^N \exp(-\beta U_N\{N\}), \tag{9b.12}$$

which determines P by the requirement that

$$\sum_{N\geq 0} \frac{1}{N!} \iint_V \cdots \int d\{N\}\, W_N\{N\} = 1. \tag{9b.13}$$

Now change our coordinates to dimensionless coordinates R,

$$\begin{aligned} R &= \frac{r}{L}, \\ d\mathbf{r} &= L^3/d\mathbf{R}, \\ \{N^*\} &= \mathbf{R}^{(N)}, \\ d\{N\} &= L^{3N}/d\{N^*\}, \end{aligned} \tag{9b.14}$$

for a system in a cubic box of edge L. The volume, $(V^*)^N = V^N/L^{3N}$, within which $d\{N^*\}$ is integrated is now unity. We write

$$(N!)^{-1}W_N\{N^*\} = \exp(-\beta PV^*L^3)\frac{z^N}{N!}\exp(-\beta U_N\{N^*\}), \tag{9b.15}$$

with

$$\sum_{N\geq 0} \frac{1}{N!} \underset{V^*=1}{\iint \cdots \int} d\{N^*\} L^{3N} W_N\{N^*\} = 1. \tag{9b.16}$$

Differentiate both sides of eq. (16) with respect to $\ln L$ using eq. (15). The volume V^* does not change, so we have, with pair potentials,

$$U_N = \sum\sum_{N\geq i>j\geq 1} u(r_{ij}) = \sum\sum_{N\geq i>j\geq 1} u(R_{ij}L), \tag{9b.17}$$

$$\sum_{N\geq 0} \underset{1}{\iint \cdots \int} \frac{1}{N!} dN^* L^{3N}\Bigg[3N - 3\beta PV^* L^3 - \beta \sum\sum_{N\geq i>j\geq 1} LR_{ij} \frac{du(LR_{ij})}{d(LR_{ij})}\Bigg] W_N\{N^*\}$$

$$= 3 \sum_{N\geq 0} \frac{1}{N!} \iint \cdots \int d\{N\}\Bigg[N - \beta PV - \frac{\beta}{3} \sum\sum_{N\geq i>j} r_{ij} \frac{du(r_{ij})}{dr_{ij}}\Bigg] W_N\{N\}$$

$$= 3\Bigg[\langle N\rangle - \beta PV - \tfrac{1}{6}\beta\langle N\rangle_\rho \int_0^\infty 4\pi r^2\, dr\left(r\frac{du(r)}{dr}\right) F_2(r)\Bigg] = 0, \tag{9b.18}$$

where the last line uses eq. (3b.3) for $\rho_2(r_{ij})$ and eq. (6) for F_2, remembering that there are $\frac{1}{2}N(N-1)$ terms u_{ij}. We find

$$PV = \langle N\rangle\Bigg[kT - \tfrac{1}{6}\rho \int_0^\infty 4\pi r^2\, dr\, r \frac{du_2(r)}{dr} F_2(r)\Bigg] \tag{9b.19}$$

as the expression for the PV product. This is often called the equation for the virial of Clausius.

9c. RELATIONS BETWEEN DIFFERENT ACTIVITIES

As a temporarily convenient compact notation we define a set of functions $\Psi_{\mathbf{n}}(V, \beta, \mathbf{z}; \{\mathbf{n}\})$ proportional to the probability densities $\rho_{\mathbf{n}}\{\mathbf{n}\}$, which we wish to examine at constant V, β values but at different activity set values $\mathbf{z}$. Define these by

$$\Psi_{\mathbf{n}}(V, \beta, \mathbf{z}; \{\mathbf{n}\}) = [\exp \beta P(\beta, \mathbf{z}) V] \frac{[\boldsymbol{\rho}(\beta, \mathbf{z})]^{\mathbf{n}}}{z^{\mathbf{n}}} \exp[-\beta \Omega_{\mathbf{n}}(\beta, \mathbf{z}; \{n\})]. \tag{9c.1}$$

With eq. (9b.3) multiplied on both sides by $\mathbf{z}^{-n} \exp P(\beta, \mathbf{z}) V$, and eq. (9b.9) for $\rho_{\mathbf{n}}$, we have

$$\Psi_{\mathbf{n}}(V, \beta, \mathbf{z}; \{\mathbf{n}\}) = \sum_{\mathbf{n}} \frac{\mathbf{z}^{\mathbf{N}}}{\mathbf{N}!} \underset{V}{\iint \cdots \int} d\{\mathbf{N}\} \exp(-\beta U_{\mathbf{N}+\mathbf{n}}\{\mathbf{N}+\mathbf{n}\}). \tag{9c.2}$$

Now the usefulness of the reduced probability densities $\rho_{\mathbf{n}}$ is pretty much confined to very small values of the number $n=\sum_a n_a$, and indeed only the pair function $\rho_2(r)$ for one-component systems of some monatomic molecules is really fairly accurately known. However, eq. (2) above defines $\Psi_{\mathbf{n}}$ for all values of n from $n=0$, for which $\Psi_0=\exp(-PV)$, to values approaching infinity. In the limit that $z\to 0$, that is, all $z_a\to 0$, the sum of eq. (2) has nonzero value only for $\mathbf{N}\equiv 0$, all $N_a=0$, and

$$\Psi_{\mathbf{n}}(V, \beta, \mathbf{z}\equiv 0, \{\mathbf{n}\})=\exp(-\beta U_{\mathbf{n}}\{\mathbf{n}\}). \tag{9c.3}$$

This is consistent with the definition of eq. (1) for $\Psi_{\mathbf{n}}$, since as all activities approach zero, the pressure P goes to zero,[(1)] $\exp(-\beta PV)$ to unity, each ρ_a/z_a to unity, and the potential $\Omega_{\mathbf{n}}$ of average force to the true potential $U_{\mathbf{n}}$ of the molecules specified to be at the position $\{\mathbf{n}\}$, since the concentration of others is vanishingly small. With eq. (3) we can rewrite eq. (2) as

$$\Psi_{\mathbf{K}}(V, \beta, \mathbf{z}; \{\mathbf{K}\})=\sum_{\mathbf{N}\geq 0}\frac{\mathbf{z}^{\mathbf{N}}}{\mathbf{N}!}\int\int_V\cdots\int d\{\mathbf{N}\}\Psi_{\mathbf{K+N}}(V, \beta, \mathbf{z}\equiv 0; \{\mathbf{K+N}\}). \tag{9c.4}$$

Now in general we do not expect to be able to solve the complete set of integral equations (4) for the functions $\Psi_{\mathbf{M}}(V, \beta, \mathbf{z}\equiv 0; \{\mathbf{M}\})$ in terms of the $\Psi_{\mathbf{K}}(V, \beta, \mathbf{z}; \{\mathbf{K}\})$, but it happens that the solution is simple, namely, that

$$\Psi_{\mathbf{M}}(V, \beta, \mathbf{z}\equiv 0; \{\mathbf{M}\})=\sum_{\mathbf{K}\geq 0}\frac{(-\mathbf{z})^{\mathbf{K}}}{\mathbf{K}!}\int\int_V\cdots\int d\{\mathbf{K}\}\Psi_{\mathbf{M+K}}(V, \beta, \mathbf{z}; \{\mathbf{M+K}\}). \tag{9c.5}$$

The proof that eq. (5) is correct is obtained by using it for $\Psi_{\mathbf{K+N}}(V, \beta, \mathbf{z}; \{\mathbf{K+N}\})$ in eq. (4) or, alternatively, using it for $\Psi_{\mathbf{M+K}}(V, \beta, \mathbf{z}; \{\mathbf{M+K}\})$ in eq. (5). It is necessary of course that the sums (and integrals) of both eqs. (4) and (5) converge, but this is required in any case and is ensured by the

[(1)] The critical reader may feel that there is a cheat in the justification given for the $\Psi_{\mathbf{n}}(V, \beta, \mathbf{z}\equiv 0, \{\mathbf{n}\})$ of eq. (3) when we use it, as we do in eq. (4), for n of the order of 10^{24} or more. The justification depends on a proper limiting process which is a little sophisticated. We wish to use eq. (4) for a large but finite volume, say, one for which $\langle N\rangle$ is 10^{24}. This requires that the sum on the right include N-values larger than this. We have argued that $\Psi_{\mathbf{N}}(V, \beta, \mathbf{z}\equiv 0, \{\mathbf{N}\})$, defined by eq. (1), at $\mathbf{z}\equiv 0$ has $P=0$. Now we use $\Psi_{\mathbf{N}}(\mathbf{z}\equiv 0)$ in the integral of eq. (4), only with the coordinates $\{\mathbf{N}\}$ in the original volume V of the open system, but at activity zero, hence for $N\cong 10^{24}$ it is defined for systems in a much larger volume, indeed in the limit of infinite volume for which all activities and the pressure go to zero. We can of course later let the original system size of the open ensemble become as large as we wish, but always defining $\Psi_{\mathbf{N}}(\mathbf{z}\equiv 0)$ in a volume larger by an infinite factor.

infinite repulsion between molecules at zero distance. We find, using eq. (4) in eq. (5),

$$\Psi_{\mathbf{M}}(V, \beta, \mathbf{z}\equiv 0; \{\mathbf{M}\}) = \sum_{N\geq 0, K\geq 0}\sum \frac{(-\mathbf{z})^{\mathbf{K}}}{\mathbf{K}!} \sum_{\mathbf{N}\geq 0} \frac{\mathbf{z}^{\mathbf{N}}}{\mathbf{N}!} \int\int \cdots \int d\{\mathbf{K}+\mathbf{N}\}$$
$$\times \Psi_{\mathbf{M}+\mathbf{K}+\mathbf{N}}(V, \beta, \mathbf{z}\equiv 0; \{\mathbf{M}+\mathbf{K}+\mathbf{N}\}). \quad (9c.6)$$

For given $\mathbf{K}+\mathbf{N}=\mathbf{L}$ the sum in front of the integral is

$$\frac{1}{\mathbf{L}!}\sum_{\mathbf{K}=0}^{\mathbf{L}} \frac{\mathbf{L}!}{(\mathbf{L}-\mathbf{K})!\mathbf{K}!}[\mathbf{z}^{\mathbf{L}-\mathbf{K}}(-\mathbf{z})^{\mathbf{K}}] = \prod_a \frac{1}{L_a!}\sum_{K_a=0}^{L_a} \frac{L_a!}{(L_a-K_a)!K_a!}[z_a^{L_a-K_a}(-z_a)^{K_a}]$$
$$= \prod_a \frac{1}{L_a!}(z_a - z_a)^{L_a} = 0, \qquad \text{if any } L_a > 0. \quad (9c.7)$$

Only the term with all $L_a = 0$ has a nonzero coefficient; the equation is an identity.

If we now use eq. (5) with an activity set $\mathbf{y} = y_a, y_b, \ldots$ on the right in eq. (4) and with an activity set $\mathbf{z}$ on the left, we find

$$\Psi_{\mathbf{K}}(V, \beta, \mathbf{z}; \mathbf{K}) = \sum_{M\geq 0, N\geq 0}\sum \frac{\mathbf{z}^{\mathbf{N}}}{N!} \sum_{\mathbf{M}\geq 0} \frac{(-\mathbf{y})^{\mathbf{M}}}{\mathbf{M}!} \int\int \cdots \int d\{\mathbf{N}+\mathbf{M}\}$$
$$\times \Psi_{\mathbf{K}+\mathbf{N}+\mathbf{M}}(V, \beta, \mathbf{y}; \{\mathbf{K}+\mathbf{N}+\mathbf{M}\}), \quad (9c.8)$$

and for $\mathbf{N}+\mathbf{M}=\mathbf{L}$, using the steps of eq. (7),

$$\Psi_{\mathbf{K}}(V, \beta, \mathbf{z}; \{\mathbf{K}\}) = \sum_{\mathbf{L}\geq 0} \frac{(\mathbf{z}-\mathbf{y})^{\mathbf{L}}}{\mathbf{L}!} \int\int \cdot_V \cdot \int d\{\mathbf{L}\} \Psi_{\mathbf{K}+\mathbf{a}}(V, \beta, \mathbf{y}; \{\mathbf{K}+\mathbf{L}\}). \quad (9c.8')$$

This equation expresses a relation between the two sets of potentials of average force $\Omega_{\mathbf{N}}(\beta, \mathbf{z}; \{\mathbf{N}\})$ and $\Omega_{\mathbf{N}}(\beta, \mathbf{y}; \{\mathbf{N}\})$ at any two different activity sets, $\mathbf{z}$ and $\mathbf{y}$, at the same value of $\beta = 1/kT$. The relation is valid even if the activities $\mathbf{z}$ and $\mathbf{y}$ are such that two different thermodynamic phases, gas and liquid, gas and crystal, or liquid and crystal, are at equilibrium. Equation (8′) includes both eqs. (4) and (5) as special cases, the former if $\mathbf{y}\equiv 0$, and the latter if $\mathbf{z}\equiv 0$, $\mathbf{z}-\mathbf{y}\equiv -\mathbf{y}$.

Of course, although the sum on the right-hand side of eq. (8′) always converges for real molecules with repulsive short-range forces, the convergence is not such that a numerically tractable number of terms is sufficient for computation. In general, if $\mathbf{z}$ and $\mathbf{y}$ really differ by an experimentally significant amount, the largest terms in the sum on the right will be those for which the numbers $L_a, L_b, \ldots$ of molecules are macroscopically significant, namely, equal to many powers of 10. In Sec.

9f we go to a cluster-type development such as that of Sec. 8d or 8f in order to compute numerically, and this type of development is not convergent between activities $\mathbf{z}$ and $\mathbf{y}$ which are those of two different thermodynamic phases.

The treatments of this and the preceding chapter have assumed that the system obeys the laws of classical mechanics and that quantum effects are negligible. Except for helium and hydrogen this is empirically a good assumption, although the next lightest noble gas, neon, shows a small but significant deviation for the law of corresponding states (Sec. 8b, end), which is almost certainly due to quantum behavior.[(1)] However, most of the formalism of this chapter, as well as much of that of Chapter 8, is directly applicable to systems with very strong quantum effects.

In Chapter 13 we discuss a formal treatment of quantum-mechanical systems using the density matrix. This matrix can be expressed in various representations in the same way a physical variable can be written in different functional forms in different coordinate systems. From this density matrix a function, the Wigner function,[(2)] which is a function of the momentum coordinate phase space, is very nearly interpretable as being proportional to a probability density. It is a real function of $\mathbf{p}_1$, $\mathbf{q}_1, \ldots, \mathbf{p}_N, \mathbf{q}_N$, and its integral over either all momenta or all coordinates is strictly proportional to the probability density in the conjugate variable space. It can be normalized to make its sum and integral unity. The one feature of this function in both $\mathbf{p}^{(N)}$, $\mathbf{q}^{(N)}$ that demonstrates is *cannot* be a true probability density is that, unlike the latter, it *can* take negative values for some local parts of the phase space.

If the classical function $\exp[-\beta H(\mathbf{p}^{(N)}, \mathbf{q}^{(N)})]$ in eq. (8c.1) is replaced by the representation in $\mathbf{p}^{(N)}\mathbf{q}^{(N)}$ space of the *matrix* $\exp[-\beta\mathcal{H}(\mathbf{p}^N, \mathbf{q}^{(N)})]$, with $\mathcal{H}$ the quantum-mechanical Hamiltonian operator, the probability density $W_{\mathbf{N}}\{\mathbf{N}\}$ of eq. (9b.1) will be

$$W_{\mathbf{N}}\{\mathbf{N}\} = \{\exp[-\beta P(\beta, \mathbf{z}) V]\}\mathbf{z}^{\mathbf{N}} \int\int d\mathbf{p}^{(\mathbf{N})})\{\exp[-\beta\mathcal{H}(\mathbf{p}^{\mathbf{N}}, \mathbf{q}^{(\mathbf{N})})\}\mathbf{q}^{(N)}\mathbf{p}^{(N)}. \tag{9c.9}$$

Then all the equations of this chapter for the reduced probability densities $\rho_{\mathbf{n}}\{\mathbf{n}\}$ are valid, *except that* $\partial \ln \rho_{\mathbf{n}} \partial x_i$ *has not the simple physical interpretation of being* $-\beta$ *times an average force* [eq. (9b.5)].

To summarize we use eq. (1) for $\Psi_{\mathbf{n}}$ to write the relation of eq. (8′) in more conventional notation. We keep $\beta = 1/kT$ fixed and consider two sets, $\mathbf{z} \equiv z_a, z_b, \ldots$ and $\mathbf{y} \equiv y_a, y_b, \ldots$, of activities and number densities,

(1) J. de Boer, *Physica*, **14,** 139 (1948).

(2) E. Wigner, *Phys. Rev.* **40,** 749 (1930). See also Moyal, *Proc. Cambridge Phil. Soc.*, **45,** 99 (1949).

$\boldsymbol{\rho}(\mathbf{z}) \equiv \rho_a(\mathbf{z}),\ \rho_b(\mathbf{z}), \ldots$ and $\boldsymbol{\rho}(\mathbf{y}) \equiv \rho_a(\mathbf{y}),\ \rho_b(\mathbf{y}), \ldots,$ for species a, b, ... of molecules. Multiply eq. (8′) on both sides by $(\prod_a z_a^{n_a})\exp[-\beta P(\mathbf{y})V]$ and, with eq. (1) for $\Psi_{\mathbf{n}}(V, \beta, \mathbf{z}, \{\mathbf{n}\})$,

$$\{\exp \beta V[P(\mathbf{z}) - P(\mathbf{y})]\}\left[\prod_a \rho_a(\mathbf{z})\right]\exp[-\beta\Omega_{\mathbf{n}}(\beta, \mathbf{z}, \{\mathbf{n}\})]$$
$$= \{\exp \beta V[P(\mathbf{z}) - P(\mathbf{y})]\}\rho_{\mathbf{n}}(\mathbf{z}, \{\mathbf{n}\})$$
$$= \prod_a \left[\frac{z_a \rho_a(\mathbf{y})}{y_a}\right]^{n_a}\left\{\sum_{N \geq 0} \prod_a \left[\frac{(z_a - y_a)\rho_a(\mathbf{y})}{y_a}\right]^{N_a}(N_a!)^{-1}\right\}$$
$$\times \int\!\!\int \cdot \underset{V}{\cdot \cdot} \int d\{\mathbf{N}\}\exp[-\beta\Omega_{\mathbf{N}+\mathbf{n}}\beta, \mathbf{y}\{\mathbf{N}+\mathbf{n}\})]. \quad (9c.10)$$

The term activity (or fugacity) of species a, ζ_a is used by chemists for a quantity proportional to $\exp[\beta\mu_a] = \exp[N_0\mu_a/RT]$, with N_0 Avogadro's number, and R the conventional gas constant, $R = N_0 k$, multiplied by some unit $C_a^{(0)}$ of concentration. In the absolute activities, z_a and y_a, in eq. (10) the unit of concentration, $C_a^{(0)} = \rho_a$, is molecules per unit volume, and μ_a is measured from a zero such that the activity z_a is equal to ρ_a in the perfect gas of infinite dilution of *all* molecular species, $\lim(\rho_a, \rho_b, \ldots \rightarrow 0)$. If cgs units are used, this will be well satisfied at unit values of $\rho_a, \rho_b, \ldots,$ and the activity will be equal to ρ_a for quite large numerical values of the number densities $\rho_a, \rho_b, \ldots$ but, if $10^{-8}\,\text{cm} = 1\,\text{Å}$ is the length unit, the activity will differ greatly from ρ_a at unit number density. The *activity coefficient* is a dimensionless quantity γ_a defined so that the activity ζ_a is $\gamma_a C_a$. Innumerable conventions as to the zero from which the chemical potential is measured and for the concentration $C_a^{(0)}$ are possible, and most have been used at some time or other. For anyone expecting to use the general equations to compare with experimentally observed results, the following discussion may be helpful, but is not fundamental, to the general principle of the equations. For gases the concentration unit $C^{(0)}$ is often P_0/kT_0 or P_0/RT, with P_0 some standard pressure, and then the term fugacity is usually employed. For solutions two conventions are common, one for solutes at relatively low concentration and one for a solvent whose mole fraction does not differ too much from unity.

For solutes a, b, ... in solvents $\nu, \mu, \ldots$ we define a general activity ζ_a of species a of solute in a solution whose concentrations are C_a, C_b in a solvent of activity set $\boldsymbol{\eta}$. The concentration units are often moles per liter or sometimes moles per some weight of solvent. The activities are defined by

$$\zeta_a(\mathbf{C}, \boldsymbol{\eta}) = \lim_{C' \rightarrow 0} \text{all}\{C'_a \exp[\beta\mu_a(\mathbf{C}, \boldsymbol{\eta}) - \beta\mu_a(\mathbf{C}' \rightarrow 0, \boldsymbol{\eta})]\}, \quad (9c.11)$$

so that

$$\lim_{C_a, C_b, \ldots \to 0} [\zeta_a(C_a, C_b, \ldots)] = C_a. \qquad (9c.11')$$

The activity of the pure solvent at fixed $T = (k\beta)^{-1}$ and some standard pressure P_0, such as 1 atm or the vapor pressure (and normally there is little dependence on which is used), is often taken to be unity. The activity η_ν can then be *defined* to be

$$\eta_\nu(\mu, \beta) = C_\nu^{(0)}(P_0, \beta) \exp \beta[\mu - \mu(P_0, \beta)], \qquad (9c.12)$$

where $C_\nu^{(0)}(P_0, \beta)$ has dimensions of reciprocal volume but may be numerically set equal to unity, that is, the volume unit as the volume of a mole, or perhaps a unit weight of pure solvent. More generally, for a multicomponent system of composition $\mathbf{C} \equiv C_a, C_b, \ldots$ in any unit of concentration we *could* define an activity η_a for each species in terms of a concentration set $\mathbf{C}^{(0)} = C_a^{(0)}, C_b^{(0)}, \ldots$ by

$$\eta_a(\mathbf{C}, \mathbf{C}^{(0)}, \beta) = C_a^{(0)} \exp[\beta\mu(\mathbf{C}, \beta) - \beta\mu(\mathbf{C}^{(0)}, \beta)], \qquad (9c.13)$$

but activities so defined are so specialized at arbitrary concentrations $\mathbf{C}^{(0)}$ that they have little significance unless $\mathbf{C}^{(0)}$ is some concentration set of special significance.

The absolute activities z_a and y_a defined for a one-component system by eq. (8c.12) may be written for a multicomponent system as

$$z_a(\boldsymbol{\rho}, \beta) = \lim_{\mathbf{z}' = \boldsymbol{\rho}' \to 0} \text{all}\{\rho_a' \exp[\beta\mu_a(z\beta) - \beta\mu_a(\mathbf{z}', \beta)]\}, \qquad (9c.14)$$

so that the ratio

$$\frac{z_a}{y_a} = \exp[\beta\mu_a(\mathbf{z}, \beta) - \beta\mu_a(\mathbf{y}, \beta)], \qquad (9c.14')$$

and

$$\frac{z_a}{y_a} \rho_a(\mathbf{y}) = \rho_a(\mathbf{y}) \exp[\beta\mu(\mathbf{z}, \beta) - \beta\mu(\mathbf{y}, \beta)] \qquad (9c.14'')$$

is an activity per molecule equal to the concentration $\rho_a(y)$ in molecules per unit volume if $\mathbf{z} \equiv \mathbf{y}$. If we change the concentration unit to $\mathbf{C}$, for instance, to moles per liter, with N_0 Avogadro's number and ρ_a in cm^{-3},

$$C_a(\text{mol liter}^{-1}) = \frac{\rho_a(\text{molec cm}^{-3})}{10^{-3} N_0} = A_a \rho_a, \qquad (9c.16)$$

which defines a constant A_a *for this particular unit of concentration* as

$$A_a = 10^3 N_0^{-1} = 1.660 \times 10^{-21} = (6.022 \times 10^{20})^{-1}. \qquad (9c.16')$$

In general, we define

$$A_a = \frac{C_a(\mathbf{z}, \beta)}{\rho_a(\mathbf{z}, \beta)} \tag{9c.17}$$

so that the activity ζ_a in the units C_a of concentration has A_a as a factor. Then $\zeta_a = A_a z_a$ *if* the activity is measured in such a way that it becomes equal to the concentration C_a at infinite dilution,

$$\lim_{\zeta_a \to 0} \left| \frac{\zeta_a}{C_a} \right| = 1. \tag{9c.18}$$

We use this in Sec. 9d on osmotic systems for the solutes.

The activity coefficient γ_a is defined by the ratio

$$\gamma_a = C_a^{-1} \zeta_a(\mathbf{C}, \beta) = \rho_a^{-1} z_a(\boldsymbol{\rho}, \beta), \tag{9c.19}$$

where, in a solution, the activities ζ_a or z_a depend on all the concentrations, $\mathbf{C} = C_a, \ldots, C_\nu, \ldots$ or $\boldsymbol{\rho} = \rho_a, \ldots, \rho_\nu, \ldots$.

If, however, as is usual for a dilute solution in which a single component solvent ν is close to unit mole fraction, one defines the activity coefficient to be unity for the pure solvent; as in eq. (12) one has

$$\gamma_\nu = C_\nu^{-1} \eta_\nu(\mathbf{C}, \beta), \tag{9c.20}$$

and

$$\lim_{C_a, C_b, \ldots = 0} C_\nu = C_\nu^{(0)}(\gamma_\nu) = 1. \tag{9c.20′}$$

$$C_a, C_b, \ldots = 0, \qquad [\gamma_\nu = 1]$$

One must then correct eq. (19) by a factor $\rho_\nu^{(0)}/z(\rho_\nu^{(0)}, \beta)$ where $\rho_\nu^{(0)} = A_\nu^{-1} C^{(0)}$, and $C_\nu^{(0)}$ is whatever concentration unit is used for the solvent. One has

$$\gamma_\nu = \frac{z_\nu(\boldsymbol{\rho}, \beta)}{\rho_\nu} \frac{\rho_\nu^{(0)}}{z_\nu[\rho_\nu^{(0)}, (\rho_a, \rho_b, \ldots = 0), \beta]} \tag{9c.21}$$

$$\eta_\nu = \gamma_\nu C_\nu = \frac{z_\nu(\boldsymbol{\rho}, \beta)}{z_\nu[\rho_\nu^{(0)}, (\rho_a, \rho_b, \ldots = 0), \beta]} C_\nu^{(0)}. \tag{9c.21′}$$

9d. OSMOTIC RELATIONS

One important concept of physical chemistry is that of systems in osmotic equilibrium. The experimental idealization is a solution in which all species of molecules are labeled as belonging to either solvent or solute. There may be one or more species of molecules for both labels, but most often the solvent is one pure chemical component of mole fraction not far from unity. This solution is in equilibrium at the same temperature with the pure solvent. The two phases, solution and pure solvent, are separated by a rigid semipermeable membrane through which solvent molecules can pass, but which is impermeable to the solute. The activity of the solvent is then equal at equilibrium in both phases. The solution has a lower pressure than the pure solvent, and the (positive) pressure difference is the osmotic pressure, namely, P (solvent) $-$ P (solution) $=$ $\Delta P =$ osmotic pressure.

Perfect semipermeable membranes are not easy to come by in general, but if the solvent has an appreciable vapor pressure and the solute none, as in the case of a salt or a sugar, a rigid porous material can be used to establish the pressure difference, and a vapor space between solution and solute serves as a perfect semipermeable membrane. In this case the solution is, at equilibrium, kept at a negative pressure, and the solvent at its vapor pressure. An inert gas, such as helium, can be used to raise both pressures, but lowers the rate of attaining equilibrium.

Consider the two-component case of a solute a and a solvent ν with the absolute activities $\mathbf{z}$, $\mathbf{y}$ of eq. (9c.10) to be $y_a = 0$, and $z_\nu = y_\nu$ to be the absolute activity of the pure solvent at some standard state P_0, T. Let z_a be the absolute activity of solute a in the solution, with solvent absolute activity z_ν. Let us examine the value of $(z_a - y_a)y_a^{-1}\rho_a(\mathbf{y}, \beta)$ for this case when expressed in conventional macroscopic concentration units, $\mathbf{C} = C_\nu$, C_a, $C_b, \ldots$, and activities ζ_ν, $\zeta_a, \ldots$. The concentration C_a of solute a is $C_a = A_a\rho_a$, where A_a is defined by eq. (9c.17) and has the numerical value of eq. (9c.16′) if C_a is in moles per liter. The ratio $(z_a - y_a)y_a^{-1}$ as $y_a \to 0$ is $a_z y_a^{-1}$, the same as the ratio of the conventional activities,

$$z_a y_a^{-1} = \zeta_a\,(\text{solution})\,[\zeta_a\,(\text{infinite dilution})]^{-1}. \tag{9d.1}$$

We have then, for the solute activities,

$$(z_a - y_a)y_a)y_a^{-1}\rho_a(\mathbf{y}, \beta) = A_a\zeta_a \lim_{C_a, C_b, \ldots \to 0.} [\zeta_a^{-1} C_a(C, \beta)] = A_a\zeta_a, \tag{9d.2}$$

where ζ_a is the conventional activity of a solute in units C_a of concentration.

In eq. (9c.10) we restrict ourselves to the case in which on the left-hand side we are interested only in reduced probability densities $\rho_{\mathbf{n}}$, with $n_\nu = 0$, namely, those of the n_a solute molecules, $\rho_{na}\{n_a\}$. We use

$$\Delta P(\zeta_a) = -P(\zeta_a, y_\nu) + P(\zeta_a \to 0, y_\nu) \tag{9d.3}$$

for the osmotic pressure. Since now in eq. (9c.10) the factor $(y_\nu - y_\nu)^{N_\nu}$ is in all terms of the sum over N_ν, only the term $N_\nu = 0$ remains. We have, with eq. (2),

$$[\exp(\beta V\,\Delta P)]\rho_{n_a}(\zeta_a, y_\nu, \{n_a\}) = (A_a\zeta_a)^{n_a} \sum_{N_a\geq 0} \frac{(A_a\zeta_a)^{N_a}}{N_a!} \times \iint\cdots\int d\{N_a\}\exp[-\beta\Omega_{N_a+n_a}(y_\nu, y_a\rightarrow 0, \{N_a+n_a\})]. \quad (9d.4)$$

In particular, if $n_a = 0$,

$$\exp(\beta V\,\Delta P) = \sum_{N_a\geq 0} \frac{(A_a\zeta_a)^{N_a}}{N_a!} \iint\underset{V}{\cdots}\int d\{N_a\}\exp[-\beta\Omega_{N_a}(y_\nu, y_a\rightarrow 0, \{N_a\})]. \quad (9d.5)$$

Equation (4) for $e^{\rho V\Delta P}\rho_n$, with $A_a\zeta_a = z$, is in exactly the same form as eq. (9b.3) for ρ_n in a single-component system, with ΔP replacing the pressure and Ω_{n+N}, which is the potential of average force between the solute molecules at infinite dilution in the solvent, replacing the true potential energy U_{n+N} in a vacuum. At sufficiently small ζ_a the large terms in the sum on the right of eq. (5) have low values of $N/V = \rho$, and for the greater part of the coordinate space $\Omega_N \rightarrow 0$. We find the integrals are V^N and

$$\lim_{z\rightarrow 0}[\exp \beta V\,\Delta P(z)] = \sum_{N\geq 0} \frac{(V_z)^N}{N!} = \exp zV,$$

$$\lim_{z\rightarrow 0}[\Delta P(z)] = kTz = kTA_a\zeta_a \quad (9d.6)$$

and, since $[\partial(\beta P)/\partial \ln z]_\beta = \rho$,

$$\lim_{z\rightarrow 0}[\rho(z)] = z = A\zeta = AC(\zeta), \quad (9d.7)$$

$$\lim_{\zeta\rightarrow 0}\left(\frac{\zeta}{C}\right) = 1,$$

namely, at infinite dilution $\beta\Delta P$ is equal to the number density of solute molecules, and the activity is equal to the concentration. This is the perfect solution limit.

The extension to several solute species follows the method of Sec. 9b.

There is a useful equation for the osmotic pressure that we use in Sec. 9j, which follows from eq. (5) by the same method as was used in eqs. (9b.11) to (9b.18) to obtain the expression (9b.18) for PV in terms of the virial of Clausius. Use eq. (5) for $\beta V\Delta P(\zeta_a)$ instead of eq. (9b.11) and go through the change in scale of eqs. (9b.13) to (9b.17). The only difference is that now we have $\beta\Omega_N(\zeta_a\rightarrow 0, \{N\})$ replacing $U_N\{N\}$ and, if we make the Kirkwood closure,

$$\beta\Omega(\zeta_a \to 0)\{N\} = \sum_{N \geq i > j \geq i} \sum \beta\omega_2(\zeta_a \to 0, r_{ij}) \tag{9d.8}$$

we find, instead of eq. (9b.18),

$$\beta P \rho^{-1} = 1 - \tfrac{1}{6}\rho \int_0^\infty 4\pi r^2\, dr\, r \frac{d\beta\omega_2(\zeta_a \to 0)}{dr} \rho_2(r) \tag{9d.9}$$

for several components, b, c, . . . ,

$$\beta\,\Delta P \rho^{-1} = 1 - \tfrac{1}{6}\rho^{-1} \sum_{b,c\cdots} \sum \int_0^\infty 4\pi r^2\, dr\, r \frac{d\beta\omega_{bc}(\zeta_b, \zeta_c \to 0)}{dr} \rho_{bc}(r) \tag{9d.9$'$}$$

9e. FORMAL CLUSTER DEVELOPMENT

The cluster function development of Sec. 8d was introduced for an ensemble of one-component monatomic molecular systems whose potential energy U_N was a sum of pair terms, $U_N = \sum\sum_{1 \leq i \leq j \leq N} u(r_{\nu j})$, but it was remarked later that it could be used more generally. Actually, one can define completely formal relations between different sets of functions corresponding to the relations between probability densities and cluster functions with very few restrictions on the nature of the functions or of the variables on which the functions depend. Since these relations are useful for a wide variety of cases, especially in the treatment of non-equilibrium systems (which we do not discuss in detail in this book) we present briefly the general relationships in this section.

Let $q^{(n)}$ be a set of n possibly multiply indexed variables,

$$\begin{gathered} q^{(ni)} \equiv \mathbf{q}_i = \mathbf{q}_{1i}, \mathbf{q}_{2i}, \ldots, \mathbf{q}_{i}, \ldots, \mathbf{q}_{(ni)i} \\ \mathbf{q}^{(n)} = \mathbf{q}_1^{(n1)}, \mathbf{q}_2^{(n2)}, \ldots, \mathbf{q}_i^{(ni)}, \end{gathered} \tag{9e.1}$$

and let $\Theta_{\mathbf{n}}(q_1^{(ni)}, q_2^{(n2)}, \ldots, q_i^{(ni)}, \ldots)$, for $0 \leq n_1 \leq N_1, \ldots, 0 \leq n_j \leq N_j, \ldots$, be known real single-valued analytic functions, regular in a real domain V, which are symmetric in exchange of any pair $q_{\lambda i}$, $q_{\kappa i}$, for every i.

The variables may be coordinates, or coordinates and conjugated momenta of molecules, including internal degrees of freedom, but may also include a time variable. However, we necessarily assume that for each i there is at least one component of $\mathbf{q}_{\lambda i}$ having the character of a position coordinate such that there is one (or more) defined distance(s), $\Delta ij = |x_i - x_j|$, between entities λi and κi or κj.

We use a symbol $\{\nu\}_n$ for a specified subset of ν indices out of a set $\{n\}$, with $l \leq \lambda i \leq n$, and *also* for the subset $\{\nu\}_n \equiv \mathbf{q}^{\{\nu\}_n}$ of variables of the original set, $\{n\} \equiv \mathbf{q}^{(n)}$. For given numerical value of n and ν there will be $n!/(n-\nu)!\nu!$ different subsets $\{\nu\}_n$. Unless all the variable sets $\mathbf{q}_i$ are alike for all i, that is, unless there is only one kind of molecule, we should use the boldface notation, $\mathbf{n} \equiv n_a, n_b, \ldots, \{\mathbf{n}\} \equiv \{n_a\}, \{n_b\}, \ldots$, so that for the subset $\{\boldsymbol{\nu}\}_{\mathbf{n}}$ there are $\mathbf{n}/(\mathbf{n}-\boldsymbol{\nu})!\boldsymbol{\nu}! \equiv \prod_a [n_a!/(n_a-\nu_a)!\nu_a!]$ different subsets.

We then assume that the functions Θ_n have a property which is common to the logarithms of probability density functions of molecules in all real physical one-phase thermodynamic situations (except exactly at critical points), namely, that there be no correlations at infinite distance between two subsets $\{\{\mathbf{n}\}-\{\boldsymbol{\nu}\}_{\mathbf{n}}\}$ and $\{\boldsymbol{\nu}\}_{\mathbf{n}}$, so that

$$\lim_{\Delta_{ij}\to\infty} [\Theta_{\mathbf{n}}\{\mathbf{n}\}] = \Theta_{\mathbf{n}-\boldsymbol{\nu}}\{\{\mathbf{n}-\{\boldsymbol{\nu}\}_{\mathbf{n}}\} + \Theta_{\boldsymbol{\nu}}\{\nu_{\mathbf{n}}\} \tag{9e.2}$$

if $i \in \{\{\mathbf{n}\}-\{\boldsymbol{\nu}\}_{\mathbf{n}}\}$, $j \in \{\boldsymbol{\nu}\}_{\mathbf{n}}$. Define functions

$$\theta_n(\{\mathbf{n}\}) = \sum_{\nu\geq 1}^{\nu\leq n} \sum_{\{\nu\}_n} (-)^{n-\nu} \Theta_{\boldsymbol{\nu}}\{\boldsymbol{\nu}\}_{\mathbf{n}}, \tag{9e.3}$$

where the lowercase θ_n's for small n are

$$\begin{aligned}
&\theta_1(\mathbf{q}_i) = \Theta_1(\mathbf{q}_i),\\
&\theta_2(\mathbf{q}_i\mathbf{q}_j) = \Theta_2(\mathbf{q}_i, \mathbf{q}_j) - \Theta_1(\mathbf{q}_i) - \Theta_1(\mathbf{q}_j),\\
&\theta_3(\mathbf{q}_i, \mathbf{q}_j, \mathbf{q}_k) = \Theta_3(\mathbf{q}_i, \mathbf{q}_j, \mathbf{q}_k) - \Theta_2(\mathbf{q}_j\mathbf{q}_k),\\
&\theta_2(\mathbf{q}_k\mathbf{q}_i) - \Theta_2(\mathbf{q}_i\mathbf{q}_j) + \Theta_1(\mathbf{q}_i) + \Theta_1(\mathbf{q}_j) + \Theta_1(\mathbf{q}_k).
\end{aligned} \tag{9e.4}$$

The solution of eq. (3) for the Θ_n's in terms of the θ_ν's is

$$\Theta_{\mathbf{n}}\{\mathbf{n}\} = \sum_{\nu\geq 1}^{\nu\leq n} \sum_{\{\boldsymbol{\nu}\}_{\mathbf{n}}} \theta_{\boldsymbol{\nu}}\{\boldsymbol{\nu}\}_{\mathbf{n}}, \tag{9e.5}$$

which can be proved by using eq. (3) for $\Theta_{\mathbf{n}}$ in eq. (5) and obtaining an identity, since the coefficient of $\Theta_{\mathbf{n}-\boldsymbol{\nu}}\{\{\mathbf{n}\}-\{\nu\}_n\}$ in eq. (5) is found to be

$$\prod_a \left[\sum_{\mu_a\geq 0}^{\mu_a=\nu_a} (-)^{\mu_a} \frac{\nu_a!}{(\nu_a-\mu_a)!\mu_a!}\right] = \prod_a (1-1)^{\nu_a} = 0, \qquad \text{unless all } \nu_a = 0. \tag{9e.6}$$

The relation (2), with eq. (5) for $\Theta_{\mathbf{n}}$, requires that

$$\lim_{\substack{\Delta ij\to\infty\\ i\in\{\{n\}-\{\nu\}_n\}, j\in\{\nu_n\}}} (\Theta_n\{n\}) = \lim_{\Delta ij\to\infty} \sum_{\boldsymbol{\nu}\geq 1}^{\boldsymbol{\nu}\leq n} \sum_{\{\boldsymbol{\nu}\}_{\mathbf{n}}} \theta_{\boldsymbol{\nu}}\{\boldsymbol{\nu}\}_{\mathbf{n}} = \sum_{\mu\geq 1}^{\boldsymbol{\mu}\leq n-\boldsymbol{\nu}} \sum_{\{\boldsymbol{\mu}\}_{\mathbf{n}-\boldsymbol{\nu}}} \theta_{\boldsymbol{\mu}}\{\boldsymbol{\mu}\}_{n-\nu}$$

$$+ \sum_{\boldsymbol{\lambda}>1}^{\boldsymbol{\lambda}\leq\boldsymbol{\nu}} \sum_{\{\boldsymbol{\lambda}\}_{\boldsymbol{\nu}}} \theta_{\boldsymbol{\lambda}}\{\boldsymbol{\lambda}\}_{\boldsymbol{\nu}}, \tag{9e.7}$$

which can be valid for all finite sets $\{\mathbf{n}\}$ and $\{\boldsymbol{\nu}\}_{\mathbf{n}}$ only if

$$\lim_{\Delta ij \to \infty} \text{any}(\theta_{\mathbf{n}}\{\mathbf{n}\}) \to 0, \tag{9e.8}$$

in which case, for finite n, any also means all. We therefore see that the functions $\theta_{\mathbf{n}}\{\mathbf{n}\}$ are compact functions, nonzero only if all the molecules of the set $\mathbf{n}$ are close together.

Now turn to functions defined by

$$F_n\{n\} = \exp(-\Theta_n\{n\}). \tag{9e.9}$$

In view of the assumption of eq. (2) we have

$$\lim_{\Delta ij \to \infty} (F_n\{n\}) = F_{n-\nu}\{n - \nu_n\} F_\nu\{\nu\}_n,$$

if

$$i \in (\{n\} - \{\nu\}_n), \qquad j \in \{\nu\}_n. \tag{9e.10}$$

Define cluster functions $g_{\mathbf{n}}\{\mathbf{n}\}$ as sums of products of functions $F_{\boldsymbol{\nu}}\{\boldsymbol{\nu}\}_{\mathbf{n}}$; these cluster functions are compact in the same sense as the θ_n's. To do so in general we have to introduce a rather cumbersome notation, but since the general definitions are necessarily complicated, something equivalent is necessary.

We use the notation $\{k\{\mathbf{n}_\alpha\}_N\}_u$ to indicate a partition of the set N of numbered molecules into k unconnected subsets, $\mathbf{n}_\alpha\, 1 \le \alpha \le k$, with coordinates $\{\mathbf{n}_\alpha\}_N$, such that

$$\sum_{\alpha=1}^{\alpha=k} \{\mathbf{n}_\alpha\}_N \equiv \{\mathbf{N}\}, \tag{9e.11}$$

so that every numbered molecule occurs in one and only one subset $\{\mathbf{n}\}_N$. Use

$$\sum \{k\{\mathbf{n}_\alpha\}_N\} u = \text{summation over all partitions} \tag{9e.12}$$

from the partition with $k = 1$, $\{\mathbf{n}_\alpha\}_N \equiv \{\mathbf{N}\}$, to the partition with $k = N$, where every subset $\{n_\alpha\}_N$ contains a single molecule only.

The cluster functions $g_{\mathbf{n}}\{\mathbf{n}_\alpha\}_N$ are then defined implicitly by the equation

$$F_N = \left(\sum \{k\{\mathbf{n}_\alpha\}\mathbf{N}\}_u\right) \prod_{\alpha=1}^{K} g_{\mathbf{n}_\alpha}\{\mathbf{n}_\alpha\}_N. \tag{9e.13}$$

The inverse equation giving g_n explicitly is

$$g_{\mathbf{n}} = \left(\sum \{k\{\boldsymbol{\nu}_\alpha\}_n\}_u\right) (-)^{k-1}(k-1)! \prod_{\alpha=1}^{k} F_{\nu_\alpha}\{\boldsymbol{\nu}_\alpha\}_n. \tag{9e.14}$$

The proof that eqs. (13) and (14) are self-consistent is, as usual, shown by replacing the terms in the right-hand sum of either of the two equations by the expression given by the other equation and noting that the resulting double summation results in an identity.

9f. CLUSTER EXPANSION AROUND ACTIVITY SET z

For an ensemble of equilibrium systems composed of monatomic molecules, or molecules for which the internal and translational Hamiltonians are separable, we may define the distribution functions $F_{\mathbf{n}}(\mathbf{z}; \{\mathbf{n}\})$ at temperature $T=(k\beta)^{-1}$ and activity set $\mathbf{z}=z_a, z_b, \ldots$ of the set $\mathbf{n}=n_a, n_b, \ldots$ of molecules of species $a, b, \ldots$ as

$$F_{\mathbf{n}}(\mathbf{z}; \{\mathbf{n}\}) = \exp[-\beta\Omega_{\mathbf{n}}(\mathbf{z}; \{\mathbf{n}\})], \tag{9f.1}$$

where $\Omega_{\mathbf{n}}$ is the potential of average force of the set $\mathbf{n}$ of molecules at coordinate positions $\{\mathbf{n}\}=\mathbf{r}_1, \mathbf{r}_2, \ldots, \mathbf{r}_n$ of their centers of mass. The cluster functions $g_{\mathbf{n}}(\mathbf{z}; \{\mathbf{n}\})$ are then defined in a fluid by eq. (9e.14), so that

$$\begin{aligned} g_a(\mathbf{r}_i) &= F_a(\mathbf{r}_i) = 1, \\ g_{ab}(\mathbf{r}_i, \mathbf{r}_j) &= F_{ab}(\mathbf{r}_i, \mathbf{r}_j) - 1, \\ g_{abc}(\mathbf{r}_i, \mathbf{r}_j\mathbf{r}_k) &= F_{abc}(\mathbf{r}_i, \mathbf{r}_j\mathbf{r}_k) - F_{bc}(\mathbf{r}_j, \mathbf{r}_k) - F_{ca}(\mathbf{r}_k, \mathbf{r}_i) - F_{ab}(\mathbf{r}_i, \mathbf{r}_j) + 2, \end{aligned} \tag{9f.2}$$

and g_4 consists of F_{abcd} minus four terms differing in the missing index from F_{bcd}, minus three different terms of type $F_{ab}F_{bc}$, plus six terms $2F_{ab}$, $2F_{ac}, \ldots$, minus 6. The first few $F_{\mathbf{n}}$'s are then, from eq. (9e.13),

$$\begin{aligned} F_a &= 1, \\ F_{ab} &= g_{ab} + 1, \\ F_{abc} &= g_{abc} + g_{bc} + g_{ca} + g_{ab} + 1, \\ F_{abcd} &= g_{abcd} + \text{four different } g_{bcd} \text{ terms} + \cdots \\ &\quad + \text{three different } g_{ab}g_{bc} \text{ terms} + \text{six } g_{ab} \text{ terms} + 1. \end{aligned} \tag{9f.3}$$

The general condition that $F_{\mathbf{n}}$ approaches $F_{\mathbf{n}-\boldsymbol{\nu}}F_{\boldsymbol{\nu}}$ in value if all the members of the coordinate set $\mathbf{n}-\boldsymbol{\nu}$ are very far from all of the set $\boldsymbol{\nu}$ [eq. (9e.10)] requires that all the cluster functions $g_{\mathbf{n}}$ approach zero value if the distance between any two subsets $\mathbf{n}-\boldsymbol{\nu}$ and $\boldsymbol{\nu}$ becomes large. The cluster integrals $b_{\mathbf{n}}(\mathbf{z})$,

$$b_{\mathbf{n}}(\mathbf{z}) = \frac{1}{\mathbf{n}!V}\int\int\cdots\int d\{\mathbf{n}\}\, g_{\mathbf{n}}(\mathbf{z}, \{\mathbf{n}\}), \tag{9f.4}$$

with $\mathbf{n}! = n_a!, n_b!, \ldots$ for finite $n = n_a + n_b + \cdots$ then go to a finite limit independent of V for sufficiently large V, provided the pair potentials of average force $\Omega_{ab}(|\mathbf{r}_i-\mathbf{r}_j|)$ fall off more rapidly than $|\mathbf{r}_i-\mathbf{r}_j|^{-3}$ (see Sec. 8n for r^{-1} potentials). In the case of dipoles, the potential energy of a pair on a line in the lowest energy orientation of the same vector direction of the moments is proportional to the product of their dipole moments divided by $|\mathbf{r}_i-\mathbf{r}_j|^3$. The integrals of eq. (4) diverge logarithmically but, when averaged over the angular orientation at finite temperature, they converge.

We now turn to eq. (9c.10) relating $\{\exp \beta V[P(\mathbf{z})-P(\mathbf{y})]\}\rho_{\mathbf{n}}(\mathbf{z}, \{\mathbf{n}\})$ to a sum of integrals of $\exp[-\beta\Omega_{\mathbf{N}}(\mathbf{y}; \{\mathbf{N}\})]$, with z and y two different activity sets, and look at the case in which the set $\mathbf{n}$ is zero, $\rho_{\mathbf{n}=0}\equiv 1$. With

$$\Delta P = P(\mathbf{z})-P(\mathbf{y}), \tag{9f.5}$$

$$\psi_a = (z_a-y_a)^{-1}\rho_a(\mathbf{y}), \tag{9f.6}$$

and eq. (1), we have

$$\exp(-\beta V\,\Delta P) = \sum_{\mathbf{N}\geq 0}\frac{\boldsymbol{\Psi}^{\mathbf{N}}}{\mathbf{N}!}\int\!\!\int\cdots\int d\{\mathbf{N}\}F_{\mathbf{N}}(\mathbf{y}; \{\mathbf{N}\}), \tag{9f.7}$$

where, as previously, $\boldsymbol{\Psi}^{\mathbf{N}}\equiv\psi_a^{N_a}\psi_b^{N_b}\cdots$, and $\mathbf{N}!\equiv N_a!, N_b!, \ldots$. The equation is exactly the same in form as eq. (8g.5), with eq. (8g.4) for a gas, where ΔP replaces the pressure P, ψ_a replaces the absolute activity z_a, and the potential of average force $\Omega_{\mathbf{N}}(\mathbf{y}, \{\mathbf{N}\})$ replaces the potential energy function $U_{\mathbf{N}}\{\mathbf{N}\}$ at zero activity. We can immediately invoke the mathematics of Sec. 8d and write, analogously to eq. (8*l*.2),

$$\beta[P(\mathbf{z})-P(\mathbf{y})] = \sum_{\mathbf{n}\geq 1} b_{\mathbf{n}}(\mathbf{y})\boldsymbol{\psi}^{\mathbf{n}}. \tag{9f.8}$$

If one sets all $y_a=y_b=\cdots=0$, so that $\rho_a(y)y_a^{-1}=1$, $\psi_a=z_a$, and $P(\mathbf{y})=0$, eq. 8 is just eq. (8*l*.2) with $b_{\mathbf{n}}(\mathbf{y}\to 0)$, the cluster integral at zero activity, with the potential of average force the true potential in the absence of other molecules. If, however, one sets $z_a=z_b=\cdots=0$, ψ_a becomes $-\rho_a$ and $\boldsymbol{\psi}=-\boldsymbol{\rho}$. One then has

$$P(\mathbf{y}) = \sum_{\mathbf{n}\geq 1}(-)^{n-1}b_{\mathbf{n}}(\mathbf{y})\boldsymbol{\rho}^{\mathbf{n}}. \tag{9f.8$'$}$$

This of course is not a true series development in that the $b_{\mathbf{n}}$'s depend on $\mathbf{y}$. Also, since this series does not converge across a phase transition, it is valueless for a low-temperature liquid but is valid for an imperfect gas.

Since

$$\frac{\partial[\beta P(\mathbf{z})]}{\partial \ln z_a} = \rho_a(\mathbf{z}), \tag{9f.9}$$

and, from eq. (6),

$$\frac{\partial \ln \psi_a}{\partial \ln z_a} = z_a(z_a-y_a)^{-1}, \tag{9f.10}$$

we have

$$\rho_a(\mathbf{z}) = z_a(z_a - y_a)^{-1}\frac{\partial}{\partial \ln \psi_a}\sum b_{\mathbf{n}}(\mathbf{y})\boldsymbol{\psi}^{\mathbf{n}}$$
$$= z_a(z_a - y_a)^{-1}\sum_{\mathbf{n}} n_a b_{\mathbf{n}}(\mathbf{y})\boldsymbol{\psi}^{\mathbf{n}}. \tag{9f.11}$$

The first term of the sum on the right of eq. (11), with $n_a = 1$, $n_b = n_c = \cdots = 0$, and $b_a = 1$, is, from eq. (6) for ψ_a, equal to $\rho_a(\mathbf{y})$. One has

$$\rho_a(\mathbf{z}) - \rho_a(\mathbf{y}) = \sum_{\substack{n_a \geq 1\, n_b,\ldots \geq 0 \\ n \geq 2}}\sum n_a b_{n_a, n_b, \ldots}(\mathbf{y})\rho_a(\mathbf{y})\psi_a^{n_a - 1}\psi_b^{n_b}\cdots. \tag{9f.12}$$

We can now introduce the irreducible functions at activity set **y** and their integrals defined at zero activity set for a sum of pair potentials by eq. (8*l*.4). The general definition of the irreducible function for many components is very complicated. As in Sec. 8*l*, we treat a one-component system and then generalize. The implicit relation for the B_m's in terms of the b_n's [eq. (8i.22) and derivation in Sec. 8j] is, for $n \geq 2$,

$$n^2 b_n = \left[\sum_{\nu_m \geq 1} \frac{\prod (nB_m)^{\nu_m}}{\nu_m!}\right]_{\sum_m (m-1)\nu_n = n-1}. \tag{9f.13}$$

The explicit inverse is, for $m \geq 2$,

$$B_m = \left\{\sum_{\nu_n \geq 0} (-)^{(1+\sum \nu_n)} \frac{[m - 2 + \sum \nu_n]}{m!} \prod_{n \geq 2} \frac{(nb_n)^{\nu_n}}{\nu_n!}\right\}_{\sum (n-1)\nu_n = m-1}. \tag{9f.14}$$

Various thermodynamic quantities are then given by series with the $B_m(y)$'s as coefficients of the quantity φ raised to the power m,

$$\psi\frac{\partial}{\partial\psi}[\beta P(z) - \beta P(y)] = (z - y)z^{-1}\rho(z) = \varphi, \tag{9f.15}$$

in particular from eq. (8i.24), with $B_m = \beta_{k-1}$,

$$\beta[P(z) - P(y)] = \varphi - \sum_{m \geq 2}(m-1)B_m(y)\varphi^m. \tag{9f.16}$$

The extension to a multicomponent system follows the same reasoning employed in Sec. 8*l*. With $\varphi_a = (z_a - y_a)z_a^{-1}\rho_a(\mathbf{z})$, $\boldsymbol{\phi}^{\mathbf{m}} = \varphi_a^{m_a}\varphi_b^{m_b}\ldots$, $m = m_a + m_b + \cdots$,

$$\beta[P(\mathbf{z}) - P(\mathbf{y})] = \sum_a \varphi_a - (m-1)B_{\mathbf{m}}(\mathbf{y})\boldsymbol{\phi}^{\mathbf{m}}. \tag{9f.17}$$

Again, if we set $y_a = y_b = \cdots = 0$ and $P(\mathbf{y}) = 0$, we have the development around the perfect-gas state, with $\varphi_a = \rho_a$ of eq. (8*l*.9) for the pressure

$P(z)$ as a power series in the number densities $\boldsymbol{\rho}^{\mathbf{m}}$. If, however, we set $z_a = z_b = \cdots = 0$, since $\lim(\rho_a/z_a) = 1$ at $\mathbf{z} = 0$, we find $\varphi_a = -y_a$ and

$$P(\mathbf{y}) = \sum_a y_a + \sum_{m\geq 2} \sum_{\mathbf{m}} (m-1)(-)^m B_{\mathbf{m}}(\mathbf{y})\mathbf{y}^{\mathbf{m}}. \tag{9f.18}$$

We now turn to the equation for the osmotic pressure, with $y_a = y_b = \cdots = 0$ for all *solute* species, but with $z_\nu = y_\nu$, $z_\mu = y_\mu, \ldots$ for the solvent. All terms in the sum for nonzero numbers of solvent molecules now vanish, so that the equations have no *explicit* dependence on the properties of the solvent except of course implicitly, and crucially, in that the potentials of average force between the solute molecules are those at infinite dilution in the presence of the pure solvent.

As all $y_a \to 0$, from eq. (6) for ψ_a, since $y_a^{-1}\rho_a(\mathbf{y}) \to 1$ and $z_a - y_a \to z_a$, we have $\psi_a \to z_a$ and the osmotic pressure ΔP is, from eq. (8),

$$\Delta P = \beta^{-1} \sum_{\mathbf{n}\geq 1} b_{\mathbf{n}} \mathbf{z}^{\mathbf{n}}, \tag{9f.19}$$

with z_a the absolute activity proportional to the conventional activity but in units of molecules per unit volume. From eq. (15) for φ_a, since $(z_a - y_a)z_a^{-1}$ goes to unity when $y_a \to 0$, with $\rho = \sum \rho_a$, we have from eq. (17) that

$$\Delta P = \beta^{-1}\left[\rho - \sum_{m\geq 2} (m-1) B_{\mathbf{m}} \boldsymbol{\rho}^{\mathbf{m}}\right]. \tag{9f.20}$$

The relation of eq. (8*l*.11) for $\ln z_a$ as a power series in the number densities ρ_a is unchanged, namely,

$$\ln z_a = \ln \rho_a - \rho_a^{-1} \sum_{\mathbf{m}} m_a B_{\mathbf{m}} \boldsymbol{\rho}^{\mathbf{m}}. \tag{9f.21}$$

9g. INTEGRAL EQUATIONS

There exists a plethora of integral equations relating the probability densities $\rho_{\mathbf{n}}$ and the dimensionless $F_{\mathbf{n}}$'s at two different activity sets, $\mathbf{z}$ and $\mathbf{y}$, used most generally with either $\mathbf{z}$ or $\mathbf{y}$ identically zero so that $F_{\mathbf{n}}$ is $\exp(-\beta U_{\mathbf{n}})$. These equations are valid, in principle at least, across a phase transition, although some sort of approximation is always necessary in order to use them numerically. They can be divided into two classes, the first, and earliest, of these is the Yvon–Born–Green hierarchy[(1)] of equations, which was also independently developed by Bogolyubov.[(2)]

(1) J. Yvon, *La Théorie Statistique des Fluides et l'Équation d'État.* (Actualitēs scientifiques et industrielles, no. 203), Paris, 1935; M. Born and H. S. Green, *Proc. R. Soc. London, Ser. A*, **188**, 10 (1946).

(2) N. N. Bogolyubov, *Problemy Dinamicheskoi Theorii v Statisticheskoi Fzike*, (Problems of Dynamic Theory in Statistical Physics), State Technical Press, 1946.

Along with these, although at first examination appearing to be different, is the Kirkwood[(1)] equation. Mayer later showed[(2)] the Kirkwood equation to be essentially identical to the others in its complete form so that, if used without any approximation, it would necessarily give the same rigorously exact result. The complete set of equations is often called the YBG, and sometimes the YBGK or YBBGK hierarchy. The second class includes the Percus–Yevick and hypernetted chain equations.

Sarolea and Mayer[(3)] later showed that an attempt to use the complete YBBGK hierarchy without approximation leads to a series that represents a function with a singularity at the activity of condensation. This does *not* mean that the equations could not be asymptotically valid when used with the type of approximation commonly employed, namely, the extended Kirkwood closure which assumes the functions $\theta_{\mathbf{n}}$ of eq. (9e.3) to be zero for some $n \geq n_{\max}$.

The approach of Yvon is straightforward and depends on the theorem proved in Sec. 9b stating that $\Omega_{\mathbf{n}} = \beta^{-1} \ln F_{\mathbf{n}}$ is a potential of average force. If the potential energy $U_{\mathbf{n}}$ is a sum of pair terms $u(r_{ij})$, the average force on one of a pair of molecules i and j, $\langle F_i^{(2)}(r)\rangle$, at a distance r apart is the direct force due the other $\mathbf{f}_i^{(2)}(\mathbf{r}_i, \mathbf{r}_j) = -\nabla_i u(|\mathbf{r}_i - \mathbf{r}_j|)$ plus an averaged term due to a third molecule. The average is given by integration over the conditional probability density for the third molecule, given the position of the pair considered. Thus, for a single-component fluid of monatomic molecules with only pair forces, one can write a single equation for the average force $-\partial\Omega_2(r)/\partial r$ *as* $-du/dr$ plus an integral involving $-du/dr$ and $\rho_3\rho_2^{-1}$. From symmetry considerations in a fluid the force is always directed along the axis connecting the two molecules.

Use the following coordinate scheme. We seek the average force on molecule 1 when

$$x_1 = y_1 = z_1 = 0, \qquad x_2 = r > 0, \qquad y_2 = z_2 = 0, \tag{9g.1}$$

which force will be directed along the x-axis. For the third molecule under the integral we use

$$-\infty \leq x = x_3 \leq \infty, \qquad \zeta = (y_3^2 + z_3^2)^{1/2}, \qquad 0 \leq \varphi = \cos^{-1}\frac{y_3}{\zeta} \leq 2\pi,$$

$$\eta = (x^2 + \zeta^2)^{1/2}, \qquad \chi = [(x-r)^2 + \zeta^2]^{1/2}, \tag{9g.1$'$}$$

so that

$$\int dr_3 = dx_3\, dy_3\, dz_3 = \int_0^{2\pi} d\varphi \int_0^{\infty} \zeta\, d\zeta \int_{-\infty}^{+\infty} dx. \tag{9g.1$''$}$$

(1) J. G. Kirkwood, *J. Chem. Phys.*, **7,** 919 (1939); Kirkwood and E. Monroe *J. Chem. Phys.*, **9,** 514 (1941), **10,** 394 (1942).
(2) J. E. Mayer, *J. Chem. Phys.*, **15,** 187 (1947).
(3) L. Sarolea and J. E. Mayer, *Phys. Rev.*, **101,** 1627 (1956).

The direct force along the positive x-axis on molecule 1 due to molecule 2 is

$$f_{x1}\text{ (due to 2)} = -\frac{du(r)}{dr}, \tag{9g.2}$$

and the x-component of that due to molecule 3 is

$$f_{x1}\text{ (due to 3)} = \frac{x}{\eta}\frac{du(\eta)}{d\eta}. \tag{9g.2$'$}$$

The conditional probability density that, if molecules 1 and 2 are at $\mathbf{r}_1$ and $\mathbf{r}_2$, molecule 3 will be at $\mathbf{r}_3$, is $\rho_3(\mathbf{r}_1, \mathbf{r}_2, \mathbf{r}_3)[\rho_2(\mathbf{r}_1, \mathbf{r}_2)]^{-1}$. We use functions $\omega_n(\{n\})$ defined in terms of the $\Omega_n\{n\}$'s in the same way as the $\theta_n\{n\}$'s of eq. (9e.3) are defined for the Θ_n's,

$$\omega_n\{n\} = \sum_{\nu\geq 1}^{\nu\leq n} \sum_{\{\nu\}_n} (-)^{n-\nu}\Omega_\nu\{\nu\}_n, \tag{9g.3}$$

with $\omega_1 \equiv 0$, so that

$$\rho_3\rho_2^{-1} = \rho \exp - \beta[\omega_2(1, 3) + \omega_2(2, 3) + \omega_3(1, 2, 3)]. \tag{9g.3$'$}$$

Since the variable φ does not appear in the argument of the integrand, the integral over φ gives only a factor 2π. One finds

$$\frac{d\omega_2(r)}{dr} = \frac{du(r)}{dr} + 2\pi\rho\int_0^\infty d\zeta\int_{-\infty}^\infty dx\frac{x\zeta}{\eta}\frac{du(\nu)}{d\nu} \times \exp\{-\beta[\omega_2(\eta) + \omega_2(\chi) + \omega_3(x, \eta, \chi)]\}. \tag{9g.4}$$

The equation is a nonlinear integrodifferential equation for ω_2, but the kernel under the integral contains an unknown function ω_3. Obviously, a similar nonlinear integrodifferential equation can be written for $\partial\omega_3/\partial r$ involving ω_2 and ω_3 in the kernel, but also the unknown ω_4 and, in general, a hierarchy of equations up to ω_N. For numerical solutions a closure at some stage is necessary, and this was done originally by the restricted Kirkwood closure which assumes that $\omega_3 \equiv 0$, writing

$$\frac{d\omega_2(r)}{dr} = \frac{du(r)}{dr} + 2\pi\rho\int_0^\infty d\zeta\int_{-\infty}^{+\infty} dx\frac{x\zeta}{\eta}\frac{du(\eta)}{d\eta}\exp\{-\beta[\omega_2(\eta) + \omega_2(\chi)]\}. \tag{9g.4$'$}$$

At the time the equation was first used, before the advent of electronic computers, it posed prodigious problems for numerical solution. Born and Green linearized the equation by using

$$\exp\{-\beta[\omega_2(\eta) + \omega_2(\chi)]\} \cong 1 - \beta[\omega_2(\eta) + \omega_2(\chi)], \tag{9g.4$''$}$$

which was still far from easy numerically, and for low-temperature liquids is a rather drastic approximation.

Since, from symmetry in x and $-x$,

$$\int_{-\infty}^{+\infty} dx \frac{x\zeta}{\eta}\frac{du(\eta)}{d\eta} = 0, \tag{9g.4'''}$$

one can drop the unity in eq. (4″) and also write the exponent minus unity for the exponent in eqs. (4) and (4′). This is usually done.

Numerical computer solutions using a Lennard–Jones 6–12 pair potential give results that look qualitatively very similar to those derived from X-ray or neutron diffraction measurements. The pressures calculated from eq. (9b.18) with the theoretical pair probability density are completely wrong, but this is hardly surprising. The small pressure is due to the almost complete cancelation of two very large terms, that due to close-range repulsion, the **b** term in the van der Waals equation (Sec. 8b) and the **a** term due to the attractive forces of the more distant molecules.

Actually, by using eq. (9c.5) in which $\exp(-\beta U_M)$ is given as a sum of integrals involving $\exp(-\beta\Omega_{M+K})$ for all K, one can obtain a linear integrodifferential equation for ω_2 if again one makes the assumption that ω_3 and ω_n, for all $n \geq 3$, are zero. Equation (9c.5), with $M=2$ for a one-component system, with eq. (9c.3) for $\Psi_2(V, \beta, \mathbf{z}\equiv 0)$, with eq. (9c.1) for $\Psi_n(V, \beta, \mathbf{z})$, and with the coordinates of eqs. (1) and (1′), is

$$e^{-\beta u_2(r)} = \sum_{N\geq 0} \int\!\!\int \overset{\cdot}{\underset{V}{\cdot\,\cdot}} \int d\{N\} e^{\beta PV} \frac{(-z)^N}{N!} \left(\frac{\rho}{z}\right)^{N+2} e^{-\beta\Omega_{N+2}}. \tag{9g.5}$$

Differentiate both sides with respect to the x-coordinate of molecule 1 at the origin and multiply both sides by $-e^{\beta u_2}$. One finds

$$\frac{du_2(r)}{dr} = \frac{d\omega_2(r)}{dr} - 2\rho z \int_0^{\infty} d\zeta \int_{-\infty}^{+\infty} dx \frac{x\zeta}{\eta} \frac{d\omega_2(\eta)}{d\eta} \exp\{-\beta[u_2(\eta)+u_2(\chi)]\} \tag{9g.6}$$

instead of eq. (4′). Equation (6) is now a linear equation in the unknown function $d\omega_2(r)/dr$.

The two kernels $\exp\{-\beta[\omega_2(\eta)+\omega_2(\chi)]\}$ of eq. (4′) and $\exp\{-\beta[u_2(\eta)+u_2(\chi)]\}$ of eq. (6) stand to each other in the same relation as reciprocal kernels in the simple theory of Fredholm inhomogeneous linear integral equations with symmetric kernels. Namely, if $K(x, y) = K(y, x)$,

$$f(x) = g(x) + \lambda \int dy\, f(x) K(x, y), \tag{9g.7}$$

and certain other conditions are fulfilled, then there exists a kernel

$K^*(x, y)$ said to be reciprocal to K, such that

$$g(x) = f(x) - \lambda \int dy\, g(y) K^*(x, y). \tag{9g.7'}$$

The reciprocal kernel K^* depends only on K and not on f or g.

The assumption that $\omega_3 \equiv 0$ is of course incorrect, even, or particularly, for the case in which one assumes $u_3 = u_4 = \cdots \equiv 0$, and it cannot be assumed that eqs. (4′) and (6) give the same answers for $\omega_2(r)$; indeed they do not. If we forego the simplification of assuming $\omega_3 = \omega_4 = \cdots \equiv 0$, eqs. (4′) and (6) become an infinite hierarchy of equations each with an infinite number of unknown functions $\omega_3, \omega_4, \ldots$, but the two sets in eqs. (4′) and (6) look quite different. In eq. (4′) we now have ω_3 in the kernel and a new equation for $\nabla_{\mathbf{r}_i}\omega_3$ with the new unknown ω_4 in the kernel and, *seriata*, an infinite set of equations for the successive unknown functions ω_n. However, eq. (6) remains a single equation, but one involving an infinite number of integrals over successive terms involving $\nabla_{r_0}\omega_n$ and more and more complicated kernels. Nevertheless, it is possible to show[(1)] that one can formally eliminate all but the kernels involving $\exp(-\beta U_n)$ and $\exp(-\beta\Omega_n)$, and that these stand in a relationship to each other equivalent to that between two simple reciprocal symmetric Fredholm kernels.

The Kirkwood integral equation was developed independently of the earlier Yvon work of which Kirkwood was unaware. The starting point looks quite different, but the same kernels are involved. This is the basis of the assertion made earlier that the YBG and Kirkwood equations are essentially identical before truncation of assuming $\omega_3 \equiv 0$, although they give different numerical results when this approximation is made.

Kirkwood starts with two types of molecules, normal molecules of type a with pair potential $u(r)$, and molecules of type b whose interaction potential $u_{ab}(r)$ with molecules of type a is $u_{ab}(r) = \lambda u(r)$, where λ varies in the range $0 \le \lambda \le 1$ and $\omega_{ab}(r)$ depends on λ, $\omega_{ab}(r) = \omega(\lambda, r)$, varying from zero at $\lambda = 0$, where there is no interaction between a- and b-type molecules, to the same value, $\omega_{ab}(r) = \omega_{aa}(r)$, as that between two molecules of type a when $\lambda = 1$.

In the original work Kirkwood used the equations for the petite canonical ensemble of fixed N_a, N_b, β, and V but set $N_b = 1$, namely, only a single molecule of type b, so that no b–b interactions are present. The grand canonical ensemble formalism can equally well be employed, setting the limit that the activity z_b of molecules of type b goes to zero, so that b–b interactions disappear. The probability density $W_{N_a,N_b}(V, \beta, z_a, z_b;$

(1) J. E. Mayer, *J. Chem. Phys.*, **15,** 187 (1947).

$\{N_a\}\{N_b\}$) in the grand canonical ensemble has a factor $z_b^{N_b}$ in the numerator.

We want to calculate the conditional probability $\rho_b^{-1}\rho_{ab}(r_{ab})$ that if one molecule of type b is at the origin $\mathbf{r}_b \equiv 0$ there will be one of type a at a distance r from the origin for given λ, but at the limit $z_b \to 0$. This requires integration over the coordinates of all but the two molecules, one of type a and one of type b, and summation over all N_a, N_b. The quantity $\rho_b^{-1}\rho_{ab}(r)$ remains finite and of the order of ρ_a in the limit $z_b \to 0$, $\rho_b \to 0$, but the sum over $N_b \geq 1$ can be limited to the first term, $N_b = 1$, since $z_b^{N_b}\rho_b^{-1} \to 0$ for $N_b > 1$. The potential energy $U_N\{N\}$ is a sum of pair terms,

$$U_{N_b=1,N_a} = \sum_{N_a \geq i > j \geq 2}\sum u(r_{ij}) + \sum_{N_a \geq i \geq 2}\sum u(|\mathbf{r}_i|), \tag{9g.8}$$

and, with eq. (9b.3) for $\rho_{\mathbf{n}}\{\mathbf{n}\}$,

$$\rho_b^{-1}\rho_{ab}(r) = \rho_b^{-1}\sum_{N_a \geq 1} e^{-\beta PV} z_b \frac{z_a}{(N_a-1)!}\int\int\cdots\int d\{N_a\}e^{-\beta U\{N_a, r_0, r_1\}}, \tag{9g.9}$$

where

$$\rho_b^{-1}\rho_{ab}(r) = \rho_a \exp[-\beta\omega(\lambda, r)], \tag{9g.10}$$

and, with the assumption that for any set of three molecules, $\omega_3 \equiv 0$, particularly $\omega_{baa} \equiv 0$,

$$\rho_{baa}(r, r_{0a}, r_{ba})[\rho_{ab}(r)]^{-1} = \rho_a \exp[-\beta\omega(\lambda, r_{0a}) - \beta\omega(\lambda \equiv 1, r_{ab})]. \tag{9g.11}$$

Differentiate both sides of eq. (9) with respect to λ using eqs. (8), (10), and (11), multiply both sides by $-\rho_b[\beta\rho_{ab}(r)]^{-1}$, and integrate over λ from zero to unity. One obtains the Kirkwood integral equation for $\omega_{aa}(r) = \omega(\lambda \equiv 1, r)$, namely, with $\rho = \rho_a$,

$$\begin{aligned}\int_0^1 d\lambda \frac{d\omega(\lambda, r)}{d\lambda} &= \omega(\lambda = 1, r) = \omega(r)\\ &= u(r) + \int^1 d\lambda\,\rho\int d\mathbf{r}_a\,u(r_a)\exp[-\beta\omega(\lambda, r_{0a}) - \beta\omega(\lambda = 1, r_{ab})].\end{aligned} \tag{9g.12}$$

This is probably the form most likely to give the nearest approach to the correct $\omega(r)$ with the use of the closure, $\omega_3 = 0$, of the three eqs. (4), (6), and (12).

The Percus–Yevick and the hypernetted chain of integral equations have an entirely different conceptual origin than the YBBGK equations. The logical origin of these follows more closely the type of procedure used in Sec. 8m where some simply integrable diagrams were discussed.

There we introduced a function $C(r)$ [eq. (8m.13)] which was the sum over all n of integrals over the coordinates of the $n-1$ intermediate molecules, $1 \leq i \leq n-1$, of chains $f(r_{01})f(r_{12})\cdots f(r_{n-1,n})$ with fixed distance, $r = r_{0n}$, between molecules zero and n. This $C(r)$ was evaluated by two consecutive integrations of single variables. Eventually, in eq. (8m.15) we wrote an integral which summed the contribution to the virial series of an infinite number of selected diagrams. It was an integral over a single variable t of a simple algebraic function of a function $\sigma(t)$, which in turn was the Fourier transform of an $S_2(r)$ equal to $e^{-\beta u(r)+C(r)} - C(r) - 1$. Thus no single step involved more than integration over a single variable.

By classifying graphs and *defining* integrals analogous to $C(r)$ over intermediate molecules of products of $f(r_{ij})$'s, which involves impossible multidimensional integrations, one can formally write a rigorous expression for a single integral giving the virial sum for $\beta P\rho$. One can also write, in a similar manner, a formal rigorous expression for the pair distribution function,[(1)] $\rho^{-1}\rho_2(r) = e^{-\beta\omega_2(r)}$, and thus for $\omega_2(r)$. In doing this one finds that $\omega_2(r)$ itself, along with other terms such as $S_2(r)$, appears in the steps by which one arrives at the final integral. Neglecting terms that are too difficult to evaluate one can finally arrive at a tractable integral equation for $\omega_2(r)$.

The hypernetted chain equation and the Percus–Yevick equation make different neglects. The hypernetted chain includes more terms than the Percus–Yevick. However, comparison with the Monte Carlo computer results (see Sec. 9h) for hard spheres shows the Percus–Yevick equation to be remarkably good for this potential.

It is doubtful that either equation is nearly so good for a low-temperature liquid, but they form relatively simple initial approximations from which a correction can be made by perturbation methods to take care of the attractive potential.[(2)]

9h. COMPUTER EXPERIMENTS

The best numerical information we have on liquid structures for an exactly defined sum of pair potentials comes from computer calculations. The computer produces a large sample of positions of a finite but reasonably large number N of molecules in a fixed volume V enclosed by rectangular faces, by a method that weights the probability density as

[(1)] The general method was first given by J. E. Mayer and Elliott Montroll, *J. Chem. Phys.*, **9,** 2 (1941).

[(2)] See, for instance, H. C. Anderson, J. D. Weeks, and D. Chandler, *Phys. Rev.*, **A4,** 1597 (1971).

proportional to $\exp(-\beta U_N\{N\})$. From this the statistical distribution $\exp[-\beta\omega_2(r)]$ can be obtained with considerable precision, and at the expense of far greater computer time a much less precise tabulation of $\omega_3(r_{12}, r_{23}, r_{31})$ is possible, in principle at least. The difference of course is due to the fact that $\omega_2(r)$ is a function of a single variable r, whereas ω_3 is a function of three variables, and even if an infinite amount of data were available, the mere numerical tabulation of the function presents a severe problem.

The values of the number of molecules N usually lie in the range between 10^2 and 10^3. The fact that the results are reasonably independent of N indicates that there is little need to go to higher values. The computations are made with periodic boundary conditions, as follows. The walls are treated as though they offer no hindrance to the passage of molecules but, if a molecule moves out of V at one wall, it is replaced by a molecule entering V from the cell beyond the opposite wall with exactly the same trajectory relative to the neighboring cell. Thus, if V has a boundary perpendicular to the x-axis at $x=0$ and one at $x=l>0$, if a molecule moves to a position y, z, $x=l+x'$, with $0\leq x'\leq l$, it is replaced by one at y, z, x'. Thus the system treated is essentially one that is one unit cell of an infinite perfect periodic crystal of cells containing N molecules in the volume V, all cells being identical in configuration and momenta at simultaneous times.

Two different methods are used to obtain the equilibrium distribution in phase space. The more obvious but more difficult one is known as the method of molecular dynamics. The moves of molecular motions are made by actually integrating the equations of motion for the system of N molecules. This has the great advantage that the actual time sequence is displayed; one can observe the time necessary to approach equilibrium from a given nonequilibrium distribution in phase space, including, for instance, the half-life of a fluctuation from equilibrium. Since self-diffusion coefficients can be calculated from the autocorrelation function, the conditional probability density that if a molecule has momentum $\mathbf{p}_0$ at a time $t_0=0$ it will have momentum $\mathbf{p}(t)$ at time t, it is possible to use the method to obtain transport properties.

The method was first developed by Alder and Wainwright for molecules with a hard sphere potential, infinite repulsive potential between pairs up to a distance r_0, and no other forces.[(1)] For this potential the programming is greatly simplified by the fact that each molecular

[(1)] B. J. Alder and T. E. Wainwright, *J. Chem. Phys.*, **31,** 459 (1959); *Phys. Rev. Lett.*, **18,** 968; *Phys. Rev.*, **A1,** 18 (1970).

trajectory is in a straight line between collisions, but Rahman,[1] and later Verlet,[2] used the method with Lennard–Jones pair potentials.

The other method is much simpler and was developed when electronic computers were less sophisticated. It goes generally under the name Monte Carlo, since the moves are made by random numbers, such as those produced by the spin of a roulette wheel. The name was devised by John von Neumann who first pointed out the use of the technique for solving multidimensional differential equations.

The machine is coded to produce random m-digit numbers ξ, between -1 and $+1$ *ad infinitum.* It also stores, at any stage, the $3N$ positional coordinates of N molecules in V; the momenta are not needed in this method. The molecule n is moved in the νth move within the rectangular volume $V = \alpha_x \alpha_y \alpha_z$, using the periodic boundary prescription, from cartesian coordinate position $X_{n\nu}$ to $X_{n,\nu+1}$, with

$$X_{n,\nu+1} = X_{n\nu} + \alpha_x \xi_{nx,\nu}, \tag{9h.1}$$

and similarly for $Y_{n,\nu+1}$ and $Z_{n,\nu+1}$, where α_x, α_y, α_z are the box edges, and $\xi_{nx,\nu}$, $\xi_{ny,\nu}$, $\xi_{nz,\nu}$ are random numbers in the range -1 to $+1$. The change ΔE_ν in energy for the move is computed from the assumed pair potential $u(r)$ by

$$\Delta E_\nu = E_{\nu+1} - E_\nu = \sum_{i \neq n} [u(r_{i,n(\nu+1)}) - u(r_{i,n\nu})], \tag{9h.2}$$

and of course for short-range potentials only the $u(r_{in})$'s for molecules relatively close to n need be considered.

The ingenuity in the method consists in the simplicity of the formula by which the correct weighting according to $\exp(-\beta U_N)$ is obtained, given in a paper by Metropolis et al.[3] If the energy change ΔE_ν is negative, the new position $X_{n,\nu+1}$, $Y_{n,\nu+1}$, $Z_{n,\nu+1}$ is given by eq. (1). If, however, ΔE_ν is positive, the positions $X_{n,\nu+1}$, $Y_{n,\nu+1}$, $Z_{n,\nu+1}$ are determined by a random number and $\exp(-\beta \Delta E_\nu)$; a random number ξ_ν between 0 and 1 is chosen and, if $\xi_\nu < \exp(-\beta \Delta E_\nu)$, eq. (1) is used for the new positions; but if $\xi_\nu > \exp(-\beta \Delta E_\nu)$, then the position at $\nu + 1$ is taken to be that at ν but, and this is important, in counting the frequency of given coordinate positions it is counted as a new position, so that the coordinate set X_ν, Y_ν, Z_ν is given double weight. The authors prove that this gives the correct weighting, namely, that the probability density W_r of a $3N$-dimensional position r is proportional to $\exp(-\beta E_r)$, as follows.

[1] A. Rahman, *Phys. Rev.*, **136,** A405 (1964).

[2] L. Verlet, *Phys. Rev.*, **159,** 98 (1967).

[3] N. A. Metropolis, A. W. Rosenbluth, M. N. Rosenbluth, A. H. Teller, and E. Teller, *J. Chem. Phys.*, **21,** 1087 (1953).

The rules of the game ensure that the system modeled is ergodic. Every position $r \equiv \{N\}$ can be reached from every position $q \equiv \{N'\}$ after a sufficient number of moves, and the periodic boundary conditions ensure a constant number N of molecules in V. In every single move between two states r and s differing only in the coordinates of a single molecule, the prescription of eq. (1) for the new coordinates is such that the a priori probability P_{rs} of a move from r to s is the same as that, P_{sr}, of a move from s to r, before the weighting by $\exp(-\beta \Delta E_{rs})$,

$$P_{rs} = P_{sr}. \tag{9h.3}$$

If W_r and W_s are the probability densities of such states in an ensemble of such games, and if $E_r > E_s$, then the number N_{rs} of transitions from r to s is

$$N_{rs} = W_r P_{rs}, \tag{9h.4}$$

since every move is allowed, but the number N_{sr}, from s to r by the formula used, is

$$N_{sr} = W_s P_{sr} \exp[-\beta(E_r - E_s)]. \tag{9h.4$'$}$$

At equilibrium we have

$$N_{sr}^{(\mathrm{eq})} = N_{rs}^{(\mathrm{eq})}, \tag{9h.5}$$

so that, from eqs. (3) to (5),

$$\frac{W_r^{(\mathrm{eq})}}{W_s^{(\mathrm{eq})}} = \exp[-\beta(E_r - E_s)]. \tag{9h.6}$$

Since the ratio is repeated in every move, it is retained after any number of consecutive moves; and since the system is ergodic, it follows that the ratio in eq. (6) applies to any two states.

In the original paper the authors used the scheme with the Los Alamos mathematical analyzer numerical integrator and computer (MANIAC), which was a second-generation radio-tube machine. The example chosen was rather trivial, 224 molecules in two dimensions. With the advent of far faster solid-state machines with memories of far greater capacity the method has been widely used for a variety of problems. It is far more economical of machine time than the molecular dynamics computation, and more results exist, but it does not represent the time evolution of the system, hence is of no value in giving transport properties.

One unexpected result has been obtained for hard-sphere potentials. If the computed pressure is plotted against the number density ρ (which is usually obtained by changing r_0, the hard-sphere diameter, and scaling rather than by changing V), two distinct curves are found giving two number densities, ρ_g and ρ_{cr}, at the same pressure P for a considerable range, with $\rho_g < \rho_{cr}$. This is interpreted as indicating a first-order phase transition from liquid, or more properly dense gas at liquid density, to a crystalline configuration. If one starts with a low density, increasing r_0 gradually, the low-density curve ρ_g is followed, making an occasional transition rather abruptly at the higher pressures to the ρ_{cr} curve. If one decreases ρ from high density, the ρ_{cr} is followed with transitions to the ρ_g curve at the lower-pressure end. The metastability is presumably to be due to the finite N values and a "surface tension" effect[(1)] between neighboring cells. Even with $N = 10^3$ and a cubic volume V the edge is only 10 molecular distances, and 488 of the 1000 molecules would be on the "surface," $488 = 10^3 - 8^3$.

9i. MOMENTS OF FUNCTIONS, HEAT CAPACITY AT CONSTANT VOLUME

In Sec. 5a we discussed averages of certain thermodynamic extensive variables such as energy $\langle E \rangle$ and moment combinations such as $\langle E^2 \rangle - \langle E \rangle^2$. The latter is related to a heat capacity,

$$C = k^{-1}(\langle E^2 \rangle - \langle E \rangle^2) = \frac{\partial E}{\partial T}, \tag{9i.1}$$

where the particular heat capacity C depends on the variables besides T held constant in the partial derivative, and the averages $\langle E^2 \rangle - \langle E \rangle^2$ are those of the Hamiltonian $H(\mathbf{p}^\Gamma, \mathbf{q}^\Gamma)$ in the corresponding ensemble. In Sec. 5a equations were given in terms of summations over probabilities of quantum states $\mathbf{K}$ of the whole system or, in the classical case, of integration in the Γ-space.

The extensive thermodynamic variables, except V (which is a boundary condition) are averages of functions of the coordinates and momenta of molecules, such as the Hamiltonian. Although there must surely be nonzero terms in the mutual potential energy U_N, such as $u_3(r_{23}, r_{31}, r_{12})$, involving simultaneous coordinate values of three, and even more, molecules in addition to the usual sum of pair terms,

$$U_N = \sum_{N \geq i > j \geq 1} \sum u(r_{ij}), \tag{9i.2}$$

these higher terms are so poorly known that in almost all, if not all, models used to attempt numerical evaluation they are assumed to be zero. In the same spirit virtually all functions whose averages are sought

(1) J. E. Mayer and W. W. Wood, *J. Chem. Phys.*, **42,** 4268 (1965).

involve, at most, pairs of molecules. The consequence is that the averages in the equations of Sec. 5a are sums of averages of functions for single molecules or of the coordinates of a pair or, as in the case of $\langle U^2 \rangle$ with eq. (2) for U_N, terms as complicated as $u(r_{ij})u(r_{kl})$ involving four different molecules. The averages can then be evaluated by integration of such functions times probability densities $\rho_n\{n\}$ for $n=1, 2, 3$, or 4.

Because of the very small but nonzero values of $\rho_2(r)-\rho^2$ when $r \gg r_0$, of $\rho_3(r_{23}\ r_{31},\ r_{12})-\rho\rho_2(r_{12})$ when $r_{31},\ r_{23} \gg r_0$, where r_0 is the distance of the minimum pair potential, and of $\rho_4(\mathbf{r}_1, \mathbf{r}_2, \mathbf{r}_3, \mathbf{r}_4)-\rho_2(r_{12})\rho_2(r_{34})$ when $r_{13}, r_{14}, r_{23}, r_{24}$ are all very large in a closed system of fixed V, T, N, the averages of products $\langle U^2 \rangle$, $\langle N^2 \rangle$, and $\langle UN \rangle$ must always be made in the ensemble of open systems, the grand canonical ensemble of fixed V, β, $-\ln z$ (see the discussion in Sec. 9a).

If the functions to be averaged involve momenta as well as coordinates, no difficulties are involved in the classical case, since the momentum probability densities are independent factors,

$$W(p_\alpha)=\left(\frac{\beta}{\pi}\right)^{1/2} e^{-\beta p_\alpha{}^2}, \tag{9i.3}$$

for each of the $3N$ cartesian momenta, $\alpha = xi,\ yi,\ zi,\ 1 \le i \le N$.

We give here the method of calculation of the constant-volume heat capacity for monatomic molecules. It involves most of the minor complications that enter similar calculations for other thermodynamic variables and serves as an instructive example. With $\beta=(kT)^{-1}$ and $\nu=-\beta\mu$, which with V are the variables for the grand canonical ensemble of one-component open systems, we write

$$k^{-1}C_V = k^{-1}\left(\frac{\partial E}{\partial T}\right)_{V,N} = -\beta^2\left(\frac{\partial E}{\partial \beta}\right)_{V,N} = -\beta^2\left(\frac{\partial E}{\partial \beta}\right)_{V,\nu} - \beta^2\left(\frac{\partial E}{\partial \nu}\right)_{V,\beta}\left(\frac{\partial \nu}{\partial \beta}\right)_{V,N}, \tag{9i.4}$$

or for constant V, N, since $dN(V, \beta, \nu)=0=(\partial N/\partial \beta)_{V,\nu}\, d\beta+(\partial N/\partial \nu)_{V,\beta}\, d\nu$, we have

$$\left(\frac{\partial \nu}{\partial \beta}\right)_{V,N} = -\left(\frac{\partial N}{\partial \beta}\right)_{V,\nu}\left(\frac{\partial N}{\partial \nu}\right)_{V,\beta}^{-1}, \tag{9i.5}$$

and

$$k^{-1}C_V = -\beta^2\left(\frac{\partial E}{\partial \beta}\right)_{V,\nu} + \beta^2\left(\frac{\partial E}{\partial \nu}\right)_{V,\beta}\left(\frac{\partial N}{\partial \beta}\right)_{V,\beta}\left(\frac{\partial N}{\partial \nu}\right)_{V,\beta}^{-1}. \tag{9i.6}$$

We now have C_V in terms of derivatives of E and N with respect to ν and β at constant V, the appropriate variables for the grand canonical ensemble.

For an ensemble of systems whose thermodynamic state is fixed by one extensive variable X_k and two defined intensive variables x_i, x_j, if $e^{-\theta}$ is the normalizing factor in the probability density, the extensive variables X_i, X_j are

$$X_i = -\left(\frac{\partial\theta}{\partial x_i}\right)_{X_k,x_j} \qquad X_j = -\left(\frac{\partial\theta}{\partial x_j}\right)_{V,x_i}. \tag{9i.7}$$

Further differentiation gives

$$\left(\frac{\partial X_i}{\partial x_j}\right)_{X_k x_i} = \left(\frac{\partial X_j}{\partial x_i}\right)_{X_k,x_j} = -\left(\frac{\partial^2\theta}{\partial x_i\,\partial x_j}\right)_{X_k}. \tag{9i.8}$$

We now use eq. (5a.20), which gives the derivative of eq. (8) in terms of moments,

$$\left(\frac{\partial X_i}{\partial x_j}\right)_{X_k,x_i} = \left(\frac{\partial X_j}{\partial x_i}\right)_{x_k,x_j} = -(\langle X_iX_j\rangle - \langle X_i\rangle\langle X_j\rangle). \tag{9i.9}$$

In this identify the various quantities as the variables in the grand canonical ensemble, $\theta = \beta PV$, $X_k = V$, $x_i = \beta$, $x_j = \nu$, and $X_i = E$, $X_j = N$, so that, from eq. (8),

$$\left(\frac{\partial E}{\partial \nu}\right)_{V,\beta} = \left(\frac{\partial N}{\partial \beta}\right)_{V,\nu}, \tag{9i.10}$$

and eq. (6) becomes

$$k^{-1}C_V = -\beta^2\left(\frac{\partial E}{\partial \beta}\right)_{V,\nu} + \beta^2\left(\frac{\partial E}{\partial \nu}\right)^2_{V,\beta}\left(\frac{\partial N}{\partial \nu}\right)^{-1}_{V,\beta}. \tag{9i.11}$$

Shorten our notation for the combination of moments,

$$\begin{aligned}
-\beta^2\left(\frac{\partial E}{\partial \beta}\right)_{V,\nu} &= \langle\beta^2E^2\rangle - \langle\beta E\rangle^2 = I(\beta^2E^2),\\
-\beta\left(\frac{\partial E}{\partial \nu}\right)_{V,\beta} &= -\beta\frac{\partial N}{\partial \beta} = \langle N\beta E\rangle - \langle N\rangle\langle\beta E\rangle = I(N\beta E),\\
-\left(\frac{\partial N}{\partial \nu}\right)_{V,\beta} &= \langle N^2\rangle - \langle N\rangle^2 = I(N^2),
\end{aligned} \tag{9i.12}$$

so that

$$k^{-1}C_V = I(\beta^2E^2) - [I(N\beta E)]^2[I(N^2)]^{-1}. \tag{9i.13}$$

In this, use that the energy E is the sum of a potential energy U and a kinetic energy $\frac{3}{2}NkT$, so that $\beta E = \beta U + \frac{3}{2}N$,

$$\begin{aligned}
I(\beta^2E^2) &= I(\beta^2U^2) + 3I(N\beta U) + \tfrac{9}{4}I(N^2),\\
I(N\beta E) &= I(N\beta U) + \tfrac{3}{2}I(N^2),
\end{aligned}$$

and

$$k^{-1}C_V = I(\beta^2 U^2) - [I(N\beta U)]^2[I(N^2)]^{-1}. \tag{9i.14}$$

The dimensionless I's of eq. (14) involve averages of functions of the coordinates of molecules,

$$\beta U_N = \sum_{N\geq i>j\geq 2}\sum \beta u(r_{ij}), \tag{9i.15}$$

$$N = \sum_{N\geq i\geq 1} 1(\mathbf{r}_i), \tag{9i.16}$$

where the function N to be averaged is simply unity for all positions $\mathbf{r}_i$ of the coordinate of molecule i. We also need averages of

$$\begin{aligned}(\beta U_N)^2 = &\sum_{N\geq i>j\geq 1}\sum \beta^2 u^2(r_{ij}) \\ &+ \sum_{N\geq i>j>k\geq 1}\sum\sum [\beta u(r_{ij})\beta u(r_{ik}) + \beta u(r_{ij})\beta u(r_{jk}) + \beta u(r_{ik})\beta u(r_{jk})] \\ &+ \sum_{N\geq i>j>k>l\geq 1}\cdots\sum [\beta u(r_{ij})\beta u(r_{kl}) + \text{five other terms}],\end{aligned} \tag{9i.17}$$

$$\begin{aligned}N\beta U_n = &\sum_{N\geq i>j\geq 1}\sum [1(r_i) + 1(r_j)]\beta u(r_{ij}) \\ &+ \sum_{N\geq i>j>k\geq 1}\sum\sum [1(\mathbf{r}_i)\beta u(r_{jk}) + 1(\mathbf{r}_j)\beta u(r_{jk}) + 1(\mathbf{r}_k)\beta u(r_{ij})],\end{aligned} \tag{9i.18}$$

$$N^2 = \sum_{N\geq i\geq 1} 1^2(\mathbf{r}_i) + \sum_{N\geq i>j\geq 1}\sum [1(\mathbf{r}_i)\times 1(\mathbf{r}_j)]. \tag{9i.19}$$

With the reduced probability densities $\rho_n(\mathbf{r}_1, \ldots, \mathbf{r}_n)$ of eq. (9b.3) and the functions

$$F_n(\mathbf{r}_1, \ldots, \mathbf{r}_n) = \rho^{-n}\rho_n(\mathbf{r}_1, \ldots, \mathbf{r}_n) = \exp[-\beta\Omega_n(\mathbf{r}_1, \ldots, \mathbf{r}_n)], \tag{9i.20}$$

for which, from eq. (9b.10) for one component,

$$\int\cdots\int_V d\mathbf{r}^{(n)}\,\rho_n(\mathbf{r}^{(n)}) = \left\langle \frac{N!}{(N-n)!}\right\rangle, \tag{9i.21}$$

we have, with eqs. (16) and (19),

$$\langle N\rangle = \int_V d\mathbf{r}\,\rho(\mathbf{r}_1) = \int_V d\mathbf{r}\,\rho, \tag{9i.22}$$

$$\langle N^2\rangle = \int_V d\mathbf{r}\,\rho + \iint_V d\mathbf{r}_1\,d\mathbf{r}_2\,\rho_2(\mathbf{r}_1, \mathbf{r}_2), \tag{9i.23}$$

in agreement with eq. (21). Write

$$\langle N\rangle^2 = \iint\limits_V d\mathbf{r}_1\, d\mathbf{r}_2\, \rho(\mathbf{r}_1)\rho(\mathbf{r}_2), \tag{9i.24}$$

and subtract it from eq. (23) to find

$$\begin{aligned} I(N^2) = \langle N^2\rangle - \langle N\rangle^2 &= \iint d\mathbf{r}_1[\rho_2(\mathbf{r}_1, \mathbf{r}_2) - \rho(\mathbf{r}_1)\rho(\mathbf{r}_2)] + \langle N\rangle \\ &= \int_V d\mathbf{r}\,\rho + \int_V d\mathbf{r}' \int_0^\infty \rho^2 4\pi r^2\, dr[F_2(r) - 1]. \end{aligned} \tag{9i.25}$$

As discussed in Sec. (9a), $\rho_2(\mathbf{r}_1, \mathbf{r}_2) - \rho(\mathbf{r}_1)\rho(\mathbf{r}_2)$ goes sufficiently rapidly to zero only in the ensemble of open systems for the correct result to be found by integration of eq. (25) over only the short-range part of $4\pi r^2[F_2(r)-1]$. Similarly, the last terms in eqs. (17) and (18) require grand canonical ensemble averages to obtain correct values of $I(\beta^2 U^2)$ and $I(N\beta U)$.

Finally, to avoid one very long equation, define seven integrals,

$$\begin{aligned}
\psi_1(N) &= \int d\mathbf{r}\,\rho = \langle N\rangle = N, \\
\psi_2(N) &= \rho\int_0^\infty 4\pi r^2\, dr[F_2(r) - 1], \\
\psi_2(\beta U) &= \rho\int_0^\infty 4\pi r^2\, dr\,\beta u(r) F_2(r), \\
\psi_2(\beta^2 U^2) &= \rho\int_0^\infty 4\pi r^2\, dr\,\beta^2 u^2(r) F_2(r), \\
\psi_3(N\beta U) &= \rho^2 \iint d\mathbf{r}_2\, d\mathbf{r}_3\, \beta u(r_{12})[F_3(r_{23}\, r_{31}, r_{12}) - F_2(r_{12})], \\
\psi_3(\beta^2 U^2) &= \rho^3 \iiint d\mathbf{r}_2\, d\mathbf{r}_3\, d\mathbf{r}_4\, \beta u(r_{12})\beta u(r_{13}) F_3(r_{23}, r_{31}, r_{12}), \\
\psi_4(\beta^2 U^2) &= \rho^4 \iint\cdots\int d\mathbf{r}_2\, d\mathbf{r}_3\, d\mathbf{r}_4\, \beta u(r_{12})\beta u(r_{34})[F_4(r_{12}, \ldots, r_{34}) \\
&\qquad - F_2(r_{12})F_2(r_{34})].
\end{aligned} \tag{9i.26}$$

All these integrals are short-range and dimensionless and have numerical values of the order of magnitude of unity except near the critical point. With computer values for $F_2(r)$ the three ψ_2's represent no difficulty in evaluation. If the Kirkwood superposition, $\omega_3 \equiv 0$, is used, the integrals ψ_3 will not be too horrendous, but the integration of ψ_4 would involve a considerable amount of computer time. It is doubtful that the ψ_4 numerical value keeping only $\omega_2(r)$ would be reliable, since the integrand has fairly large positive and negative parts in configuration space. The total value would then be a fairly small difference of rather large inaccurately known contributions. With the Kirkwood superposition approximation we have

$$F_3(r_{23}, r_{31}, r_{12}) = F_2(r_{23})F_2(r_{31})F_2(r_{12}), \tag{9i.27}$$

$$F_3(r_{23}, r_{31}, r_{12}) - F_2(r_{12}) = F_2(r_{12})[F_2(r_{23})F_2(r_{31}) - 1], \tag{9i.27$'$}$$

$$\begin{aligned} F_4(r_{12}, r_{13}, r_{14}, r_{23}, r_{24}, r_{34}) &- F_2(r_{12})F_2(r_{34}) \\ &= F_2(r_{12})F_2(r_{34})]F_2(r_{13})F_2(r_{14})F_2(r_{23})F_2(r_{24}) - 1]. \end{aligned} \tag{9i.28}$$

With eqs. (15) to (19) the I functions of eq. (14) for $k^{-1}C_V$ are given by the integrals of eq. (26) multiplied by the integral over the volume V of ρ; $\int_V d\mathbf{r}\,\rho = N$, so that

$$\begin{aligned} I(N^2) &= N[\psi_2(N) + 1], \\ I(\beta^2 U^2) &= N[\tfrac{1}{2}\psi_2(\beta^2 U^2) + \psi_3(\beta^2 u^2) + \tfrac{1}{4}\psi_4(\beta^2 U^2)], \\ I(N\beta U) &= N[\psi_2(\beta U) + \tfrac{1}{2}\psi_3(N\beta U)], \end{aligned} \tag{9i.29}$$

and

$$\begin{aligned} K^{-1}C_V = N\{&\tfrac{1}{2}\psi_2(\beta^2 U^2) + \psi_3(\beta^2 U^2) + \tfrac{1}{4}\psi_4(\beta^2 U^2) \\ &+ [\psi_2(N) + 1]^{-1}[\psi_2(\beta U) + \tfrac{1}{2}\psi_3(N\beta U)]^2\}. \end{aligned} \tag{9i.30}$$

9j. ELECTROLYTIC SOLUTIONS, DEBYE–HÜCKEL EQUATION

In Sec. 8n we discussed the treatment of a plasma, a gas composed of charged ions. The treatment of an electrolytic solution is often made with what is frequently called the primitive model, namely, an idealization of hard-sphere ions swimming in a continuous medium of dielectric constant D so that the electric long-range potential energy between pairs of type a and b of charge $z_a\epsilon$ and $z_b\epsilon$, respectively, with ϵ the electronic charge magnitude, is $z_a z_b \epsilon^2/Dr_{ab}$ outside the hard core,

$$\begin{aligned} \beta u_{ab}(r_{ab}) &= (\beta z_a z_b \epsilon^2 D^{-1}) r_{ab}^{-1}, & r_{ab} &> r_{ab}^{(0)}, \\ &= \infty. & r_{ab} &\le r_{ab}^{(0)}, \end{aligned} \tag{9j.1}$$

The equations are then the same as those for a plasma of temperature DT. Since the dielectric constant for water is close to $D = 80$, the behavior of a water solution of ions at 300 K is similar to that of a plasma of the same concentration at 24,000 K.

The first derivation of what is known as the Debye-Hückel limiting law[1] was an ingenious one published in 1923. The argument can be presented as follows. Around an isolated single ion of charge $z_a\epsilon$ at a distance r in a pure water solution there is an electrical potential, $\psi_a^{(0)}(r) = z_a\epsilon/Dr$, but in a solution of other ions at average number densities $\rho_a, \rho_b, \ldots$ and charge $z_a\epsilon, z_b\epsilon, \ldots$ the potential $\psi_a(r)$ is partly screened by a concentration of the oppositely charged ions greater than the concentration of those of the same charge. At a distance r the local average number density of ions of type b is altered by a factor $\exp[-(\beta z_a z_b\epsilon^2 D^{-1}\psi_a(r)]$. We linearize this to be approximately

$$\exp[-(\beta z_a z_b\epsilon^2 D^{-1})\psi_a(r)] \cong 1 - (\beta z_a z_b \epsilon^2 D^{-1})\psi_a(r), \tag{9j.2}$$

so that the total average local electric charge density $q_a(r)$ at distance r from an ion of type a is

$$q_a(r) = \sum_b \rho_b z_b \epsilon - \sum_b (\rho_b \beta z_b^2 z_a \epsilon^3 D^{-1})\psi_a(r). \tag{9j.3}$$

However, the condition of electrical neutrality of the bulk solution is that

$$\sum_b \rho_b z_b \epsilon = 0. \tag{9j.4}$$

At this stage we introduce the Debye *kappa* of the dimension of reciprocal length,

$$\kappa^2 = \sum_b 4\pi\beta\rho_b z_b^2 \epsilon^2 D^{-1}, \tag{9j.5}$$

so that eq. (3), with eqs. (4) and (5), becomes

$$q_a(r) = -\frac{1}{4\pi}\kappa^2 \psi_a(r) z_a \epsilon \tag{9j.4}$$

for the average electric charge density $q_a(r)$ at distance r from an ion of type a. The electric potential $z\epsilon\psi(r)$ in a continuous medium with a spherically symmetric net electric charge density $q(r)$ obeys the Poisson equation

$$\begin{aligned} z\epsilon\nabla^2\psi(r) &\equiv z\epsilon\left(\frac{\partial^2}{\partial x^2} + \frac{\partial^2}{\partial y^2} + \frac{\partial^2}{\partial z^2}\right)\psi(r) \\ &= z\epsilon\left(2r^{-1}\frac{d\psi}{dr} + \frac{d^2\psi}{dr^2}\right) = -4\pi q(r), \end{aligned} \tag{9j.5}$$

so that the potential $z_a\epsilon\psi_a(r)$ around ion a yields, from eq. (4),

$$2r^{-1}\frac{d\psi_a(r)}{dr} + \frac{d^2\psi_a(r)}{dr} = \kappa^2\psi_a(r). \tag{9j.6}$$

[1] P. Debye and E. Hückel, *Phys. Z.*, **24,** 185 (1923).

The general solution for the differential equation of the form of eq. (6) is that $\psi_a = Ar^{-1}e^{-\kappa r} + Br^{-1}e^{\kappa r}$, and it is obvious that, since we must have $\psi(r)$ approach zero for large r, $B=0$. The constant A is determined by the condition that at sufficiently small r the quantity $z_a \epsilon r \psi_a(r)$ must approach $z_a \epsilon D^{-1}$ in value. We then have

$$\psi_a(r) = (Dr)^{-1} e^{-\kappa r}, \tag{9j.7}$$

and the pair probability density $\rho_{ab}(r)$ between ions of type a and b at large values of r is

$$\rho_{ab}(r) = \rho_a \rho_b [1 - \beta z_a \epsilon z_b \epsilon (Dr)^{-1} e^{-\kappa r}]. \tag{9j.8}$$

Use this equation in eq. (9d.9′) for $\beta \Delta P \rho^{-1}$ with ΔP the osmotic pressure, namely, with $\rho = \sum \rho_a$,

$$\beta \Delta P^{-1} = 1 - \tfrac{1}{6}\rho^{-1} \sum_{ab}\sum \int_0^\infty 4\pi r^2 \, dr \left(r \frac{d\omega_{ab}^{(0)}(r)}{dr} \right) \rho_{ab}(r), \tag{9j.9}$$

with $\beta \omega_{ab}^{(0)}(r) = \beta z_a \epsilon z_b \epsilon (Dr)^{-1}$, and

$$\begin{aligned} \sum_{ab}\sum r \frac{d\omega_{ab}}{dr} \rho_{ab} &= \frac{D\beta}{r^2}\left(\sum_a \rho_a z_a \epsilon \right)^2 + \left[\sum_a \rho_a \beta z_a^2 \epsilon^2 (Dr)^{-1} \right]^2 e^{-\kappa r} \\ &= \kappa^4 (4\pi)^{-2} r^{-2} e^{-\kappa r}, \end{aligned} \tag{9j.10}$$

since from electric neutrality [eq. (4)], $\sum \rho_a z_a \epsilon = 0$. With eq. (10) in eq. (9), neglecting the hard core at r_0,

$$\begin{aligned} \beta \, \Delta P \rho^{-1} - 1 &= -\rho^{-1} \frac{\kappa^3}{24\pi} \int_0^\infty d(\kappa r) 1 e^{-\kappa r}, \\ &= -\rho^{-1} \frac{\kappa^3}{24\pi} \end{aligned} \tag{9j.11}$$

in agreement with eq. (8n.31) and proportional to the concentration ρ to the one-half power. The hard-sphere term has a smaller value at low concentration.

The numerical value of κ is of interest. Define an ionic strength I proportional to m, the concentration of the formula weight of salt in moles per liter. If the formula has ν_a ions of type a each with charge $z_a \epsilon$, define

$$I = \tfrac{1}{2} \sum_a \nu_a z_a^2 m \tag{9j.12}$$

so that the I values are those in Table (9j.1).

Table 9j.1

Salt Type	Example	I
1–1	NaCl	m
2–2	$MgSO_4$	$4m$
3–3	$AlPO_4$	$9m$
1–2	Na_2SO_4	$3m$
1–3	Na_3PO_4	$6m$
2–3	$Mg_3(PO_4)_2$	$15m$

The dielectric constant of water is 80.36 at 20°C, 78.54 at 25°C, and 76.75 at 30°C; it is quite temperature-dependent. We choose 80 as a convenient round number. Numerical evaluation of eq. (5) then gives

$$\kappa^2 = \sum_a 4\pi\beta\rho_a z_a^2 \epsilon^2 D^{-1}$$

$$= 10.538\,5\times 10^{14}\ \frac{300}{T}\frac{80}{D} I \qquad \text{cm}^{-2}, \qquad (9j.13)$$

$$\kappa = 3.246\,3\times 10^7 \left(\frac{300}{T}\right)^{1/2}\left(\frac{80}{D}\right)^{1/2} I^{1/2} \qquad \text{cm}^{-1}, \qquad (9j.13')$$

and the Debye length,

$$\kappa^{-1} = 3.080\,4\times 10^{-8} \left(\frac{T}{300}\right)^{1/2}\left(\frac{D}{80}\right)^{1/2} I^{-1/2} \qquad \text{cm}. \qquad (9j.13'')$$

At a distance $r=\kappa^{-1}$ the factor $e^{-\kappa r}$ has dropped to $e^{-1}=0.368$, and the linear approximation assumed in eq. (2) is pretty poor and much worse for $r<\kappa^{-1}$. The distance between neighboring ions in a salt crystal is of the order of 3×10^{-8} cm, so that at a concentration of $m=1$, even in a 1–1 salt, the limiting law proportional to $m^{1/2}$ is expected to fail drastically and does so. Even at $m=10^{-2}$, for which the concentration is about that of a 300 K gas at $\frac{1}{2}$ atm (2 mol of ions per mole of salt), the deviation from the $m^{1/2}$ behavior of $\beta P/\rho - 1$ is nonnegligible. For more highly charged ions, even at 10^{-3} mol liter^{-1}, the deviations from the limiting law are appreciable. Using eq. (9j.11), we find

$$\beta\,\Delta P\rho^{-1} - 1 = 0.7534\ \left(\frac{300}{T}\right)^{3/2}\left(\frac{80}{D}\right)^{3/2} I^{3/2} m^{-1}. \qquad (9j.14)$$

CHAPTER 10

CRYSTALLINE SOLIDS

10a. INTRODUCTION

In the broadest definition of statistical mechanics as the treatment of macroscopic systems from an atomic or molecular viewpoint, all of theoretical solid-state physics is a part of statistical mechanics. Since solid-state physics is the concern of a very large segment of the research physics community and the subject matter of many books published each year, it is almost arrogant to give the title "Crystalline Solids" to a short chapter such as the one that follows. In what follows we present only some of the more primitive early use of the semiclassical formalism for the structure and thermodynamic functions of simple nonconducting crystal lattices at equilibrium. The modern extension of even this limited part of solid-state physics is the subject matter of a 727-page book reporting the papers presented to the 1963 International Conference on Lattice Dynamics at Copenhagen,[(1)] as well as of very many papers since then. An equally brief discussion of some of the simpler treatments of electronic states in conducting materials is given in Sec.'s 11g and 11h.

In the solid state the distance between neighboring atoms or molecules is so small that the forces they exert on each other are considerable, even greater than in the liquid.

The investigation of the solid crystalline state is simplified by its great regularity. X-Rays show that a crystal is in a state of complete, or almost complete, order. A certain fundamental arrangement of a few atoms or molecules is repeated periodically in space. In an ideal crystal therefore the centers of the atoms form the points of a regular space lattice which is triply periodic, that is, periodic in three directions which are not in the same plane. Since the lattice arrangement is stable, the lattice points must

[(1)] R. F. Wallis, Ed., *Lattice Dynamics*, Pergamon Press, New York, 1965.

be equilibrium positions for the atoms, which implies that there must be no force acting on any atom or, in other words, that the forces on any one atom from all others must exactly cancel if all atoms are at the lattice points. Moreover, all tensions must be zero, and the equilibrium poition must be an energetic minimum.

If the forces are known, it can be determined by calculation which lattice is stable for a given substance. In practice, the opposite procedure is usually employed, and from the observed crystal structure conclusions are drawn about the forces. This, however, is a purely mechanical problem and does not interest us here. In all calculations of this chapter the lattice structure is assumed to be given.

10b. NORMAL COORDINATES AND CLASSICAL EQUATIONS

Actually, the atoms are never at rest at the lattice points but undergo small vibrations around these equilibrium positions. Let us denote the cartesian coordinates of the deviation from the equilibrium position for each atom by ξ_i, the index i running from 1 to $3N$ if the crystal contains N atoms; let m_i be the masses of the particles and $p_{\xi i} = m_i\dot{\xi}_i$ the momenta. For small deviations ξ_i from equilibrium the potential energy of the system may be developed with respect to the ξ_i's. If we choose the potential at the equilibrium, $\xi_i = 0$, as the zero of energy, the development has no constant term. The linear terms in ξ_i must vanish, since it has been assumed that no forces act on the atoms in their equilibrium positions. The first terms arising are therefore quadratic in the ξ_i's, and the energy may be written as

$$H(p, \xi) = \sum_{i=1}^{3N} \frac{1}{2m_i} p_{\xi i}^2 + \frac{1}{2} \sum_{i,j=1}^{3N} a_{ij}\xi_i\xi_i. \tag{10b.1}$$

For small enough displacements cubic and higher-order terms may be neglected. The mechanical task of calculating the force constants a_{ij} from the forces between the atoms is beyond the scope of this book.

If, instead of the deviations ξ_i of individual atoms, linear combinations of ξ_i's are used as coordinates, the potential energy in eq. (1) expressed in these new coordinates remains a quadratic expression, and the kinetic energy a quadratic expression in the corresponding momenta. It is a theorem of mechanics, or rather of mathematics, that there exist $3N$ special independent linear combinations of the $3N$ ξ_i's, which are called the normal coordinates and are designated by q_i. They have the property that, if eq. (1) is expressed in the q_i's and the corresponding momenta p_i, the potential energy contains no cross terms between two different q_i's, while the kinetic energy is a sum of squares,

$$H(p, q) = \frac{1}{2} \sum_{i=1}^{3N} (p_i^2 + 4\pi^2 \nu_i^2 q_i^2). \tag{10b.2}$$

These normal coordinates, the q_i,s, are of course the same in principle as those discussed in Sec. 7b for polyatomic molecules, except that in the case of the crystal the number of them is $3N \sim 10^{24}$. The problem of finding the distribution of the ν_i^2's is essentially that of diagonalizing a matrix, but in general no direct exact analytic solution can be found for a matrix of 10^{24} rows and columns. The Hamiltonian of eq. (2) differs from that of eq. (7b.2) in that by a change in scale the mass factor $\mu\lambda^{-1}$ has been eliminated from the kinetic energy in eq. (2) and $\kappa_\lambda\mu_\lambda$ has been replaced by $4\pi\nu_i^2$.

The Hamiltonian is thereby expressed as a sum of $3N$ functions, each containing one of the coordinates and the corresponding momentum only. Moreover, these functions are the well-known Hamilton functions of a harmonic oscillator of frequency ν_i. The ν_i^2's are complicated functions of the force constants a_{ij}. If the cyrstal is stable, that is, if the energy for $q_i = 0$ is a minimum, all ν_i^2's must be positive, except six which are zero.

The equations of motion are

$$\dot{q}_i = \frac{\partial H}{\partial p_i} = p_i, \qquad \ddot{q}_i = \dot{p}_i = -\frac{\partial H}{\partial q_i} = -4\pi^2 \nu_i^2 q_i,$$

and the classical solution is

$$q_i = A_i \cos(2\pi\nu_i t + \delta_i).$$

If only one normal coordinate q_i is excited, and the others constantly zero, all particles in the crystal vibrate with the same frequency ν_i but with different amplitudes, determined by the coefficients with which the displacement ξ_i enters into q_i. In a general state of vibration the motion of each particle is a complicated superposition of motions of different frequencies.

Equation (2) shows that a crystal consisting of N strongly coupled atoms is mechanically equivalent to a system of $3N$ independent oscillators. The terms of third and higher orders in q_i in the development of the potential energy, which were neglected in eqs. (1) and (2), introduce a coupling between the oscillators and make an exchange of energy between them possible. These deviations from harmonicity in the vibrations establish the equilibrium distribution of energy between the oscillators. They play therefore much the same role here that the collisions do for the perfect gas. With increasing temperature, as the amplitudes of vibration become larger, it may not be correct to neglect these terms. Their influence on the specific heat, and so on, of the crystal can then be calculated, at least approximately. This is analogous to the fact that at high concentrations in a gas the forces between atoms give rise to deviations from the perfect gas law.

Among the normal coordinates there are three that describe the simple translations of the crystal as a whole along the three axes, $\sum \xi_i$, and so on, and three that correspond to rotations around the center of mass. The forces counteracting these six motions are zero. It follows that six of the ν_i's in eq. (2) vanish, so that the second sum should be extended only over $3N-6$ terms. However, since six is a very small number compared to $3N$, this difference may be neglected.

We use the petite canonical ensemble of systems with fixed V, β, N, for which the partition formation is given in eq. (3i.11′); it is classically

$$Q(V, \beta, N) = \int \cdots \int d\mathbf{p}^{(3N)}\, dq^{(3N)} \frac{1}{h^{3N}N!} \exp[-\beta H(\mathbf{p}^{(3N)}\mathbf{q}^{(3N)})], \tag{10b.3}$$

$$-\beta A = \ln Q(V, \beta, N). \tag{10b.4}$$

We restrict our discussion to a crystal of one monatomic species. Since there are $N!$ different ways of placing N numbered atoms in a lattice if we integrate over a Hamiltonian that is valid only for small displacements of the atoms from their assigned lattice sites, which is the case in eq. (2), we must multiply by $N!$, which removes it from the denominator of eq. (3). With this, and eq. (2) for H, we have

$$Q(V, \beta, N) = \prod_{i=1}^{i=3N} \frac{1}{h} \int_{-\infty}^{+\infty} dp_i\, e^{(-1/2)\beta p_i^2} \int_{-\infty}^{+\infty} dq_i\, e^{-\beta 2\pi^2 \nu_i^2 q_i^2} = \prod_{i=1}^{i=3N} (\beta h \nu_i)^{-1}, \tag{10b.5}$$

or, with eq. (4) for the Helmholtz free energy A,

$$A = \sum_{i=1}^{3N} \ln (\beta h \nu_i). \tag{10b.6}$$

Since

$$\left[\frac{\partial(\beta A)}{\partial B}\right]_{V,N} = A + \beta\frac{\partial A}{\partial \beta} = A - T\left(\frac{\partial A}{\partial T}\right)_{V,N} = A + TS = E \tag{10b.7}$$

(see also Sec. 3j), one finds

$$E = \sum_{i=1}^{i=3N} \beta^{-1} = 3NkT, \tag{10b.8}$$

and

$$C_V = \left(\frac{\partial E}{\partial T}\right)_V = 3Nk. \tag{10b.9}$$

Here we encounter again the law of equipartition of energy: The system has $3N$ degrees of freedom; the energy depends quadratically on each of the $3N$ coordinates and momenta; the average energy for each coordinate and for each momentum at the temperature T is $\frac{1}{2}kT$. The heat capacity per mole at constant volume is therefore a constant for all crystals. If the substance in question is monatomic, the value of C_V for a mole of substance is $3R = 5.959$ cal deg^{-1}. This theorem is called the law of Dulong-Petit. Experimentally, direct determination of C_V is difficult. C_V is, however, connected with the easily measurable quantity C_P, the heat capacity at constant pressure, through the relation

$$C_P - C_V = \frac{9\alpha^2 VT}{\kappa}, \tag{10b.10}$$

in which α signifies the linear expansion coefficient,

$$\alpha = \frac{1}{3V}\left(\frac{\partial V}{\partial T}\right)_P, \tag{10b.11}$$

and κ the compressibility,

$$\kappa = -\frac{1}{V}\left(\frac{\partial V}{\partial P}\right)_T. \tag{10b.12}$$

The value of C_V calculated from these experimental quantities is in excellent agreement with the theoretical one [eq. (9)] for most monatomic and simple ionic crystals at room temperature. However, for all substances the heat capacity falls below this classical theoretical value at low temperatures and, indeed, approaches zero at absolute zero. Classical mechanics was unable to account for this. The explanation is given by the quantum-mechanical treatment of the vibrations (see Sec. 10c). A few monatomic crystalline elements show deviations from the law of Dulong-Petit even at room temperature. These are diamond, beryllium, and silicon, which have some unusually large vibrational frequencies. In crystals containing di- or polyatomic molecules, for instance, CO_3^{2-} groups, the vibration frequencies of the *molecule* are so high that their average energy at room temperature is less than the classical value kT.

The entropy of the crystal at room temperature is, according to eqs. (4), (5), (7), and (8),

$$S = k \ln Q + \frac{E}{T} = 3Nk\left(\ln kT + 1 - \frac{1}{3N}\sum_{i=1}^{3N} \ln h\nu_i\right). \tag{10b.13}$$

It is seen that for the calculation of the classical heat capacity of crystals the knowledge of the values of the frequencies is unnecessary. The frequencies, however, contain the volume dependence of the entropy and therefore play an important part if the equation of state of the crystal is to be determined. The product of the frequencies, which enters into the entropy, also has to be known if calculations of equilibrium between the crystal and other phases are made.

10c. QUANTUM-MECHANICAL CALCULATION

If it is assumed that the distribution of the frequency square ν_i^2's is known, the formal quantum-mechanical calculation of the thermodynamic functions will be straightforward.

A state $\mathbf{n}$ of the total crystal is determined by the numbers n_i of quanta of vibration for each normal coordinate. Its energy is

$$E_{\mathbf{n}} = \sum_{i=1}^{3N} (n_i + \tfrac{1}{2})h\nu_i. \tag{10c.1}$$

The lowest energy of the crystal, when all n_i's are zero, is $\sum \frac{1}{2}h\nu_i$. In treating the diatomic and polyatomic molecules the zero of energy was altered to give the lowest quantum state zero energy. We do *not* follow that practice here.

The partition function of the system becomes a product of the partition functions for each individual oscillator,

$$\begin{aligned} Q(V, \beta, N) &= \sum_{\mathbf{n}} e^{-\beta E_{\mathbf{n}}} = \prod_{i=1}^{i=3N} e^{(-1/2)\beta h\nu_i} \sum_{n\geq 0} e^{-n\beta h\nu_i} \\ &= \prod_{i=1}^{i=3N} e^{(-1/2)\beta h\nu_i}(1-e^{-\beta h\nu_i})^{-1}. \end{aligned} \tag{10c.2}$$

We then have

$$-\beta A = \ln Q(V, \beta, N) = -\sum_{i=1}^{i=3N} [\tfrac{1}{2}\beta h\nu_i + \ln(1-e^{-\beta h\nu_i})], \tag{10c.3}$$

and from eq. (10b.7),

$$E = \left[\frac{\partial(\beta A)}{\partial \beta}\right]_{V,N} = \sum_{i=1}^{i=3N} [\tfrac{1}{2}h\nu_i + h\nu_i(e^{\beta h\nu_i}-1)^{-1}], \tag{10c.4}$$

$$C_V = \left(\frac{\partial E}{\partial T}\right)_{V,N} = k\sum_{i=1}^{i=3N} (\beta h\nu_i)^2 e^{\beta h\nu_i}(e^{\beta h\nu_i}-1)^{-2}, \tag{10c.5}$$

and, using $S = T^{-1}(E-A)$, with eqs. (3) and (4),

$$S = k\sum_{i=1}^{i=3N} [-\ln(1-e^{-\beta h\nu_i}) + \beta h\nu_i(e^{\beta h\nu_i}-1)^{-1}]. \tag{10c.6}$$

At high temperatures, $\beta h\nu_i \ll 1$, these all go over to the classical expressions. We can develop the exponentials as in eq. (7h.28) for the vibrational energy, finding

$$E = 3NkT\left[1 + \tfrac{1}{12}(3N)^{-1}\sum_{i=1}^{i=3N}(\beta h\nu_i)^2 + \tfrac{1}{720}(3N)^{-1}\sum_{i=1}^{i=3N}(\beta h\nu_i)^4 + \cdots\right]$$

$$= 3NkT\left[1 + \frac{1}{12}\left\langle\left(\frac{h\nu_i}{kT}\right)^2\right\rangle + \frac{1}{720}\left\langle\left(\frac{h\nu_i}{kT}\right)^4\right\rangle + \cdots\right], \tag{10c.7}$$

$$C_V = 3Nk\left[1 - \frac{1}{12}\left\langle\left(\frac{h\nu_i}{kT}\right)^2\right\rangle - \frac{1}{240}\left\langle\left(\frac{h\nu_i}{kT}\right)^4\right\rangle + \cdots\right]. \tag{10c.8}$$

It is seen that the quantum-mechanical equations have the classical ones as the limiting case for high temperatures. Since it is known experimentally that the heat capacity has reached the classical value even at room temperature for most monatomic crystals, one must expect the lattice frequencies to be very low compared to molecular frequencies. The decrease in heat capacity at lower temperatures is correctly predicted by the quantum-mechanical formula. As the temperature tends to zero, the average energy content of the crystal approaches the zero-point energy, and the heat capacity approaches zero. From eq. (5) it is seen that the contribution of any one frequency to C_V decreases exponentially as kT becomes much smaller than $h\nu_i$, that is, as $\beta h\nu_i$ becomes very large. In that case the unity in the denominator may be neglected compared to $e^{\beta h\nu_i}$, and the additive part becomes $(\beta h\nu_i)^2 e^{-\beta h\nu_i}$, in which the exponential term is overwhelmingly important. However, the experimental heat capacity of the crystal does not decrease as rapidly with decreasing T at low temperature, but only with about the third power of the temperature. This shows that it is impossible to approximate the behavior of the crystal by oscillators of one or of a few frequencies only. Some of the crystal frequencies are extremely low and are classical even at very low temperature, so that the dependence of the sum of terms in eq. (5) on the temperature is different from that of any single terms. The discussion of the Debye equation in Sec. 10d is based on this.

It is interesting to note that, if all $3N$ frequencies of the crystal were assumed to be identical, the thermodynamic functions of the crystal would be the same as those for the vibrational contribution to $3N$ identical diatomic molecules. One then finds the simple equations

$$E = eNkT[\tfrac{1}{2}\beta h\nu + \beta h\nu(e^{\beta h\nu} - 1)^{-1}], \tag{10c.9}$$

$$C_V = 3Nk(\beta h\nu)^2 e^{\beta h\nu}(e^{\beta h\nu} - 1)^{-2}. \tag{10c.10}$$

These are referred to as the Einstein equations for a crystal, and eq. (10)

is the Einstein equation for the heat capacity, the same as that for the vibrational contribution of one frequency to the independent molecules of a gas. Under this simple assumption of equal numerical value for all $3N$ frequencies of the crystal Einstein[1] had first explained the observed decrease in heat capacity below the Dulong-Petit value of $3NkT$ at low temperature.

Equations (3) to (6) are exact within the limits of the development of the potential energy as a sum of quadratic terms only, and eqs. (10b.1) and (10b.2) are applicable to crystals of more than one chemical monatomic species, or of ions. They are also applicable to molecular crystals with N the number of atoms in the crystal. In this case the frequencies ν_i includes the relatively high intramolecular vibrations which, in many cases at least, are in narrow bands centered closely at the frequencies of the gaseous molecules. In the next section we discuss an approximate method of evaluating the distribution of the frequencies, but this method is tailored for a crystal of monatomic atoms of one species, or for an ionic crystal of ions of equal or nearly equal masses.

The frequencies of a general three-dimensional lattice can be determined only with the use of various approximations. Before discussing these we wish to treat a simple crystal model which is not realized in nature, but which shows the essential characteristics of a general lattice without cumbersome mathematics.[2] This model is the so-called one-dimensional crystal.

Consider N point particles arranged on a line, the x-axis, and restricted to move along that line only. The particles are numbered according to their position, from left to right, by indices i running from 1 to N. It is assumed that neighboring particles act only on each other with a potential energy depending on the distance between them $\phi(x_{i+1}-x_i)$. The total potential energy of the N points is then given by

$$U=\sum_{i=1}^{N-1}\phi(x_{i+1}-x_i). \qquad (10c.11)$$

If the points are placed equidistant, the forces acting on any of the middle particles from both neighbors cancel. A stable equilibrium is reached only if there are also no forces acting on the two end points, a condition that determines the distance a between the points by the relation

$$\left(\frac{d\phi(r)}{dr}\right)_{r=a}=0. \qquad (10c.12)$$

(1) Albert Einstein, *Ann. Phys.* [4], **22,** 180 (1907).

(2) M. Born and Th. von Karman, *Phys. Z.*, **13,** 297 (1912).

It follows that the minimum of potential energy is obtained if the particles form a regular one-dimensional lattice. With suitable choice of the arbitrary zero point of the x-axis, the equilibrium position of the first particle is $x_1 = 0$, and of the ith particle, $x_i = (i-1)a$.

A small displacement of the ith particle from its equilibrium point is denoted by ξ_i. The potential energy [eq. (11)] may be developed with respect to these quantities. The linear terms vanish on account of eq. (12). If all terms higher than quadratic in the ξ's are neglected, the potential energy becomes

$$U = (N-1)\phi(a) + \tfrac{1}{2}\phi''(a) \sum_{i=1}^{N-1} (\xi_{i+1} - \xi_i)^2, \tag{10c.13}$$

where the symbol $\phi''(a)$ is used to denote the second derivative of ϕ with respect to r at $r = a$. We now make the further assumption that the masses of all particles are equal. The equation of motion for the ith particle in this potential is then given by

$$m\ddot{\xi}_i = -\frac{\partial U}{\partial \xi_i} = \phi''(a)(\xi_{i+1} + \xi_{i-1} - 2\xi_i). \tag{10c.14}$$

For the two particles at the ends, $i = 1$ and $i = N$, the equations are somewhat different.

Equation (14) for different values of i is satisfied by a motion corresponding to a standing wave, namely,

$$\begin{aligned}\xi_i &= \sin\left(\frac{2\pi x_i}{\lambda} + \alpha\right) A \sin 2\pi\nu t \\ &= \sin\left[\frac{2\pi a}{\lambda}(i-1) + \alpha\right] A \sin 2\pi\nu t.\end{aligned} \tag{10c.15}$$

Inserting eq. (15) in eq. (14) leads to

$$-4\pi^2\nu^2 m\xi_i = \phi''(a)\left(2\cos\frac{2\pi a}{\lambda} - 2\right)\xi_i = -\phi''(a) 4\sin^2\left(\frac{\pi a}{\lambda}\right)\xi_i,$$

or

$$\nu = \frac{1}{\pi}\left(\frac{\phi''(a)}{m}\right)^{1/2} \sin\frac{\pi a}{\lambda}. \tag{10c.16}$$

Not the whole course of the wave function [eq. (15)], but only its values at isolated points $x_i = a(i-1)$, are of physical significance. It is seen that waves of wavelength λ', with $a/\lambda' = (a/\lambda) + 1$ or with $a/\lambda' = 1 - (a/\lambda)$, give rise to exactly the same displacements ξ_i at every lattice point as those of wavelength λ (in the latter case with opposite sign and different phase). From eq. (16) it follows that they also have the same value of ν^2. It is therefore sufficient to restrict ourselves to wavelengths with $a/\lambda \leq \frac{1}{2}$, of

$\lambda \geq 2a$. The frequencies take on all the possible values the function in eq. (16) can have as λ goes from infinity to $2a$. To the shortest wavelength, $\lambda = 2a$, corresponds the highest frequency, $\nu = (1/\pi)[\phi''(a)/m]^{1/2}$. The mode of this vibration is such that neighboring particles have opposite amplitudes; the distance between nodes is $\frac{1}{2}\lambda = a$, equal to the distance between particles. The occurrence of this smallest wavelength is characteristic for the lattice structure, in contradistinction to a continuum for which no lowest wavelength exists.

To avoid the complications arising from the ends we assume that the two end points are fixed in their equilibrium positions, that is, $\xi_1 = \xi_N = 0$. As in the case of the vibrating rod, to which this problem has great similarity, this border condition of clamped ends influences the types of vibration, but not the distribution of the frequencies. Besides, it reduces the number of degrees of freedom by two, which is of little importance, since N is assumed to be a very large number. One of the types of motion excluded this way is the simple translation along the x-axis.

The border condition that the waves have nodes at the ends is taken care of at $i = 1$, $x_i = 0$, by choice of the phase factor $\alpha = 0$ in eq. (15). At $i = N$ it imposes

$$\sin \frac{2\pi a}{\lambda}(N-1) = 0, \qquad \frac{2(N-1)a}{\lambda} = n,$$

where n is a whole number. The distance between nodes, that is, half the wavelength λ, must be equal to the length of the lattice $(N-1)a$ divided by an integral number.

For a one-dimensional continuum of length L all possible displacements, subject to the condition that the ends remain fixed, can be represented by linear combinations of functions $\sin(\pi nx/L)$, where n goes from 1 to infinity. The amplitude factors of these functions represent the components of the Fourier development of the displacement. They may therefore be used as coordinates for the description of the system. If the mass is discretely distributed, according to the previous discussion, the functions $\sin[\pi n(i-1)/a(N-1)]$, with $1 \leq n \leq N-1$, suffice; every conceivable displacement of the $N-2$ inner mass points may be expressed as a linear combination of these $N-2$ functions. The amplitudes of the wave functions may therefore be used as coordinates, and the equations of motion show that they vary periodically in time. The normal coordinates are

$$\begin{aligned} q_n &= \left(\frac{2m}{N-1}\right)^{1/2} \sum_{i=1}^{N} \sin\left[\frac{\pi n(i-1)}{(N-1)}\right]\xi_i \\ &= \left[\frac{(N-1)m}{2}\right]^{1/2} A_n \sin 2\pi\nu_n t, \end{aligned}$$

where A_n is the amplitude in eq. (15). The normalization factor in front has been chosen such as to make the kinetic energy a sum of squares in the momenta $p_n = \dot{q}_n$ without mass factors, in agreement with eq. (10b.2).

The $N-2$ frequencies are

$$\nu_n = \frac{1}{\pi}\left[\frac{\phi''(a)}{m}\right]^{1/2} \sin\frac{\pi n}{2(N-1)}, \qquad 1 \le n \le N-2. \tag{10c.17}$$

For the very low frequencies one may replace the sine by the argument in eq. (16) or (17) and obtain

$$\nu\lambda = a\left[\frac{\phi''(a)}{m}\right]^{1/2}. \tag{10c.18}$$

It is seen that in a large lattice there exist vibrations of extremely low frequency. Physically, these motions represent elastic or acoustic waves. The quantity on the right-hand side is then the velocity of sound. For decreasing wavelength, when eq. (18) ceases to be valid, the velocity of sound is dependent on the wavelength. However, if one uses eq. (18) up to the shortest wavelength, an approximation that has to be made for the general crystal, the error made would not be very large. The highest frequency would then be $\frac{1}{2}[\phi''(a)/m]^{1/2}$ instead of $(1/\pi)[\phi''(a)/m]^{1/2}$, as obtained from the correct formula in eq. (17).

10d. THE FREQUENCIES OF SIMPLE ISOTROPIC LATTICES

An approximative formula for the distribution of the frequencies of a simple three-dimensional lattice may be obtained following a method of Debye.[(1)] Let us at first neglect the lattice structure entirely and treat the crystal as an isotropic elastic continuum. If, for simplicity, the crystal is assumed to have the shape of a cube with major axes parallel to the x-, y-, z-axis, the proper vibrations of this block of elastic matter are standing waves, the displacements at any point x, y, z being proportional to

$$u(x, y, z) = \sin 2\pi\tau_x x \sin 2\pi\tau_y y \sin 2\pi\tau_z z. \tag{10d.1}$$

The waves are characterized by vectors $\boldsymbol{\tau}$ with positive components τ_x, τ_y, τ_z whose magnitude is inversely proportional to the wavelength, $|\boldsymbol{\tau}| = \lambda^{-1}$. The standing wave is a superposition of the eight progressive waves of the same wavelength but different directions of propagation given by the eight vectors $(\pm\tau_x, \pm\tau_y, \pm\tau_z)$, which have the same magnitude, but different sign of components as the vector $\boldsymbol{\tau}$. The nodal planes of the waves are parallel to the surface planes. The distance between the nodal planes normal to the x-axis is $(2\tau_x)^{-1}$.

The border conditions at the surface determine the possible vectors $\boldsymbol{\tau}$. We may either stipulate that there is no motion at all at the surface, that is, the surface particles are rigidly fixed, so that the wave function [eq. (1)] must have a node there, or that the ends are free to vibrate and are therefore a place of maximum amplitude. In either case these conditions demand that the length L of the side of the cube, divided by the distance between the nodal planes normal to it, be an integer number n, namely,

$$\tau_z = \frac{n_x}{2L}, \qquad \tau_y = \frac{n_y}{2L}, \qquad \tau_z = \frac{n_z}{2L}. \tag{10d.2}$$

To every wave vector $\boldsymbol{\tau}$ [eq. (2)] and function u [eq. (1)] there belong three different types of vibration, since the displacements may be in any direction in space. The three waves originate from two transversal and one longitudinal progressive wave. In an isotropic medium the frequencies of the vibrations depend on the wavelength through the relation

$$\nu_1\lambda = \nu_2\lambda = c_t, \qquad \nu_3\lambda = c_l, \tag{10d.3}$$

where c_t, c_l are the velocities of the transversal and longitudinal elastic waves. The (compressional) longitudinal wave is the sound wave.

The number of longitudinal vibrations whose frequencies lie between ν and $\nu+\Delta\nu$ is then the same as the number of vectors $\boldsymbol{\tau}$ with $\nu/c_l \leq |\boldsymbol{\tau}| \leq (\nu+\Delta\nu)/c_l$, or the same as the number of vectors $\mathbf{n}$ with positive integer components n_x, n_y, n_z whose end points lie in a spherical shell determined by

$$\frac{2L\nu}{c_l} \leq (n_x^2+n_y^2+n_z^2)^{1/2} \leq \frac{2L(\nu+\Delta\nu)}{c_l}.$$

This latter number is asymptotically equal to one-eighth of the volume of the spherical shell, namely, $4\pi(L/c_l)^3\nu^2\,\Delta\nu$. The number of transversal waves in the same frequency range is twice the same expression, but with c_t in the place of c_l. The total number of vibrations with frequencies between ν and $\nu+\Delta\nu$ is then, considering that $L^3 = V$, the volume,

$$N(\nu)\,\Delta\nu = 4\pi V\left(\frac{2}{c_t^3}+\frac{1}{c_l^3}\right)\nu^2\,\Delta\nu. \tag{10d.4}$$

In a lattice the form of the vibrations is essentially the same as in a continuum. The displacements from equilibrium of a lattice particle in a simple harmonic vibration are proportional to the value of a standing wave u [eq. (1)] at the equilibrium position. An essential difference enters, however, owing to the fact that now not the whole course of the function u, but only its value at discrete points, is of physical importance. This introduces an upper limit for the components of the vector $\boldsymbol{\tau}$, that is, a lower limit for the wavelength. Motions with nodes closer together than the distance between neighboring particles may just as well be described as motions with smaller τ, greater distance between nodes, for the same reason we found it in the one-dimensional lattice (Sec. 10c). If the lattice contains N particles, there exist precisely N different wave vectors $\boldsymbol{\tau}$ which give rise to functions u which are different from each other at the N lattice points. This means that there exist $3N$ different modes of harmonic vibration, as many as the number of degrees of freedom.

For motions with long wavelength the lattice structure plays no important part, so that eq. (3) is still fulfilled. Only, now the velocity of the elastic waves is in general somewhat different for different directions. For wavelengths comparable with the distance between particles, eq. (3) breaks down completely.

An approximate distribution of the frequencies may be obtained by assuming eq. (3) to hold for all permissible wavelengths and directions of propagation. The velocity of the elastic waves must then be replaced by some average over the different directions. The distribution of frequencies is still given by eq. (4). The lattice structure is now taken into account by cutting eq. (4) at a highest frequency ν_m, determined in such a way that the total number of vibrations has the correct value $3N$, namely,

$$\begin{aligned} 3N &= 4\pi V\left(\frac{2}{c_t^3}+\frac{1}{c_l^3}\right)\int_0^{\nu_m} \nu^2\, d\nu \\ &= \frac{4\pi V}{3}\left(\frac{2}{c_t^3}+\frac{1}{c_l^3}\right)\nu_m^3. \end{aligned} \tag{10d.5}$$

This may be used to eliminate the sound velocities from eq. (33), leading to

$$N(\nu)\,\Delta\nu = 9N\frac{\nu^2}{\nu_m^3}\,\Delta\nu, \qquad \nu \leq \nu_m. \tag{10d.6}$$

Obviously, the cut should be made independently at $2N$ for the transverse waves and at N for the longitudinal ones. This is at least logical but, since the sound velocities in any case depend strongly on the wave numbers, the approximation of using a single cutoff, $\nu \leq \nu_m$, is probably less influenced by the difference between c_t and c_l than by the frequency dependence of both.

10e. THE DEBYE FORMULA

With this distribution [eq. (10d.6)] of the frequencies of the oscillators, the average energy of the lattice at any temperature T can immediately be calculated from eq. (10c.7). The average energy of one oscillator is multiplied with the number $N(\nu)\,d\nu$ of oscillators with frequencies in the range $d\nu$, and the product integrated over ν from zero to ν_m,

$$E=\frac{9N}{\nu_m^3}\int_0^{\nu_m}\left(\tfrac{1}{2}h\nu+\frac{h\nu}{e^{h\nu/kT}-1}\right)\nu^2\,d\nu. \tag{10e.1}$$

The integral of the second term is abbreviated by the use of the symbol $D(u)$ for the Debye function,

$$D(u)=\frac{3}{u^3}\int_0^u\frac{x^3\,dx}{e^x-1}, \tag{10e.2}$$

in which

$$u=\frac{h\nu_m}{kT}=\frac{\theta}{T}, \qquad \theta=\frac{h\nu_m}{k}. \tag{10e.3}$$

The quantity θ is called the Debye temperature, or characteristic temperature, of the lattice. Using eq. (2) in the equation for the energy, one obtains

$$E=\tfrac{9}{8}Nh\nu_m+3NkT\cdot D\left(\frac{h\nu_m}{kT}\right). \tag{10e.4}$$

The first term is the zero-point energy of the oscillators.

For purposes of numerical calculation one can find two series approximations to eq. (2), one of which is valid at high temperatures and the other at low temperatures. The ranges of temperature for which the two series converge overlap. At high temperatures u is small and, since the integral extends from zero to u, x is small throughout the range of integration. One can develop

$$\begin{aligned}\frac{3}{u^3}\frac{x^3}{e^x-1}&=\frac{3x^3}{u^3(x+x^2/2+x^3/6+x^4/24+x^5/120+\cdots)}\\&=\frac{1}{u^3}\left(3x^2-\frac{3}{2}x^3+\frac{x^4}{4}-\frac{x^6}{240}\cdots\right),\end{aligned}$$

and then integrate, obtaining

$$D(u)=1-\tfrac{3}{8}u+\tfrac{1}{20}u^2-\tfrac{1}{1680}u^4, \qquad u\le 1. \tag{10e.5}$$

For low temperatures, u is large, and it is more convenient to transform to

$$D(u)=\frac{3}{u^3}\left(\int_0^\infty\frac{x^3\,dx}{e^x-1}-\int_u^\infty\frac{x^3\,dx}{e^x-1}\right).$$

The first integral is a definite integral having the value $\pi^4/15$. In the second term x is large throughout the range of integration, and one can develop

$$\frac{x^3}{e^x-1}=\frac{x^3e^{-x}}{1-e^{-x}}=x^3e^{-x}+x^3e^{-2x}+\cdots.$$

Integration of this leads to

$$D(u)=\frac{\pi^4}{5}\frac{1}{u^3}-\left(3+\frac{9}{u}+\frac{18}{u^2}+\frac{18}{u^3}\right)e^{-u}$$
$$-\left(\frac{3}{2}+\frac{9}{4u}+\frac{9}{4u^2}+\frac{9}{8u^3}\right)e^{-2u}, \qquad (10e.6)$$

The energy of the crystal may then be written as

$$E=\tfrac{9}{8}Nh\nu_m+3NkT\cdot D\left(\frac{\theta}{T}\right)$$
$$=3NkT\left[1+\frac{1}{20}\left(\frac{\theta}{T}\right)^2-\frac{1}{1680}\left(\frac{\theta}{T}\right)^4+\cdots\right]$$
$$=\tfrac{9}{8}Nh\nu_m+3NkT\left[\frac{\pi^4}{5}\left(\frac{T}{\theta}\right)^3-\cdots\right] \qquad (10e.7)$$

for high and low temperatures respectively.

The heat capacity at constant volume is the derivative of the energy with respect to temperature.

$$C_V=3Nk\cdot D\left(\frac{\theta}{T}\right)+3NkT\frac{d}{dT}D\left(\frac{\theta}{T}\right).$$

The derivative of the Debye function is

$$\frac{d}{dT}D\left(\frac{\theta}{T}\right)=-\frac{1}{T}u\frac{d}{du}D(u)$$

and, from eq. (2),

$$\frac{d}{du}D(u)=-\frac{3}{u}\cdot D(u)+\frac{3}{e^u-1},$$

so that

$$C_V=3Nk\left[4D\left(\frac{\theta}{T}\right)-\frac{3(\theta/T)}{e^{\theta/T}-1}\right]. \qquad (10e.8)$$

At high temperatures the approximative formula is

$$C_V=3Nk\left[1-\frac{1}{20}\left(\frac{\theta}{T}\right)^2+\frac{1}{560}\left(\frac{\theta}{T}\right)^4\cdots\right], \qquad (10e.8')$$

which goes over into the classical $3Nk$ at sufficiently high temperatures. At low temperatures,

$$C_V = 3Nk\left[\frac{4\pi^4}{5}\left(\frac{T}{\theta}\right)^3 + \cdots\right]. \tag{10e.8''}$$

Experimental measurements at very low temperatures bear out this proportionality of the specific heat to the cube of the temperature.

The entropy, according to eqs. (10c.6) and (10d.6), is determined by

$$S = 9Nk\left(\frac{\theta}{T}\right)^3 \int_0^{\theta/T}\left[-\ln(1-e^{-x}) + \frac{x}{e^x - 1}\right]x^2\, dx.$$

Partial integration of the first term brings this into the form

$$S = 3Nk\left[\frac{4}{3}\cdot D\left(\frac{\theta}{T}\right) - \ln(1-e^{-\theta/T})\right]. \tag{10e.9}$$

At high temperatures this becomes

$$S = 3Nk\left[\frac{4}{3} - \ln\left(\frac{\theta}{T}\right) + \frac{1}{40}\left(\frac{\theta}{T}\right)^2 - \frac{1}{2240}\left(\frac{\theta}{T}\right)^4 + \cdots\right]. \tag{10e.9'}$$

At low temperatures the approximation is

$$S = 3Nk\left[\frac{4\pi^4}{15}\left(\frac{T}{\theta}\right)^3 + \cdots\right]. \tag{10e.9''}$$

The Helmholz free energy $A = E = TS$ takes the form

$$\begin{aligned} A &= \tfrac{9}{8}Nh\nu_m + 3NkT\left[\ln(1-e^{-\theta/T}) - \tfrac{1}{3}\cdot D\left(\frac{\theta}{T}\right)\right] \\ &= 3NkT\left[\ln\left(\frac{\theta}{T}\right) - \tfrac{1}{3} + \tfrac{1}{40}\left(\frac{\theta}{T}\right)^2 - \tfrac{1}{6720}\left(\frac{\theta}{T}\right)^4 + \cdots\right] \\ &= \tfrac{9}{8}Nh\nu_m - 3NkT\left[\frac{\pi^4}{15}\left(\frac{T}{\theta}\right)^3 + \cdots\right], \end{aligned} \tag{10e.10}$$

which may also be derived from the relation $A = -kT\ln Q$.

The Debye temperature θ depends on the interaction of the atoms and therefore on the volume V of the crystal. The pressure,

$$P = -\left(\frac{\partial E}{\partial V}\right)_S = -\left(\frac{\partial A}{\partial V}\right)_T,$$

therefore contains $d\theta/dV$, a quantity that is not easily evaluated. The heat content H and the Gibbs free energy G differ from E and A, respectively, by the term PV. However, the volume per molecule V/N in a crystal is very much smaller, indeed about one-thousandth of that in a gas at room temperature and 1 atm pressure. Except at high pressure the term PV is therefore about one-thousandth of RT per mole, and of little importance.

It is seen that the heat capacity C_V, as well as E/NkT, depend only on θ/T. By shifting the temperature scale the curves for various substances can be brought into coincidence.

For isotropic nonatomic crystals the course of the experimentally observed specific heats is moderately well represented by a formula of the type in eq. (8). The values of θ, and therefore ν_m, determined from thermal data, are in fair agreement with those obtained from elastic constants. The Debye temperature for most substances lies below room temperature. This is in agreement with the observed and calculated fact that the lattice frequencies, as a result of small force constants and large vibrating masses, are very small.

However, careful comparison of the experimental heat capacity curves with those predicted by eq. (8) shows consistent, although small, discrepancies. Actually, the proportionality of the heat capacity to T^3 is observed for higher temperatures than would be predicted from the Debye equation. Blackman[(1)] was probably the first to make a detailed calculation of the number density $N(\nu)$ of frequencies. His method was an extension of work by Born and von Karman[(2)] who analyzed the linear lattice model numerically (Sec. 10c). He found that the function $N(\nu)$ is fairly complicated, deviating very considerably from the ν^2 law of eq. (10d.6) and even showing a maximum at a frequency considerably lower than ν_m, as well as the higher maximum at about ν_m predicted in eq. (10d.6). It appears that the agreement with experiment obtained with the Debye function is partly fortuitous. Since then, far more detailed studies making use of computer calculations have been made.[(3)]

For crystals that are not monatomic and isotropic, other, more complicated, types of behavior occur. Qualitatively, some of these are predictable. They give rise to heat capacity functions considerably different from Debye's.

The assumption was made, however, that the crystal be isotropic. If it is not, that is, if the velocity of sound is very different for propagation in different directions, the distribution of frequencies is necessarily more complicated than eq. (10d.6) and the Debye formula must be expected to fail. This indeed happens.

(1) M. Blackman, *Proc. R. Soc. London*, Ser. A, **159,** 416 (1937).
(2) M. Born and Th. von Karman, *Phys. Z.*, **13,** 297 (1912).
(3) See, for instance, R. F. Wallis, Ed., *Lattice Dynamics*, Pergamon Press, New York, 1965.

Simple isotropic monatomic lattices, although Blackman's detailed calculation shows that the distribution of frequencies is not given very well by a formula of the type in eq. (10d.6), still obey the Debye equation rather remarkably.

The formulas of the last section may be used with some success for many crystals of simple ions, like the alkali halides. These substances form lattices in which each ion is surrounded by a rather large number (six or eight) of equivalent neighbors of opposite charge. The forces acting on the positive and negative ions are essentially the same, so that the modes of vibration are of the same type as in the monatomic crystal.

The vibration with smallest wavelength and largest frequency in the lattice is that where neighboring particles vibrate with opposite phase. In ionic crystals like the alkali halides, neighboring particles are oppositely charged. This fastest oscillation corresponds then to a large vibrating electric dipole moment and should therefore be capable of emitting and absorbing light. This is indeed the fact. These optically active frequencies of salt crystals are usually called reststrahlen, or residual ray frequencies, after the method by which they were first observed. They are so far in the infrared (optical wavelength between 20 and 150 μm) that optical absorption methods failed for some time to disclose them. Rubens[(1)] discovered them originally by studying the selective reflection of salt crystals.[(2)]

If the masses of the different ions in the crystal are equal, or nearly equal, as for instance in KCl, the optical frequencies of increasing wavelength go continuously over into the elastic frequencies. The optical frequency ν_0 is then simply the highest frequency ν_m and may be used in the Debye formula (8) to calculate the heat capacity. The total number of degrees of freedom is of course three times the total number of ions, or six times the total number of KCl molecules.

If the masses of the constituent particles are different, however, the situation is more complicated. One then finds gaps in the frequencies, that is, ranges of frequency between the optical and the elastic frequencies, which correspond to no lattice vibration. In the alkali halides, for instance, where one reststrahlen frequency is observed, the lattice frequencies form two branches, each containing half as many vibrations as there are degrees of freedom in the lattice. The distribution of frequencies in the lower or elastic branch is given rather well by eq. (10d.6), replacing the number of ions N by $\frac{1}{2}N$. The frequencies of the upper or optical

(1) H. Rubens and E. F. Nichols, *Wied. Ann.*, **60,** 45 (1897), and numerous later papers.

(2) The laws of optics show that the frequency of the light that is selectively reflected is somewhat different from that which is most strongly absorbed; the latter is the true frequency of the vibrational motion in the crystal.

branch are all higher than the largest one of the elastic branch, and the highest of them is the reststrahlen frequency. If the masses of the ions are very different, all $3N/2$ optical vibrations have almost the same frequency.

The reason for this effect lies not in a difference of force constants but in a difference in the vibrating masses. For the motion corresponding to the reststrahlen frequency the vibrating mass is the reduced mass of the two kinds of particles $m_1m_2/(m_1+m_2)$ which is smaller than the mass of either particle, and nearly equal to the mass of the lighter particle if m_1 and m_2 are very different. The vibrations of the optical branch consist primarily of vibration of the different kinds of particles against each other, down to the vibration of the lighter particles alone. The elastic branch consists of vibrations in which neighbors move in phase, up to the motions of the heavier particles alone.

This can be seen very clearly in the simple one-dimensional lattice. Assume the particles to have alternately different masses, m_1 and m_2, with $m_1 > m_2$. The break in the frequencies occurs at the wavelength $\lambda = 4a$, where every second particle is standing still. To this wavelength correspond two motions of exactly the same force constant but different frequency; for the lower one the heavier, for the higher one the lighter, particles vibrate alone. The frequencies are given by $\nu_1 = (1/\pi)[\phi''(a)/2m_1]^{1/2}$ and $\nu_2 = (1/\pi)[\phi''(a)/2m_2]^{1/2}$, respectively, since $\sin(\pi/4) = 2^{-1/2}$. There are $\frac{1}{2}N$ vibrations with longer wavelength, which have lower frequencies than ν_1, and $\frac{1}{2}N$ with shorter wavelength and higher frequency than ν_2, up to the highest with $\nu_0 = (1/\pi)[\phi''(a)(m_1+m_2)/2m_1m_2]^{1/2}$.

In this case, a fair approximation for the specific heat may be obtained by representing the heat capacity of the oscillators of the elastic branch by a Debye formula containing $3N/2$ vibrations. The frequencies of the optical branch may be considered to be identical to the reststrahlen frequency ν_0, so that their heat capacity is given by $3N/2$ times the average heat capacity of one oscillator. The energy of the crystal is similarly calculated as the sum of two parts, each contributed to be $3N/2$ degrees of freedom,

$$E = \tfrac{9}{16}Nh\nu_m + \tfrac{3}{2}NkT \cdot D\left(\frac{\theta}{T}\right) + \tfrac{3}{4}Nh\nu_0 + \tfrac{3}{2}NkT\frac{(h\nu_0/kT)}{e^{h\nu_0/kT}-1}. \tag{10e.11}$$

The reststrahlen frequency is of the same order of magnitude as ν_m, so that the heat capacity of these substances at room temperature has the classical value, kT times the number of ions.

Still different are molecular crystals, or crystals that contain molecular ions, for instance, CO_3^{2-} groups. The forces within the molecules are usually much stronger than those between the molecules. The shape of the molecule or ion complex therefore remains, essentially the same as in the gas or in solution, and the molecular frequencies, at least the higher ones corresponding to the stretching of bonds and not to bending, are little influenced by the fact that the molecule is cemented into the lattice. The molecular frequencies are so much higher than the lattice frequencies that their contribution to the heat capacity at room temperature is very small. The contribution of the true lattice frequencies may again be represented by a Debye curve.

The advent of high-speed computers has radically changed the extent of our knowledge of the distribution of frequencies in crystals. Blackman's 1937 results already showed that the actual spectral distribution was far more complicated than the simple Debye $\nu^2\,d\nu$ equation. What might be called brute force computations take about a large fraction of an hour but give highly precise functions for a given model of assumed nearest-neighbor interactions. The actual distribution functions show singularities in the slope and even discontinuities in the frequency density distribution. More modern mathematical techniques using modified moments[(1)] have cut the required computer time to fractions of a minute with equal precision of results.

[(1)] See, for instance, John C. Wheeler, *Phyus. Rev.*, **B10,** 2429 (1974).

CHAPTER 11

BOSE–EINSTEIN AND FERMI–DIRAC PERFECT GASES

11a. INTRODUCTION

In the previous four chapters we consistently used Boltzmann counting and to a great extent classical mechanics. In this chapter we examine a few problems which require Bose-Einstein counting or Fermi-Dirac counting, particularly involving perfect gases, and present a more qualitative discussion of some of the real systems for which the perfect gas treatment is applicable as an approximation.

11b. THE BOSE–EINSTEIN PERFECT GAS

We treat a perfect one-component monatomic gas composed of Bosons which demand symmetric wave functions. The standard example is helium, ^{4}He. In Sec. 7a we derived the general equation for this case from the grand canonical ensemble to be [eq. (7a.13)]

$$\beta PV = \sum_{\mathbf{k}} -\ln(1 - e^{\beta(\mu - \epsilon_{\mathbf{k}})}), \tag{11b.1}$$

where $\beta = 1/kT$, $\mathbf{k}$ are the quantum numbers of the single molecules having energy $\epsilon_{\mathbf{k}}$, and μ is the chemical potential.

In eq. (1) we expand the logarithm, $-\ln(1-x)=\sum_{\nu\geq 1}(x^{\nu}/\nu)$, to write

$$\beta PVB=\sum_{\mathbf{k}}\sum_{\nu\geq 1}\nu^{-1}e^{\nu\beta(\mu-\epsilon_{\mathbf{k}})}, \tag{11b.2}$$

and make use of the fact that for the translational quantum states $\mathbf{k}=k_x$, k_y, k_z in any macroscopic volume they are so close together in energy that we can replace summation by integration in phase space [see eq. (7b.13)],

$$\sum_{\mathbf{k}}=\sum_{k_x}\sum_{k_y}\sum_{k_z}\Rightarrow Vh^{-3}\iiint dp_x\,dp_y\,dp_z. \tag{11b.3}$$

With the kinetic energy $\epsilon_{\mathbf{k}}$ replaced by the classical expression $\epsilon_{\mathbf{k}}\Rightarrow p^2/2m$, we have

$$\beta P=\sum_{\nu\geq 1}\nu^{-1}e^{\nu\beta\mu}\left(\int_{-\infty}^{+\infty}h^{-1}\,dp\,e^{-\nu(\beta/2m)p^2}\right)^3. \tag{11b.4}$$

Into this introduce the de Broglie wavelength of eq. (7b.16) (discussed at some length in Sec. 7d) as

$$\lambda=\left(\frac{h^2}{2\pi mkT}\right)^{1/2}=\left(\frac{\beta h^2}{2\pi m}\right)^{1/2}, \tag{11b.5}$$

so that, with a dummy variable y,

$$\frac{\nu\beta p^2}{2m}=\frac{\beta h^2}{2\pi m}(\nu^{1/2}\pi^{1/2}ph^{-1})^2$$
$$=(\nu^{1/2}\lambda\pi^{1/2}ph^{-1})^2=y^2,$$

we can write eq. (4) as

$$\beta P=\sum_{\nu\geq 1}\nu^{-5/2}e^{\nu\beta\mu}\left(\pi^{-1/2}\lambda^{-1}\int_{-\infty}^{+\infty}dy\,e^{-y^2}\right)^3$$
$$=\lambda^{-3}\sum_{\nu\geq 1}\nu^{-5/2}e^{\nu\beta\mu}. \tag{11b.6}$$

We can differentiate either eq. (1) or V times its series development [eq. (6)] with respect to $\beta\mu$ at constant V, β and with respect to β at constant V, $\beta\mu$ using the thermodynamic values for the derivatives which are N (the number of molecules) and E (the energy), respectively. These thermodynamic relations are discussed in Sec. 3h, and the equation for βPV is eq. (3h.27). Using eq. (1) we have

$$\left[\frac{\partial(\beta PV)}{\partial(\beta\mu)}\right]_{V,\beta}=N=\sum_{\mathbf{k}\geq 0}\frac{1}{e^{\beta(\epsilon_{\mathbf{k}}-\mu)}-1}, \tag{11b.7}$$

$$\left[\frac{\partial(\beta PV)}{\partial\beta}\right]_{V,\beta\mu}=E=\sum_{\mathbf{k}\geq 0}\frac{\epsilon_{\mathbf{k}}}{e^{\beta(\epsilon_{\mathbf{k}}-\mu)}-1}, \tag{11b.8}$$

or, if we use V times eq. (6), we find

$$N = V\lambda^{-3} \sum_{\nu \geq 1} \nu^{-3/2} e^{\nu\beta\mu}, \tag{11b.7'}$$

$$E = \tfrac{3}{2} V\beta^{-1}\lambda^{-3} \sum_{\nu \geq 1} \nu^{-5/2} e^{\nu\beta\mu}, \tag{11b.8'}$$

$$E = \tfrac{3}{2}(\beta PV). \tag{11b.9}$$

The zero of the chemical potential μ must be measured from the zero used for the quantum states $\epsilon_{\mathbf{k}}$. We choose the state $\mathbf{k} = k_x, k_y, k_z \equiv 0$ with the lowest kinetic energy to be zero energy, $\epsilon_0 = 0$, so that, since $N_{\mathbf{k}}$ or rather the average number, $\langle N_{\mathbf{k}} \rangle = (e^{\beta(\epsilon_{\mathbf{k}} - \mu)} - 1)^{-1}$, in each quantum state $\mathbf{k}$ cannot be negative and for the zero state is

$$\langle N_0 \rangle = (e^{-\beta\mu} - 1)^{-1}, \tag{11b.10}$$

we see that necessarily

$$\beta\mu \leq 0, \tag{11b.11}$$

with $\langle N_0 \rangle$ infinity for $\beta\mu$ identically zero.

In all the previous chapters in Part 3 of this book we used Boltzmann counting which is valid if $-\beta\mu$ is large enough so that $e^{-\beta\mu} \gg 1$ and $\langle N_{\mathbf{k}} \rangle \cong e^{\beta(\mu - \epsilon_{\mathbf{k}})} \ll 1$; there are then very few molecules per quantum state. We now examine the case in which $0 \leq (-\beta\mu) \ll 1$ and $\langle N_0 \rangle \gg 1$. With $-\beta\mu$ sufficiently small, we develop the exponential of eq. (9) as $e^{-\beta\mu} = 1 - \beta\mu \cdots$ to find

$$\langle N_0 \rangle \cong (-\beta\mu)^{-1} \gg 1. \tag{11b.12}$$

The numerical value of the sums over ν that are in eqs. (7′) and (8′), when $\beta\mu \equiv 0$, $e^{\nu\beta\mu} \equiv 1$, are

$$\sum_{\nu \geq 1} \nu^{-3/2} = 2.612, \tag{11b.13}$$

$$\sum_{\nu \geq 1} \nu^{-5/2} = 1.341,$$

so that eq. (7′) for N gives a maximum value for

$$\langle N \rangle = \sum_{\mathbf{k} > 0} \langle N_{\mathbf{k}} \rangle = 2.612\ V\lambda^{-3} \tag{11b.14}$$

as $-\beta\mu \Rightarrow 0$. The contradiction is not difficult to explain. In deriving eq. (7′) we used eq. (3), expressing the sum over quantum states as an integral in the momentum space,

$$\sum_{\mathbf{k}} \Rightarrow Vh^{-3} \iiint_{-\infty}^{+\infty} dp_x\, dp_y\, dp_z = \int_0^{\infty} 4\pi p^2\, dp,$$

which is, for all macroscopically appreciable kinetic energies, extremely good but gives zero weight to the state at $|\mathbf{p}|=0$. For this state, and presumably also for a few quantum states of immeasurably small momenta above zero, the series in eq. (7′) is invalid and we must use eq. (7).

The conclusion we draw is that we have what is called Bose-Einstein condensation, a condensation of molecules into the zero kinetic energy state at very low temperatures and high number density. In Sec. 13*l* we show that this is accompanied by a clustering of molecules in configuration space as well, and that the development of eqs. (6), (7′), and (8′) as a power series in an absolute activity $z=\lambda^{-3}e^{\beta\mu}$ yields coefficients b_ν and νb_ν which are integrals of cluster functions in configuration space strictly analogous to the classical development of Chapter 8.

Bose-Einstein condensation begins when, at fixed temperature T, the number density ρ reaches a value, from eqs. (7′) and (13), of

$$\rho_0(T) = 2.612\lambda^{-3} = 2.612\left(\frac{2\pi mkT}{h^2}\right)^{3/2}. \tag{11b.15}$$

In terms of the molecular weight is when the volume V per mole decreases to

$$V_0(T) = 1227(MT)^{-3/2} \qquad \text{cm}^3, \tag{11b.15′}$$

with M the molecular weight. The pressure, as condensation starts, is given by eq. (8′), with eq. (13), as $\frac{2}{3}$ of E/V to be

$$P_0(T) = \beta^{-1} 1.341\lambda^{-3} = \frac{2}{3}\frac{E}{V}, \tag{11b.16}$$

so that, from this and eq. (15),

$$\frac{\beta P_0}{\rho_0} = \frac{PV}{RT} = 0.5134. \tag{11b.17}$$

The energy and the pressure stay constant at constant temperature if the density is increased. Since we have used the fictitious model of a perfect gas with absolutely no forces between molecules, the equations predict it will stay constant to zero volume. If the density is held constant and the temperature is lowered below that of condensation, the energy and pressure will decrease proportionally to $T^{5/2}$.

The condensation line in a number density temperature plane follows the line $\rho_0 T_0^{-3/2}$ constant,

$$\rho_0 T_0^{-3/2} = 2.612 \left(\frac{2\pi mk}{h^2}\right)^{3/2} \tag{11b.18}$$

$$V_0 T_0^{3/2} = 1227 M^{-3/2} \qquad \text{cm}^3\,\text{deg}^{-3/2}, \tag{11b.18'}$$

with M the molecular weight,

$$V_0^{2/3} T_0 = 1146 M^{-1}. \tag{11b.18''}$$

At temperatures higher than T_0 or densities lower than ρ_0 the thermodynamic functions of the system can be developed in convergent power series of the state defining variables. In Sec. 8c we developed equations for the imperfect gas as power series of the absolute activity [eq. (8c.5) and later discussion],

$$z = \left(\frac{2\pi mkT}{h^2}\right)^{3/2} e^{\beta\mu} = \lambda^{-3} e^{\beta\mu}, \tag{11b.19}$$

using the notation [eq. (8d.10′)],

$$\beta P = \sum_{\nu \geq 1} b_\nu z^\nu, \tag{11b.20}$$

with $b_1 \equiv 1$. From eq. (8d.14) for the number density $\rho = N/V$ we had

$$\rho = \left[\frac{\partial(\beta P)}{\partial \ln z}\right]_\beta = \sum_{\nu \geq 1} \nu b_\nu z^\nu. \tag{11b.21}$$

From eq. (6) for βP and eq. (7′) for ρ, we find that for the Bose-Einstein perfect gas these are valid, with

$$b_\nu = \nu^{-5/2}\left(\frac{h^2}{2\pi mkT}\right)^{3/2(\nu-1)} = \nu^{-5/2} \lambda^{3(\nu-1)}. \tag{11b.22}$$

The activity z can be found by inverting the series of eq. (10) as a series for z in powers of ρ, as was done at the beginning of Sec. 8i.[1] One finds

$$\begin{aligned} z = \rho[1 - 0.35355(\lambda^3\rho) + 0.05755(\lambda^3\rho)^2 - 0.00576(\lambda^3\rho)^3 \\ + 0.000402(\lambda^3\rho)^4 - \cdots], \end{aligned} \tag{11b.23}$$

[1] The formula for this inversion, which becomes tedious, is given to the seventh power in the *Handbook of Mathematical Tables*, 1st ed., (supplement to the *Handbook of Chemistry and Physics*), Chemical Rubber Publishing Co., Cleveland Ohio, p. 399.

or in terms of $(\rho/\rho_0) = 2.612(\lambda^3\rho)$,

$$z = \rho\left[1 - 0.92348\frac{\rho}{\rho_0} + 0.39248\left(\frac{\rho}{\rho_0}\right)^2 - 0.10272\left(\frac{\rho}{\rho_0}\right)^3 + 0.01871\left(\frac{\rho}{\rho_0}\right)^4 - \cdots\right]. \tag{11b.23'}$$

With M the molecular weight,

$$\lambda^3\rho = 3206[V(\text{cm}^3\ \text{mol})^{-1}]^{-1}(MT)^{-3/2}. \tag{11b.23''}$$

The latter series at $\rho/\rho_0 = 1$ gives $z/\rho = 1/2.612$ if completed, and up to the term $(\rho/\rho_0)^4$ of eq. (23′) is within 1% of the limiting value.

The development for $\beta\mu$ is

$$\begin{aligned}\beta\mu &= \ln(\lambda^3\rho) - 0.35355(\lambda^3\rho) - 0.00492(\lambda^3\rho)^2 - 0.00014(\lambda^3\rho)^3 - \cdots \\ &= 0.958907 + \ln\frac{\rho}{\rho_0} - 0.9235\frac{\rho}{\rho_0} - 0.00336\left(\frac{\rho}{\rho_0}\right)^2 - 0.0024\left(\frac{\rho}{\rho_0}\right)^3 - \cdots,\end{aligned} \tag{11b.24}$$

where $0.958907 = \ln 2.612$.

From eq. (10) for βP, with eq. (12) for b_ν, we have

$$\beta P = \sum_{\nu \geq 1} \nu^{-5/2}\lambda^{3(\nu-1)}z^\nu, \tag{11b.25}$$

and if eqs. (23) and (23′) are used in this, we find with a little tedious algebra that, with the enthalpy $H = E + PV$,

$$\begin{aligned}\beta P &= \rho[1 - 0.17678(\lambda^3\rho) - 0.00330(\lambda^3\rho)^2 - 0.00011(\lambda^3\rho)^3 - \cdots] \\ &= \rho\left[1 - 0.4618\left(\frac{\rho}{\rho_0}\right) - 0.00225\left(\frac{\rho}{\rho_0}\right)^2 - 0.00196\left(\frac{\rho}{\rho_0}\right)^3 - \cdots\right] \\ &= \frac{2}{3}\frac{\beta E}{V} = \frac{2}{5}\frac{\beta H}{V}.\end{aligned} \tag{11b.26}$$

The value of the entropy divided by Boltzmann's constant k per molecule is

$$\begin{aligned}\frac{S}{k} = \frac{S}{kN} &= \frac{\beta H}{N} - \beta\mu \\ &= \tfrac{5}{2}\frac{\beta P}{\rho} - \beta\mu.\end{aligned} \tag{11b.27}$$

With eqs. (24) and (26),

$$\frac{S}{k} = \tfrac{5}{2} - \ln(\lambda^3\rho) - 0.08840(\lambda^3\rho) - 0.00333(\lambda^3\rho)^2 - 0.000135(\lambda^3\rho)^3 - \cdots$$

$$= 1.54109 - 0.2309\frac{\rho}{\rho_0} - 0.0225\left(\frac{\rho}{\rho_0}\right)^2 - 0.0023\left(\frac{\rho}{\rho_0}\right)^3 - \cdots. \tag{11b.28}$$

The heat capacity at constant volume has a cusp at the condensation temperature T_0. From eqs. (25) and (26) the energy per molecule at the condensation line ρ_0, T_0, is

$$\epsilon_0 = \frac{E}{N} = \frac{3}{2}\frac{P}{\rho} = \frac{3}{2}\,0.5134kT_0, \tag{11b.29}$$

and below this is $\epsilon(T) = \frac{3}{2}\,0.5134kT(T/T_0)^{3/2}$, $T \le T_0$. The heat capacity at constant volume per molecule, $C_V = C_V/N$, divided by Boltzmann's constant k, is then

$$\begin{aligned}\frac{C_V}{k} &= \tfrac{15}{4}0.5134\left(\frac{T}{T_0}\right)^{3/2}, \qquad T \le T_0,\\ &= 1.925\left(\frac{T}{T_0}\right)^{3/2} = 1.925\frac{\rho}{\rho_0},\end{aligned} \tag{11b.30}$$

and its derivative with respect to temperature is

$$\left[\frac{\partial(C_V/k)}{\partial T}\right]_V = \tfrac{45}{8}0.5134T^{-1}\left(\frac{T}{T_0}\right)^{3/2}, \tag{11b.31}$$

so that, as the condensation temperature is approached from below,

$$\begin{aligned}\lim_{T-T_0\to 0} -T_0\left[\frac{\partial(C_V/k)}{\partial T}\right]_V^{(T-T_0=<0)} &= \tfrac{45}{8}0.5134T_0^{-1}\\ &= 2.888T_0^{-1}.\end{aligned} \tag{11b.31'}$$

For temperatures greater than T_0 we use $\epsilon = 3P/2\rho$ with eq. (26) for βP,

$$\epsilon = \tfrac{3}{2}kT[1 - 0.17678(\lambda^3\rho) - 0.00330(\lambda^3\rho)^2 - 0.00011(\lambda^3\rho)^3 - \cdots],$$

to find

$$\begin{aligned}\frac{C_V}{k} = \left(\frac{\partial\varepsilon}{\partial T}\right)_\rho &= \tfrac{3}{2}[1 + \tfrac{1}{2}0.1768(\lambda^3\rho) + 2.000330(\lambda^3\rho)^2\\ &\quad + \tfrac{7}{2}0.00011(\lambda^3\rho)^3 + \cdots]\\ &= \tfrac{3}{2}[kT\ 1 + 0.08809(\lambda^3\rho) + 0.00660(\lambda^3\rho)^2 + 0.00686(\lambda^3\rho)^3 + \cdots]\\ &= \frac{3}{2}\left[1 + 0.2301\frac{\rho}{\rho_0} + 0.04503\left(\frac{\rho}{\rho_0}\right)^2 + 0.00686\left(\frac{\rho}{\rho_0}\right)^3 + \cdots\right], \qquad T > T_0,\end{aligned} \tag{11b.32}$$

or at $T=T_0$, $\rho=\rho_0$, $(C_V/k)_{T=T_0}\cong 1.923$. If eq. (32) is continued to higher powers, it becomes identical with the value of eq. (30) approached from $T<T_0$.

The value of C_V/k for the classical perfect gas is $\frac{3}{2}$. That of the Bose-Einstein perfect gas is zero at $T=T_0$, increases to the value 1.925, which is 28% greater than that of the classical value at the condensation line, and then decreases to the classical value of $\frac{3}{2}$.

If C_V/k of eq. (32) is differentiated again with respect to T, one finds

$$\left(\frac{\partial(C_V/k)}{\partial T}\right)_V = -\tfrac{3}{2}T^{-1}\left[0.345\frac{\rho}{\rho_0} + 0.135\left(\frac{\rho}{\rho_0}\right)^2 + 0.031\left(\frac{\rho}{\rho_0}\right)^3 + \cdots\right], \qquad T>T_0, \quad (11b.33)$$

or has a limit

$$\lim_{\epsilon\to 0^+}\left(\frac{\partial(C_V/k)}{\partial T}\right)_V^{(T-T_0=\epsilon>0)} = -(0.511+\cdots)T_0^{-1} \qquad (11b.33')$$

Helium-saturated vapor at its boiling point of 4.2 K, were it a perfect gas, would deviate from the classical value of unity for $\beta P/\rho$, as a result of its Bose-Einstein statistics, by about 4% which is of the same order as the deviation due to the mutual attraction of the molecules. At the density of the liquid for which the volume V per mole is equal to 27.6 cm^3, the condensation temperature of a Bose-Einstein perfect gas is 3.14 K. Liquid helium undergoes a sharp transition at 2.174 K to a fluid having many remarkable properties, known as liquid helium II. This transition is without doubt essentially related to the Bose-Einstein perfect gas condensation.

The condensation is into a state in which the condensed fraction of the molecules moves with the same momentum, and they can flow through the normal molecules having random kinetic energy apparently without friction. The condensed fraction is superfluid, showing zero viscosity in most of the experimental methods of measuring viscosity. The heat capacity, unlike that predicted for the perfect Bose-Einstein gas, apparently becomes infinite at the transition, diverging as $-\ln|T-T_0|$ or possibly as $|T-T_0|^{-\epsilon}$, with ϵ extremely small.

In the next section we discuss the equilibrium gas of photons, the extreme example of a perfect Bose-Einstein gas.

11c. BLACKBODY RADIATION, CLASSICAL THEORY

One of the most esthetic accomplishments of late nineteenth century physical theory was the development of the thermodynamic properties of blackbody radiation based on a pure analysis of the laws of electromagnetic theory plus those of thermodynamics.

It is an everyday experience that a solid body, if heated, emits light, the intensity and color of which change with temperature. At any one temperature, the intensity and spectral distribution of the emitted radiation is a characteristic of the body, which can be determined theoretically only by a detailed investigation of the process of light emission. Statistical calculations, however, permit one to deduce the energy density at different frequencies of the radiation in equilibrium with the body.

The application of the second law of thermodynamics to radiation processes enables one to derive the fact that a body capable of emitting light of a certain frequency must also be capable of absorbing it, and furthermore that the radiation with which the body is in equilibrium, which is described by the intensity of radiation of different frequencies, is a function of the temperature only, and independent of the body in question. This is demonstrated here.

We wish to introduce the device of a box with perfectly reflecting walls, that is, with walls that neither emit nor absorb radiation. Light contained in this box is then effectively insulated from the outside; if the box is otherwise empty, neither intensity nor frequency distribution changes with time. In this box two different objects, a and b, are placed, and the temperature of both is maintained at the same value T. Between the objects, dividing the box into two unconnected parts, a screen is introduced, which has the property of perfectly reflecting light of all frequencies except one, ν, for which it is transparent. Although this is certainly an idealizing assumption, in practice screens can be found that approximate the qualities stipulated here to a certain degree.

If either one of the objects a or b were *alone* in its box, it would emit and absorb radiation until it came to equilibrium, that is, until the intensity of light surrounding it became so high that it would absorb as much light of each frequency as it emitted. Let us assume that the density of light of frequency ν in equilibrium with the bodies is higher for object a than for b. If now the two objects are separated by the screen, transparent for this frequency ν only, each body will tend to create the intensity of light in its surroundings with which it is in equilibrium. There will be a flow of light through the screen from side a to side b. Since this will decrease the density of light on side a, body a will emit more energy than it absorbs, whereas the converse would be true for b. The effect would be a net flow of energy between two objects at the same temperature, without the intervention of work, a result in disagreement with the second law of thermodynamics.

It follows that the density of light of each frequency must be the same for the radiation in equilibrium with any two bodies a and b at the same temperature, independent of the nature of the objects. The energy per unit volume of light of frequencies between ν and $\nu+\Delta\nu$ in equilibrium with a body of temperature T is termed $u(\nu, T)\Delta\nu$. We have reached the conclusion that $u(\nu, T)$ is a universal function of frequency and temperature alone.

A body that absorbs all the light falling on it is called black. The energy of radiation absorbed per second by such a body is easily calculated from the radiation in equilibrium with it. Of the light of frequency ν, a fraction $d\Omega/4\pi$ has a direction located in one solid angle $d\Omega$. An element of surface is struck, in the time interval dt, by the fraction of light contained within a hemisphere of radius $c\,dt$, which is directed towards it. The velocity of light c is independent of the frequency. The energy arriving per unit time, for each frequency, is therefore simply proportional to $u(\nu, T)$. The proportionality constant, calculated in the same way as for molecules hitting a wall, is $c/4$ for unit area and unit time. For a black surface all this energy is absorbed, and the energy emitted per second, unit area, and frequency range $\Delta\nu$, must be the same, namely, $(c/4)u(\nu, T)\ \Delta\nu$. For this reason the function $u(\nu, T)$ is called the blackbody distribution function.

No real bodies are truly black for all frequencies. One can conclude, however, that, if r_ν is the reflection coefficient for the frequency ν at the temperature T, that is, $1-r_\nu$ the fraction absorbed of the light of frequency ν striking the surface, the emission from the surface is $1-r_\nu$ times the emission from a blackbody at the same temperature. This is known as Kirchhoff's law. The intensity of radiation emitted by a blackbody represents the upper limit attainable from any surface of a given temperature.

The function $u(\nu, T)$ might be determined by calculating the rate of emission from a blackbody. Since, however, $u(\nu, T)$ is the density of light in equilibrium with *any* body at temperature T, it must be an inherent property of the radiation field itself and subject to a simpler statistical derivation.

We present here a rather short summary of the classical theory of blackbody radiation thermodynamics. A beautiful and very careful analysis was made by Planck in his book, *Wärmestrahlung*. The first classical step requires the analysis of electromagnetic theory to derive the radiation pressure. The conclusion, derived first by Maxwell but later examined by many others[(1)], is that

$$P=\tfrac{1}{3}u_t(T)=\tfrac{1}{3}\int_0^\infty u(\nu, T)\,d\nu. \tag{11c.1}$$

(1) The difficulties in understanding the failure of the Rayleigh-Jeans law, [eq. (22)] led even Rayleigh [John William Strutt, *Phil. Mag.*, **XLV**, 522–525 (1898)] to question the validity of eq. (1).

We omit the classical derivation. It can be derived simply quantum mechanically by the same consideration one uses to derive that the pressure of a perfect gas due to the change in momentum of the molecules at the wall per unit time and unit area is $P = \frac{2}{3}V^{-1}E_{\text{kin}}$, rather than the $\frac{1}{3}$ in (eq. 1). The difference of a factor of 2 arises from the fact that the photon energy is $\epsilon = h\nu$ but its momentum is $h\nu/c$ which is the energy divided by the velocity c. For a nonrelativistic molecule the momentum $p = mv = \epsilon/2v$, since the kinetic energy is $E = \frac{1}{2}mv^2$.

Now consider a cylinder of perfectly reflecting walls with an absolutely tight-fitting perfectly reflecting frictionless piston at the top and with a blackbody base. The temperature of the base can be controlled by contact with a reservoir at temperature T. At equilibrium the total energy of the radiation in the cylinder is

$$E = Vu_t(T) = 3P(T)V, \tag{11c.2}$$

where V is the volume of the cylinder.

Let the piston rise reversibly, doing the work $P\,dV = \frac{1}{3}u_t(T)\,dV$, but maintain the temperature of the black bottom at T so that a heat flow δq is transmitted from the reservoir to the radiation. Conservation of energy requires that

$$dE + P\,dV - \delta q = 0. \tag{11c.3}$$

The radiation has a definite entropy S. The whole expansion is done reversibly and so slowly that the energy density of the radiation remains at equilibrium $u(T)$. The increase dS_{rad} of the radiation plus the loss $-\delta q/T$ of entropy of the reservoir is zero for this reversible process, so that

$$dS_{\text{rad}} = T^{-1}(\delta q)_{\text{rev}} = T^{-1}(dE + P\,dV), \tag{11c.4}$$

from eq. (3).

Now the thermodynamic state of the radiation is completely determined by any two of the variables E, V, u, T, or P provided at least one is extensive, E or V. We choose the volume V and temperature T so that with $dE = d(u_tV) = u_t(T)\,dV + V(du/dT)\,dT$ and, from eq. (2), $P = \frac{1}{3}u_t$, where $u_t = \int d\nu\, u(\nu, T)$, we have

$$dS = VT^{-1}\frac{du_t}{dT}\,dT + \tfrac{4}{3}u_tT^{-1}\,dV, \tag{11c.5}$$

$$\left(\frac{\partial S}{\partial T}\right)_V = VT^{-1}\frac{du_t}{dT}, \tag{11c.5′}$$

$$\left(\frac{\partial S}{\partial V}\right)_T = \tfrac{4}{3}u_tT^{-1}. \tag{11c.5″}$$

The second derivative of S with respect to V and T is the partial of eq. (5′) with respect to V, equal to the partial derivative of eq. (5″) with respect to T,

$$\left(\frac{\partial^2 S}{\partial V\,\partial T}\right) = T^{-1}\frac{du_t}{dT} = \tfrac{4}{3}T^{-1}\frac{du_t}{dT} - \tfrac{4}{3}u_t T^{-2}, \tag{11c.6}$$

from which we find

$$u_t^{-1}\,du_t = d\ln u_t = 4T^{-1}\,dT = 4d\ln T \tag{11c.6'}$$

or, since $u_t(T=0)=0$,

$$u_t = aT^4, \tag{11c.6''}$$

and for the total radiant energy,

$$E = aT^4V = 3PV. \tag{11c.7}$$

The total entropy S of blackbody radiation is found by integrating eq. (5) to be

$$S = \tfrac{4}{3}aT^3V. \tag{11c.8}$$

The Helmholz free energy $A-TS$ is

$$A = -\tfrac{1}{3}aT^4V, \tag{11c.9}$$

and the Gibbs free energy $G = A + PV$ is

$$G = E - TS + PV = 0. \tag{11c.10}$$

Equation (7), for the energy of blackbody radiation, is known as the Stefan-Boltzmann law, first established experimentally by Stefan[(1)] on the basis of rather rough measurements and later deduced on a thermodynamic basis similar to that given above by Boltzmann.[(2)] A much more careful and detailed derivation than that given here is in Planck's *Wärmestrahlung.*[(3)]

The next step in the classical derivation goes beyond pure thermodynamics in that it states an important conclusion about the mathematical form of the dependence of $u(\nu, T)$ on frequency and results in Wien's displacement law,[(4)] namely, that

$$u(\nu, T)\,d\nu = T^3 f\left(\frac{\nu}{T}\right) d\left(\frac{\nu}{T}\right), \tag{11c.11}$$

where f is a function of the ratio ν/T only.

(1) J. Stefan, *Sitzungsber. Acad. Wiss. Wien* **79,** 391 (1879).

(2) L. Boltzmann, *Ann. Phys.* (*Wiedemann*) **22,** 291 (1884).

(3) An English translation, *The Theory of Heat Radiation*, of the second edition of *Wärmestrahlung* (M. Masius, Ed.) has been published in paperback by Dover (New York, 1959). The pertinent material is on pp. 49–102 which also cover Wien's displacement law.

(4) W. Wien, *Ann. Phys.* (*Wiedemann*) **58,** 662 (1896).

The argument we give here may seem too brief to be convincing to the critical student, but it serves to outline the essentials of the steps. The first of these is to prove that radiation having the equilibrium frequency energy density $u(\nu, T_1)$, if compressed reversibly and adiabatically in a container with perfectly reflecting walls, will always remain in the equilibrium energy density spectral distribution $u(\nu, T)$ with increasing T, VT^3 constant. If the container includes a blackbody object of negligible size and negligible heat capacity, and the compression is done so slowly as to maintain blackbody equilibrium, the reversible process with no heat flow will be at constant entropy and, from eq. (8), VT^3 will be constant. Now trap blackbody radiation $u(\nu, T_3)$ in a volume V_1 and *remove* the black object of temperature T_1 necessary to establish equilibrium. Then compress to V_2. Since the pressure depends only on $u_t = \int d\nu\, u(\nu, T)$, the work done and the total energy change are the same as with an infinitesimal black dust particle, the process is reversible, there is no heat flow, and the entropy is constant. Now insert the infinitesimal black object. Any change must involve an increase in entropy if the blackbody distribution of $u(\nu, T)$ is assumed to be a thermodynamically stable distribution determined by having a maximum entropy at a fixed thermodynamic state E_2, V_2.

Unless the expansion without the presence of the black dust particle retains the equilibrium energy density distribution $u(\nu, T)$, there is a conflict with the second-law principle that all reversible adiabatic paths between two thermodynamically defined states show the same zero change in entropy.

The next step is to consider the same reversible adiabatic compression carried out on a small region containing at V_1, T_1 the radiation $u(\nu, T_1)$ between two frequencies, ν_1 and $\nu_1 + \Delta_1\nu$, only. This can be established by emission into a totally reflecting cylinder with the usual fictitious totally reflecting tight frictionless piston of radiation passing through a screen transparent only between ν_1 and $\nu_1 + \Delta_1\nu$ from a black object at temperature T_1. The screen is then replaced by a total reflector.

In the process of compression one then concludes that the frequencies change; both ν and $\Delta\nu$ are proportional to $V^{-1/3}$ and therefore to T, so that ν/T is constant. The classical argument that νL stays constant for a *one*-dimensional standing wave along the x-axis trapped between two perfect reflectors normal to the x-axis a distance L apart is simple to understand. The standing wave has k half-wavelengths in the distance L, its amplitude is proportional to $\sin(\pi k/L)x$, with k an integer, and is zero at each reflector. If L is decreased sufficiently slowly k does not change. The wavelength is $\lambda = k/2L$, and the frequency is

$$\nu = c\lambda^{-1} = \frac{ck}{2L}. \tag{11c.12}$$

The extension to three dimensions that $\nu V^{1/3}$ is constant requires considerably more analysis and discussion to be absolutely convincing, it is given 33 pages in Planck's *Wärmestrahlung* (Dover English translation).

The equivalent theorem in the quantum formulation is much simpler. The quanta of energy $h\nu$ are independent, and each adds to its independent contribution to the energy density, $u_i = h\nu/V$, and to the time-average pressure, $\langle p_i \rangle_t = \frac{1}{3}h\nu_i/V$. On compression the work done on the quantum i by sufficiently slow compression increases its energy by

$$\begin{aligned} d(h\nu_i) &= -\langle p_i \rangle_t\, dV = \tfrac{1}{3}h\nu_i V^{-1}\, dV, \\ \nu_i^{-1}\, d\nu_i &= d \ln \nu_i = -\tfrac{1}{3}V^{-1}\, dV = -\tfrac{1}{3}\, d \ln V, \\ \nu_i V^{1/3} &= \text{constant}, \qquad dS = 0, \\ \frac{\nu_i}{T} &= \text{constant}, \qquad dS = 0. \end{aligned} \tag{11c.13}$$

The total energy density $u_t(T)$ is a sum of independent contributions $u(\nu, T)\,\Delta\nu$, and in each of these the dependence of frequency on temperature at equilibrium is such that ν/T is constant. We can write

$$u_t(T) = F(T) \int_0^\infty d\left(\frac{\nu}{T}\right) f\left(\frac{\nu}{T}\right). \tag{11c.14}$$

The integral

$$\int_0^\infty d\left(\frac{\nu}{T}\right) f\left(\frac{\nu}{T}\right) = A = T^{-1} \int_0^\infty d\nu\, f\left(\frac{\nu}{T}\right) \tag{11c.15}$$

is a constant A independent of temperature, but from eq. (6″), where $u_t(T)$ is proportional to T^4, we derive eq. (11) with the added relation that the proportionality constant a of eq. (6″), $u_t = aT^4$, is the same as the A of eq. (15),

$$A\,[\text{eq. (15)}] = a\,[\text{eq. (6″)}]. \tag{11c.16}$$

Independently of the Wien displacement law we can calculate the number of independent standing electromagnetic waves in a hohlraum, an evacuated space containing only electromagnetic radiation. The calculation follows that used for the number of translational kinetic energy functions of a gas in Sec. 7b. With a total magnitude k,

$$k = (k_x^2 + k_y^2 + k_z^2)^{1/2}, \tag{11c.17}$$

the number density $n(k)$ of standing *compressional* waves is given by equating the number of positive integer k_x, k_y, k_z values between k and

$k+\Delta k$ to $n(k)\,\Delta k$. One finds

$$n(k)\,dk = \tfrac{1}{8}4k^2\,dk, \tag{11c.18}$$

namely, the area of the positive octant surface of the sphere of radius k. With eq. (12), in which $\nu = ck/2L$,

$$k^2\,dk = 8L^3c^{-3}\nu^2\,d\nu,$$

and for the number density $N^*(\nu)$ in frequency space in a cubic hohlraum, $V = L^3$,

$$N^*(\nu)\,d\nu = Vc^{-3}4\pi\nu^2\,d\nu.$$

This counts k-values alone and refers to compressional waves of frequency ν. The electromagnetic waves are transverse, and there are two waves of different polarization for each set k_x, k_y, k_z of wave numbers. For electromagnetic radiation the number of standing waves in a volume V between ν and $\nu + d\nu$ divided by $d\nu$ is then

$$N(\nu) = V8nc^{-3}\nu^2. \tag{11c.19}$$

If $\langle\epsilon(\nu)\rangle_T$ is the average energy per standing wave at temperature T, the energy density $u(\nu, T)$ per unit frequency range per unit volume will be

$$u(\nu, T) = 8\pi c^{-3}\nu^2\langle\epsilon(\nu)\rangle_T, \tag{11c.20}$$

and the total energy density will be

$$u_t(T) = c^{-3}\int_0^\infty 8\pi\nu^2\,d\nu\,\langle\epsilon(\nu)\rangle_T. \tag{11c.21}$$

We are now all set to find disaster, the famous "ultraviolet catastrophy" or "white death" derived near the end of the nineteenth century by the rigorous classical equations. An electromagnetic wave is a classical oscillator, and its average energy $\langle\epsilon(\omega)\rangle_T$ is kT. The Rayleigh-Jeans equation

$$\begin{aligned} u(\nu, T) &= 8\pi c^{-3}\nu^2 kT \\ &= kT^3 8\pi c^{-3}\left(\frac{\nu}{T}\right)^2 \end{aligned} \tag{11c.22}$$

is the classical result. It obeys the Wien displacement law, which is consoling, but the integral of eq. (21) diverges horribly and u_t is predicted to be infinite at all finite temperatures.

The Rayleigh-Jeans form of eq. (22) was confirmed experimentally by Rubens and Kurlbaum[(1)] at very long wavelengths. This was really after

[(1)] H. Rubens and F. Kurlbaum, *Drude Ann.*, **IV,** 649 (1901).

the Planck equation had been proposed, but one could hardly say accepted. Wien had proposed a form

$$u(\nu, T) \sim \nu^3 e^{-\alpha(\nu/T)}, \tag{11c.23}$$

which was known to fit the experimental data for optical frequencies quite well.

This was the state of the theoretical understanding of blackbody radiation at the end of the nineteenth century. There were similar difficulties with the known heat capacities of diatomic gases, which were lower than theory demanded. The latter led to considerable skepticism about the reality of the atomic molecular theory. However, Maxwell's laws of electromagnetic radiation were firmly believed. The discrepancies were, to say the least, disturbing.

11d. QUANTUM THEORY OF BLACKBODY RADIATION

It was Max Planck who, in 1900, had the genius and temerity to solve the problem of the "white death" by taking the radical step of assuming a deviation from the fundamental classical laws of mechanics for sufficiently small objects. This of course was the origin of quantum mechanics. The equation for $\langle\epsilon(\nu)\rangle_T$, the energy of a single oscillator ($\beta = 1/kT$),

$$\langle\epsilon(\nu)\rangle_T = h\nu(e^{\beta h\nu} - 1)^{-1}, \tag{11d.1}$$

in eq. (11c.20) for $u(\nu, T)$,

$$u(\nu, T) = 8\pi c^{-3} h\nu^3 (e^{\beta h\nu} - 1)^{-1}, \tag{11d.2}$$

gives Wien's law of eq. (11c.23) for high frequencies, $\beta h\nu \gg 1$, when the unity in the denominator of eq. (2) can be neglected. It also gives the Rayleigh-Jeans expression [eq. (11c.22)] when $\beta h\nu \ll 1$ and $e^{\beta h\nu} - 1 \cong \beta h\nu$. Planck originally proposed that mechanical dipole oscillators could absorb or emit radiation only in energy changes by an amount $h\nu$. It was really Einstein in 1905 who completely accepted the now usual point of view that electromagnetic radiation *consists* of quanta of energy $h\nu$ propagated by a probability density field according to the laws of classical electromagnetic radiation. Bose, in 1923, sent to Einstein a derivation of the blackbody equation based on essentially the method we used in Sec. 7a for the grand canonical ensemble treatment of a perfect gas. Einstein submitted Bose's paper[(1)] for publication with his blessing.

The photons, each of energy $h\nu$, form a completely degenerate perfect

[(1)] S. N. Bose, *Z. Phys.*, **26,** 178 (1924); A. Einstein, *Sitzungsber. Preuss. Akad. Wiss.*, 261 (1924).

Bose-Einstein gas. We have derived eq. (11c.10) that for blackbody radiation the Gibbs free energy G is zero, so that the chemical potential μ is zero. If $\mu = 0$ is substituted in eq. (11b.8) for the energy E of a Bose-Einstein perfect gas, we find, with $\beta = 1/kT$,

$$E = \sum_{\mathbf{k}\geq 0} \epsilon_{\mathbf{k}}(e^{\beta\epsilon_{\mathbf{k}}} - 1)^{-1}. \tag{11d.3}$$

With eq. (11c.19) for the number density, $N(\nu) = V8\pi c^{-3}\nu^2$, of photon states per unit frequency range, and $\epsilon_{\mathbf{k}} = h\nu$, we find for the total energy E in a hohlraum of volume V at temperature T the integral of the Planck blackbody energy density $u(\nu, T)$ over frequency,

$$u_t(T) = \int_0^\infty d\nu\, u(\nu, T) = 8\pi c^{-3} h \int_0^\infty \nu^3 (e^{\beta h\nu} - 1)^{-1}\, d\nu \tag{11d.4}$$

If the integrand is written as a function of $\beta h\nu$, $d\nu = (\beta h)^{-1}\, d(\beta h\nu)$ in accord with Wien's displacement law of eq. (11c.11), we have

$$u(T) = \frac{8\pi(kT)^4}{c^3 h^3} \int_0^\infty \frac{(\beta h\nu)^3}{e^{\beta h\nu} - 1}\, d(\beta h\nu). \tag{11d.5}$$

The definite integral $I = \int_0^\infty dx\, x^3(e^x - 1)^{-1}$ can be evaluated exactly by developing the denominator, $(e^x - 1)^{-1} = \sum_{n\geq 1} e^{-nx}$, and using

$$\int_0^\infty dx\, x^3 e^{-nx} = n^{-4} \int_0^\infty dy\, y^3 e^{-y} = 6n^{-4},$$

from which, with eq. (5),

$$E = V \frac{48\pi(kT)^4}{(hc)^3} \sum_{n\geq 1}^{\infty} n^{-4}. \tag{11d.6}$$

The infinite series is given in tables as

$$\sum_{n=1}^{\infty} n^{-4} = \frac{\pi^4}{90}, \tag{11d.6$'$}$$

so that, with eq. (11c.6″), where $u_t = aT^4$,

$$u_t(T) = \frac{E}{V} = \frac{8\pi^5 k^4 T^4}{15(hc)^3} = aT^4. \tag{11d.6$''$}$$

The Stefan-Boltzmann constant, which is the radiation energy emitted per unit time and unit area divided by T^4, is

$$\sigma = \tfrac{1}{4}ca. \tag{11d.7}$$

Its numerical value is listed in many tables of the important physical constants. It is

$$\sigma = \frac{2\pi^5 k^4}{15h^3 c^{-2}} = 5.6696 \times 10^{-12}\,\mathrm{J\,sec^{-1}\,cm^{-2}\,K^{-4}} \tag{11d.8}$$

At 1000 K the radiant energy emitted by a square centimeter surface is over a calorie per second. The value of a is

$$\begin{aligned} a = \frac{4\sigma}{c} &= 7.57 \times 10^{-12}\,\mathrm{erg\,liter^{-1}\,K^{-4}} \\ &= 0.00757 \times 10^{-16}\,\mathrm{J\,liter^{-1}\,K^{-4}}, \end{aligned} \tag{11d.9}$$

an extremely small energy density at 1000 K, but at 10^5 K it has about the energy in 1 liter equal to the kinetic energy of 3 mol of a perfect monatomic gas at room temperature, 300 K. The mass density $\rho(T)$ is $c^{-2}a$, or 0.84×10^{32} g liter^{-1} K^{-4}, which at 10^8 K is 0.84 g liter but at 10^9 K is 8.4 Kg liter^{-1}. The thermodynamic functions are listed in eqs. (11c.7) to (11c.10).

11e. THE PERFECT FERMI–DIRAC GAS AT $T = 0$

Atoms, molecules, and fundamental particles demanding antisymmetric state functions form Fermi-Dirac systems. The standard example is the gas of electrons which, because of their very low mass, show extreme deviation from Boltzmann counting at high concentrations. Obviously, a high concentration of electrons alone in macroscopic amounts is not a common laboratory system, but many of the simple properties of metals are given fairly well by treating the metal as a container, which by having positively charged ions in a crystalline array neutralizes the electrons but traps the mobile valence electrons within the metal. The zeroth order approximation is to treat the potential for the electrons within the metal as constant and, because of the high screening due to the positively charged ions, to neglect the electrostatic repulsion between the electrons. The model is then of a perfect Fermi-Dirac gas of very light mass at a very high number density, namely, of 1 mol of electrons for a univalent metal, sodium, potassium, and so on, in the volume of 1 mol of metal.

One other laboratory system, liquid helium isotope of mass 3, consisting of two protons, two electrons, and one neutron, has an odd number of fundamental fermions, has a half-integer nuclear spin, and obeys Fermi-Dirac statistical counting of states. It is of course not a perfect gas, and its treatment has all the complexities of the liquid state. An almost perfect Fermi-Dirac gas, even at quite high density, would be a neutron gas, as yet not a common laboratory system, although presumably existing in some stars at fantastically high densities of 10^9 times that at the earth's core.

We start with eq. (7a.13) for the Fermi-Dirac system,

$$\beta PV = \sum_{\mathbf{k}\geq 0} \ln\left(1+e^{\beta(\mu-\epsilon_{\mathbf{k}})}\right), \tag{11e.1}$$

and use the thermodynamic relation [eq. (3h.27)]

$$\left[\frac{\partial(\beta PV)}{\partial(\beta\mu)}\right]_{V,\beta} = N = \sum_{\mathbf{k}\geq 0} \frac{1}{e^{\beta(\epsilon_{\mathbf{k}}-\mu)}+1}. \tag{11e.2}$$

If each metal atom contributes one, or as many electrons as its valency, to the electron gas, the density of particles in the gas will be very high. The molal volume V of the electron gas is the atomic volume of the metal divided by a small integer. Atomic volumes of metals are of the order of 10 cm^3. The classical distribution function is that in which the unity in the denominator of this equation is omitted. The temperature above which this neglect is justified is determined by the condition that $-\mu/kT$ be considerably larger than unity. Evaluation of this quantity with omission of the unity leads to $-\mu/kT = \ln(0.000634M^{3/2}VT^{3/2})$. If the electron atomic weight $M=\frac{1}{1840}$, and $V=10\text{ cm}^3$ is inserted, $-\mu/kT= \ln(8\times 10^{-8}T^{3/2})$. This shows that the classical equation can be applied successfully only above 10^6 deg, a temperature at which no metal is solid. At room temperature the classical distribution function is not even suitable as a starting point for an approximate calculation.

In view of this it is interesting to study first the opposite extreme, namely, the properties of the gas at $T=0$, which gives a much better approximation for room temperature than the classical distribution. This calculation can be done without recourse to the distribution function.

The lowest energy of a classical or a Bose-Einstein gas at $T=0$ is $E_0=0$. At zero temperature all particles crowd into the lowest state and lose all kinetic energy. For a Fermi-Dirac gas this is not possible. The particles in it are subject to the Pauli exclusion principle: No more than one may be in one quantum state. The lowest energy of a gas of N particles is therefore obtained if the N quantum states of lowest energy are filled with one particle in each. The energy E_0 of the gas at $T=0$ is therefore different from zero.

This quantity E_0 can be calculated easily. We designate by μ_0 the energy of an electron in the highest quantum state still filled at $T=0$. At zero temperature all quantum states of energy below μ_0 are occupied, and all quantum states with energy above μ_0 are empty. There must, then, exist precisely N states with energy lower than or equal to μ_0, and this condition is sufficient to determine μ_0. Since the volume is macroscopic, the translational states lie close together in momentum space, and we can replace summation over the translational quantum state **k** by integration over the classical phase space divided by h^3 (see Sec. 7b),

$$\sum_{\mathbf{k}} \Rightarrow gh^{-3}\int_V dr\int 4\pi p^2\, dp = Vgh^{-3}\int 4\pi p^2\, dp, \tag{11e.3}$$

where g is the number of internal quantum states of zero internal energy. This number is $g=2$ for electrons of spin $\frac{1}{2}$. Integrating eq. (3) from $p=0$ to p_0, the magnitude of momentum of the highest energy $\mu_0=(2m)^{-1}p_0^2$ filled at $T=0$, and equating this to N, one finds, with $\rho=N/V$,

$$N = Vgh^{-3}\frac{4\pi}{3}p_0^3 = Vgh^{-3}\frac{4\pi}{3}(2m\mu_0)^{3/2},$$

$$p_0 = \left(\frac{3}{4\pi g\rho}\right)^{1/3} h, \tag{11e.4}$$

$$\mu_0 = (2m)^{-1}p_0^2 = (2m)^{-1}h^2\left(\frac{3}{4\pi g\rho}\right)^{2/3}, \tag{11e.5}$$

or for electrons with $g=2$,

$$\mu_0 = \frac{h^2}{8m}\left(\frac{3\rho}{\pi}\right)^{2/3}, \qquad g=2. \tag{11e.5$'$}$$

This quantity μ_0, the uppermost energy of the filled cells, is called the Fermi energy.

The use of eq. (5′) in eq. (3) permits one to write an alternate form for the sum over quantum states as an integral over the energy, $\epsilon=(2m)^{-1}p^2$, namely,

$$\sum_{\mathbf{k}} \Rightarrow \tfrac{3}{2}N\int\left(\frac{\epsilon}{\mu_0}\right)^{1/2} d\frac{\epsilon}{\mu_0}. \tag{11e.3$'$}$$

It is to be noted that we define a quantum state by both the translational quantum numbers k_x, k_y, k_z and the internal quantum numbers of the particle, which in this case consist of the two spin directions. One sometimes defines a state by the translational quantum numbers only and says that *two* electrons, of opposite spin, may occupy this state. The difference, obviously, is only one of nomenclature.

The total energy of the N particles in this distribution, namely, the energy E_0 of the Fermi gas at $T=0$, is given by

$$E_0 = \tfrac{3}{2}N\mu_0\int_0^\infty\left(\frac{\epsilon}{\mu_0}\right)^{3/2} d\frac{\epsilon}{\mu_0} = \tfrac{3}{5}N\mu_0. \tag{11e.6}$$

The average energy per electron in the Fermi gas at $T=0$ is three-fifths of that of the energetically highest particle, or three-fifths of the Fermi energy μ_0.

The energy μ_0 depends inversely on the mass of the particles. By inserting for m the mass of the electron, for N_0 Avogadro's number, and for V the molal volume V in cubic centimeters of the electron gas, which last quantity is equal to the atomic volume of the metal divided by the number of valence electrons, one obtains

$$\begin{aligned} \mu_0 &= 4.1\times 10^{-11} V_{\mathrm{cm}}^{-2/3} \quad \text{erg molec}^{-1} \\ &= 26 V_{\mathrm{cm}^3}^{-2/3} \quad \text{eV}, \\ N_0\mu_0 &\cong 600\, V_{\mathrm{cm}^3}^{-2/3} \quad \text{kcal}, \end{aligned} \tag{11e.7}$$

and μ_0/k, a temperature, has the value

$$\frac{\mu_0}{k} = 3.02\times 10^5\, V_{\mathrm{cm}}^{-2/3} \quad \text{K}$$

Since V is about 10 cm^3 for most metals, it is seen that the Fermi energy of an electron gas is extremely high. In the next section it is shown that the thermodynamic properties of the gas above $T=0$ can be obtained as a power series in kT/μ_0. A series of this type must be expected to converge very rapidly, so that the behavior of the electron gas at room temperature is not greatly different from that at $T=0$.

Equation (5′) shows that both the small mass and the high density of the electron gas favor this high value of μ_0. Atoms or molecules have masses more than 2000 times that of an electron, so that the value of μ_0 for a chemical Fermi-Dirac gas, even at the same density, is very much smaller. A development with respect to kT/μ_0 for a chemical gas obeying the Pauli principle leads to a series that converges at very low temperatures only, and at room temperature the thermodynamic functions are radically different from those at $T=0$, in accordance with the discussions of Chapter 7.

The same results for the electron gas at zero temperature could have been obtained of course from the distribution function [eq. (2)]. In eq. (2) the quantity $\mu(T)$, a function of the temperature, is determined by the condition that the total number of particles is fixed. At $T=0$, $\beta=\infty$, eq. (2) before summation is zero if $\epsilon_{\mathbf{k}}>\mu(0)$; it is equal to unity if $\epsilon_{\mathbf{k}}<\mu(0)$. The distribution function states, then, that all states with energy lower than $\mu(0)$ are filled and all states with higher energy are empty. At $\epsilon=\mu(0)$ the function has a discontinuity, as it drops suddenly from unity to zero. It is seen then that the Fermi energy μ_0 of the filled level of highest energy is equal to the value μ occurring in eq. (2) at $T=0$. In Chapter 7 it was shown in general that the quantity μ in the distribution function is the chemical potential which is $1/N$ times the free energy G. The free energy of the electron gas at $T=0$ is therefore $G_0 = N\mu_0$. We verify this result by direct calculation of the various thermodynamic functions at $T=0$.

A quantum state of the total gas is described by giving the number of electrons in each quantum state. It is clear, then, that the distribution at $T=0$, when all cells of energy $\epsilon < \mu_0$ are full and all cells with $\epsilon > \mu_0$ are empty, can be realized by one state (or, owing to the possible small degeneracy of the very highest level with $\epsilon = \mu_0$, by a very few states) of the total system only. The entropy, $S = k \ln \Omega$, of the gas at $T=0$ is therefore practically zero, since Ω is equal to a small integer.

The work function, $A = E = TS$, is $A_0 = E_0$ for the gas in the lowest energy state. The pressure of the gas, $P = -(\partial A/\partial V)_T$, is obtained by using eq. (5′) in eq. (6) for $E_0 = A_0$,

$$P_0 = -\frac{\partial E_0}{\partial V} = -\tfrac{3}{5}N\frac{\partial \mu_0}{\partial V} = \frac{2}{5}\frac{N}{V}\mu_0 = \frac{2}{3}\frac{E_0}{V}. \qquad (11e.8)$$

It is seen that the relation $P_0 V = \tfrac{2}{3}E_0$ is precisely the same as that for the classical gas. A difference arises here from the fact that at $T=0$ neither P_0 nor E_0 is zero. The numerical value of μ_0 [eq. (7)] shows that

$$\begin{aligned} P_0 &= 10\times 10^{12} V^{-5/3} \quad \text{dyne/cm}^2 \\ &= 9.9\times 10^{6} V^{-5/3} \quad \text{atm} \end{aligned} \qquad (11e.8')$$

Remembering that V is of the order of 10 cm^3 for metals, one finds that the zero-point pressure in metals is very high. A gas at such high pressures can be contained only in a very strong box. For the electron gas, the metal itself provides this "box." On account of the strong attraction between the electrons and the remaining positive metal ions the potential energy of the electrons inside the metal is much lower than outside, so that the electrons remain confined to the metal. The electron pressure, tending to increase the volume, is balanced by the attraction between the ions and the electrons, which tends to decrease it.

The free energy, $G = A + PV$, is, at $T=0$, $G_0 = E_0 + P_0 V$. Using eq. (6) for E_0 and eq. (8) for P_0, one obtains

$$G_0 = E_0 + P_0 V = N\mu_0,$$

in agreement with the general identification of μ with G/N.

At temperatures above $T=0$ the distribution function equation (2) must be used for evaluation of the thermodynamic functions. Since the integrals cannot be evaluated in closed form, an approximation method is developed in the next section.

11f. THE ELECTRON GAS AT FINITE TEMPERATURE

For finite $\beta = 1/kT$ the number density of electrons in energy space $N(\epsilon)$ is obtained by multiplying eq. (11e.3′), $\sum_{\mathbf{k}} \Rightarrow \frac{3}{2}N\mu_0^{-3/2}\int \epsilon^{1/2}\, d\epsilon$, for the quantum state density, by $(e^{\beta(\epsilon-\mu)}+1)^{-1}$, for the number of electrons per quantum state,

$$N(\epsilon) = \tfrac{3}{2}N\mu_0^{-3/2}\epsilon^{1/2}(e^{\beta(\epsilon-\mu)}+1)^{-1}, \tag{11f.1}$$

in which μ_0, defined by eq. (11e.5′), is the chemical potential at $T = 0$, and μ is the chemical potential at the temperature in question.

The integration of eq. (1) over all values of ϵ serves to determine μ as a function of temperature by equating the result $\int_0^\infty N(\epsilon)\, d\epsilon$ to the total number of particles N. It is to be observed from eq. (1) for $N(\epsilon)$ that μ necessarily results as a function of μ_0 and β alone.

The energy may be determined from the equation

$$E = \int_0^\infty \epsilon N(\epsilon)\, d\epsilon.$$

It is seen that in general one is confronted with the problem of making integrations of the type

$$I = \int_0^\infty f(\epsilon)g(\epsilon)\, d\epsilon, \tag{11f.2}$$

where the function $f(\epsilon)$ is some simple continuous function of ϵ such as $\epsilon^{1/2}$ or $\epsilon^{3/2}$, and

$$g(\epsilon) = (e^{\beta(\epsilon-\mu)}+1)^{-1}. \tag{11f.2′}$$

We have already seen that, at $T = 0$, $\beta = \infty$, the function $g(\epsilon)$ is a step function, unity for $\epsilon < \mu_0$ and zero for $\epsilon > \mu_0$. Also at $T = 0$ the value of μ, $\mu = \mu_0$, is extremely high, with μ_0/k of the order of magnitude of 5×10^4 to 10^5 K for most metals.

It is not to be expected that μ will decrease enormously with temperature, so that at room temperature μ/kT is of the order of magnitude of 10^2, $e^{-\mu/kT} \approx 10^{-40}$. The approximation method used for *classical* gases is valid only if μ is negative and $e^{-\mu/kT}$ greater than unity. In this section integrals of the type in eq. (2) are evaluated under the assumption that $\mu/kT \gg 1$, and their values are obtained in terms of a power series in the small quantity kT/μ_0. The result will show, a posteriori, that for electrons in metals this assumption, $\mu/kT \gg 1$, is justified up to temperatures above those at which metals melt.

In order to integrate equations of the type in eq. (2) it is necessary to resort to a trick. Since, as we have found, $e^{-\mu/kT} \approx 10^{-40}$, $g(\epsilon)$ (eq. (2′)] is practically unity at $\epsilon = 0$ and decreases monotonically to zero at $\epsilon = \infty$. Its derivative $g'(\epsilon) = dg/d\epsilon$ is always negative, but has a single sharp minimum at $\epsilon = \mu$ as long as μ is positive. For $\mu/kT \gg 1$ this maximum of $-g'(\epsilon)$ is very sharp, and the function $-g'(\epsilon)$ is negligibly small for all values of ϵ differing greatly from $\epsilon = \mu$. By partial integration eq. (2) can be transformed into an integral over $-F(\epsilon)g'(\epsilon)$ and, because of the form of $-g'(\epsilon)$, only the values in the neighborhood of $\epsilon = \mu$ contribute to the integral. The limits of integration are actually from $\epsilon = 0$ to $\epsilon = \infty$ but, since $-g'(\epsilon)$ is practically zero for $\epsilon \leq 0$, no significant error is introduced by changing the limits of integration to minus infinity and plus infinity. With these limits the integration can be performed by developing the function $F(\epsilon)$ as a Taylor's series in powers of $\epsilon - \mu$ about the place of maximum $-g'(\epsilon)$.

The first and second derivatives of the function $g(\epsilon)$ in eq. (2′) are

$$g' = \frac{dg(\epsilon)}{d\epsilon} = -\frac{\beta e^{\beta(\epsilon-\mu)}}{(e^{\beta(\epsilon-\mu)}+1)^2} \tag{11f.3}$$

and

$$\begin{aligned} g''(\epsilon) = \frac{d^2 g(\epsilon)}{d\epsilon} &= -\frac{\beta^2 e^{(\epsilon-\mu)}}{(e^{\beta(\epsilon-\mu)}+1)^2} + \frac{2\beta^2 e^{2\beta(\epsilon-\mu)}}{(e^{\beta(\epsilon-\mu)}+1)^3} \\ &= \frac{\beta^2 e^{\beta(\epsilon-\mu)}(e^{\beta(\epsilon-\mu)}-1)}{(e^{\beta(\epsilon-\mu)}+1)^3} \end{aligned} \tag{11f.4}$$

The first derivative is always negative. The second derivative is zero when

$$e^{\beta(\epsilon-\mu)} - 1 = 0, \qquad \epsilon = \mu. \tag{11f.5}$$

At $\epsilon = \mu$ the function $-g'(\epsilon)$ has a maximum which is sharper the lower the temperature. The negative of the slope of the original function is greatest at this point.

Using $\epsilon = \mu$ in eqs. (2′) and (4), one finds that the value of the function $g(\epsilon)$ at this point is

$$g(\mu) = \tfrac{1}{2},$$

and that of its derivative is

$$g'(\mu) = -\tfrac{1}{4}\beta.$$

The logarithmic decrease of $g(\epsilon)$ with $\ln \epsilon$ is

$$-\frac{d \ln g(\epsilon)}{d \ln \epsilon} = \tfrac{1}{2}\beta\mu,$$

so that the relative abruptness of the descent of $g(\epsilon)$ from almost unity to almost zero increases with the value of $\beta\mu$.

The functions $g(\epsilon)$ and $g'(\epsilon)$ are plotted in Fig. 11f.1 for various values of kT.

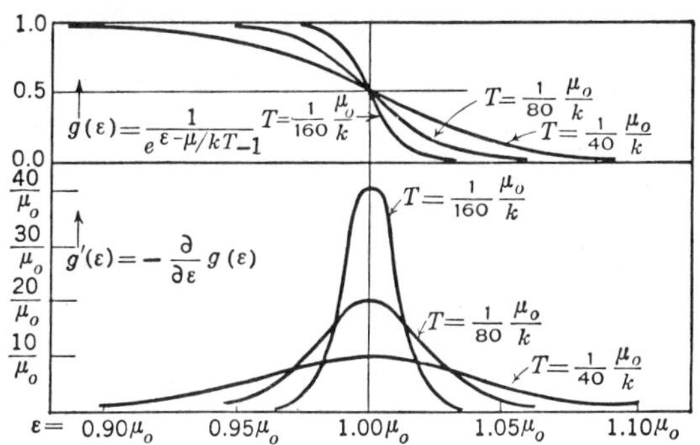

Figure 11f.1

By partial integration of the integral I of eq. (2) one finds

$$I=\int_0^\infty f(\epsilon)g(\epsilon)\,d\epsilon=F(\infty)g(\infty)-F(0)g(0)-\int_0^\infty F(\epsilon)g'(\epsilon)\,d\epsilon,$$

where

$$F(\epsilon)=\int_0^\varepsilon f(\epsilon')\,d\epsilon'. \tag{11f.6}$$

If $f(\epsilon)$ is not infinity at $\epsilon=0$, then $F(0)$ and the product $F(0)g(0)$ are zero. If $f(\epsilon)$ does not go exponentially to infinity with ϵ, the product $F(\infty)g(\infty)$ will be zero, since $g(\epsilon)$ approaches zero as $e^{-\beta\epsilon}$ with increasing ϵ. One may consequently write

$$I=\int_0^\infty f(\epsilon)g(\epsilon)\,d\epsilon=-\int_0^\infty F(\epsilon)g'(\epsilon)\,d\epsilon. \tag{11f.7}$$

We now transform to the new variable

$$x=\beta(\epsilon-\mu) \tag{11f.8}$$

and develop the function $F(x)$ as a power series of x,

$$F(x)=\sum_{n=0}^{\infty}\frac{x^n}{n!}F^{(n)}(x=0), \tag{11f.9}$$

where, in the old variable,

$$F(x=0)=\int_0^\mu f(\epsilon)\,d\epsilon,$$

$$F^{(n)}(x=0)=\beta^{-n}\left[\frac{d^nF(\epsilon)}{d\epsilon^n}\right]_{\epsilon=\mu}=\beta^{-n}\left[\frac{d^{n-1}f(\epsilon)}{d\epsilon^{n-1}}\right]_{\epsilon=\mu}$$
$$=\beta^{-n}f^{(n-1)}(\mu). \tag{11f.10}$$

By introducing this development into eq. (7) one may write

$$I = -F(0)\int_0^\infty f'(\epsilon)\,d\epsilon - \sum_{n=1}^{\infty} \frac{\beta^{-n}}{n!} f^{(n-1)}(\mu)\int_{x=-\mu/kT}^{\infty} x^n g'(x)\,dx. \quad (11f.11)$$

The integral of the first term is

$$-\int_0^\infty g'(\epsilon)\,d\epsilon = g(0) - g(\infty) = (1+e^{-\beta\mu})^{-1} \cong 1. \quad (11f.12)$$

The function $g'(x)$ is obtained by using the expression from eq. (8) for x in eq. (3),

$$g'(x) = -\frac{\beta e^x}{(e^x+1)^2} = -\frac{\beta}{(e^x+1)(e^{-x}+1)}. \quad (11f.13)$$

This function is completely symmetric in x, that is, $g'(x) = g'(-x)$. The function approaches zero exponentially as x approaches minus infinity. If $\beta\mu$ is large, the value of the function will be already negligible at the lower limit, $x = -\beta\mu$, of the integral in eq. (11). No error is introduced, consequently, by changing the limits of integration of the terms in the sum of eq. (11) to $x = -\infty$ and $x = +\infty$.

We must now evaluate the integrals

$$\int_{-\infty}^{+\infty} \frac{x^n}{(e^x+1)(e^{-x}+1)}\,dx.$$

From the symmetry of the denominator it is seen that the integrand is antisymmetric in x if n is odd, that is, it changes sign if x is replaced by $-x$, and the integral is therefore zero for odd n. For even values of n we may integrate from zero to infinity and multiply by 2, since then the integrand is symmetric in x.

By developing

$$\frac{1}{(e^x+1)(e^{-x}+1)} = \frac{e^{\ x}}{(1+e^{-x})^2} = e^{-x} - 2e^{-2x} + 3e^{-3x} - \cdots$$

$$= -\sum_{m=1}^{\infty} (-1)^m m e^{-mx},$$

the integration may be performed as

$$\int_{-\infty}^{+\infty} \frac{x^n}{(e^x+1)(e^{-x}+1)}\,dx = -2\sum_{m=1}^{\infty} (-1)^m m \int_0^\infty x^n e^{-mx}\,dx$$

$$= -2n! \sum_{m=1}^{\infty} \frac{(-1)^m}{m^n}, \qquad n \text{ even.} \quad (11f.14)$$

Using eqs. (13) and (12) in eq. (11), with eqs. (10) and (14), one finally arrives at

$$I=\int_0^{\infty} f(\epsilon)g(\epsilon)\,d\epsilon=-\int_0^{\infty} F(\epsilon)g'(\epsilon)\,d\epsilon$$

$$=\int_0^{\mu} f(\epsilon)\,d\epsilon-2\sum_{n=1}^{\infty}(kT)^{2n}f^{(2n-1)}(\mu)\sum_{m=1}^{\infty}\frac{(-1)^m}{m^{2n}},$$

$$f^{(2n-1)}(\mu)=\left[\frac{d^{2n-1}f(\epsilon)}{d\epsilon^{2n-1}}\right]_{\epsilon=\mu}. \tag{11f.15}$$

The sums occurring have the numerical values

$$-\sum_{m=1}^{\infty}\frac{(-1)^m}{m^2}=\frac{\pi^2}{12},\qquad -\sum_{m=1}^{\infty}\frac{(-1)^m}{m^4}=\frac{7\pi^4}{720}, \tag{11f.16}$$

so that

$$I=\int_0^{\infty}\frac{f(\epsilon)}{e^{(\epsilon-\mu)/kT}+1}\,d\epsilon$$

$$=\int_0^{\mu} f(\epsilon)\,d\epsilon+\frac{\pi^2}{6}\beta^{-2}\left(\frac{df}{d\epsilon}\right)_{\epsilon=\mu}+\frac{7\pi^4}{360}\beta^{-4}\left(\frac{d^3f}{d\epsilon^3}\right)_{\epsilon=\mu}+\cdots. \tag{11f.15$'$}$$

Equation (15′) is now applied to calculate μ. Using eq. (1) and

$$N=\int_0^{\infty} N(\epsilon)\,d\epsilon=\frac{3N}{2\mu_0^{3/2}}\int_0^{\infty}\frac{\epsilon^{1/2}\,d\epsilon}{e^{(\epsilon-\mu)/kT}+1}, \tag{11f.17}$$

one finds that $f(\epsilon)=\epsilon^{1/2}$ in this problem. The integral $F(\mu)$ is

$$\int_0^{\mu} f(\epsilon)\,d\epsilon=\int_0^{\mu}\epsilon^{1/2}\,d\epsilon=\tfrac{2}{3}\mu^{3/2}.$$

The derivatives are

$$\left(\frac{df}{d\epsilon}\right)_{\epsilon=\mu}=\tfrac{1}{2}\mu^{-1/2},\qquad \left(\frac{d^3f}{d\epsilon}\right)_{\epsilon=\mu}=\tfrac{3}{8}\mu^{-5/2}.$$

Using these with eq. (15′) in eq. (7), one finds

$$\left(\frac{\mu}{\mu_0}\right)^{3/2}\left[1+\frac{\pi^2}{8}(\beta\mu)^{-2}+\frac{7\pi^4}{640}\beta\mu)^{-4}+\cdots\right]=1, \tag{11f.18}$$

which determines μ as a function of μ_0 and T.

In order to make the equation explicit in μ the development

$$\frac{1}{(1+x)^{2/3}}=1-\frac{2x}{3}+\frac{5x^2}{9}-\cdots$$

is used to obtain the form

$$\mu=\mu_0\left[1-\frac{\pi^2}{12}(\beta\mu)^{-2}+\frac{\pi^4}{720}(\beta\mu)^{-4}+\cdots\right],$$

and in this $\mu^{-2}=\mu_0^{-2}\,[1+(\pi^2/6)(\beta\mu)^{-2}]$ is substituted in the quadratic term. In the quartic term, which is the last correction, μ_0 is simply substituted for μ. One obtains

$$\mu=\mu_0\left[1-\frac{\pi^2}{12}(\beta\mu_0)^{-2}-\frac{\pi^4}{80}(\beta\mu_0)^{-4}+\cdots\right] \tag{11f.19}$$

as an equation for μ, the chemical potential, in terms of β and μ_0, the chemical potential at $T=0$, is given in turn as a function of the volume V by eq. (11e.7).

The energy E may be calculated by using the relationship

$$E=\int_0^\infty \epsilon N(\epsilon)\,d\epsilon=\frac{3N}{2\mu_0^{3/2}}\int_0^\infty \frac{\epsilon^{3/2}\,d\epsilon}{e^{(\epsilon-\mu)/kT}+1} \tag{11f.20}$$

from eq. (1).

In this integral $f(\epsilon)=\epsilon^{3/2}$, and

$$\int_0^\mu f(\epsilon)\,d\epsilon=\tfrac{2}{5}\mu^{5/2},$$

$$\frac{df}{d\epsilon}=\tfrac{3}{2}\mu^{1/2},\qquad \frac{d^3f}{d\epsilon^3}=-\frac{3}{8}\frac{1}{\mu^{3/2}},$$

so that

$$E=\tfrac{3}{5}N\frac{\mu}{\mu_0}^{3/2}\mu\left[1+\frac{5\pi^2}{8}(\beta\mu_0)^{-2}-\frac{7\pi^4}{384}(\beta\mu_0)^{-4}+\cdots\right]$$

is obtained.

By using eq. (19) to replace μ with μ_0, one finds for the energy

$$E=\tfrac{3}{5}N\mu_0\left[1+\frac{5\pi^2}{12}(\beta\mu_0)^{-2}-\frac{\pi^4}{16}(\beta\mu_0)^{-4}+\cdots\right]. \tag{11f.21}$$

The easiest way to obtain the other thermodynamic functions is to use the relation $PV=\frac{2}{3}E_{\text{kin}}$, valid for all nonrelativistic perfect gases. This relation is most easily derived from eq. (11e.1), $\beta PV=\sum_{\mathbf{k}}\ln(1+e^{\beta(\mu-\epsilon_k)})$, and noting, see for instance eq. (11e.3′), that summation over $\mathbf{k}$ can be replaced by $\sum_{\mathbf{k}}\Rightarrow A\int\epsilon^{1/2}\,d\epsilon$ so that partial integration of PV yields

$$PV=A\int_0^\infty \epsilon^{1/2}\,d\epsilon\,\ln(1+e^{\beta(\mu-\epsilon)})=\tfrac{2}{3}A\int_0^\infty \epsilon^{3/2}(e^{\beta(\epsilon-\mu)}+1)^{-1}\,d\epsilon=\tfrac{2}{3}E \tag{11f.22}$$

from the fact that $\epsilon^{3/2}\ln(1+e^{\beta(\mu-\epsilon)})$ is zero at $\epsilon=0$ and at $\epsilon=\infty$, where $\ln(1+e^{\beta(\mu-\epsilon)})\sim e^{\beta(\mu-\epsilon)}$.

We then have, from eq. (19),

$$G = N\mu = N\mu_0\left[1-\frac{\pi^2}{12}(\beta\mu_0)^{-2}-\frac{\pi^4}{80}(\beta\mu_0)^{-4}-\cdots\right], \tag{11f.23}$$

$$H=\tfrac{5}{3}E = N\mu_0\left[1+\frac{5\pi^2}{12}(\beta\mu_0)^{-2}-\frac{\pi^4}{16}(\beta\mu_0)^{-4}+\cdots\right], \tag{11f.24}$$

$$TS = H-G = N\mu_0\left[\frac{\pi^2}{2}(\beta\mu_0)^{-2}-\frac{\pi^4}{20}(\beta\mu_0)^{-4}+\cdots\right], \tag{11f.25}$$

$$PV=\tfrac{2}{3}E=\tfrac{2}{5}N\mu_0\left[1+\frac{5\pi^2}{12}(\beta\mu_0)^{-2}-\frac{\pi^4}{16}(\beta\mu_0)^{-4}+\cdots\right], \tag{11f.26}$$

$$A = E-TS=\tfrac{3}{5}N\mu_0\left[1+\frac{5\pi^2}{12}(\beta\mu_0)^{-2}+\frac{\pi^4}{48}(\beta\mu_0)^{-4}+\cdots\right]. \tag{11f.27}$$

The heat capacity at constant volume is easily obtained from the expression for E, with $\beta=1/kT$, to be

$$C_V=\tfrac{3}{2}Nk\frac{\pi^2}{3}(\beta\mu_0)^{-1}\left[1-\frac{3\pi^2}{10}(\beta\mu_0)^{-2}+\cdots\right], \tag{11f.28}$$

and is linear in T at low temperatures. The factor $\frac{3}{2}Nk$ in eq. (28) is the classical value of C_V for a perfect monatomic gas. In eq. (11e.7) the value of μ_0/k was given as $3\times10^5\,V^{-2/3}$ K, so that

$$\frac{\pi^2}{3}(\beta\mu_0)^{-1}\cong 10^{-5}\,V_{\text{cm}^3}^{2/3}.$$

With V of the order of 10 cm^3 at room temperature, $T=300$, the heat capacity at constant volume due to the electron gas is about 1.4% of the classical value for a univalent metal.

At very low temperatures the lattice heat capacity varies as T^3, hence becomes lower than the linear contribution due to the electrons. The crossover occurs experimentally in metals at somewhere between 1 and 10 K.

11g. METALS

The postulate that metals owe their electrical conductivity and high heat conductivity to an essentially free electron gas within them is due primarily to Drude.[(1)] The great success of the idea was that it explained,

[(1)] P. Drude, *Ann. Phys.*, **1,** 566 (1900).

even semiquantitatively, the Wiedemann-Franz law that the ratio of the heat conductivity κ to the electrical conductivity σ is proportional to the temperature with the *same proportionality constant* for all metals.

The idea is that the electrons move freely within the metal, suffering occasional collisions with atomic ions in the fixed metal crystal lattice, having therefore a mean free path λ. Calculation of the heat conductivity due to the electrons and the electrical conductivity can then be made similarly to that for a perfect classical gas. The mean free path cancels in the ratio, as does the number density ρ of the electrons. A very simple but inexact calculation gives

$$\frac{\kappa}{\sigma}=\frac{\pi^2}{3}\left(\frac{k}{e}\right)^2 T, \tag{11g.1}$$

in which not only the known experimental proportionality to T is given but also the magnitude of the proportionality constant. The success was striking. The heat conductivity of a metal is enormously greater than that of a nonmetallic crystal. For instance, the heat conductivity of silver in calories per square centimeter per gradient of 1°C per centimeter is about unity, whereas that of glass is about 2×10^{-3}. The heat conductivity of an insulator is due to the lattice vibrations, which of course also conduct some of the heat in metals. However, at least in the heavy metals, this can be neglected compared to the enormously greater electronic conduction. Experimentally, the ratio κ/σ is constant for different metals at the same temperature to within about 15%, and the average ratio agrees quite well with eq. (1).

Later Lorentz[(1)] amplified the theory and pointed out severe difficulties in its acceptance. Both conductivities κ and σ should be proportional to the product of the mean free path λ and the number density ρ. Both conductivities are high, but the contribution of the free electron gas of number density ρ to the heat capacity C_V should be $\frac{3}{2}V\rho k$ for a classical gas. Since the value of C_V for metals is almost that of the Dulong-Petit rule, $3k$ per atom, one concludes that the number density of free electrons must be negligible compared to the number of metal atoms per unit volume. This had two unpleasant consequences: first, the values of the conductivities would require the mean free paths, already rather unbelievably large, to be completely incredible and, second, a low value of ρ would be understandable only if the free electrons consisted of a very small fraction of the total number of electrons that were thermally excited. In this latter case their concentration should increase and the

(1) H. A. Lorentz, *Proc. Acad. Sci. Amsterdam*, **7,** 438, 585, 684 (1905).

conductivities should increase exponentially with T. Actually, the electrical conductivity decreases with increasing temperature.

In 1928 Arnold Sommerfeld explained the solution to all the fundamental discrepancies by treating the electron gas as a perfect Fermi-Dirac system. The low heat capacity of the electron gas was immediately explained; see eq. (11f.28) and the subsequent discussion. In many respects the Fermi-Dirac gas presents an easier treatment than the classical Boltzmann gas. Because of the Pauli exclusion principle the electrons of energy appreciably below the energy μ_0 cannot be excited into translational states of nearly the same energy, since these are already filled; and, since collisions of the very light electrons with the heavy ions are close to elastic with little change in energy, these low-lying electrons play no role. Only the electrons close to the Fermi surface of energy μ_0 take part in conduction, where by close we mean those having energies not much different from μ_0 than a few kT. Since kT/μ_0 is very small, this means that their energies are all the same to within a percent or two, and their mean free paths λ, which would be expected to depend considerably on their velocities, should be nearly the same.

The only temperature-dependent quantity entering the electrical conductivity σ is the mean free path λ. Quantum-mechanical calculation shows that for a strictly periodic potential, such as that caused by the ions of a perfect crystal at absolute zero, with all the ions in their exact equilibrium positions, the mean free path of the electrons is infinite. Collisions of the electrons with the ions are due only to imperfections in the lattice. These imperfections arise from two causes: impurities in the crystal and temperature motion.

For the lattice without impurities, the interaction potential between ions and electrons is proportional to the amplitude of vibration of the ions, which in turn is proportional to $T^{1/2}$ near room temperature. The number of collisions per unit time is proportional to the square of the interaction potential and therefore to T. The effect of impurities is to cause a constant, temperature-independent contribution to the number of collisions per unit time. The mean free path is inversely proportional to the number of collisions, so that

$$\lambda = \frac{\lambda_0}{a+T}. \tag{11g.2}$$

The temperature-independent constant a is highly dependent on the amount of impurity in the metal and presumably is zero for a completely pure metal.

As a result, since σ is proportional to λ, it is seen that the temperature dependence of the electrical conductivity is given as

$$\sigma = \frac{A}{B+T}. \tag{11g.3}$$

The resistivity σ^{-1} is linear in the temperature. This is experimentally observed.

At low temperatures, at which the lattice vibrations are quantized, the conclusion that the amplitudes of vibration vary with $T^{1/2}$ is unjustified, and the temperature dependence of σ is more complicated than that of eq. (2).

Although eq. (2) leads to infinite electrical conductivity for an absolutely pure metal at $T = 0$, this prediction is not in agreement with the observed superconductivity at extremely low temperatures. In the observed phenomenon, the conductivity increases discontinuously to infinity at a sharply defined temperature. Superconductivity is due to an interaction between the electrons and the lattice ions, partly but not solely with the phonons or vibrational waves of the lattice. This in turn leads to an effective interaction between electrons having equal magnitude but opposite direction of momentum. Such an electron pair having two fermions is a boson, and the pairs obey Bose-Einstein statistics. Superconductivity has considerable analogy to Bose-Einstein condensation and superfluidity in liquid helium.

The treatment of the electron gas as moving in a completely constant potential within the metal is obviously a zeroth order approximation. The potential within the metal is periodic, with deep minima at the position of the positive ions. The problem of the states of an electron in a periodic potential was first treated by Bloch[(1)] and has since been carried out in great detail for a considerable number of metals. The wave function of the electron at each ion has the general form of the function for the same quantum number *ns*, *np*, $\cdots$ as in the free atom but with a slightly different energy, so that it connects to that in the neighboring atom function with no discontinuity in value or slope. The free-electron wave function of the metal is then a sum of these modified atomic functions with amplitudes proportional to $\sin(\pi k_x/L)$, $\sin(\pi k_y/L)$, $\sin(\pi k_z/L)$ in a cubic crystal of edge L having a crystal structure of cubic symmetry.

The energy of the longest wavelength state, $k_x = k_y = k_z = 1$, is the appropriate potential energy for use in the constant-potential free-electron approximation. For relatively low $\mathbf{k}$ values the distribution, especially in the monovalent alkali metals, sodium, potassium, and so on,

[(1)] Felix Bloch, *Z. Phys.*, **52,** 555 (1928).

obeys very closely that of the constant-potential form [eq. (11e.3′)] but is improved by the introduction of an effective mass m_{eff} for the electron mass slightly different from the true mass in the expression in eq. (11e.5′) for μ_0. At short wavelengths, when one of the k_α values approaches the value $N^{1/3}$ of the number of atoms along that axis, the energies become markedly perturbed and eq. (11e.5′) is significantly altered. There is a definite energy gap between the states, say, of fixed k_y, k_z, $k_x = N^{1/3} - n$ and those of k_y, k_z, $k_x = N^{1/3} + n$, with n small.

Actually, in order to treat problems involving conduction it is far more convenient, and indeed necessary, to use progressive wave functions rather than the stationary states $\sin(\pi k_x xL), \ldots$. The three-dimensional progressive waves in a cubic box are, with k_z, k_y, and k_z running from minus to plus infinity,

$$(k_x, k_y, k_z) = \exp[2\pi i L^{-1}(k_x x + k_y y + k_z z)], \tag{11g.4}$$

for which the wavelength λ_x is L/k_x, and

$$p_x = \frac{h}{2\pi i}\frac{\partial}{\partial x}\psi = \frac{h}{L}k_x = \frac{h}{\lambda_x}. \tag{11g.5}$$

The total wavelength λ is given by

$$\lambda^{-2} = (\lambda_x^2 + \lambda_y^2 + \lambda_z^2)^{-1} = h^{-2}(p_x^2 + p_y^2 + p_z^2). \tag{11g.6}$$

In the crystal of cubic symmetry the energy gaps due to the periodic potential occur along the surfaces of cubes containing N atoms, each having $2N$ electronic states; in the cube of lowest energy these surfaces occur when one or more of the k_α's are $N^{1/3}$ or $-N^{1/3}$. Since the unperturbed energy is proportional to $k_x^2 + k_y^2 + k_z^2$, the energy before the perturbation at the center of the cube face is one-half of that at the center of an edge and only one-third of that at a corner. If the gap is not very large, the lower energies of the neighboring zones overlap the highest energy of the lowest cubic zone and there is always a continuum of free-electron states at all energies. For crystals of more complicated structure there are similar cells in the momentum space with gaps at the interfaces called Brillouin zones.

The alkali metals, lithium, sodium, potassium, and so on, with a single *ns* electron, just half fill the first zone. The density of states follows the constant-potential equation quite well up to the Fermi energy μ_0. They are excellent conductors. The alkaline earth metals beryllium, magnesium, and calcium have two valence electrons, enough to completely fill the *ns* electron zone. The *np* levels are not greatly higher in energy and well overlap the *s* level in the metal. The conductivity is high. If the energy gap between the different zones is very high and the number of electrons per atom is even, there may be no overlap of energy and the material is an insulator. This is the case in the lighter elements, carbon (diamond), silicon, and germanium of the fourth column of the periodic table. In other cases, for instance, hydrogen and the halides, fluorine, chlorine, and so on, the tendency to form strong homopolar bonded diatomic molecules is so strong that the stable structure is molecular. The electronic excited states have so high an energy that the material is an insulator. The a priori calculation of which structural form will be the most stable for a given element or combination of elements is a quantum-mechanical problem of enormous complexity, for which only the very simplest cases can be said to be even approximately treated with convincing rigor. For instance, it is known with some certainty that at sufficient pressure hydrogen is metallic, but estimates of the actual P-T curve of the transition vary widely. On the other hand, with the use of some semiempirical relationships a posteriori understanding is often quite satisfactory.

11h. SEMICONDUCTORS

In view of the enormous importance of semiconductors in all modern electronic circuitry it would be a startling omission to discuss the behavior of free electrons in solid materials, that is, of electronically conducting solids, without even mentioning their existence. However, the amount of research and detailed information, and the number and complexity of the phenomena observed, are so great that it seems inappropriate in a book of this type to do much more than give a brief qualitative description of their nature.

We have already mentioned that the lighter elements of column IVa in the periodic table, carbon in the diamond lattice and silicon, conduct negligibly at room temperature, but even pure germanium is a semiconductor at room temperature. Actually, both germanium and silicon crystals of commercial quality conduct electricity considerably. Absolutely pure crystals conduct less.

In silicon and germanium the lowest energy zone is completely filled, and there is no overlap in energy with the next excited levels, although the gap is relatively small. At sufficiently high temperatures electrons are excited into the next higher conduction band, and there is electrical conduction increasing exponentially with temperature. These are *intrinsic* semiconductors. This is *not* the phenomenon of the commercially important semiconductors. The latter are carefully prepared crystals with a closely controlled very small amount of dissolved impurity, one of the elements in either column III or V of the periodic table. Germanium, for instance, has an atomic radius of 1.22 Å (10^{-8} cm). Gallium in column III has a radius of 1.25 Å, and arsenic in column V a radius of 1.21 Å; both dissolve substitutionally in germanium at low concentrations. If gallium is dissolved, there are fewer electrons than fill the lowest band and, if arsenic is present, there are more.

The result is a band of rather sharp energy of either electrons or of holes. The cases of excess electrons are known as n-type conductors. If there is a deficiency of electrons, the conduction is p-type conduction, and the holes in the band behave like positive charges. Since the properties can be carefully controlled by the composition, various exotic and extremely useful effects can be obtained especially at junctions between n- and p-type conductors, including almost perfect rectification.

The details of the band structure have been studied in great detail in a large number of cases.

11i. THERMIONIC EMISSION, THE RICHARDSON EFFECT

The numerical values of the energies in metals are high. From eq. (11e.7) we see that the Fermi energy μ_0 is several tens of electronvolts, which is several hundred kilocalories per mole. The work function w of a metal is the negative difference between the energy of an electron at the top of the Fermi sea μ_0 and that of an electron of zero kinetic energy u in the vacuum near the metal surface, $w = u - \mu_0$. The values of w vary from about 2.3 eV for the alkalis to over 5 eV for platinum and are even higher for metals having an oxide layer on the surface.

The thermionic emission, namely, the rate of emission of electrons from the surface of a heated clean metal surface, can be calculated readily under a number of assumptions which are often reasonably well satisfied and serve to give measurements of the work function.

At equilibrium the chemical potential μ of the electrons in the evacuated space just outside the metal surface is equal to that in the metal which, given by eq. (11f.19) with (11e.7) for μ_0, even at 1000 to 2000 K, is quite close to μ_0. It is then $-w$ if measured from the electric potential of a zero kinetic energy electron at the surface just outside the metal. By keeping a relatively small positive electric gradient at the surface the electrons can be prevented from building up a space charge at the surface. The electrons leaving the surface can be assumed to be equal to those that would strike the surface from a hypothetical equilibrium electron gas at zero electric potential. The most dubious assumption is that the reflection coefficient r is zero, and all but $1 - r$ "stick." The emissivity is

$1-r$ times the number calculated. For most clean metal surfaces the assumption that r lies close to zero is probably correct; quantum-mechanical calculations indicate that it should be zero.

The actual calculation is very simple with previously derived equations. From eq. (7b.1) for the low pressure, and therefore classical electron gas, with internal partition function, $Q_i = 2$, due to the two spin directions, the pressure P is

$$P = 2kT\left(\frac{2\pi mkT}{h^2}\right)^{3/2} e^{-\beta w},$$

whereas the number Z hitting the unit surface in unit time [eq. (2d.4)] is

$$Z = \frac{P}{(2\pi mkT)^{1/2}},$$

so that the current $I = Ze$ is

$$I = 4\pi em\left(\frac{kT}{h}\right)^2 e^{-\beta w}.$$

Numerical evaluation leads to

$$I = 120\ T^2 e^{-\beta w} \qquad \text{amp cm}^{-2},$$

$$\log_{10} I\ (\text{amp cm}^{-2}) = 2.08 + 2\log_{10} T - \frac{w}{2.303kT}$$

$$= 8.08 + 2\log_{10}\frac{T}{1000} - 5.03\,w(\text{eV})\left(\frac{T}{1000}\right)^{-1}.$$

For tungsten with a work function listed as $4.58 + 0.015(T/1000)$ at 2000 K the emission is about 1.5×10^{-3} amps cm^{-2}.

CHAPTER 12

ELECTRIC AND MAGNETIC FIELDS

12a. INTRODUCTION

In all problems considered up to now the only variables of the system, besides the numbers of molecules or of atoms, were two in number and could be chosen as E and V. Instead of the extensive properties E and V one could use intensive variables, $\beta = 1/kT$, or βP, as well as replacing the numbers $N_a, \ldots$ of molecules by $\beta\mu_a$, provided only that one extensive variable was kept to define the size of the system. What we call intensive variables are related to what are frequently called thermodynamic forces (see Sec. 3h).

In general, a system may be influenced by various other variables corresponding to extensive properties other than energy and volume. For instance, a rigid crystalline solid may be subjected to various tensions which result in an alteration of its shape, without necessarily involving a change in volume. The amount of force applied is related to the displacement in these cases by one of the several elastic constants of the body.

In this chapter we are concerned with the effect of electric and magnetic forces applied to the systems by means of external electric or magnetic fields. The calculations are at first made in the simplest possible manner. In later sections the more general method of development by which more complicated problems might be treated is indicated.

Maxwell's equations for the vector electric field $\mathscr{E}$ and electronic displacement vector $\mathscr{D}$, combined with those for the corresponding magnetic vector field quantities $\mathscr{H}$ and $\mathscr{B}$, are

$$\mathbf{curl}\ \mathscr{E} = \frac{-c^{-1}\,\partial\mathscr{B}}{\partial t}, \qquad \mathbf{div}\ \mathscr{D} = 4\pi\rho, \tag{12a.1}$$

with c the light speed in vacuum, and ρ the charge density, and with $\mathbf{i}$ the vector current density,

$$\mathbf{curl}\ \mathscr{H} = c^{-1}\left(4\pi\mathbf{i} + \frac{\partial\mathscr{D}}{\partial t}\right), \qquad \mathbf{div}\ \mathscr{B} = 0. \tag{12a.2}$$

These equations require two experimentally determined constants dependent on the medium in which the fields propagate. The two constants are the dielectric constant $\boldsymbol{\epsilon}$ and the magnetic permeability $\boldsymbol{\mu}$. The two necessary equations are

$$\mathscr{D} = \boldsymbol{\epsilon}\mathscr{E} \qquad \mathscr{H} = \boldsymbol{\mu}\mathscr{B}. \tag{12a.3}$$

In vacuum both $\boldsymbol{\epsilon}$ and $\boldsymbol{\mu}$ are unity, so that $\mathscr{E} \equiv \mathscr{D}$ and $\mathscr{H} \equiv \mathscr{B}$.

The equations take no cognizance of the atomic or molecular nature of the medium. The fact that matter is composed of charged particles, nuclei and electrons, was of course unknown to Maxwell. Lorentz[(1)] was the first, or one of the first, to attempt the more modern viewpoint of defining $\boldsymbol{\epsilon}$ and $\boldsymbol{\mu}$ by eq. (3) and, since the space between the point charges is a vacuum, to use microscopic electric and magnetic fields $\mathbf{e}$ and $\mathbf{h}$ which are functions of position $\mathbf{r}$ in the material, and to derive the values of $\boldsymbol{\epsilon}$ and $\boldsymbol{\mu}$ from the action of these fields in displacing the charged particles within the molecules. One of the best more modern treatises is still probably that of Van Vleck.[(2)]

A system composed of independent molecules possessing rigid permanent dipoles and subjected to an electric field $\mathscr{E}$ is treated first.

The electric field, of dimensions force divided by charge, which is the same as charge divided by length squared, gives the magnitude and direction of the electric force exerted on unit charge. $\mathscr{E}$ is therefore a vector quantity, but since only relatively simple problems are dealt with, in which the field direction is kept constant, and in isotropic media for

[(1)] See, for instance, H. A. Lorentz, *The Theory of Electrons*, Lectures delivered at Columbia University, New York, spring 1906. Published in 1909 by B. G. Teubner, Leipzig and Columbia University Press.

[(2)] J. H. Van Vleck, *Electric and Magnetic Susceptibilities*, Oxford University Press, 1932.

which the effects are independent of the direction of the field, the equations used involve only the magnitude of the field, which is written $\mathscr{E}$.

The force acting, per unit charge, on the charged part of a molecule in a dense medium is not in general the same as the external applied field $\mathscr{E}$. At first we limit ourselves to a dilute gas, where the difference between the local field $\mathscr{E}_{\text{loc}}$ and the external field $\mathscr{E}$ is negligible. The correction for this effect is discussed later.

12b. RIGID DIPOLES IN A CONSTANT AND UNIFORM ELECTRIC FIELD

The dipole moment of a molecule is defined as a vector pointing from the center of negative charge to that of positive charge. The magnitude p_0 of the dipole moment has the dimensions charge times length. In atoms, which consist of a positive nucleus and a spherically symmetric distribution of electrons, the centers of the positive and negative charges are located at the same place. The dipole moment is zero. This is not true of nonsymmetric molecules. The total amount of negative and positive charge in a neutral molecule must of course be equal, namely, ze, where z signifies the number of electrons and e the magnitude of electronic charge. If the distance between the centers of positive and negative charge is called l, the magnitude of the dipole moment p_0 is defined as the product lze. Actually, a dipole field is defined as the field around such an object in the limit $l \to 0$, lze constant. If l is finite, the molecule will also have quadruple and higher moments, but these are normally small.

The electric field tends to orient the dipoles in space. The potential energy of the molecule in the field does not depend on the position of the center of mass, but only on the angle of orientation. If θ is the angle between the axis of the molecule and the direction of the field, so chosen that when $\theta = 0$ the positive end of the molecule is directed toward the negative plate producing the field, then the potential energy of the molecule in the field $\mathscr{E}$ will be

$$u(\theta) = -lze\mathscr{E} \cos \theta = -p_0\mathscr{E} \cos \theta. \tag{12b.1}$$

The magnitude of the dipole moment p_0 in a real molecule depends somewhat, but in most diatomic molecules at least not greatly, on the amplitude of vibration or on the quantum state of vibration. This effect is neglected at present.

In the absence of a field the probability that any axis of the molecule lies in a certain solid angle range $d\Omega$ with respect to fixed coordinates in space is just proportional to the solid angle range $d\Omega$. The solid angle range $d\Omega$ corresponding to values of θ between θ and $\theta + d\theta$ is proportional to $\sin\theta\, d\theta$. The angle θ may vary between 0 and π. The integral of $\sin\theta\, d\theta$ from 0 to π is 2,

$$\int_0^\pi \sin\theta\, d\theta = 2,$$

so that, if $w(\theta)\, d\theta$ is used for the probability that the angle θ lies between θ and $\theta + d\theta$, one may write

$$w(\theta)\, d\theta = \tfrac{1}{2}\sin\theta\, d\theta \qquad \text{(zero field)}. \tag{12b.2}$$

The probability $w(\theta)\, d\theta$ of the orientation is altered in the presence of the field by the dependence of the energy of the molecule on the angle θ [eq. (1)]. The average density of molecules in the element dq of the configuration space is proportional to $e^{-\beta u(q)}\, dq$, where $u(q)$ is the potential energy as a function of the position q in the coordinate space. The probability $w(\theta)$ that a molecule will have the orientation θ is then proportional to the configuration volume corresponding to that angle [eq. (2)] times the exponential $e^{-\beta u(q)}$. In this, $u(\theta)$, the part of the potential energy that depends on θ and θ alone, is given by eq. (1). Since $w(\theta)\, d\theta$ must be unity when integrated over all angles θ from 0 to π, the equation has to be normalized by division with this integral.

One may therefore write, in the field $\mathscr{E}$,

$$w(\theta)\, d\theta = \tfrac{1}{2}\sin\theta\, d\theta e^{-\beta u(\theta)}\left(\int_0^\pi \tfrac{1}{2}\sin\theta\, d\theta e^{-\beta u(\theta)}\right)^{-1}. \tag{12b.3}$$

Expression (3) gives the classical probability that the angle θ between the dipole moment p_0 and the field $\mathscr{E}$ lies between θ and $\theta + d\theta$, if $u(\theta)$ is given by eq. (1), independently of the complications of the molecular structure, provided only that the dipole is rigid, that is, not stretched or altered in magnitude by the field.

The integral in the denominator of eq. (3) may be evaluated exactly, using eq. (1) and $\cos\theta = \zeta$, $\sin\theta\, d\theta = -d\zeta$,

$$\begin{aligned}\int_0^\pi e^{\beta p_0 \mathscr{E}\cos\theta}\tfrac{1}{2}\sin\theta\, d\theta &= \tfrac{1}{2}\int_{-1}^{+1} e^{\beta p_0\mathscr{E}\zeta}\, d\zeta \\ &= (2\beta p_0\mathscr{E})^{-1}(e^{\beta p_0\mathscr{E}} - e^{-\beta p_0\mathscr{E}}) \\ &= (\beta p_0\mathscr{E})^{-1}\sinh(\beta p_0\mathscr{E}).\end{aligned}$$

Expanding $e^x = 1 + x + \frac{1}{2}x^2 + \cdots$ the result may be written

$$\begin{aligned}\int_0^\pi e^{-\beta u(\theta)}\tfrac{1}{2}\sin\theta\, d\theta &= (\beta p_0\mathscr{E})^{-1}\sinh(\beta p_0\mathscr{E}) \\ &\cong 1 + \tfrac{1}{6}(\beta p_0\mathscr{E})^2 + \cdots.\end{aligned} \tag{12b.4}$$

The average projection of the dipole moment along the field (positive end of the dipole toward the negative end of the plates producing the field) is

$$\langle p \rangle = \int_0^{\pi} p_0 \cos\theta w(\theta)\, d\theta. \qquad (12b.5)$$

One must integrate, using again $\cos\theta = \zeta$,

$$\frac{p_0}{2}\int_{-1}^{1} \zeta e^{\beta p_0 \mathscr{E}\zeta} d\zeta = \frac{p_0}{2}(\beta p_0 \mathscr{E})^{-1}(e^{\beta p_0 \mathscr{E}} + e^{-\beta p_0 \mathscr{E}})$$

$$-\frac{p_0}{2}(\beta p_0 \mathscr{E})^{-2}(e^{\beta p_0 \mathscr{E}} - e^{-\beta p_0 \mathscr{E}})$$

$$\cong \frac{p_0}{3}(\beta p_0 \mathscr{E})[1 + \tfrac{1}{10}(\beta p_0 \mathscr{E})^2 + \cdots].$$

Dividing this by eq. (4), one finds

$$\langle p \rangle = p_0 \frac{e^{\beta p_0 \mathscr{E}} + e^{-\beta p_0 \mathscr{E}}}{e^{\beta p_0 \mathscr{E}} - e^{-\beta p_0 \mathscr{E}}} - p_0(\beta p_0 \mathscr{E})^{-1} = p_0[\coth(\beta p_0 \mathscr{E}) - (\beta p_0 \mathscr{E})^{-1}]$$

$$\cong \tfrac{1}{3}\beta p_0^2 \mathscr{E}[1 - \tfrac{1}{15}(\beta p_0 \mathscr{E})^2 + \cdots]. \qquad (12b.6)$$

The function of eq. (6), $\coth x - x^{-1}$, called the Langevin function $L(x)$, is plotted in Fig. 12h.2 (curve for $j = \infty$). It increases monotonically with x, from zero at $x = 0$, to unity as x approaches infinity. From the expansion it is seen that $L(x) = x/3$, if $x \ll 1$. For increasing argument the slope of $L(x)$ decreases, approaching zero as $x = \infty$, $L(x) = 1$. It is seen that $\langle p \rangle$ behaves qualitatively as one would expect. For small fields $\langle p \rangle$ is proportional to the field $\mathscr{E}$; as $x = \beta p_0 \mathscr{E}$ increases, the ratio of $\langle p \rangle / \mathscr{E}$ decreases until, for large values of x, that is, high fields or very low temperatures, $\langle p \rangle$ becomes independent of $\mathscr{E}$, $\langle p \rangle = p_0$. The system has then reached saturation; $\langle p \rangle$ has reached its maximum value, and all the dipoles are oriented in the direction of the field.

For all practical cases with electric fields the approximate result

$$\langle p \rangle = \tfrac{1}{3}\beta p_0^2 \mathscr{E} \qquad (12b.6')$$

is of sufficient accuracy, since the quantity $\beta p_0 \mathscr{E}$ is very small compared to unity. The exact eq. (6) is referred to later in dealing with magnetic effects.

12c. THE DIELECTRIC CONSTANT

In the previous section it was found that the application of an electric field of magnitude $\mathscr{E}$ to a dilute gas composed of molecules having a permanent rigid dipole moment $\not{p}_0$ results in a net average orientation of the dipoles in the direction of the field. The average projection $\langle \not{p} \rangle$ of the dipole moment on the field is given by eq. (12b.6′).

The product of the average oriented dipole moment $\langle \not{p} \rangle$ per molecule and the total number of molecules is a quantity of the dimensions of charge times distance. The polarization $\mathcal{P}$ of the gas is the value of this quantity divided by the volume, the total (oriented) net dipole moment, in the direction of the field, per unit volume,

$$\mathcal{P}_0 = \frac{n}{V}\langle \not{p} \rangle = \rho \langle \not{p} \rangle, \tag{12c.1}$$

with ρ the number density, and where in the general case both $\boldsymbol{\mathcal{P}}$ and $\boldsymbol{\not{p}}$ are vectors and, if the field is not static, depend on the frequency.

The dielectric constant ϵ of the gas may be defined in various ways. One of the definitions is

$$\frac{\epsilon - 1}{4\pi} = \frac{\mathcal{P}_0}{\mathcal{E}}. \tag{12c.2}$$

If this is compared with eq. (12a.3), one will find for the electric displacement $\boldsymbol{\mathcal{D}}$,

$$\boldsymbol{\mathcal{D}} = \boldsymbol{\mathcal{E}} + 4\pi \boldsymbol{\mathcal{P}}. \tag{12c.3}$$

In an isotropic medium $\boldsymbol{\mathcal{E}}$ and $\boldsymbol{\mathcal{P}}$, and therefore $\boldsymbol{\mathcal{D}}$, are parallel or even possibly antiparallel vectors at high frequencies. This is not necessarily true in a crystal and in this case $\boldsymbol{\epsilon}$, the dielectric constant, may be a symmetric tensor consisting of six different individual single numbers.

In the isotropic dilute gas, with eq. (12b.6′) in eq. (1) and eq. (1) in eq. (2), one finds, with ρ the number density N/V,

$$\epsilon - 1 = \frac{4\pi\mathcal{P}}{\mathcal{E}} - \frac{4\pi}{3}\rho\beta\langle \not{p}_0^2 \rangle, \tag{12c.4}$$

and the magnitude $\mathcal{D}$ of the vector $\boldsymbol{\mathcal{D}}$ is

$$\mathcal{D} = \mathcal{E}\left[1 + \frac{4\pi}{3}\rho\beta\langle \not{p}_0^2 \rangle\right]. \tag{12c.4′}$$

Since, at least in static and very low-frequency fields, the direction of polarization is *parallel* to the direction of the electric field, the dielectric constant is greater than unity.

Qualitatively, the result of eq. (4) is easily interpreted. A medium of high dielectric placed between the plates of a condenser increases the capacity. The increase in capacity is caused by an induced polarization in the material, bringing the negatively charged ends of the molecules toward the positive plate, which partially neutralizes the field, requiring a greater applied charge to produce the same voltage difference between the plates. The amount of net orientation of the molecules is proportional to their dipole moments p_0 and to the field $\mathscr{E}$, and inversely proportional to the temperature, which tends to keep their orientation random. The amount of polarization produced is proportional to the density and to the product of the dipole moment and the degree of orientation, therefore to the square of the moment p_0.

12d. TEMPERATURE-INDEPENDENT POLARIZATION

Monatomic molecules and symmetric di- or polyatomic molecules, like He, Ne, H_2, N_2, CH_4, and CCl_4, possess no permanent dipole moment p_0. Their gases, however, have dielectric constants ϵ differing from unity, although usually smaller than those of gases composed of nonsymmetric molecules like HCl, CH_3Cl, and NH_3.

The value of $\epsilon - 1$ in gases of symmetric molecules is due to the fact that the electric field induces a dipole moment in the molecule. This may be regarded as being caused by the equal and opposite forces of the field acting on the negatively and positively charged parts of the molecule, tending to separate their centers which are coincident in the absence of the field.

This effect is primarily electronic, namely, due to the displacement of the electron clouds around the nuclei in the electric field. The displacement, and therefore the polarization, are proportional to the field. The proportionality constant α is called the polarizability of the molecule,

$$p_i = \alpha \mathscr{E}, \tag{12d.1}$$

where p_i is the induced dipole moment along the field. In molecules with permanent dipoles, this effect is superimposed on that of orientation discussed in Sec. 12b.

Actually, α is not a simple number. If the molecule is not spherically symmetric, the value of α will depend on the direction of the field with respect to the various axes of the molecule. In addition, if the field does not act parallel to a symmetry axis of the molecule, it may produce perpendicular components of polarization. These components at right angles to the field always average to zero in a gas, although they are of importance in a nonisotropic crystal.

The observed average induced dipole moment in the gas is always in the direction of the electric field and is given by eq. (1), where α is the value of the polarization averaged over all orientations of the molecule with respect to the field. If the molecule possesses a permanent dipole moment, certain orientations are preferred if a field is applied. This has some, but little, influence on the interpretation of α as the average over all orientations.

The general equation for the total average projection of the dipole moment along the direction of the field is then, instead of eq. (12b.6′),

$$\langle \not{p} \rangle = (\alpha + \tfrac{1}{3}\beta \not{p}_0^2)\mathscr{E} \tag{12d.2}$$

where α is always positive and nonzero, but $\not{p}_0$ is zero for symmetric molecules.

Using eq. (2) in eq. (2c.2), with eq. (12c.1), one obtains

$$\epsilon - 1 = 4\pi \frac{\mathscr{P}}{\mathscr{E}} = 4\pi\rho(\alpha + \tfrac{1}{3}\beta \not{p}_0^2). \tag{12d.3}$$

This equation is known as the Debye equation for the dielectric constant. It is found to give the temperature dependence of ϵ excellently for most simple gases. Experimentally, $\epsilon - 1$ is proportional to the density at low densities. In Sec. 12f this equation is further improved as regards its density dependence.

At the beginning of this section it was stated that the polarizability α is primarily an electronic effect. This statement is not always true. The qualitative distinction between electronic and atomic polarization may be described as follows.

An increase in the distance between the two atoms of a symmetric diatomic molecule, such as H_2 or N_2, does not in general produce a dipole moment. The induced polarization caused by the electric field may be described as being due to the displacement of the negative electrons with respect to the nuclei. However, a change in the distance of separation of the two nuclei in a molecule like HBr in general results in a change in dipole moment. It is to be observed that an increase in the distance between nuclear centers by no means necessarily increases the dipole moment. The reverse may take place. Inversely, the force of the field tends to change the distance between the atomic centers, and the induced polarization in the field is partly due to a change in the internal vibrational coordinates of the molecule. A related, but at first thought apparently different, effect arises from the change in the dipole moment with temperature due to increased vibration.

As long as the temperature is low enough to use the approximate quadratic terms in the normal coordinates for the potential energy, one finds, using the classical equations, that the average square of the dipole moment is that of the square of the dipole moment at the lowest potential configuration plus a term linear to T. Using this in eq. (2), one again obtains the form eq. (3) for the dielectric constant.

In a molecule such as $CH_2Cl—CH_2Cl$ there exists a normal coordinate which measures the angle of the equal and opposite rotation of the two CH_2Cl groups around the C—C bond. The potential energy up to about kT at room temperature is *not* given well by a purely quadratic term in this displacement. The vibration of this coordinate deviates considerably from harmonic behavior. Only in cases such as this does one except a more complicated dependence on ϵ of T in gases than that given by eq. (3).

However, the part of α due to molecular vibration is in one respect experimentally distinguishable from the electronic polarization term. This part is due to the displacements of the heavy nuclei by the electric field, and these do not follow an alternating field of much higher frequency than the natural vibrational frequencies of the molecule. The extrapolation to zero frequency of index-of-refraction measurements made with visible light, which are unaffected by the term consequently leads to values for the *electronic* polarization and does not include the temperature-independent contribution to eq. (3) due to the atomic displacement. This latter term is connected with the intensity of the infrared vibrational bands of the molecule. The reader interested in the relations between the various properties of the molecules and the theory of molecular structure is referred to J. H. Van Vleck, *The Theory of Electric and Magnetic Susceptibilities*, Clarendon Press, Oxford, 1932.

12e. COMPARISON WITH REFRACTIVE INDEX AND EXPERIMENTAL DATA

The dielectric constant ϵ depends on the frequency ν of the field used, if alternating fields are applied. Equation (12d.3) was derived only for static or low-frequency fields. Connected with ϵ is the index of refraction $n(\nu)$ of light of frequency ν. The index of refraction $n(\nu)$ is determined by the refraction of a beam of light crossing from a vacuum through a sharp boundary into the medium in question. It is defined as the ratio of the sines of the angles of the incident and refracted beams to the direction normal to the plane of the surface.

Neglecting the usually truly insignificant difference between the magnetic permeability and unity, one may identify the dielectric constant ϵ with the square of the index of refraction for the *same* frequency,

$$\epsilon(\nu)=[n(\nu)]^2. \qquad (12e.1)$$

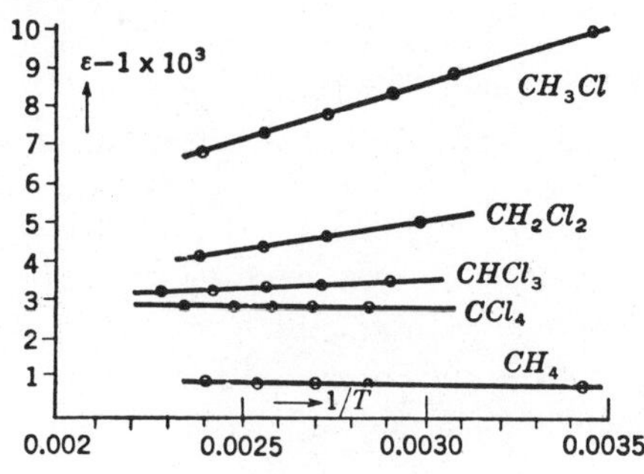

Figure 12e.1

At the frequency of visible light the term in eq. (12d.3) due to the orientation of the permanent dipole moment $\beta \not{p}_0^2$ contributes nothing to the index of refraction. This may be made plausible by the observation that the frequency 10^{14} sec^{-1} of visible light is far higher than the frequency 10^{10} sec^{-1} of rotation of most molecules. The molecules do not have time to orient themselves before the direction of the electric field due to the light has changed sign. They cannot follow a field of high frequency. The polarizability α in eq. (12d.3) may then be obtained by either of two methods. It is the intercept on the ordinate $1/T = 0$, $T = \infty$, of a plot of $V(\epsilon - 1)/4\pi N$ for ϵ at very low frequencies against the reciprocal temperature. It may also be determined by an extrapolation to $\nu = 0$ of the index of refraction from the values for various frequencies in the visible region. The extrapolation of $[n(\nu)]^2$ for visible frequencies gives only the electronic contribution. The extrapolation of $\epsilon(\nu)$ for low frequency to $1/T$ equal to zero includes the nuclear displacement contribution in the lowest state.

One of the standard examples for the interpretation of the experimental data, and of the information that has been obtained from them, is the series of compounds CH_4, CH_3Cl, CH_2Cl_2, $CHCl_3$, and CCl_4. Figure 12e.1 shows Sänger's[1] observations on the dielectric constants of these gases as functions of temperature at constant density.

It is seen that the curves for CH_4 and CCl_4 are both horizontal, showing zero permanent dipole moments. The molecules must consequently be symmetric, and either a tetrahedral or plane square arrangement of the Cl and H atoms about the carbon is demanded. The polarizability of CCl_4 is seen to be about three times that of CH_4, owing to the far greater polarizability of the Cl atoms than those of the carbon

[1] R. Sänger, *Phys. Z.*, **27,** 556 (1926).

and hydrogen. The value of $\epsilon - 1 = 0.00096$ for CH_4 found here is fairly close to the value 0.00086 found for $n^2 - 1$ by extrapolation from the measured index of refraction in the visible region. The discrepancy is presumably due to the contributions of the infrared vibration bands to the index of refraction $n(0)$ which is not obtained by extrapolating measurements made with visible light.

This addition to n^2 at $\nu = 0$ from the infrared vibrational bands is usually much smaller in diatomic molecules than in those of methane. It is the part of the polarizability α due to the displacement of the atomic nuclei by the field, as discussed in Sec. 12d.

The appreciable slope of the lines for the other chlorides in Fig. 12e.1 indicates that they have permanent dipole moments. The values of p_0 calculated from the slopes are

$$1.86 \times 10^{-18} \text{ esu for } CHCl_3,$$
$$1.59 \times 10^{-18} \text{ esu for } CH_2Cl_2,$$

and

$$0.95 \times 10^{-18} \text{ esu for } CHCl_3.$$

The dipole moment of a molecule containing several rigidly connected dipoles is the vector sum of their several moments. By making the unwarranged assumption that the dipoles along the C—H bonds are zero, the value of p_0 for CH_3Cl gives directly the moment of the C—Cl bond. By assuming this dipole moment for the bond to be constant, one may use the laws of vector addition to obtain the angle between the C—Cl bonds in CH_2Cl_2 and $CHCl_3$.

For CH_2Cl_2 the angle ϕ between the two C—Cl bonds is found by setting $\frac{1}{2} \times 1.59/1.86 = \cos \frac{1}{2}\phi$, which leads to $\phi = 128°$. For $CHCl_3$ the angle ψ between the CH bond and any one of the C—Cl bonds (assuming a distorted tetrahedral model so that all three angles are equal) is obtained by $\frac{1}{3} \times 0.95/1.86 = \sin(\psi - 90°)$, $\psi = 96°$. Of course, far more precise methods are now available for obtaining this kind of information on the molecular structure.

The numerical values of the dipole moments themselves are not without interest. Using the value of the elementary charge, $e = 4.8 \times 10^{-10}$ esu, one sees that the dipole moment, 1.86×10^{-18} esu, of CH_3Cl corresponds to a positive and a negative elementary charge separated by $1.86/4.8 = 0.4$ Å. This is a comparatively large dipole moment. That of HCl is only 1.02×10^{-18} esu, and that of HI only 0.38×10^{-18} esu.

For HI, for example, band spectral data show the separation of the H and I nuclei to be 1.41 Å. The separation of two elementary charges corresponding to the dipole moment is only 0.08 Å.

12f. THE LORENTZ–LORENZ FORCE

The electric field $\mathscr{E}$ within a gas or any other material having a dielectric constant ϵ not unity is different from the value of the field outside the material. In a vacuum between two condenser plates the electric displacement $\mathscr{D}$ and the electric field $\mathscr{E}$ are identical. If the charge on the condenser plates is kept constant, the displacement $\mathscr{D}$ is unchanged by the introduction of the material between the plates, provided that the plates are large compared to the distance between them. The field $\mathscr{E}$, however, is reduced within the (isotropic) material to the value $\mathscr{E}=\mathscr{D}/\epsilon$. This reduction in the field $\mathscr{E}$ is due to polarization of the material, the induced charge brought to the surface of the material tending to cancel the effect of the charges on the condenser plates. The average electric force on an infinitesimal charge δe, averaged over all positions in the material, is given by $\mathscr{E}\,\delta e$.

However, the electric force acting on the charged parts of a single molecule is different from $\mathscr{E}$. This is because the average field of the molecule itself must be subtracted in making such a calculation.

The local field $\mathscr{E}_{\text{loc}}$ acting on one molecule is therefore not $\mathscr{E}$, but differs from it by a term which may be calculated[(1)] from the average polarization density $\mathscr{P}$ in the material. This additional force is known as the Lorentz–Lorenz force. Its magnitude depends on the distribution of the molecules in space.

If the molecules are randomly distributed, the local field acting on a molecule $\mathscr{E}_{\text{loc}}$ is given by the Clausius-Mossotti formula,

$$\mathscr{E}_{\text{loc}}=\mathscr{E}+\frac{4\pi}{3}\mathscr{P}, \tag{12f.1}$$

which is not derived here.[(2)] By using eq. (1) for $\mathscr{E}_{\text{loc}}$ instead of $\mathscr{E}$ in eqs. (2c.1) to (2d.2),

$$\mathscr{P}=\rho(\alpha+\tfrac{1}{3}\beta\,p_0^2)\left(\mathscr{E}+\frac{4\pi}{3}\mathscr{P}\right),$$

$$(\epsilon-1)=4\pi\frac{\mathscr{P}}{\mathscr{E}}=4\pi\rho(\alpha+\tfrac{1}{3}\beta\,p_0^2)\left[1-\frac{4\pi}{3}\rho(\alpha+\tfrac{1}{3}\beta\,p_0^2)\right]^{-1},$$

or finally,

$$\frac{\epsilon-1}{\epsilon+2}=\frac{4\pi}{3}\rho(\alpha+\tfrac{1}{3}\beta\,p_0^2). \tag{12f.2}$$

(1) See, for instance, the discussion in the first chapter of J. H. Van Vleck, *The Theory of Electric and Magnetic Susceptibilities*, Clarendon Press, Oxford, 1932.

(2) The derivation is given in H. A. Lorentz, *The Theory of Electrons*, Columbia University Press, New York, 1909, Section 117 and Note 54.

In the limit that N/V is small, ϵ approaches unity, and $3/(\epsilon+2)\cong 1$ this becomes eq. (12d.3).

The quantity $\rho^{-1}(\epsilon-1)/(\epsilon+2)$ is found to be independent of density up to very high pressure in gases. For instance, Magri[1] found, in measurements on air, that it remains constant within the experimental error of about 0.4% up to 180 times normal density, whereas $(V/N)(\epsilon-1)$ increases by 4% in this range.

In fact, for nonpolar molecules, $\not{p}_0=0$, the quantity $\rho^{-1}(\epsilon-1)/(\epsilon+2)$ is constant to within about 10% in going from the vapor to the liquid phase. This is an extremely severe test, since the method of derivation is scarcely applicable to liquids.

12g. PARA- AND DIAMAGNETISM

The equations for para- and diamagnetic materials in magnetic fields are analogous to those for molecules with and without permanent dipoles in electric fields.

Assume every molecule in the gas to have the same permanent magnetic moment μ_0. In the magnetic field $\mathcal{H}$ the potential energy is

$$u(\theta)=-\mu_0\mathcal{H}\cos\theta, \tag{12g.1}$$

analogously to eq. (12b.1), where θ is the angle between the magnetic moment and the direction of the field. The subsequent calculations of Sec. 12b are exactly valid, and one finds, for the average component of the magnetic moment $\langle\mu\rangle$ in the direction of the field,

$$\langle\mu\rangle=\mu_0[\coth(\beta\mu_0\mathcal{H})-(\beta\mu_0\mathcal{H})^{-1}]\cong\tfrac{1}{3}\beta\mu_0^2\mathcal{H}. \tag{12g.2}$$

The magnetic polarization (per unit volume) $\mathcal{M}$ is

$$\mathcal{M}=\rho\langle\mu\rangle, \tag{12g.3}$$

and magnetic susceptibility χ may be defined as

$$\chi=\frac{\mathcal{M}}{\mathcal{H}}=\tfrac{1}{3}\rho\beta\mu_0^2, \tag{12g.4}$$

in analogy with eq. (12c.4). This equation holds for paramagnetic substances. The fact that the paramagnetic susceptibility is inversely proportional to temperature is called Curie's law.

Just as in Sec. 12d we discussed the fact that the electric field induces a dipole in molecules, so molecules possessing no permanent magnetic moment acquire an induced moment in the presence of a magnetic field.

[1] L. Margi, *Phys. Z.*, **6**, 629 (1905).

However, this induced magnetic moment is always in the direction opposite the field, that is, it opposes the field. The equation obtained is then analogous to eq. (12d.3), except that the temperature-independent term occurs with a negative sign in the magnetic case. That the impressed magnetic moment opposes the field is a consequence of electrodynamics. The current induced in a closed electric circuit by the imposition of a magnetic field is such as to creaste a moment opposite the field.

There is one other distinct difference between the influence of electric and magnetic fields. In eq. (12d.3) the two terms, α and $\beta \not{p}_0^2$, are both of the same order of magnitude. The paramagnetic susceptibility (orientation effect) given by eq. (4), if present at all, that is, if μ_0 is not zero, is about hundred- to several thousandfold larger than the effect of the induced magnetic moment, called the diamagnetic susceptibility. As a result, one usually neglects the diamagnetic effect in dealing with paramagnetic substances and uses eq. (4) without the negligible diamagnetic term.

It is not in the province of this book to treat the relation between the diamagnetic constant and the structure of the molecule. However, it may be mentioned that for free atoms the diamagnetic susceptibility is related to the average square of the distance of the electrons from the nucleus, which we write $\langle r^2 \rangle$.

The equation

$$\chi = -\frac{e^2}{6mc^2} \rho \sum_i \langle r_i^2 \rangle \tag{12g.5}$$

gives the diamagnetic susceptibility χ in which e and m are the charge and mass of the electron, respectively, and c is the velocity of light. The sum $\sum \langle r_i^2 \rangle$ is that over all electrons i of their average squared distances r_i^2 from the nucleus.

The method employed in this section is a hybrid of classical and quantum mechanics. The assumption has been made that every molecule (of the same kind) in the gas has the same magnetic moment μ_0. Actually, if an attempt were made to explain the magnetic moment by the motion of electrons under the influence of the electric field of the nuclei, the application of classical statistics would lead to a variety of magnetic moments each weighted with a certain probability dependent on T. Van Leeuwin[(1)] has shown that a purely classical system of electric point charges should exhibit zero magnetic susceptibility. The reason, however, is intimately connected with the complete inability of classical mechanics to account for the properties of atoms if they are each composed of one point mass nucleus and of electrons.

[(1)] J. H. van Leeuwin, Dissertation, University of Leiden, 1919; also *J. Phys.* [6], **2,** 361 (1921).

Whereas in the electric example only the approximation to eq. (2) is experimentally significant, magnetic susceptibilities in cases to which these equations apply can be measured at extremely low temperatures, and the complete eq. (2) can be checked experimentally (Sec. 12). Although the approximation for low values of $\mathcal{H}\beta\mu_0$ is the same for classical and quantum-mechanical systems, the complete equation differs slightly for the two methods of calculation.

We consequently, in Sec. 12h, make a somewhat more logical development, taking cognizance of the quantum-mechanical nature of the phenomenon.

12h. PARAMAGNETISM IN QUANTUM MECHANICS

A complete discussion of magnetic phenomena involves comparatively complicated considerations of the applications of quantum mechanics to atomic and molecular structure, which are not treated in this book. Even after limiting the discussion to independent atoms and ions, there remains a considerable number of different cases which must be considered separately.

For instance, the behavior of the atoms depends on whether only one electronic level is excited at the temperature considered, or whether the multiplet separation of the energy levels is small compared to kT. The effect of weak and of strong fields must also be distinguished. For a complete description of the phenomena encountered the reader is again referred to Van Vleck's *Theory of Electric and Magnetic Susceptibilities.*

The discussion in this section is limited to the example in which the energy of separation of the lowest and first excited electronic levels in the atom or ion is large compared both to the energy of interaction with the magnetic field and to kT. Furthermore, diamagnetic effects are neglected. In addition, it is assumed that all the ions or atoms considered are monatomic, and that the directions of their magnetic moments are independent of one another.

The last condition does not, however, limit the applicability of the equations derived to gases. The reason for this is that the electrostatic and exchange repulsion forces that operate between many ions and their environment in water solution or in ionic crystals do not involve *primarily* the electronic spin responsible for paramagnetism. The orientation of the magnetic moment in the ion does not affect its interaction with neighboring particles. The effect of the magnetic field on the ions may therefore be treated as if the ions were completely independent of their surroundings, that is, as if they composed a monatomic perfect gas.

The equations derived by this method are found to apply excellently to relatively concentrated water solutions of paramagnetic ions. They also fit well the observations made on many crystals, provided only that the distance between the magnetically active ions is not too small. This is realized in practice by using crystals containing considerable water of hydration, or dilute solid solutions of magnetic ions in crystals of non-magnetic salts. Ions having the completed octet electronic structure, a lowest $^1S_0$ state, are diamagnetic. Paramagnetism is observed only in ions of the transition elements and the rare earths. These ions have the lowest electronic levels possessing an angular momentum ($j \neq 0$).

Since these ions also have several low electronic levels, the limitation to cases in which the lowest excited level has a high energy compared to kT is not an unimportant restriction.

The magnetic moments of isolated atoms or ions are dependent on their angular momenta. The connection between these two quantities is not utterly simple but may be calculated correctly with the help of the so-called vector model treated below for Russel-Saunders coupling. The angular momentum of the ion in a given electronic level is due to the rotation of the electrons. This rotation has two different components. One of these is called the orbital angular momentum and is due to the motion of the center of mass of the electrons about the nucleus. The second one is due to the spin of the electrons about their own axes.

In Russell-Saunders coupling the orbital angular momenta of the different electrons are strongly coupled, resulting in a net vectorial angular momentum of the atom, which is determined by an integer quantum number l in such a way that the square of the orbital angular momentum is $(h/2\pi)^2 l(l+1)$. The energy of the level is determined primarily by l, where $l=0$ is an S level, $l=1$ a P level, $l=2$ a D level, $l=3$ an F level, and so on. The vector sum of the electrons spins has a smaller effect on the energy, and the relative vector orientation of orbit and spin still less. The total spin angular momentum square is $(h/2\pi)^2 s(s+1)$, where s is the total spin quantum number. The total angular momentum squared is

$$\left(\frac{h}{2\pi}\right)^2 j(j+1),$$

where j is the quantum number of total angular momentum, namely, the magnitude of the vector sum of orbital and spin angular momenta. The projection of the total angular momentum on a fixed axis in space may take values $mh/2\pi$, where $j \geq m \geq -j$, so that m takes $2j+1$ values.

The magnetic moment μ_0 is likewise determined by the rotation of the electrons.

The symbol b_0 is introduced for the Bohr magneton, $b_0 = he/4\pi mc$, where e and m are the charge and mass of the electron, and c is the velocity of light. The contribution to the magnetic moment due to the orbital angular momentum is just $\sqrt{l(l+1)}b_0$, but the moment is directed along the axis of the orbital angular momentum.

The contribution to the magnetic moment due to spin is $2\sqrt{s(s+1)}b_0$, along the spin axis. The difference in the case of orbit and spin is a relativistic effect but may be crudely and very naively described as being due to the fact that in orbits the mass and charge of the electron move together, while in spin the charge, located on the surface of the electron, rotates with a greater average radius than the mass, which is distributed throughout the particle.

The square of the magnetic moment of the atom or ion may be calculated as follows.

We consider the two orbital and spin vectors of magnitude $\sqrt{l(l+1)}$ and $\sqrt{s(s+1)}$, adding vectorially to make the total angular momentum $\sqrt{j(j+1)}$ (Fig. 12h.1). The orbital magnetic moment $\sqrt{l(l+1)}b_0$ and the spin moment $2\sqrt{s(s+1)}b_0$ are vectors of the same direction as the respective angular momenta but are of different relative lengths. The two vectors, orbital and spin, precess rapidly about the total angular momentum vector. The magnetic moment square is found by squaring the projection μ_0 of the vector sum of $\sqrt{l(l+1)}b_0$ and $2\sqrt{s(s+1)}b_0$ on the total angular momentum vector.

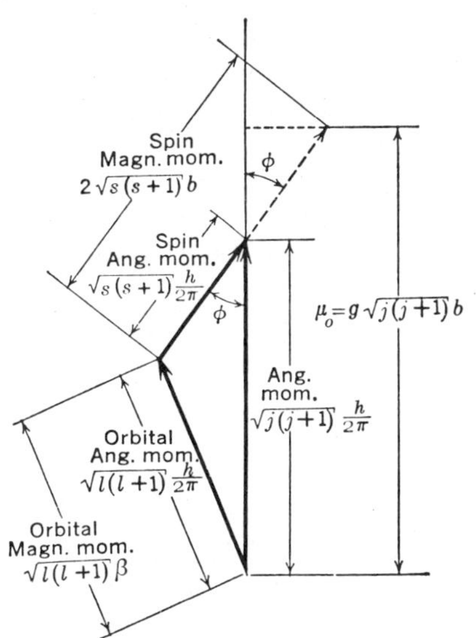

Figure 12h.1 Diagram for the g-factor in Russell-Saunders coupling.

The problem is one of simple trigonometry. The cosine of the angle ϕ in the figure is given by

$$\cos\phi = \frac{s(s+1)+j(j+1)-l(l+1)}{2\sqrt{j(j+1)}\sqrt{s(s+1)}}.$$

From the figure it is seen that, if $\mu_0 = \sqrt{j(j+1)}gb_0$, the factor g is

$$g = 1\frac{\sqrt{s(s+1)}}{\sqrt{j(j+1)}}\cos\phi,$$

$$g = 1+\frac{s(s+1)+j(j+1)-l(l+1)}{2j(j+1)}. \tag{12h.1}$$

The square of the magnetic moment μ_0^2 of an atom or ion in an electronic level with given quantum number j of the angular momentum is

$$\mu_0^2 = j(j+1)g^2 b_0^2, \tag{12h.2}$$

where b_0 signifies the Bohr magneton,

$$b_0 = \frac{eh}{4\pi mc} = 0.9274\times 10^{-20}\text{ emu} \tag{12h.3}$$

The Landé g-factor in eq. (2) is given by eq. (1) in terms of the quantum numbers j, l, and s for Russell-Saunders coupling.

In what is known as j-j coupling the orbital angular momentum and spin of each electron i couple to a net vectorial angular momentum characterized by a quantum number j_i (which is always a half-integer). The vector sum of these results in an angular momentum square of $(h/2\pi)^2 j(j+1)$. In either case, as well as in more complicated cases, the square of the magnetic moment at zero field of the lowest electronic level can always be given by eq. (2) which defines a number g. At sufficiently low magnetic fields $\mathcal{H}$, the analysis that follows can be used and g determined experimentally. Its value can then be used with other information to find the character of the lowest electronic level.

If, however, there is a second electronic level with a low energy $\Delta\varepsilon$ above the lowest when the magnetic field becomes of the order of $\mathcal{H} \sim \Delta\varepsilon/gb_0$ [see eq. (5) below], both levels contribute. The splitting in energy due to the magnetic field is more complicated than that discussed below.

The level, of given j, is degenerate and consists of $2j+1$ states differing in the value of m, the quantum number determining the projection of the angular momentum on the direction of the field. In the absence of a field, $\mathcal{H} \to 0$, these have the same energy. The projection μ_m of the magnetic moment along the axis of a field is

$$\mu_m = mgb_0, \qquad j \geq m \geq -j, \tag{12h.4}$$

for the state m, and the energy due to the magnetic field has the form

$$\varepsilon_m = -\mu_m \mathcal{H} = -mgb_0\mathcal{H}. \tag{12h.5}$$

The probability that an ion is in the state m is proportional to $e^{-\beta\varepsilon_m}$, so that the average projection $\langle\mu\rangle$ of the magnetic moment along the field, for all ions, is

$$\langle\mu\rangle = \frac{\sum_{m=-j}^{m=+j} mgb_0 \exp(+\mathcal{H}mg\beta b_0)}{\sum_{m=-j}^{m=+j} \exp(+\mathcal{H}mg\beta b_0)}. \tag{12h.6}$$

The first approximation of eq. (6) for low values of $g\beta\mathcal{H}/kT$ is exactly the classical result eq. (12g.2), with eq. (21) for μ_0^2. This may be seen as follows.

The exponential functions in the numerator and denominator of eq. (6) are expanded in the usual manner as $e^y = 1 + y + \cdots$, and all but the first nonzero term after summation is neglected in both numerator and denominator. The first term in the denominator is obtained by using unity for the exponential and is just the number of terms $2j+1$. The term arising from the unity of the expansion in the numerator is zero, since $\sum m$ extended over both positive and negative m values from $-j$ to $+j$ is zero. The next term does not vanish and is linear in T^{-1}. Since

$$\mathcal{H}\frac{g^2b_0^2}{kT}\sum_{m=-j}^{m=+j} m^2 = \mathcal{H}g^2\beta b_0^2\frac{(j+1)j(2j+1)}{3},$$

one obtains

$$\langle\mu\rangle = \frac{1}{3}\frac{g^2b_0^2 j(j+1)}{kT}\mathcal{H} = \tfrac{1}{3}\beta\mu_0^2\mathcal{H}, \tag{12h.7}$$

with eq. (2) for μ_0^2.

Equation (6) may be brought into an explicit form without recourse to the series development of the exponential. By writing $e^{g\beta h_0\mathcal{H}} = x$,

$$\langle\mu\rangle = g\beta\frac{\sum_{m=-j}^{m=+j} mx^m}{\sum_{m=-j}^{m=+j} x^m}$$

$$= g\beta x\frac{d}{dx}\left(\ln\sum_{m=-j}^{m=+j} x^m\right). \tag{12h.6'}$$

One uses the relations

$$\sum_{m=-j}^{m=+j} x^m = x^{-j} \sum_{n=0}^{n=2j} x^n = \frac{x^{-j} - x^{j+1}}{1-x} = \frac{x^{j+1/2} - x^{j+1/2}}{x^{1/2} - x^{-1/2}}$$

and

$$x\frac{d}{dx}\left(\ln \frac{x^{j+1/2} - x^{-(j+1/2)}}{x^{1/2} - x^{-1/2}}\right) = (j+\tfrac{1}{2})\frac{x^{j+1/2} + x^{-(j+1/2)}}{x^{j+1/2} - x^{-(j+1/2)}} - \frac{1}{2}\frac{x^{1/2} + x^{-1/2}}{x^{1/2} - x^{-1/2}}.$$

If now the Brillouin function $B_j(y)$ is defined as

$$B_j(y) = \frac{j+\frac{1}{2}}{j}\coth\left(\frac{j+\frac{1}{2}}{j}y\right) - \frac{1}{2j}\coth\frac{1}{2j}y$$
$$= \frac{j+\frac{1}{2}}{j}\frac{e^{(j+1/2)y/j} + e^{-(j+1/2)y/j}}{e^{(j+1/2)y/j} - e^{-(j+1/2)y/j}} - \frac{1}{2j}\frac{e^{-y/2j} + e^{-y/2j}}{e^{y/2j} - e^{-y/2j}}, \qquad (12h.8)$$

it is seen that eq. (6) becomes

$$\langle\mu\rangle = jgb_0 B_j(jg\beta b_0 \mathscr{H}). \qquad (12h.9)$$

The imposing function in eq. (8) is plotted in Fig. 12h.2 for several j values. It has properties similar to those of the classical $L(x) = \coth x - x^{-1}$ [eq. (12b.6)]. For $y \ll 1$ the function may be developed, and one finds

$$B_j(y) \cong \tfrac{1}{3}j(j+1)\frac{y}{j}, \qquad y \ll 1,$$

which, on substitution of $y = jg\beta b_0\mathscr{H}$ in eq. (9), leads to eq. (7).

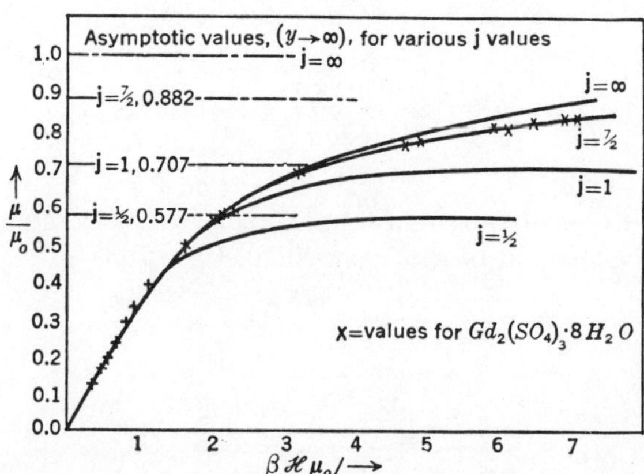

Figure 12h.2 Plot of $\langle\mu\rangle/\mu_0 = j(j+1)B_j(y)$ versus $\beta\mathscr{H}\mu_0$ for various values of j. From van Vleck, *The Theory of Electric and Magnetic Susceptibilities*, Clarendon Press, Oxford, 1932.

For $y \gg 1$, $e^y \gg e^{-y}$, both hyperbolic cotangents become unity, and

$$B_j(y) \cong 1, \qquad y \gg 1,$$

so that, on substitution in eq. (9), one finds

$$\langle \mu \rangle = jgb_0, \qquad jg\beta b_0 \mathcal{H} \gg 1. \tag{12h.10}$$

The saturation effect is reached when all the ions are pressed, by the field, into the state $m = j$, where the projections of the magnetic moments on the direction of the field have their maximum values, $\mu_m = jgb_0$. This saturation moment, however, is *not* the square root of μ_0^2 given by eq. (2). The value of the square of the magnetic moment obtained by measurements with weak fields, using eq. (7), leads to the $j(j+1)$ [eq. (2)], whereas the saturated magnetic moment obtained with a strong field is only $jg\beta$.

For large j values the difference between $j(j+1)$ and j^2 is negligible. The Brillouin function has the property that

$$\lim_{j\to\infty}[B_j(y)] = \coth y - \frac{1}{y} = \frac{e^y + e^{-y}}{e^y - e^{-y}} - \frac{1}{y}.$$

The classical eq. (12g.2) is obtained if j becomes large, in which event the magnetic moment μ_0 is given by jgb_0.

Actually, at high temperatures, that is, at about room temperature, saturation or even very appreciable deviations from the approximate equation (7) are not observable in the laboratory, except with ferromagnetic substances, discussed in Sec. 12i. The same is true of electrical polarization produced by electric fields.

However, whereas the equations derived in this chapter for electrical polarization are strictly applicable to gases only and, less rigorously, to dilute liquid solutions, those for magnetic fields are applicable to certain crystals. The equations derived in this section may therefore be used in comparison with experiments performed at extremely low temperatures. Since the behavior of the material depends on $\beta\mathcal{H}$, a lowering of the temperature corresponds to an increase in the magnetic field.

Approximate saturation was obtained, in the 1920s with hydrated gadolinium sulfate, $Gd_2(SO_4)_3 \cdot 8H_2O$, at 1.3 K, in Leiden.[(1)] The gadolinium ion, Gd^{3+}, the lowest level of which is $^8S_{7/2}$, is the magnetically active ion. Since the angular momentum of this ion is entirely due to spin (S state, $l = 0$), the g factor is 2. The results agree perfectly with the prediction using $j = \frac{7}{2}$, $g = 2$, $y = 7\beta b_0 \mathcal{H}$ in eq. (9). In Fig. 12h.2 the crosses indicate the experimental values for this salt.

(1) H. R. Woltjer and H. Kamerling-Onnes, *Leiden Commun.* 167c or *Versl. Amst. Akad.* **32,** 772 (1923).

Measurements of magnetic moments as a function of temperature and magnetic field have played an essential role in the elucidation of the electronic configuration of ions of the salts of the transition elements and of the rare earths. The advent of magnetic resonance spectroscopy has given us a much more powerful tool nowadays.

12i. FERROMAGNETISM

In Sec. 12f it was found that the electric force acting on a molecule in the material treated was not given simply by the electric field, but by the macroscopic field $\mathscr{E}$ plus a term proportional to the polarization density $\mathscr{P}$. The proportionality constant was $4\pi/3$ in the Clausius-Mossotti equation [eq. (12f.1)].

If it is assumed that the forces acting on one ion contain a term due to the magnetic polarization $\mathscr{M}$, so that eq. (12g.1) or (12h.5) has the form

$$u(\theta) = -\mu_0(\mathscr{H} + a\mathscr{M})\cos\theta \qquad \text{or} \qquad \varepsilon_m = -\mu_m(\mathscr{H} + a\mathscr{M}), \quad (12\text{i}.1)$$

one obtains equations predicting properties similar to those of ferromagnetic substances by using a sufficiently large value of a. Without discussing the possible origin of the strong interaction term $a\mathscr{M}$ between the magnetic moments in eq. (1) one may investigate the effect it has on the macroscopic behavior of the substance.

Since the classical equation (12g.2) and the quantum-mechanical equation (12h.9) have the same qualitative features, it is somewhat simpler, and not essentially different, to treat the classical case. The substitution of $\mathscr{H} + a\mathscr{M}$ for $\mathscr{H}$ in eq. (12g.2), and multiplication of $\langle\mu\rangle$ by N/V to obtain $\mathscr{M}$ [eq. (12g.3)], leads to

$$\mathscr{M} = \rho\mu_0\{\coth[\beta\mu_0(\mathscr{H} + a\mathscr{M})] - [\beta\mu_0(\mathscr{H} + a\mathscr{M}]^{-1}\}. \quad (12\text{i}.2)$$

Equation (2) determines the magnetic polarization $\mathscr{M}$ (per unit volume) in terms of the applied magnetic field $\mathscr{H}$ at any temperature. The task of analytically solving eq. (2) for $\mathscr{M}$ is not an easy one. The solution may, however, be undertaken readily by a graphical method.

The term in brackets on the right-hand side of eq. (2), plotted as a function of

$$x = \beta\mu_0(\mathscr{H} + a\mathscr{M}),$$

is the Langevin function $L(x)$ of Sec. 12b, a monotonically increasing function starting from zero x at $=0$ with a slope of $\frac{1}{3}$ and approaching unity asymptotically as x goes to infinity. The quantity $(\rho\mu_0)^{-1}\mathscr{M}$ which, according to eq. (2), should be equal to the bracket, plotted as a function of the *same* argument x, is a straight line,

$$(\rho\mu_0)^{-1}\mathcal{M} = -(a\rho\mu_0)^{-1}\mathcal{H} + (a\beta\rho\mu_0^2)^{-1}x. \tag{12i.3}$$

The slope of this line is independent of the magnetic field. With increasing field $\mathcal{H}$ the line moves parallel to itself downward.

For given values of $\mathcal{H}$ and T the magnetic polarization $\mathcal{M}$ is determined by the equality of the right- and left-hand sides of eq. (2) or, in the plot, by the intersection of the straight line of eq. (3) with the curve $\coth x - 1/x$ (see Fig. 12h.2, curve for $j=\infty$). The ordinate of the intersection point determines $\mathcal{M}V/\mu_0 N$. It is seen that, with increasing field but constant temperature, both the abscissa and ordinate of the intersection point increase monotonically.

The temperature T determines the slope of eq. (3) but not the intersection with the axis $x=0$. For constant $\mathcal{H}$ an increase in T increases the slope and thereby decreases the ordinate of the intersection point.

For zero magnetic field the straight line of eq. (3) goes through the origin $x=0$, $\mathcal{M}=0$, and there intersects the curve $L(x)$. It is then apparent that for $\mathcal{H}=0$ two different cases may occur, depending on the inclination of eq. (3). If the line in eq. (3) is steeper than the slope of the curve at the origin, that is,

$$(a\beta\rho\mu_0^2)^{-1} > \tfrac{1}{3}, \qquad kT > \tfrac{1}{3}a\rho\mu_0^2,$$

this is the *only* intersection point. If this holds, as $\mathcal{H}$ goes to zero, the magnetic polarization $\mathcal{M}$ vanishes and for small enough values of the field $\mathcal{M}$ is proportional to $\mathcal{H}$. Qualitatively, the substance behaves like a paramagnetic one.

In the other case, if the slope of the line is smaller than $\frac{1}{3}$,

$$kT < \tfrac{1}{3}a\rho\mu_0^2,$$

there exists a second intersection point for $\mathcal{H}=0$ with a nonvanishing value of $\mathcal{M}$. If the magnetic field is gradually reduced to zero, $\mathcal{M}$ approaches this finite value. The substance has then a remanent magnetic moment.

The temperature T_c above which this remanent polarization disappears,

$$\beta_c = 3(a\rho\mu_0^2), \qquad T_c = \frac{1}{3}\frac{a\rho\mu_0^2}{k}, \tag{12i.4}$$

is called the Curie temperature.

For $T \gg T_c$, $\beta \ll \beta_c$, eq. (2) may be developed as a power series in β,

$$\mathcal{M} = \tfrac{1}{3}(\beta\rho\mu_0^2)(\mathcal{H}+a\mathcal{M})[1-\tfrac{1}{15}(\beta\mu)^2\mathcal{H}+a\mathcal{M})+\cdots]. \tag{12i.2′}$$

By neglecting all but the first term this leads to

$$\mathcal{M}=\frac{\mathcal{H}}{3}(\beta\rho\mu_0^2[1-\tfrac{1}{3}a\beta\rho\mu_0^2)]^{-1}=\frac{\mathcal{H}}{3}\frac{\rho\mu_0^2}{k(T-T_c)}, \tag{12i.5}$$

which breaks down as T approaches T_c.

Qualitatively, the Curie temperature corresponds to a sort of critical temperature. The term with the constant a in eq. (1) is an interaction between the atoms, which reduces the energy of one of them if it orients itself in the same direction as the preponderant one of the others, which is measured by $\mathcal{M}$. The orientation of all the atomic magnets of the system in the same direction, even at zero magnetic field, results then in a considerable decrease in energy, but also in a decrease in the available phase space, a decrease in entropy. Above the Curie temperature T_c the temperature motion is sufficient to maintain a random orientation. If a field is applied, however, the interaction tends to aid the magnetic field, so that for paramagnetic substances $T-T_c$ appears in the denominator of eq. (5) instead of T. As the temperature is lowered, and T_c approached from above, the interaction force becomes more and more predominant. Even a weak field, instead of orienting only a few atoms, merely supplies the initiative. The force a brings about *almost* a landslide, although, as long as T is greater than T_c, the number oriented is still proportional to the magnetic field for very weak fields. Below T_c there exists a real landslide. The temperature motion of the atoms is not sufficient to counteract the energy decrease produced by a common preferential orientation.

The description supplied by eq. (2) is essentially in agreement with the experimental observations on ferromagnetic materials. From the measured Curie temperatures of the remanent magnetic substances one may then calculate backward the magnitude of a, the strength of the interaction. For iron, the Curie temperature is about 1000 K. The volume per atom V/N is 11.8×10^{-24} cm^3. If, for μ_0, 1 Bohr magneton, $\mu_0\cong b_0=10^{-20}$, is substituted, one obtains for the dimensionless quantity a the value 10^4, an entirely different order of magnitude compared to the $4\pi/3$ occurring in the interaction of electric dipoles.

It might perhaps be mentioned here that not all samples of iron, which at room temperature are all below the Curie point, show permanent magnetism. This is because the macroscopic material is composed of domains of microscopic size. Presumably, below T_c there exist individual domains which have permanent magnetic moments, but the orientation of the moments of the individual domains is random until a slight field is applied to bring them into alignment.

In substances that are paramagnetic at room temperature, the peculiarly strong interaction of magnetic moments is missing. However, the equivalent of the Clausius-Mossotti formula [eq. (12f.1)] still predicts an interaction of the type considered here with $a=4\pi/3$. Paramagnetic substances would then have a Curie point, according to eq. (4), which depends on ρ but lies, for most salts, well below 1 K. Below this temperature the substances would be ferromagnetic. Actually, there are probably always interactions between the magnetic spins and orbital states of the electrons, which are responsible for the large values of a in the case of ferromagnetic materials, although these may be, and are in many cases, very small. It should also be noted that antiferromagnetic crystals exist. For these there is an interaction term corresponding essentially to a negative a in eq. (1). In general, the value of a can never be expected to be $4\pi/3$, except by a numerical accident, but this value may be regarded as giving a lower-limit order of magnitude.

The discussion of ferromagnetism given in this section is practically the classical theory of Weiss[(1)] which so far as it goes may be regarded as essentially correct. The chief contribution of quantum mechanics has not been to alter the equations, except unimportantly by the substitution of the Brillouin function for eq. (2), but to explain the occurrence of a in eq. (1). The nature of the strong interaction potential in ferromagnetic substances had long been a mystery. Of course, the force need not be of magnetic origin. It is entirely unessential that the energy change is written as though it depends on the magnetic moment μ_0. Both μ_0 and a are characteristics of the material, and the force tending to orient the magnetic moments of the atoms (or, rather, the spins of the electrons in the metal) may have nothing to do with magnetism. The only reason that it is connected at all with the magnetic field $\mathcal{H}$ is that both $\mathcal{H}$ and this force tend to accomplish the same result, a net orientation of the electrons in the atoms in such a way as to tend to point all their magnetic moments in the same direction.

It remained for Heisenberg[(2)] to explain the orienting force introduced by a in eq. (1) as due to the exchange forces between electrons. The resultant energy of one atom includes terms proportional to higher powers of $\mathcal{M}$ than the first, which alone is considered in eq. (1), and the complete quantum-mechanical development is not as simple as that given here.

The origin of the orienting force is intimately connected with the necessity for using antisymmetric eigenfunctions for electrons. If two electrons have parallel spins, their spin function is necessarily symmetric. The orbital function of their positions must then be antisymmetric.

The antisymmetric and symmetric orbital functions differ primarily in the fact that the probability of the electrons being spatially close together

[(1)] P. Weiss, *J. Phys.*, **6,** 667 (1907).

[(2)] W. Heisenberg, *Z. Phys.*, **49,** 619 (1928).

is less for the antisymmetric function. Since the electrons repel each other, these functions, other things being equal, have lower energy. The result is a certain force tending to line up the electron spins in the same direction so as to create the low-energy antisymmetric orbital functions.

12j. MAGNETIC COOLING

Low temperatures are ordinarily produced by use of the Joule-Thomson effect in gases. If the lowest-boiling gas, helium, is so liquified and then evaporated under reduced pressure, temperatures somewhat under 1 K can be produced. There exists, however, a practical lower limit to the temperature that may be obtained in this manner, since the rate of evaporation, which decreases with decreasing temperature, finally becomes as low as the heat leak into the apparatus. Recourse has been taken to adiabatic demagnetization to cool even further.

The principle of the method is simple. In the absence of a magnetic field the random orientation of the angular momentum vectors of the (magnetically active) ions introduces an entropy $R \ln(2j+1)$ per gram-atom in excess of that due to lattice vibrations. The application of a strong magnetic field, while the material is kept at constant temperature, tends to orient all the ions in the energetically lower directions, reducing the entropy. This reduction in entropy at constant temperature in the strong field demands a flow of heat, $q = \Delta S/T$, out of the material (calculated as if the field were applied reversibly). If the material, while under the influence of the field, is insulated so that no heat can flow in or out, and the field removed slowly enough so that the whole process takes place reversibly, no change in the total entropy of the material occurs. The material is therefore cooled, since entropy, and consequently heat, flow from the lattice vibrations to raise the orientation entropy of the magnetic moments to their original value of $R \ln(2j+1)$.

The process is usually carried out by applying the field while the salt containing the magnetically active ions is kept in contact with liquid helium boiling at reduced pressure. The initial temperature T_i in the process is then about 1 K if the cooling is done by pumping on ^{4}He. More recently, dilution refrigeration has been used, which utilizes the properties of a two-phase mixture of ^{3}He in ^{4}He, by which temperatures below 0.1 K can be obtained. The entropy due to the lattice vibrations at these initial temperatures is extremely low, so that the method is very efficient. Temperatures as low as about 0.004 K have been attained in this manner. Indeed, it is quite practicable at 1 K to reduce the entropy of some salts in a magnetic field by a greater amount than the total lattice vibrational entropy at the initial temperature, and of course even more so if the initial temperature is 0.1 K. The simple consideration given above leads one to expect cooling to 0 K in such a case, and of course there is obviously a flaw in the argument that leads to such a prediction.

The error made lies in the assumption that a random orientation of the magnetic moments prevails at all temperatures in the absence of a field. If any interaction at all exists between the moments, either through a true magnetic force or owing to an interaction with the electric crystalline field, a particular specified orientation of zero entropy will be stable at 0 K. Essentialy, this just means that every paramagnetic material becomes ferromagnetic or antiferromagnetic below some critical temperature. The entropy of orientation of the magnets is greater than zero above 0 K but approaches the high-temperature value of $R \ln(2j+1)$ only at temperatures for which kT exceeds the energy of interaction.

It is seen that the effectiveness of the magnetic cooling depends peculiarly on the substance used. If the material has a high density of ions with great magnetic moment, that is, if it has a high magnetic susceptibility, the change in entropy ΔS at the initial temperature T_i with a given magnetic field is high. In this event a relatively high initial temperature, or a relatively low magnetic field, may be used to attain a given final temperature. However, high magnetic susceptibility in the salt favors a large interaction betwen the magnetic moments, and if low initial temperatures and high fields are available, the final temperature reached may be appreciably lower if a salt of low magnetic susceptibility is used. For this reason salts with relatively small susceptibilities, that is, with ions not having extraordinarily high moments, and very much diluted with water of crystallization or other inert ions, are used to attain the lowest temperatures.

The interactions that cause deviations from the ideal behavior calculated here for ions with magnetic moments completely unaffected by their surroundings are of two kinds. That which usually has the higher interaction energy, and therefore becomes important at the higher temperature, is the interaction between the magnetic moment and the field of the crystal due to the surrounding (magnetically inactive) molecules or ions. This tends to favor one of the possible axes of orientation of the magnetic moment of each ion relative to that of its neighbors, giving it a lower energy than the others. However, the crystalline field never distinguishes between the two possible directions along this axis, that is, it always leaves two of the $2j+1$ states with the same energy. As a result, if this perturbation alone were present, the entropy at 0 K would be $R \ln 2$. There remains the true magnetic interaction which tends to line up all the spins in the same direction along the axis favored by the crystal field, and which reduces the entropy at 0 K to zero. The purely dipole magnetic interaction is, at least approximately, given by $a = 4\pi/3$, but the other effects are usually larger.

The actual temperatures obtained are somewhat difficult to determine. One might, for instance, measure the magnetic susceptibility in a very weak magnetic field [eq. (12g.4)] and attempt to use Curie's law to determine the temperature. However, marked deviations from eq. (12g.4) are to be expected at these low temperatures. This method, however is used to determine a qualitative temperature scale, the T^* scale. By experiment, then, one observes that adiabatic demagnetization from the same initial temperature T_i, but from different applied fields $\mathcal{H}$, results in different end temperatures T^* on this scale. In short, one can correlate the magnetic field $\mathcal{H}$ used in the cooling with the final qualitative temperature T^*, always starting from the same initial temperature.

In order to determine the thermodynamic temperature T as a function of T^* one proceeds as follows. The entropy decrease at T_i in going from zero magnetic field to the field $\mathcal{H}$ may be measured by the heat evolved or calculated exactly by equations given later in this section, since at this relatively high temperature, 1 K, the perturbing influences already discussed play no role. After the adiabatic removal of the field, the amount by which the entropy is below that at T_i is then known in terms of the qualitative scale T^*. One knows, then, S as a function of T^*, and in particular one knows $(\partial S/\partial T^*)_{V,\mathcal{H}}$ as a function of T^* at zero field.

The next measurement is that of heat capacity as a function of this qualitative temperature scale T^*. This may be accomplished by determining the rate of heating (on a T^* scale) as the material absorbs γ-rays of known intensity. One then knows $(\partial E/\partial T^*)_{V,\mathcal{H}}$ at zero field as a function of T^*. From the thermodynamic relationship

$$(dS)_{V,\mathcal{H}} = \frac{1}{T}(dE)_{V,\mathcal{H}},$$

one now finds the true temperature T as a function of T^* by

$$T = \frac{(\partial E/\partial T^*)_{V,\mathcal{H}}}{(\partial S/\partial T^*)_{V,\mathcal{H}}}.$$

The entropy $S_{\mathcal{H}}$ at a given field, due to the orientations of the ions, may be readily calculated. It is, from eq. (7e.6) which gives the contribution to the entropy due to the internal partition function Q_i of independent molecules,

$$S_{\mathcal{H}} = R\frac{d}{dT}(T \ln Q_{\mathcal{H}}) \tag{12j.1}$$

per gram atom of magnetically active ions, and $Q_{\mathscr{H}}$ is the sum

$$Q_{\mathscr{H}} = \sum_{m=-j}^{m=+j} e^{mg\beta b_0 \mathscr{H}}. \tag{12j.2}$$

Carrying out the differentiation one finds

$$S_{\mathscr{H}} = R\left(\ln Q_{\mathscr{H}} - \frac{\sum_{m=-j}^{m=+j} mg\beta_0 \mathscr{H} e^{mg\beta b_0 \mathscr{H}}}{\sum_{m=-j}^{m=+j} e^{mg\beta b_0 \mathscr{H}}} \right).$$

The fraction is just $\beta\langle\mu\rangle$, as is seen by comparison with eq. (12h.6) which is, with eq. (12h.9), $(jg\beta b_0\mathscr{H})B_j(jg\beta b_0\mathscr{H})$. By the same method employed to obtain eq. (12h.9) it is seen that

$$\begin{aligned} \ln Q_{\mathscr{H}} &= \ln(e^{(j+1/2)g\beta b_0 \mathscr{H}} - e^{-(j+1/2)g\beta b_0 \mathscr{H}}) - \ln(e^{(1/2)g\beta b_0 \mathscr{H}} - e^{(-1/2)g\beta b_0 \mathscr{H}}) \\ &= \ln \sinh\left(\frac{j+\frac{1}{2}}{j}\, y\right) - \ln \sinh\left(\frac{1}{2j}\, y\right), \end{aligned} \tag{12j.3}$$

with

$$y = jg\beta b_0 \mathscr{H} \tag{12j.4}$$

Using this in eq. (2) one obtains

$$S_{\mathscr{H}} = R\left[\ln \sinh\left(\frac{j+\frac{1}{2}}{j}\, y\right) - \ln \sinh\left(\frac{1}{2j}\, y\right) - yB_j(y)\right]. \tag{12j.5}$$

$S_{\mathscr{H}}$ varies from $R\ln(2j+1)$ at $y=0$ to zero as y approaches infinity. This function then represents the dependence of the entropy on the magnetic field $\mathscr{H}$ at constant temperature. Other more direct measurements of the absolute temperatures using nmr are now available.

Using this technique with electronic magnetic moments temperatures of the order of 3 to 4 mdeg absolute (3 to 4×10^{-3} K) are attainable. The decrease of a little over 1 K from the 1.2 K obtainable by pumping on liquid helium may sound unimpressive. It sounds more spectacular if one talks in terms of the variable $\beta = 1/kT$ which is increased several hundredfold. The properties of superconductors and superfluids, and particularly of the isotope of helium of mass 3, are of great interest in this temperature range.

Far lower temperatures are obtainable by using the same procedure on nuclear magnetic moments. The Bohr magneton, $b_0 = eh/2\pi m_e c$ (with m_e the electronic mass), of 0.9274×10^{-20} emu [eq. (12h.3)] enters all the significant equations of this section in the combination $g\beta b_0\mathscr{H}$. The unit of nuclear magnetic moment, $b_n = eh/2\pi m_n c$ (with m_n the nuclear mass), is nearly 2000-fold lower than the Bohr magneton. The factor g, which also appears for different nuclear states is, as it is for the electronic levels, not much greater than unity. For the same magnetic field $\mathscr{H}$, then, the equations for nuclear magnetic cooling involving $g\beta b_n \mathscr{H}$ have their effective range with β about 1000-fold or more greater and temperatures about 1000-fold lower. Starting with a sample already cooled into the millidegree range by electronic magnetic cooling one can attain nuclear spin temperatures of the order of microdegrees, $\sim 10^{-6}$ K.

These low spin temperatures, although spectacular, are not of much value in the investigation of the properties of other materials such as liquid or solid ^{3}He until the helium is cooled. Both the actual attainment of the low nuclear spin temperatures and their use in cooling other materials require remarkably clever and ingenious cryogenic experimental techniques. One of these is to use thin wires of a superconducting metal from which the superconductivity can be removed by applying a sufficient magnetic field. In the normal (nonsuperconducting) state at these low temperatures the conductivity is good, and the wires make good thermal contact between the electronically magnetic material and the electronically nonmagnetic matter having only nuclear magnetic properties. It is possible to attain good thermal equilibration even at fairly large spatial separation. Removal of the magnetic field leaves the wires in the (electrically) superconducting state. They then have negligible thermal conductivity.

The extremely low spin temperatures are impressive and an interesting accomplishment by themselves, but they alone do not lead to useful information about the properties of other materials at very low temperatures (high values of $\beta = 1/kT$). Two related obstacles remain. First, the lattice vibrational heat capacity due to the phonons, proportional to T^3, is extremely small at these temperatures. For a lattice Debye temperature of 100 K, at $T = 10^{-3}$ K the value of $u = \theta/T$ is 10^5. From eq. (10e.8), $C_V = 3Nk4D(u)$ for $u \gg 1$, and eq. (10e.6), in which, for $u \gg 1$, $D(u) = (\pi^4/5)u^{-3}$, we find that, with $\pi^4 \sim 10^2$, C_V is 10^{-13} times the Dulong-Petit value of $3Nk = 6$ cal/g-atom, extremely small. Second, the coupling between nuclear spin and phonons is extraordinarily weak, as it would also be between the lattice phonons and any other interesting material such as ^{3}He. It would take an inordinately long time to use the low spin temperature to cool the ^{3}He to an equlibrium value from an insulating crystal. The solution lies in using a metal for the nuclear spin cooling. From eq. (11f.28) for a metal the contribution C_V (electronic) to the heat capacity is $3Nk(\pi^2/6)(kT/\mu_0)$, where the Fermi energy μ_0, divided by k from eq. (11e.7) is $\mu_0/k \cong 3 \times 10^5 V^{-2/3}$ K, with V the metallic volume in cubic centimeters per mole of valence electrons. If $V = 30$ cm^3, $V^{2/3} \cong 10$ cm^2, and we have at $T = 10^{-3}$ K $kT/\mu_0 \cong 3 \times 10^{-8}$ and C_V

(electronic)$\cong 3Nk \times (5 \times 10^{-8})$. The electronic heat capacity is about 5×10^5-fold greater than that due to the lattice vibrations.

The coupling between nuclear spin and the electrons is better than that with the phonons, but still not rapid. The time required to reach equilibrium within the metal is given by the Korringer relation, $\tau = \kappa/T$, where τ is the relaxation time, that necessary to reduce the deviation from equilibrium by the fraction e^{-1}, and κ is an empirical constant for each metal. Its value for copper is $\kappa = 1.27$ sec K. Using copper as metal, temperatures in the metal below 4×10^{-4} K have been obtained, and in ^{3}He absorbed in a copper sponge below 10^{-3} K.(1)

12k. THERMODYNAMIC EQUATIONS IN ELECTRIC FIELDS

Throughout this chapter, up to the last section, statistical mechanics was employed solely with the use of the Boltzmann factor $e^{-u/kT}$ to weight the various orientations of the molecules, atoms, and ions in order to ascertain the total electric and magnetic polarization. Only in the last section was a common thermodynamic function S calculated. This was chiefly occasioned by the fact that the thermodynamics of materials in magnetic and electric fields is not a familiar subject.

In previous chapters, the experimental properties of the systems treated might have been calculated without explicit use of the word thermodynamics or explicit naming of the various functions that occur and are also common to the general thermodynamic treatment.

The material of this chapter might have been presented by first developing the equations for the thermodynamic properties of macroscopic systems in electric and magnetic fields and subsequently calculating the thermodynamic functions by statistical mechanics. This method is indicated here.

In order to illustrate, rather than to prove, the thermodynamic equations for a system under the influence of an electric field, a simple plate condenser is considered, in which all linear dimensions of the plates are large compared to their distance apart. Particularly in static fields, because of the long range of the electric and magnetic forces, the behavior of the system depends critically on its size and shape. At high frequencies, when the wavelength is much smaller than the system dimensions, this is not the case.

Two conducting parallel plates of area A each, at a distance l apart in a

(1) J. M. Dundon, J. M. Goodkind, *Phys. Rev. Lett.*, **32,** 1343 (1974); J. M. Dundon, D. L. Stoffa, and J. M. Goodkind, *Phys. Rev. Lett.*, **30,** 843 (1973); E. B. Osgood and J. M. Goodkind, *Phys. Rev. Lett.*, **18,** 894 (1967).

vacuum, form a condenser of capacity

$$C=\frac{\mathscr{q}}{\mathscr{V}}=\frac{A}{4\pi l}. \tag{12k.1}$$

The ratio of the charge $\mathscr{q}$ on the plates to the voltage $\mathscr{V}$ between them is C. The electric work of charging the plates gives the free energy increase as a function of $\mathscr{q}$ as

$$w=\int_0^{\mathscr{q}}\mathscr{V}\,d\mathscr{q}'=\int_0\frac{4\pi l}{A}\mathscr{q}'\,d\mathscr{q}'=\frac{4\pi l}{A}\frac{\mathscr{q}^2}{2}. \tag{12k.2}$$

The electrical displacement $\mathfrak{D}$ between the plates is

$$\mathfrak{D}=4\pi\frac{\mathscr{q}}{A}=4\pi\frac{\mathscr{q}l}{V}, \tag{12k.3}$$

where in the right-hand expression the volume, $V=Al$, of the field between the plates is introduced.

The field $\mathscr{E}$ is

$$\mathscr{E}=\frac{\mathscr{V}}{l}, \tag{12k.4}$$

which is equal to $\mathfrak{D}$ when no material is placed between the condenser plates, that is, $\mathfrak{D}=\mathscr{E}$ in a vacuum. Equation (2) may then be rewritten as

$$w=\frac{V}{4\pi}\int_0^{\mathfrak{D}}\mathscr{E}\,d\mathfrak{D}'=\frac{V}{8\pi}\mathfrak{D}^2. \tag{12k.5}$$

Of course, the final form of eqs. (2) and (5) agree if relation (3) is used. Both eq. (3) for $\mathfrak{D}$ and eq. (4) for $\mathscr{E}$ are valid if a material of dielectric constant ϵ is introduced between the plates of the condenser.

$$C=\frac{\mathscr{q}}{\mathscr{V}}=\frac{A\epsilon}{4\pi l}, \tag{12k.6}$$

so that

$$\mathscr{E}=\frac{1}{\epsilon}\mathfrak{D}. \tag{12k.7}$$

The field $\mathscr{E}$ and displacement $\mathfrak{D}$ are not equal in a material of dielectric constant ϵ differing from unity.

The work of charging the condenser is still $\int_0^{\mathscr{q}}\mathscr{V}\,d\mathscr{q}$, which may be written

$$w=\frac{V}{4\pi}\int_0^{\mathfrak{D}}\mathscr{E}\,d\mathfrak{D}'=\frac{V}{8\pi}\frac{\mathfrak{D}^2}{\epsilon}=\frac{V}{8\pi}\mathscr{E}\mathfrak{D}. \tag{12k.5'}$$

If the process is carried out at constant temperature and pressure, the work done on the system determines its increase in Gibbs free energy G.

One now makes the distinction that the free energy of the *field* (and its energy, since the entropy of an electric field in a vacuum is zero) is still given by eq. (5), and the difference between eqs. (5′) and (5) lies in the free-energy change of the material (at constant temperature) as the field is applied. For ΔG, the change in free energy of the material, as a function of the electrical displacement $\mathscr{D}$, one has

$$\Delta G(\mathscr{D}) = \frac{V}{8\pi}\mathscr{D}^2\left(\frac{1}{\epsilon}-1\right) = -\frac{V}{8\pi}\mathscr{D}^2\frac{\epsilon-1}{\epsilon}. \tag{12k.8}$$

It was assumed, in integrating eq. (5′), that ϵ is actually a constant. This method was employed since the equations for a condenser in terms of ϵ are more familiar than those in terms of the polarization.

One may proceed somewhat more generally. The application of the charge q to the plates induces a polarization $\mathscr{P}$ in the material between them. The dimensions of polarization are dipole moment per unit volume, that is, charge times (vector) length in the direction normal to the plates divided by volume. $\mathscr{P}$ has therefore the same dimensions as $\mathscr{E}$ or $\mathscr{D}$. This polarization in the material brings an induced charge of opposite sign up to the plates, partially neutralizing the applied charge q, and the field $\mathscr{E}$ is reduced to

$$\mathscr{E} = 4\pi\left(\frac{ql}{V}-\mathscr{P}\right) = \mathscr{D} - 4\pi\mathscr{P} \tag{12k.9}$$

The use of the definition [eq. (12c.2)] of ϵ by $\epsilon - 1 = 4\pi P/\mathscr{E}$ is seen to result in eq. (7), $\mathscr{E} = \mathscr{D}/\epsilon$.

The voltage $\mathscr{V}$ produced on the plates is $\mathscr{V} = \mathscr{E}l$, so that the work of charging $\int \mathscr{V}\, dq$ is, as always, given by eq. (5) as $(V/4\pi)\int \mathscr{E}\, d\mathscr{D}$. Using eq. (9) for $\mathscr{E}$ one has

$$(dG)_{T,P} = -V\mathscr{P}\,(d\mathscr{D})_{T,P}, \qquad \left(\frac{\partial G}{\partial \mathscr{D}}\right)_{T,P} = -V\mathscr{P} \tag{12k.10}$$

It is to be noted that, if a dielectric constant exists, that is, if $\mathscr{P}$ is always proportional to ϵ, as is normally the case, $\mathscr{P}\, d\mathscr{D}$ and $\mathscr{D}\, d\mathscr{P}$ are equal and may be used interchangeably.

If one now wishes to calculate the free energy G for a gas from statistical mechanics, the Hamiltonian or energy of each molecule must be expressed as a function of the electrical displacement $\mathscr{D}$ (which gives the effect of the true charges on the plates of the condenser) plus terms due to the interaction of the molecules. The interaction energy of the molecules is different in the presence of the field than in its absence, since the field tends to orient the permanent dipoles in one direction.

If one wishes to obtain an equation equivalent to eq. (12f.2) in which the Lorentz–Lorenz force has been taken into account, the statistical treatment must include the interaction between the molecules, which are therefore not independent, and the method of calculation for the perfect gas may not be rigorously employed.

One may, however, use a rather illogical method which leads to almost the correct results. In this it is simply assumed that the averaging over the interaction betwen the molecules leads to eq. (12f.1) for the local electric force acting on one molecule. The method is, then, logically no whit superior to that used in Sec. 12f, being indeed somewhat inferior, but is followed here to demonstrate the use of eq. (10).

It is first necessary to investigate the effect of the electric force $\mathcal{F}$ on the energy of the molecule due to the polarization term α. It can be shown that the energy change u_α in the molecule due to this term is

$$u_\alpha = -\tfrac{1}{2}\alpha\mathcal{F}^2. \tag{12k.11}$$

This energy is made up of two terms of opposite sign: one, the electric energy in the field, is $-\alpha\mathcal{F}^2$; the second is the internal potential energy of the molecule, which is $+\frac{1}{2}\alpha\mathcal{F}^2$. The electric energy is just $-p_i\mathcal{F}$, since the induced dipole is directed along the field, and from eq. (12d.1) $p_i = \alpha\mathcal{F}$, so that one finds this energy to be $-\alpha\mathcal{F}^2$. The increased internal potential energy of the molecule is due to a restoring force f tending to keep the positive and negative charges ze and $-ze$ from being displaced. The displacement λ of their centers is $\lambda = p_i/ze = \alpha\mathcal{F}/ze$ when the electric force tending to pull them apart, which must be balanced by f, is $ze\mathcal{F}$. Using $f = -ze\mathcal{F}$, $d\lambda = (\alpha/ze)\,d\mathcal{F}$, and integrating

$$-\int f\,d\lambda = \int_0^{\mathcal{F}} \alpha\mathcal{F}'\,d\mathcal{F}' = \tfrac{1}{2}\alpha\mathcal{F}^2,$$

one obtains the internal potential as half the negative of the electric energy.

One may therefore write for the potential energy of the single molecule in the field $\mathcal{F}$ the equation

$$u = -\mathcal{F}p_0 \cos\theta - \tfrac{1}{2}\alpha\mathcal{F}^2. \tag{12k.12}$$

This equation is satisfactory enough. The difficulty that arises is only what to substitute for the electric force $\mathcal{F}$.

If the electrical displacement $\mathcal{D}$ is substituted for $\mathcal{F}$, one neglects entirely the interaction between the molecules, not only that due to the Lorentz–Lorenz force making $\mathcal{E}_{\text{loc}}$ different from $\mathcal{E}$ [eq. (12f.1)], but also the surface polarization which accounts for the difference between the electric field $\mathscr{E}$ and the displacement $\mathscr{D}$. Logically, one should use $\mathscr{D}$ for $\mathscr{F}$ in eq. (12) and add interaction terms between the dipoles of the molecules.

We adopt an easy way out, which leads to approximately correct results.

Since from eq. (9) $\mathscr{E}=\mathscr{D}-4\pi\mathscr{P}$, and from eq. (12f.1) $\mathscr{E}_{\text{loc}}=\mathscr{E}+4\pi\mathscr{P}/3$, we find $\mathscr{E}_{\text{loc}}=\mathscr{D}-8\pi\mathscr{P}/3$. The correction term $-8\pi\mathscr{P}/3$ represents the averaged contribution to the electric force acting on one molecule due to the others. This must be halved, since it is a mutual force acting between two molecules, and we may approximately account for it by assigning half the term to each molecule, writing

$$\mathscr{F}=\mathscr{D}-\frac{4\pi}{3}\mathscr{P}. \tag{12k.13}$$

It is now a question of straight substitution into familiar equations to arrive at the equivalent of eq. (12f.2). The potential u in eq. (12) is a function of the angle θ alone. In calculating the classical factor of the partition function Q due to the angle for the molecule, one has, without an electric field, only the integral over the volume element $\int_0^\pi \frac{1}{2}\sin\theta\, d\theta = 1$. One must now replace this by

$$\int_0^\pi \tfrac{1}{2}\sin\theta e^{-u/kT}\, d\theta = Q_{\mathscr{F}}. \tag{12k.14}$$

In this, $Q_{\mathscr{F}}$ gives the factor of the partition function Q affected by the field $\mathscr{F}$ and normalized in such a way that $Q_{\mathscr{F}}=1$ when $\mathscr{F}=0$.

The additional term in the free energy G due to the electric field is $-NkT\ln Q_{\mathscr{F}}$, so that

$$G=G_0-NkT\ln Q_{\mathscr{F}}, \tag{12k.15}$$

where G_0 is the free energy in zero field.

Substituting eq. (12) for u into eq. (14) for $Q\mathscr{F}$, it is seen that, except for the factor $e^{\alpha\mathscr{F}^2/2kT}$, which does not contain θ, the integral is that already evaluated in eq. (12b.4), and

$$\begin{aligned} Q_{\mathscr{F}} &= e^{\alpha\beta\mathscr{F}^2/2}(\beta p_0\mathscr{F})^{-1}\sinh(\beta p_0\mathscr{F}) \\ &\cong e^{\alpha\beta\mathscr{F}^2/2}[1+\tfrac{1}{6}(\alpha\beta p_0\mathscr{F})^2+\cdots]. \end{aligned} \tag{12k.16}$$

Using this in eq. (15),

$$G=G_0-N\left(\frac{\alpha}{2}+\tfrac{1}{6}\beta p_0^2\right)\mathscr{F}^2. \tag{12k.17}$$

There remains still a considerable amount of juggling to arrive at an equation for the dielectric constant. One uses eq. (10) to find the polarization $\mathcal{P} = -(\partial G/\partial \mathcal{D})_{T,P}/V$, with eq. (13) for $\mathcal{F}$ in eq. (17). The dielectric constant ϵ is defined by eq. (12c.2) as $\epsilon - 1 = 4\pi\mathcal{P}/\mathcal{E}$, which with eq. (7) where $\mathcal{E} = \mathcal{D}/\epsilon$ gives $\mathcal{P}/\mathcal{D} = (\epsilon - 1)/4\pi\epsilon$. Using this in eq. (13),

$$\mathcal{F} = \mathcal{D}\left(1 - \frac{\epsilon - 1}{3\epsilon}\right) = \mathcal{D}\frac{2\epsilon + 1}{3\epsilon}, \qquad \frac{\partial \mathcal{F}}{\partial \mathcal{D}} = \frac{2\epsilon + 1}{3\epsilon}. \tag{12k.18}$$

With eqs. (10) and (17),

$$\begin{aligned} \mathcal{P} = -\frac{1}{V}\left(\frac{\partial G}{\partial \mathcal{D}}\right)_{T,P} &= -\frac{1}{V}\left(\frac{\partial G}{\partial \mathcal{F}}\right)_{T,P} \frac{d\mathcal{F}}{d\mathcal{D}} \\ &= \frac{N}{V}(\alpha + \tfrac{1}{3}\beta \not{p}_0^2)\mathcal{D}\left(\frac{2\epsilon + 1}{3\epsilon}\right)^2 \end{aligned} \tag{12k.19}$$

is obtained. Finally, noting that $(2\epsilon + 1)^2 = 3\epsilon(\epsilon + 2) + (\epsilon - 1)^2$, so that $(2\epsilon + 1)^2/9\epsilon^2 = (\epsilon + 2)/3\epsilon + (\epsilon - 1)^2/9\epsilon^2$, and neglecting $(\epsilon - 1)^2$ as a second-order correction, one arrives at eq. (12f.2) by using

$$\frac{\mathcal{P}}{\mathcal{D}} = \frac{\epsilon - 1}{4\pi\epsilon} = \frac{N}{V}(\alpha + \tfrac{1}{2}\beta \not{p}_0^2)\frac{\epsilon + 2}{3\epsilon},$$

$$\frac{\epsilon - 1}{\epsilon + 2} = \frac{4\pi}{3}\frac{N}{V}(\alpha + \tfrac{1}{3}\beta \not{p}_0^2). \tag{12k.20}$$

The method used in this section is rather awkward. Actually, the calculation of the behavior in electric fields is one of the few examples in which the statistical method is applied most easily without following an essentially thermodynamic method. Of course, the development in Sec. 12f is also by no means rigorous, and as far as we are aware there is no strict statistical method by which eq. (12f.2) has been derived.

A careful and completely satisfactory method would use the electrical displacement $\mathcal{D}$ for $\mathcal{F}$ in eq. (12) and, with the equations for a system of dependent particles, insert the interaction between the electric dipoles as an additional term in the Hamiltonian.

CHAPTER 13

THE DENSITY MATRIX

13a. INTRODUCTION

In the previous chapters all the fundamental equations started with a summation over the quantum states indexed by a set $\mathbf{K}$ of the quantum numbers of the system. The quantity summed was the probability $W_{\mathbf{K}}$ that a member of an ensemble will be found in the state $\mathbf{K}$. In many cases summation over quantum states could be approximately represented by integration over the classical phase $\mathbf{q}^{(\Gamma)}\mathbf{p}^{(\Gamma)}$ divided by $h^{\Gamma}N!$. In all other discussions of specific problems it was assumed that the states of quantum number $\mathbf{K}$ were the stationary states characterized by a state function $\Psi_{\mathbf{K}}(\mathbf{q}^{(\Gamma)})$ which is a solution of the time-independent Schrödinger equation

$$\mathscr{H}\Psi_{\mathbf{K}} = E_{\mathbf{K}}\Psi_{\mathbf{K}}, \tag{13a.1}$$

in which $\mathscr{H}$ is the quantum-mechanical Hamiltonian operator and, $E_{\mathbf{K}}$ is energy (see Sec. 3b).

Actually, as pointed out, we always use a fictitious approximate Hamiltonian operator $\mathscr{H}_0$ in eq. (1) and the perturbation $\Delta\mathscr{H}$. With the relation

$$\mathscr{H}_{\text{true}} = \mathscr{H}_0 + \Delta\mathscr{H}, \tag{13a.2}$$

the perturbing $\Delta\mathscr{H}$ causes transitions between the approximate solutions $\Psi_{\mathbf{K}}^{(0)}$ to eq. (1), with $\mathscr{H}_0$ replacing $\mathscr{H}$. Nevertheless, our formalism always required the use of eq. (1) with a reasonable approximation $\mathscr{H}_0$ to the true $\mathscr{H}$ to solve for the $E_{\mathbf{K}}^{(0)}$'s, and *then* the use of these in the partition functions $\sum e^{-\beta E_{\mathbf{K}}}$ (Sec. 3j).

Needless to say, the solution to eq. (1) for $\Psi_{\mathbf{K}}(\mathbf{q}^{(\Gamma)})$ and $E_{\mathbf{K}}$ is completely impractical unless a reasonable approximate $\mathcal{H}_0$ is used that is separable (Sec. 3d) into a sum of terms each depending on a subset $\mathbf{q}_i^{(\gamma_i)}$ of only a very small number of coordinates, preferably a single coordinate. This is possible, with some approximations, for even relatively complicated molecules in a perfect gas, since the assumption of no mutual potential between different molecules automatically makes the Hamiltonian separable into the coordinates and momenta of single molecules. Similarly, for crystals the truncation of the potential energy into a power series of coordinates giving the displacements from an equilibrium lattice array at the quadratic terms, the harmonic approximation, permits an analysis in normal coordinates. The Hamiltonian is then separable into single coordinates each of harmonic oscillator form with different frequencies ν_i, $1 \leq \nu_i \leq 3N$, with N the number of atoms. The analysis of the frequency distribution $N(\nu)$ is still a formidable problem but has been accomplished with some precision in relatively simple cases. For the treatment of dense gases and of liquids we used the assumption that classical mechanics was adequate.

In this chapter we discuss a formal method which obviates the necessity for ever solving the time-independent Schrödinger equation [eq. (1)]. The formalism permits the introduction of a function in the $\mathbf{p}^{(\Gamma)}\mathbf{q}^{(\Gamma)}$ classical phase space, the gamma space, which is the quantum-mechanical analogue of the classical probability density $W_Q(\mathbf{p}^{(\Gamma)}, \mathbf{q}^{(\Gamma)})$. It is the Wigner function, obtained from the elements of the density matrix.

The Wigner function for an ensemble has the following properties. It is a real function, not complex. Its integral over the momenta $\mathbf{p}^{(\Gamma)}$ is the probability density $W(\mathbf{q}^{(\Gamma)})$ in the coordinate space. Its integral over the coordinate space $\mathbf{q}^{(\Gamma)}$ is the probability density $W(\mathbf{p}^{(\Gamma)})$ in the momentum space. In the limit that Planck's constant h goes to zero it becomes equal to the classical probability density $W_{\text{cl}}(\mathbf{p}^{(\Gamma)}, \mathbf{q}^{(\Gamma)})$. We prove this in Sec. 13i. In Sec. 13i we also give the mathematical proof of the assertion made earlier that integration over $(N!)^{-1}h^{-\Gamma}\, d\mathbf{p}^{(\Gamma)}\, d\mathbf{q}^{(\Gamma)}$ is the semiclassical equivalent of counting solutions to the time-independent Schrödinger equation [eq. (1)].

The unintegrated Wigner function of both $\mathbf{p}^{(\Gamma)}\mathbf{q}^{(\Gamma)}$ is, however, not a true probability density as evidenced by the fact that it may, and often does, take negative values at places in the gamma space. These are in the regions of nonclassical finite $W(\mathbf{q}^{(\Gamma)})$, where the eigenfunctions $\Psi_{\mathbf{K}}(\mathbf{q}^{(\Gamma)})$ at energies near the classical energy at $\mathbf{p}^{(\Gamma)}\mathbf{q}^{(\Gamma)}$ of the systems extend into portions of the configuration space for which the potential energy $U(\mathbf{q}^{(\Gamma)})$ is greater than the total energy $E_{\mathbf{K}}$. In these portions of the configuration space the kinetic energy must be negative. However, since integration over either momenta $\mathbf{p}^{(\Gamma)}$ or coordinates $\mathbf{q}^{(\Gamma)}$ gives the true always positive probability densities of the other ($\mathbf{q}^{(\Gamma)}$ or $\mathbf{p}^{(\Gamma)}$, respectively), it follows that integration of any classical function $f(\mathbf{q}^{(\Gamma)})$ or $g(\mathbf{p}^{(\Gamma)})$, or any sum of two such functions multiplied by the Wigner function and integrated over both $\mathbf{p}^{(\Gamma)}$ and $\mathbf{q}^{(\Gamma)}$, gives the average value of the function in the systems of the ensemble.

The reader not thoroughly acquainted with the use of matrices in quantum mechanics is advised to consult Appendices VII and VIII.

13b. THE COORDINATE REPRESENTATION

The quantum state of a system i that is a member of an ensemble may be described by a single-valued square integrable continuous function $\Psi_i(\mathbf{q}^{(\Gamma)})$ of the coordinate space $\mathbf{q}^{(\Gamma)}$ defined in the volume V of the system. We assume that the function is always normalized to unity. To be continuous and zero outside V it must be zero at the boundaries of V, and to have finite energy everywhere its second derivative must be finite wherever Ψ_i is nonzero. For an ensemble of M systems one might think that an average function defined in the limit $M \to \infty$,

$$\langle \Psi_{\text{ens}}(\mathbf{q}^{(\Gamma)}) \rangle = \lim_{M\to\infty} \left[M^{-1} \sum_{i=1}^{M} \Psi_i(\mathbf{q}^{(\Gamma)}) \right],$$

where Ψ_i is the state function of the ith system, could have some utility, but this is *not* the case. The reason is apparent if one remembers that average values $\langle f \rangle$ of physical observables $f(\mathbf{p}^{(\Gamma)}\mathbf{q}^{(\Gamma)})$ are given by forming the quantum-mechanical Hermitian operator $\mathscr{F}$ and, for each system i,

$$\langle f_i \rangle = \int d\mathbf{q}^{(\Gamma)}\, \Psi_i^*(\mathbf{q}^{(\Gamma)}) \mathscr{F} \Psi_i(\mathbf{q}) = \int dq^{(\Gamma)} \Psi_i(q) \mathscr{F} \Psi_i^*(q), \qquad (13b.1)$$

since $\langle f_i \rangle$ is real. Alternatively, we can use

$$\langle f_i \rangle = \int dq^{(\Gamma)} \left[\Psi_i^*(\mathbf{q}^{(\Gamma)}) \tfrac{1}{2} \mathscr{F} \Psi_i(\mathbf{q}^{(\Gamma)}) + \Psi_i(\mathbf{q}^{(\Gamma)}) \tfrac{1}{2} \mathscr{F} \Psi_i^*(\mathbf{q}^{(\Gamma)}) \right]. \qquad (13b.1')$$

For the ensemble one has

$$\langle f \rangle_{\text{ens}} = \lim_{M\to\infty} \left(M^{-1} \sum_{i=1}^{M} \langle f_i \rangle \right). \qquad (13b.2)$$

If $\langle \Psi_{\text{ens}}(\mathbf{q}^{(\Gamma)}) \rangle$ were used in eq. (1) there would be completely meaningless terms due to integrals of $\Psi_i^* \Psi_j$, with $i \neq j$ contributing spurious interactions between independent systems of the ensemble. Indeed, since Ψ_i and Ψ_j would not in general be orthogonal; the function $\langle \Psi_{\text{ens}}(\mathbf{q}^{(\Gamma)}) \rangle$ would not, in general, be normalized.

Since in eq. (1) the integration is carried out with the same value, $\mathbf{q}^{(\Gamma)}$, in both Ψ_i^* and $\mathscr{F}\Psi_i$, one might try to use

$$W(\mathbf{q}^{(\Gamma)}) = \lim_{M\to\infty}\left[M^{-1}\sum_{i=1}^{M}\Psi_i^*(\mathbf{q}_i^{(\Gamma)})\Psi_i(\mathbf{q}_i^{(\Gamma)})\right] \tag{13b.3}$$

as sufficient to describe the ensemble, and this is adequate to give average values of any function $f(\mathbf{q}^{(\Gamma)})$ of the coordinates alone,

$$\langle f_{\text{ens}}\rangle = \int dq^{(\Gamma)}\, W(\mathbf{q}^{(\Gamma)})f(\mathbf{q}^{(\Gamma)}), \tag{13b.4}$$

but tells us nothing about a function of the momenta. That this latter statement is true is evidenced by the simple example of a perfect gas of N monatomic atoms for which $W(\mathbf{q}^{(\Gamma)})$ has the constant value V^{-N} for all values of $\mathbf{q}^{(\Gamma)}$, independently of temperature or average kinetic energy.

To obviate this difficulty we introduce the coordinate representation of the density matrix $\mathbf{\Omega}$, namely, the matrix of elements,

$$\omega(\mathbf{q}'^{(\Gamma)}, \mathbf{q}''^{(\Gamma)}) = \lim_{M\to\infty}\left[M^{-1}\sum_{i=1}^{M}\Psi_i(\mathbf{q}'^{(\Gamma)})\Psi_i^*(\mathbf{q}''^{(\Gamma)})\right], \tag{13b.5}$$

for now we *define* the operation of an operator $\mathscr{F}$ on the element ω to be

$$\begin{aligned}\mathscr{F}\omega &= \mathscr{F}(\mathbf{q}'^{(\Gamma)}, \mathbf{q}''^{(\Gamma)})\omega(\mathbf{q}'^{(\Gamma)}, q''^{(\Gamma)})\\ &\equiv \lim_{M\to\infty}\left\{M^{-1}\sum_{i=1}^{M}\tfrac{1}{2}[\Psi_i^*(q'')\mathscr{F}\Psi_i(q') + \Psi_i(\mathbf{q}')\mathscr{F}(q'')\Psi_i^*(q_i'')]\right\},\end{aligned} \tag{13b.6}$$

so that for the ensemble average we have

$$\begin{aligned}\langle f_{\text{ens}}\rangle &= \int dq^{(\Gamma)}\,\mathscr{F}(\mathbf{q}^{(\Gamma)})\omega(\mathbf{q}^{(\Gamma)}, \mathbf{q}^{(\Gamma)})\\ &\equiv \text{tr}(\mathscr{F}\mathbf{\Omega}),\end{aligned} \tag{13b.7}$$

where (see Appendix VII) $\text{tr}(\mathscr{F}\mathbf{\Omega})$ is the trace of the matrix $\mathscr{F}\mathbf{\Omega}$.

At this stage all we appear to have accomplished is some semantic obfuscation by introducing a new phrase, density matrix, and by the use of the somewhat awkward definition of eq. (6) have shortened the expression for the ensemble average of a physically observable function of $\mathbf{q}^{(\Gamma)}\mathbf{p}^{(\Gamma)}$. Actually, we have opened the door to the use of the powerful mathematical techniques of matrix algebra. In particular, the fact that the trace of a Hermitian matrix is invariant in the representation is most useful (see Sec. 13e).

For the microcanonical and petite canonical ensembles of fixed numbers, $\mathbf{N} = N_a, N_b, \ldots,$ of molecules the coordinate space $\mathbf{q}'^{(\Gamma)}\mathbf{q}''^{(\Gamma)}$ is of course fixed. For a grand canonical ensemble the density matrix is not strictly a single matrix but rather a sum of independent matrices differently weighted for each set $\mathbf{N}$ of molecules with representations in different dimensional coordinate spaces with different numbers $\Gamma_{\mathbf{N}}$ of degrees of freedom. In any actual use of the matrices to evaluate average values of observables only the reduced matrices, which we discuss in Sec. 13d, are used. These are matrices of a fixed and small number γ of degrees of freedom obtained by integration of the diagonal elements of the $(\Gamma-\gamma)\mathbf{q}^{(\Gamma-\gamma)}\mathbf{q}^{(\Gamma-\gamma)}$ coordinates. The awkwardness of the grand canonical ensemble density matrices then disappears.

The density matrix of element $\omega(\mathbf{q}'^{(\Gamma)}\mathbf{q}''^{(\Gamma)})$ given by eq. (5) is Hermitian,

$$\omega(\mathbf{q}''^{(\Gamma)}, \mathbf{q}'^{(\Gamma)}) = [\omega(\mathbf{q}'^{(\Gamma)}, \mathbf{q}''^{(\Gamma)})]^*, \tag{13b.8}$$

and since the operators $\mathcal{F}$ of physical observables are Hermitian [eq. (AVII.5)], the $\mathcal{F}\mathbf{\Omega}$'s of elements given by eq. (6) are also Hermitian.

We have chosen to normalize the density matrix so that its trace is unity, a normalization that is convenient in comparing its equations with those of the classical probability density. Other normalizations differing by a constant factor are frequent in the literature.

13c. DENSITY MATRIX OF A SINGLE SYSTEM

Although the density matrix $\mathbf{\Omega}_i$ of a single system i in a definite normalized quantum state $\Psi_i(\mathbf{q}^{(\Gamma)})$ can be defined and has elements

$$\omega_i(\mathbf{q}'^{(\Gamma)}, \mathbf{q}''^{(\Gamma)}) = \Psi_i(\mathbf{q}'^{(\Gamma)})\Psi_i^*(\mathbf{q}''^{(\Gamma)}) \ . \tag{13c.1}$$

in the coordinate representation, it is too trivial a matrix to have properties that make it useful. The matrix Ω_i^2 has elements

$$\begin{aligned}
\omega_i^2(\mathbf{q}', \mathbf{q}'') &= \int d\mathbf{q}''' \ \omega_i(\mathbf{q}', \mathbf{q}''')\omega_i(\mathbf{q}''', \mathbf{q}'') \\
&= \Psi_i(\mathbf{q}')\left[\int d\mathbf{q}''' \ \Psi_i^*(\mathbf{q}''')\Psi_i(\mathbf{q}''')\right]\Psi_i^*(\mathbf{q}'') \\
&= \Psi_i(\mathbf{q}')\Psi_i^*(\mathbf{q}'') = \omega_i(\mathbf{q}', \mathbf{q}'')
\end{aligned} \tag{13c.2}$$

or, for the matrix product,

$$\mathbf{\Omega}_i\mathbf{\Omega}_i = \mathbf{\Omega}_i. \tag{13c.2′}$$

Matrices for which eq. (2′) holds are said to be *idempotent.*

The matrix $\mathbf{\Omega}_i$, like all Hermitian matrices, can be brought into a diagonal representation. Let $\Phi_\nu(\mathbf{q}^{(\Gamma)})$ be one of an infinite number of orthonormal functions, $1 \leq \nu \leq \infty$, which are all orthogonal to Ψ_i and which form a complete set with Ψ_i. This can actually be accomplished by projecting out of each $\Phi'_\nu(\mathbf{q})$ of a complete set that part which is not orthogonal to Ψ_i and then making the resultant functions normalized and mutually orthogonal. Without examining the details of the process suppose that it is accomplished. We then have a complete set of functions labeled by state numbers, $1 \leq \nu \leq \infty$, and i, which with their conjugate complexes Ψ_i^* and Φ_ν^* can be regarded as elements (see Appendix VII) of a unitary matrix $\mathbf{U}$ with its adjoint $\mathbf{U}^\dagger$,

$$\mathbf{U}\mathbf{U}^\dagger = \mathbf{U}^\dagger\mathbf{U} = \mathbf{1} \tag{13c.3}$$

of elements, $1 \leq \nu \leq \infty$,

$$\begin{aligned} u(\mathbf{q}, i) &\equiv \Psi_i(\mathbf{q}), & u(\mathbf{q}, \nu) &\equiv \Phi_\nu(\mathbf{q}), \\ u^\dagger(i, \mathbf{q}) &\equiv \Psi_i^*(\mathbf{q}), & u^\dagger(\nu, \mathbf{q}) &\equiv \Phi_\nu^*(\mathbf{q}). \end{aligned} \tag{13c.3$'$}$$

This unitary matrix, with its adjoint, transforms any Hermitian matrix from the $\mathbf{q}'$, $\mathbf{q}''$ representation to the state number representation, and vice versa. In particular, for the density matrix $\mathbf{\Omega}_i$ of the single system in the quantum state Ψ_i,

$$\mathbf{\Omega}^{\text{state number}} = \mathbf{U}\,\mathbf{\Omega}^{(\mathbf{q})}\mathbf{U} \tag{13c.4}$$

$$\mathbf{\Omega}^{(\mathbf{q})} = \mathbf{U}\mathbf{\Omega}^{\text{state number}}\,\mathbf{U}\,. \tag{13c.4$'$}$$

Examine the elements ω_{ii}, $\omega_{i\nu}$, $\omega_{\nu\nu}$, and $\omega_{\nu\mu}$ of Ω^{number}, using the orthonormality conditions $\int dq\, \Psi_i^*\Psi_i = 1$, $\int dq\, \Phi_\nu^*\Psi_i = 0$, to find

$$\omega_{ii} = \left(\int d\mathbf{q}\, \Psi_i^*\Psi_i\right) = 1,$$

$$\omega_{i\nu} = \omega_{\nu i}^* = \left(\int d\mathbf{q}\, \Psi_i^*\Psi_i\right)\left(\int d\mathbf{q}\, \Psi_i^*\Phi_\nu\right) = 0, \tag{13c.5}$$

and

$$\omega_{\nu\nu} = \omega_{\nu\mu} = \omega_{\mu\nu} = 0.$$

In this state number representation the density matrix of a single system is diagonal with only one diagonal element, and that element is unity.

One could of course have deduced this from the fact that the trace of Ω_i, which is the sum of the diagonal elements, is unity and the idempotency condition of eq. (2′) which requires $\mathbf{\Omega}_i^n = \mathbf{\Omega}_i$, $\text{tr}(\mathbf{\Omega}_i^n) = 1$. In the diagonal representation the trace of the nth power of a matrix is the sum of the nth power of the diagonal elements of the original matrix. The only way this can be unity for all n is for the sum to consist of one number only, namely, unity.

There is no requirement that the function $\Psi_i(\mathbf{q}^{(\Gamma)})$ be independent of time, nor in general that the ensemble density matrix be time-independent, although of course that for an equilibrium system necessarily is. The function $\Psi_i(t, \mathbf{q}^{(\Gamma)})$ can always be developed as a linear combination with time-dependent coefficients of the eigenfunctions $\Psi_{\mathbf{K}}(\mathbf{q}^{(\Gamma)})$ of the Hamiltonian operator $\mathcal{H}$,

$$\mathcal{H}\Psi_{\mathbf{K}} = E_{\mathbf{K}}\Psi_{\mathbf{K}}, \tag{13c.6}$$

$$\Psi_i(t, \mathbf{q}) = \sum_{\mathbf{K}} \alpha_{\mathbf{K}}(t)\Psi_{\mathbf{K}}(\mathbf{q}), \tag{13c.7}$$

where (Appendix VII)

$$\alpha_{\mathbf{K}}(t) = \int dq\, \Psi^*_{\mathbf{K}}(\mathbf{q})\Psi_i(t, \mathbf{q}) \tag{13c.7$'$}$$

and, since Ψ_i is normalized and the $\Psi_{\mathbf{K}}$'s are orthonormal,

$$\sum_{\mathbf{K}} \alpha_{\mathbf{K}}\alpha^*_{\mathbf{K}} = 1. \tag{13c.7$''$}$$

Use the time-dependent Schrödinger equation,

$$\frac{\partial\Psi_i}{\partial t} = \frac{2\pi i}{h}\mathcal{H}\Psi_i, \tag{13c.8}$$

to obtain from eqs. (6) and (7) that

$$\begin{aligned}\frac{\partial\Psi_i}{\partial t} &= \sum_{\mathbf{K}'} \frac{\partial\alpha_{\mathbf{K}'}}{\partial t}\Psi_{\mathbf{K}'}(\mathbf{q}) \\ &= \sum_{\mathbf{K}'} \alpha_{\mathbf{K}'}\frac{2\pi i}{h}\mathcal{H}\Psi_{\mathbf{K}'} = \sum_{\mathbf{K}'} \alpha_{\mathbf{K}'}\frac{2\pi i}{h}E_{\mathbf{K}'}\Psi_{\mathbf{K}'}.\end{aligned} \tag{13c.9}$$

Multiply by $\Psi^*_{\mathbf{K}}(\mathbf{q})$ and integrate over $\mathbf{q}$ using the orthonormality condition to find, for any arbitrary $\mathbf{K}$-value,

$$\frac{d\alpha_{\mathbf{K}}}{dt} = \frac{2\pi i E_{\mathbf{K}}}{h}\alpha_{\mathbf{K}}(t). \tag{13c.10}$$

The general solution is

$$\alpha_{\mathbf{K}}(t) = a_{\mathbf{K}} \exp\left[\frac{2\pi i E_{\mathbf{K}}}{h}(t - \delta_{\mathbf{K}})\right], \tag{13c.11}$$

where, by adjusting the phases $\delta_{\mathbf{K}}$, we can make $a_{\mathbf{K}}$ real and positive.

The energy of the system is now an average,

$$\langle E\rangle = \int d\mathbf{q}\, [\mathcal{H}\Psi_i(\mathbf{q})]\Psi^*_i(\mathbf{q}), \tag{13c.12}$$

which, with eqs. (6) and (7) and the orthonormality condition, is

$$\langle E\rangle = \sum_{\mathbf{K}'\mathbf{K}}\sum E_{\mathbf{K}'}\alpha_{\mathbf{K}'}(t)\alpha^*_{\mathbf{K}}(t)\int d\mathbf{q}\,\Psi_{\mathbf{K}'}\Psi^*_{\mathbf{K}}$$

$$= \sum_{\mathbf{K}} \alpha_{\mathbf{K}}(t)\alpha^*_{\mathbf{K}}(t)E_{\mathbf{K}} = \sum_{\mathbf{K}} a^2_{\mathbf{K}}E_{\mathbf{K}}, \qquad (13c.12')$$

where the last expression follows from eq. (11). The energy is constant in time, provided the systems of the ensemble have no contact with external reservoirs, that is, if it is a microcanonical ensemble.

The density matrix $\mathbf{\Omega}_i$ of this system given in the representation of the quantum numbers $\mathbf{K}'\mathbf{K}$ of the eigenfunctions of the Hamiltonian has elements

$$\omega_i(\mathbf{K}', \mathbf{K}'') = \iint d\mathbf{q}''\, d\mathbf{q}'\, \Psi^*_{\mathbf{K}'}(\mathbf{q}')\Psi_i(\mathbf{q}')\Psi^*_i(\mathbf{q})\Psi_{\mathbf{K}''}(\mathbf{q}). \qquad (13c.13)$$

With eq. (7) and the orthonormal $\Psi_{\mathbf{K}}$'s, this gives

$$\omega_i(\mathbf{K}', \mathbf{K}'') = \alpha_{\mathbf{K}'}(t)\alpha^*_{\mathbf{K}''}(t). \qquad (13c.14)$$

That idempotency [eq. (2′)] is preserved in this representation, as it must be, follows from the condition that $\sum \alpha^*_{\mathbf{K}}\alpha_{\mathbf{K}} = 1$ [eq. (7′)].

For a quantum-mechanically observable quantity for which the classical function is $f(\mathbf{q}, \mathbf{p})$ the corresponding operator $\mathcal{F}(\mathbf{q})$ is Hermitian, and the matrix of its representation in the quantum numbers $\mathbf{K}'\mathbf{K}''$ of eigenfunctions $\Psi_{\mathbf{K}}(\mathbf{q})$ of the Hamiltonian operator $\mathcal{H}$ has elements

$$f_{\mathbf{K}'\mathbf{K}''} = \int d\mathbf{q}\, \tfrac{1}{2}\{[\mathcal{F}\Psi_{\mathbf{K}'}(\mathbf{q})]\Psi^*_{\mathbf{K}''}(\mathbf{q}) + \Psi_{\mathbf{K}'}(\mathbf{q})\mathcal{F}\Psi_{\mathbf{K}''}(\mathbf{q})\}, \qquad (13c.15)$$

with

$$f_{\mathbf{K}'\mathbf{K}''} = f^*_{\mathbf{K}''\mathbf{K}'}. \qquad (13c.15')$$

Unless $\mathcal{F}$ commutes with the Hamiltonian, there will be nonzero off-diagonal elements with $\mathbf{K}' \neq \mathbf{K}''$. With eq. (7) for Ψ_i, the average $\langle f_i(\mathbf{q}, \mathbf{p})\rangle$ in the system i is

$$\langle f_i(\mathbf{q}, \mathbf{p})\rangle = \int d\mathbf{q}\, [\mathcal{F}\psi_i(\mathbf{q})]\Psi^*_i(\mathbf{q})$$

$$= \sum_{\mathbf{K}'\mathbf{K}''}\sum \alpha_{\mathbf{K}'}(t)\alpha^*_{\mathbf{K}''}(t)f_{\mathbf{K}'\mathbf{K}''}. \qquad (13c.16)$$

The diagonal terms $\mathbf{K}' = \mathbf{K}''$ are all real and, from eq. (11) for $\alpha_{\mathbf{K}}$, are independent of time. The paired off-diagonal terms $\alpha_{\mathbf{K}'}\alpha^*_{\mathbf{K}''}f_{\mathbf{K}'\mathbf{K}''}$ and $\alpha_{\mathbf{K}''}\alpha^*_{\mathbf{K}'}f_{\mathbf{K}''\mathbf{K}'}$ are conjugate complexes of each other, and their sum is real.

We can write eq. (16), using eq. (11) for α, as

$$\langle f_i \rangle = \sum_{\mathbf{K}} a_{\mathbf{K}}^2 f_{\mathbf{KK}} + 2\sum_{K'>K''}\sum [\alpha_{\mathbf{K}'}(t)\alpha^*_{\mathbf{K}''}(t) f_{K'K''}]_{\text{real}}. \qquad (13c.16')$$

Remembering that $e^{ix} = \cos x + i \sin x$, $e^{-ix} = \cos x - i \sin x$, and writing

$$f_{\mathbf{K}'\mathbf{K}''} = f^{\text{real}}_{\mathbf{K}'\mathbf{K}''} + i f^{\text{imag}}_{\mathbf{K}''\mathbf{K}'},$$
$$f_{\mathbf{K}''\mathbf{K}'} = f^{\text{real}}_{\mathbf{K}'\mathbf{K}''} - i f^{\text{imag}}_{\mathbf{K}''\mathbf{K}'}$$

with both f^{real} and f^{imag} real numbers, we define

$$\nu_{\mathbf{K}'\mathbf{K}''} = h^{-1}(E_{\mathbf{K}'} - E_{\mathbf{K}''}),$$
$$\Delta_{\mathbf{K}'\mathbf{K}''} = h^{-1}(E_{\mathbf{K}'}\boldsymbol{\delta}_{\mathbf{K}'} - E_{\mathbf{K}''}\delta_{\mathbf{K}''}), \qquad (13c.17)$$

and obtain

$$\langle f_i(\mathbf{q}, \mathbf{p}) \rangle = \sum_{\mathbf{K}} a_{\mathbf{K}}^2 f_{\mathbf{KK}}$$
$$+ 2\sum_{\mathbf{K}'>\mathbf{K}}\sum a_{\mathbf{K}'} a_{\mathbf{K}} \{ f^{\text{real}}_{\mathbf{K}'\mathbf{K}''} \cos[2\pi\nu_{\mathbf{K}'\mathbf{K}''}t - 2\pi\, \Delta_{\mathbf{K}'\mathbf{K}''}]$$
$$- f^{\text{imag}}_{\mathbf{K}'\mathbf{K}''} \sin(2\pi\nu_{\mathbf{K}'\mathbf{K}''}t - 2\pi\, \Delta_{\mathbf{K}'\mathbf{K}''})\}. \qquad (13c.18)$$

The average value is real, but oscillates in time if Ψ_i is not a single-energy eigenstate.

We comment here that the frequency, $\nu_{\mathbf{K}} = E_{\mathbf{K}}/h$, with which $\alpha_{\mathbf{K}}(t)$ oscillates depends on the zero from which the energy is measured. In his original introduction of plane waves in the absence of any potential, de Broglie used the only absolute energy, $E = mc^2$, which for a macroscopic system would be a pretty horrendous frequency. However, for any quantum-mechanical observable only energy *differences* enter. The arbitrary zero of energy cancels out.

13d. REDUCED DENSITY MATRICES

The ultimate use of statistical mechanics is to calculate the average value $f(\mathbf{q}^{(\Gamma)}\mathbf{p}^{(\Gamma)})$ of some observable function of coordinates and momenta in an ensemble of reproducibly defined macroscopic systems, either defined at equilibrium by the relatively few thermodynamic variables necessary to fix the macroscopic state or, in some ensembles of nonequilibrium systems, for instance, by a few functions, such as the energy density or concentrations, dependent on the position $\mathbf{r}$ in the volume of the system.

In classical mechanics any function $f(\mathbf{q}^{(\Gamma)}\mathbf{p}^{(\Gamma)})$ is supposedly observable, but there are limitations in quantum mechanics connected with the uncertainty principle. For instance, the classical function that is a sum over all molecules α, $1 \le \alpha \le N$, of $\sum_\alpha \mathbf{q}_\alpha \cdot \mathbf{p}_\alpha = \sum_\alpha \mathbf{p}_\alpha \cdot \mathbf{q}_\alpha$ is not an observable in quantum mechanics, since the quantum-mechanical operators are different for $\mathbf{q}_\alpha \cdot \mathbf{p}_\alpha$ and $\mathbf{p}_\alpha \cdot \mathbf{q}_\alpha$, [see the discussion in Sec. 3b for eqs. (3b.5) and 3b.6)]. Only those classical functions for which the classical function $f(\mathbf{q}, \mathbf{p})$, with p_v replaced by $(h/2\pi i)\partial/\partial q_v$, leads to a unique operator $\mathcal{F}[\mathbf{q}, (h/2\pi i)\nabla \cdot \mathbf{q}]$ that is Hermitian [eq. (AVII.5)] are allowed the distinction of being called observable.

There is a different practical limitation in the use of statistical mechanics on functions $f(\mathbf{q}^{(\Gamma)}, \mathbf{p}^{(\Gamma)})$, even in the case of the use of the equations of classical mechanics. This limitation is that we must be able to find a coordinate conjugate momentum phase space of the Γ degrees of freedom in which $f(\mathbf{q}^{(\Gamma)}, \mathbf{p}^{(\Gamma)})$ is a sum of terms labeled by κ, each involving only a small number, $\gamma_\kappa = 1, 2, 3, \ldots$, of molecules or of degrees of freedom,

$$f(\mathbf{q}^{(\Gamma)}, \mathbf{p}^{(\Gamma)}) = \sum_\kappa f_\kappa(\mathbf{q}^{(\gamma_\kappa)}, \mathbf{p}^{(\gamma_\kappa)}). \tag{13d.1}$$

In fluid systems of one chemical component, using center-of-mass coordinates of the molecules, there may be $(n!)^{-1}(N!)/(N-n)!$ identical terms involving the coordinates and momenta of n molecules each. However, if the systems are crystalline, one presumably starts with the $3N$ normal coordinates, $1 \le i \le 3N$, of frequency ν_i. In a higher approximation than the harmonic one might wish to examine interactions between the different normal modes of vibration. There are then $\frac{1}{2}3N(3N-1)$ interactions between pairs of normal modes, but in general the interaction terms are all different, depending on the frequencies ν_i and ν_j as well as the directions of the two wave vectors.

The quantum-mechanical operator $\mathcal{F}(\mathbf{q}^{(\Gamma)}, \mathbf{p}^{(\Gamma)})$ corresponding to the classical function $f(\mathbf{q}^{(\Gamma)}, \mathbf{p}^{(\Gamma)})$ of eq. (1) is then also a sum,

$$\mathcal{F}(\mathbf{q}^{(\Gamma)}) = \sum_\kappa \mathcal{F}_\kappa(\mathbf{q}^{(\gamma_\kappa)}), \tag{13d.1$'$}$$

of operators $\mathcal{F}_\kappa$ acting only on the γ_κ-dimensional coordinate $\mathbf{q}^{(\gamma_\kappa)}$ subset of the total Γ degrees of freedom. Since the average is

$$\langle f(\mathbf{q}^{(\Gamma)}, \mathbf{p}^{(\Gamma)})\rangle = \mathrm{tr}[\mathcal{F}(\mathbf{q}'^{(\Gamma)}, \mathbf{q}''^{(\Gamma)})\mathbf{\Omega}(\mathbf{q}'^{(\Gamma)}, \mathbf{q}''^{(\Gamma)})], \tag{13d.2}$$

from eq. (13b.7) we have

$$\langle f(\mathbf{q}^{(\Gamma)}, \mathbf{p}^{(\Gamma)}\rangle = \sum_\kappa \langle f_\kappa(\mathbf{q}^{(\gamma_\kappa)}, \mathbf{p}^{(\gamma_\kappa)}\rangle, \tag{13d.3}$$

where

$$\langle f_\kappa(\mathbf{q}^{(\gamma_\kappa)}, \mathbf{p}^{(\gamma_\kappa)})\rangle = \mathrm{tr}[\mathcal{F}_\kappa(\mathbf{q}'^{(\gamma_\kappa)}, \mathbf{q}''^{(\Gamma)})\mathbf{\Omega}(\mathbf{q}'^{(\Gamma)}, \mathbf{q}''^{(\Gamma)})]. \tag{13d.4}$$

Just as in our treatment of liquids with classical mechanics we found it convenient to introduce probability densities $\rho_n(\mathbf{r}^{(n)})$ [eq. (9b.3)] giving the average probability density of a subset $\{n\}_N$ of the N molecules in the space of their coordinates, and write equations as integrals of functions of coordinates times the ρ_n's, so also in the use of density matrices one actually always employs reduced density matrices of a limited number of coordinates. These are defined by taking the trace over all but the subset $\mathbf{q}'^{(\gamma_\kappa)}$, $\mathbf{q}''^{(\gamma)}$ coordinates of the complete matrix of Γ degrees of freedom.

We write, for the reduced density matrix of an ensemble in the coordinate representation $\mathbf{q}''^{(\gamma)}, \mathbf{q}^{(\gamma)}$,

$$\boldsymbol{\Omega}_{(\gamma)} = \mathrm{tr}(\Gamma - \gamma)\boldsymbol{\Omega}_{\mathrm{ens}}, \tag{13d.4}$$

of elements

$$\omega(\mathbf{q}'^{(\gamma)}, \mathbf{q}''^{(\gamma)}) = \int d\mathbf{q}^{(\Gamma-\gamma)}\, \omega(\mathbf{q}'^{(\gamma)} + \mathbf{q}^{(\Gamma-\gamma)}, \mathbf{q}''^{(\gamma)} + \mathbf{q}^{(\Gamma-\gamma)}), \tag{13d.4$'$}$$

so that

$$\begin{aligned}\langle f(\mathbf{q}'^{(\gamma)}, \mathbf{p}^{(\gamma)})\rangle &= \mathrm{tr}[\mathcal{F}(\mathbf{q}'^{(\gamma)}, \mathbf{q}''^{(\gamma)})\boldsymbol{\Omega}_{(\gamma)}] \\ &= \int d\mathbf{q}^{(\gamma)}\, (\mathcal{F}(\mathbf{q}'^{(\gamma)}, \mathbf{q}''^{(\gamma)})\omega(\mathbf{q}'^{(\gamma)}, \mathbf{q}''^{(\gamma)})_{\mathbf{q}'=\mathbf{q}''}.\end{aligned} \tag{13d.5}$$

13e. PHASES—RANDOM AND OTHERWISE, [$\mathcal{F}\Omega$] MATRICES

In discussions of the density matrix for an ensemble one frequently encounters this statement without further amplification: "We use the random phase approximation." The term *approximation* is unfortunate. Whether the phases are random or not depends on the problem, namely, on the macroscopic variables that characterize the systems of which the ensemble is composed. The phases for an ensemble of equilibrium systems *are* random. For other cases they often are not, and indeed for systems whose macroscopic properties are time-dependent they cannot be *strictly* random.

The phases referred to are the quantities $\delta_{\mathbf{K}}$ of the dimensions of time appearing in the expression (13.11) for $\alpha_{\mathbf{K}}(t)$, or actually the dimensionless difference term $\Delta_{\mathbf{K}'\mathbf{K}''}$ of eq. (13c.17). These were so chosen initially in eq. (13c.11) so that the factors $a_{\mathbf{K}}$ in the expression for $\alpha(t)$ are real. In general, in the ensemble, even in an ensemble of equilibrium systems, the individual systems are not expected to be independent of time. They fluctuate in the values of measurable properties away from their average value, although normally only to a small extent. For the individual systems, then, the off-diagonal elements in eq. (13c.18) contribute to $\langle f_i(\mathbf{q}, \mathbf{p})\rangle$ as shown in that equation; they oscillate in time, and their value depends on $\nu_{\mathbf{K}'\mathbf{K}}t-\Delta_{\mathbf{K}'\mathbf{K}}$. The ensemble of equilibrium systems consists of an infinite number of members i, all prepared in the same thermodynamically defined state, and average values,

$$\langle f_{\text{ens}}(\mathbf{q},\mathbf{p})\rangle = \lim_{M\to\infty} M^{-1}\sum_{i=1}^{M} \langle f_i\mathbf{q},\mathbf{p}\rangle \ , \qquad (13e.1)$$

are to be calculated corresponding to observations made on random systems at *random* times.

Since the average values of both $\cos 2\pi x$ and $\sin 2\pi x$ over the range of x-values, $0 \leq x \leq 1$, are zero, the average contribution of the off-diagonal elements is zero, and

$$\langle f_{\text{ens}}\mathbf{q},\mathbf{p}\rangle = \sum_{\mathbf{K}} W(\mathbf{K}) a_{\mathbf{K}}^2 f_{\mathbf{KK}}, \qquad (13e.2)$$

with $W(\mathbf{K})$ the equilibrium probability of the single energy state $\mathbf{K}$, proportional to $\exp(-\beta E_{\mathbf{K}})$. The randomness of the phase factors is not an *approximation* but a characteristic of the ensemble.

Nonequilibrium systems are less simple. Consider, for example, an ensemble of molecular systems in which there is thermal or molecular transport along gradients of temperature or composition, or both. One possible treatment is to consider an ensemble of open systems in contact through open walls at opposite ends with infinite reservoirs of fixed, but different, temperatures and/or concentrations of molecules. The ensemble is a steady-state ensemble with time-independent properties. One asks for average values measured on random systems at random times, and random phases are appropriate. However, if the analysis is carried out in terms of the real solutions $\Psi_{\mathbf{K}}(\mathbf{q})$ of eigenfunctions of the Hamiltonian operator $\mathcal{H}$, there will be nonzero off-diagonal elements $\omega(\mathbf{K}'\mathbf{K}'')$ with complex values like $\exp \sum 2\pi i h^{-1}\mathbf{p}_\nu \cdot \mathbf{r}_\nu$, where the sum runs over the molecules ν. These are necessary to give nonzero average transport in one direction. In principle at least these will contain phase oscillations like those in eq. (13c.18) if the ensemble is not one of steady-state systems but changes with time.

However, in problems like those of heat or molecular diffusion, in which the half-time of approach to equilibrium is characteristically 15 min ($\cong 10^3$ sec) or much longer, it may be most useful to assume that the ensemble properties are essentially constant for times of the order of 1 sec or longer. In such a case the treatment using random phases could be numerically perfectly justified but would be appropriately called an approximation, albeit a very good one.

There are now, however, experimentally sophisticated, but conceptually quite simple, cases in which the phase in the ensemble average is not random but vary precisely the same for all systems of the ensemble. Consider the example of a photochemical reaction in which a molecule A absorbs a high-energy photon to be raised into an activated state A* which can decompose in two ways; in one of these an activated fragment α^* is produced, and in the other a different activated fragment γ^*. Both fragments soon decay to their ground states. We have two different processes:

$$\mathrm{B}' + \alpha^* \rightarrow \mathrm{B}' + \alpha + \Delta\epsilon \qquad \text{(I)}$$

$$h\nu + \mathrm{A} \rightarrow \mathrm{A}^* \quad \nearrow \quad \searrow$$

$$\mathrm{B}'' + \gamma^* \rightarrow \mathrm{B}'' + \gamma + \Delta\epsilon \qquad \text{(II)}.$$

Now the activation can be performed on a dilute gas of A with a very short pulse from an intense laser source, and the concentrations of both activated fragments α^* and γ^* followed in time by their absorption spectra or by some other method of measurement precisely timed relative to the laser pulse.

The observed systems of the ensemble are then systems consisting initially of independent activated molecules A* and their subsequent histories measured at precise times t after the laser-pulse activation. The phases δ_{I} and δ_{II}, as well as $\Delta_{\mathrm{I,II}}$, of all members of the ensemble are the same. The relative concentrations of the activated fragments can be expected to oscillate in time. Many similar examples occur in high-energy physics experiments.

In summary, then, the phases in an ensemble of equilibrium or of steady-state nonequilibrium systems are random, and the off-diagonal time-dependent terms in the representation of the ensemble density matrix in the quantum states of the eigenfunctions of the Hamiltonian average to zero. In other ensembles of nonequilibrium slowly evolving systems the approximation of random phase may be a very close and useful one. There are, however, ensembles of time-dependent systems in which the time t is very precisely fixed with respect to the same definite initial state in all member systems, and in which the phase of the ensemble density matrix has a fixed single value.

13f. REPRESENTATION TRANSFORMATIONS OF Ω

For the petite canonical ensemble of fixed V, N, T the equilibrium density matrix is diagonal in the quantum numbers $\mathbf{K}$ of the eigenfunctions $\Psi_{\mathbf{K}}(\mathbf{q}^{(\Gamma)})$ of the Hamiltonian operator $\mathcal{H}$,

$$\mathcal{H}\Psi_{\mathbf{K}}(\mathbf{q}) = E_{\mathbf{K}}\Psi_{\mathbf{K}}(q). \qquad (13f.1)$$

With A the Helmholtz free energy and $\beta=(kT)^{-1}$, the diagonal elements are

$$\omega(\mathbf{KK})=\exp[-(\beta A+E_{\mathbf{K}})],$$
$$\omega(\mathbf{K'K''})=0 \qquad \text{if } \mathbf{K}'\neq\mathbf{K}. \tag{13f.2}$$

The elements of the coordinate representation $\omega(\mathbf{q'q''})$ are given by

$$\omega(\mathbf{q'q''})=\sum_{\mathbf{K'K''}}\sum \Psi_{\mathbf{K'}}(\mathbf{q'})\omega_{(\mathbf{K'},\mathbf{K''})}\Psi^*_{\mathbf{K''}}(\mathbf{q''}), \tag{13f.3}$$

and for this equilibrium ensemble, with eq. (2), by

$$\omega(\mathbf{q'q''})=\sum_{\mathbf{K}}\Psi_{\mathbf{K}}(\mathbf{q'})e^{-\beta(A-E_{\mathbf{K}})}\Psi^*_{\mathbf{K}}(\mathbf{q''}). \tag{13f.3'}$$

In matrix notation we say that there is a unitary transformation matrix $\boldsymbol{\Psi}$ of elements $\psi(\mathbf{q},\mathbf{K})\equiv\Psi_{\mathbf{K}}(\mathbf{q})$ whose adjoint matrix $\boldsymbol{\Psi}^\dagger$ has elements $\psi^\dagger(\mathbf{K},\mathbf{q})\equiv\Psi^*_{\mathbf{K}}(\mathbf{q})=\psi^*(\mathbf{q},\mathbf{K})$, for which

$$\boldsymbol{\Psi}\boldsymbol{\Psi}^\dagger=\mathbf{1}, \tag{13f.4}$$

namely, a unit matrix of diagonal elements which are products of the Dirac function $\delta(\mathbf{q}'-\mathbf{q}'')=\prod_{j=1}^{\Gamma}\delta(q_j'-q_j'')$ and

$$\boldsymbol{\Psi}^\dagger\boldsymbol{\Psi}=\mathbf{1} \tag{13f.4'}$$

of elements $\delta(\mathbf{K}'-\mathbf{K}'')$ (see Appendix VIII). The transformation of eq. (3) for the elements $\omega_{\mathbf{K'K''}}$ to those $\delta(\mathbf{q'q''})$ is then between the two representations

$$\boldsymbol{\Omega}^{(\mathbf{q'},\mathbf{q''})}=\boldsymbol{\Psi}\boldsymbol{\Omega}^{(\mathbf{K'},\mathbf{K''})}\boldsymbol{\Psi}^\dagger \tag{13f.5}$$

and

$$\boldsymbol{\Omega}^{(\mathbf{K'},\mathbf{K''})}=\boldsymbol{\Psi}^\dagger\boldsymbol{\Omega}^{(\mathbf{q'q''})}\boldsymbol{\Psi}, \tag{13f.5'}$$

which follows from eq. (5) using eqs. (4) and (4′).

The general density matrix for any ensemble of systems defined by some unique macroscopic description is defined by eq. (13b.6). We assume that this description is adequate to define the elements $\omega(\mathbf{K}',\mathbf{K}'')$ in the $\mathbf{K}'$, $\mathbf{K}''$ representation. We may possibly prefer to compute the elements $\omega(\mathbf{M}',\mathbf{M}'')$ of some other complete orthonormal set of functions $\theta_{\mathbf{M}}(\mathbf{q})$ which are eigenfunctions,

$$\mathcal{O}\theta_{\mathbf{M}}(\mathbf{q})=\theta_{\mathbf{M}}\theta_{\mathbf{M}}(\mathbf{q}), \tag{13f.6}$$

of some Hermitian operator $\mathcal{O}$.

If the representation is known, or sought in terms of the functions $\theta_{\mathbf{M}}(\mathbf{q})$ of eq. (6), we will define a unitary matrix $\boldsymbol{\theta}$ of elements, $\theta(\mathbf{q}, \mathbf{M}) = \theta_{\mathbf{M}}(\mathbf{q})$, with its adjoint $\boldsymbol{\theta}^\dagger$ of elements, $\theta^\dagger(\mathbf{M}, \mathbf{q}) = \theta^*_{\mathbf{M}}(\mathbf{q})$, with products

$$\boldsymbol{\theta}\boldsymbol{\theta}^\dagger = \mathbf{1}, \qquad \boldsymbol{\theta}^\dagger\boldsymbol{\theta} = \mathbf{1}, \tag{13f.7}$$

as in eqs. (4) and (4′), and the transformations between the $\mathbf{q}'$, $\mathbf{q}''$ and $\mathbf{M}'$, $\mathbf{M}''$ representations will be analogous to eqs. (5) and (5′).

The product matrices

$$\mathbf{U} = \boldsymbol{\theta}^\dagger\boldsymbol{\Psi}, \qquad \mathbf{U}^\dagger = \boldsymbol{\Psi}^\dagger\boldsymbol{\theta} \tag{13f.8}$$

are self-adjoint,

$$u(\mathbf{M}, \mathbf{K}) = \int d\mathbf{q}\ \theta^*_{\mathbf{M}}(\mathbf{q})\Psi_{\mathbf{K}}(\mathbf{q}) = u^\dagger(\mathbf{K}, \mathbf{M}), \tag{13f.8′}$$

and, since

$$\mathbf{U}\mathbf{U}^\dagger = \boldsymbol{\theta}^\dagger\boldsymbol{\Psi}\boldsymbol{\Psi}^\dagger\boldsymbol{\theta} = \boldsymbol{\theta}^\dagger\mathbf{1}\boldsymbol{\theta} = \boldsymbol{\theta}^\dagger\boldsymbol{\theta} = \mathbf{1}, \tag{13f.8″}$$

are unitary. They transform the density matrix Ω from the $\mathbf{K}'$, $\mathbf{K}''$ to the M', M'' representation, and vice versa, by the equations

$$\boldsymbol{\Omega}^{(\mathbf{M}',\mathbf{M}'')} = \mathbf{U}\boldsymbol{\Omega}^{(\mathbf{K}',\mathbf{K}'')}\mathbf{U}^\dagger, \tag{13f.9}$$

$$\boldsymbol{\Omega}^{(\mathbf{K}',\mathbf{K})} = \mathbf{U}^\dagger\boldsymbol{\Omega}^{(\mathbf{M}',\mathbf{M})}\mathbf{U}. \tag{13f.10}$$

13g. THE MOMENTUM REPRESENTATION $\omega(\mathbf{p}', \mathbf{p})$

We have chosen to define the quantum state of a single system i by a function $\Psi_i(\mathbf{q})$ of the coordinate set $\mathbf{q}$. In Appendix VIII we show that for a single cartesian coordinate the function Φ_ν,

$$\Phi_\nu(p) = \int dx \frac{1}{\sqrt{h}} \Psi_\nu(x) e^{(2\pi i/h)px}, \tag{13g.1}$$

for which

$$\Psi_\nu(x) = \int dp \frac{1}{\sqrt{h}} \Phi_\nu(p) e^{-(2\pi i/h)px}, \tag{13g.1′}$$

where $\Psi_\nu(x) = 0$ for $x \leq \frac{1}{2}l$, $x \geq -\frac{1}{2}l$, can be used to define the state ν in the momentum space $p = p_x$ conjugate to x. Further, we prove that, if there is a complete orthonormal set Ψ_ν, $\Psi_\mu, \ldots, 1 \leq \nu,\ \mu \leq \infty$, of functions $\Psi(x)$, the corresponding functions Φ_ν, Φ_μ form a complete orthonormal set and $\Phi_\nu(p)\Phi^*_\nu(p)$ is the probability density of the momentum p in the state ν.

In Appendix VIII we discuss the example in which the coordinate function is defined in a macroscopic length l, such as 1 cm, so that the integer k in eq. (AVIII.1) and momentum p become essentially continuous. This is used as a mathematical device to derive the change in the normalization factor $1/\sqrt{l}$ in the Fourier sum over integer k values to $1/\sqrt{h}$ in the continuous Fourier integral function $\Phi(p, x) = (1/\sqrt{h})e^{(2\pi i/h)px}$ The function $\Phi(p, x)$ is an eigenfunction of the momentum operator, $\mathcal{P} = (h/2\pi i)\,\partial/\partial x$, independently of the length l in which $\Psi_\nu(x)$ is nonzero. If l is of the order of 10^{-8} to 10^{-6} cm, say, and Ψ_ν is a real stationary wave $\sqrt{2}\sin(2\pi kx/l)$, the corresponding momentum function $\Psi_\nu(p)$ [eq. (1)] will have maximum absolute amplitudes at $p = \pm hk/l$, and the difference $\Delta p = h/l$ between consecutive maxima will be large. However, $\Phi(p)\Phi^*(p)$ will have appreciable nonzero positive values for p-values on both sides of these maxima of the order of $h/2l$, corresponding to the uncertainty rule $\Delta p\,\Delta x = \Delta pl = h$.

An ensemble of systems in the stationary states $\sqrt{2}\sin(2\pi kx/l)$ with equal probability, $W(k) = 1/k''$, $k'' \gg 1$ for all states k between k' and $k' + k''$, has equal probability densities for all momenta in the middle of the two ranges $-h(k'+k'')/l < p < -hk'/l$ and $hk'/l < p < h(k'+k'')/l$. The mathematics of this is essentially the same as that for a parallel beam of light of wavelength λ passing through narrow and wide slits. If the slit width d is of the order of λ, the beam is strongly diffracted. If the width is nd, $n \gg 1$, the pattern on a screen behind the slit can be computed from the sum of amplitudes of n adjacent slits each of width d. The result is uniform intensity in the beam of width about nd and some diffraction at the edges.

Since for a generalized coordinate q the operator for the conjugate momentum p is $\mathcal{P} = (h/2\pi i)\,\partial/\partial q$, eq. (1) holds for any coordinate, with q replacing the cartesian x. One caution should be noted. If q is an angle such that the stationary coordinate functions $\Psi_\nu(q)$ are periodic, $\Psi_\nu(q + 2\pi n) = \Psi_\nu(q)$, then p will be sharply defined; the probability density function $\Phi_\nu(p)\Phi_\nu^*(p)$ will have values proportional to Dirac functions $\delta(p)$ at allowed values of p (if the systems remain for long in the same rotational state).

The state function $\Phi_i(\mathbf{p}^{(\Gamma)})$ for a system i of Γ degrees of freedom is an obvious extension of the single-coordinate function discussed up to now. The Fourier integral transformation function is then

$$\Phi(\mathbf{p}^{(\Gamma)} \cdot \mathbf{q}^{(\Gamma)}) = h^{-\Gamma/2} \exp\left(\frac{2\pi i}{h}\mathbf{p}^{(\Gamma)} \cdot \mathbf{q}^{(\Gamma)}\right), \tag{13g.2}$$

where the notation used is

$$\mathbf{p}^{(\Gamma)} \cdot \mathbf{q}^{(\Gamma)} = \sum_{j=1}^{j=\Gamma} p_j q_j. \tag{13g.2'}$$

The unitary matrix $\mathbf{\Phi}$ of elements $\phi(\mathbf{p}^{(\Gamma)}, \mathbf{q}^{(\Gamma)}) \equiv \Phi(p, q)$ is adjoint to $\mathbf{\Phi}^\dagger$ of elements $\phi^\dagger(\mathbf{q}^{(\Gamma)}, \mathbf{p}^{(\Gamma)}) = \phi^*(\mathbf{p}, \mathbf{q})$. The products

$$\mathbf{\Phi\Phi}^\dagger = \mathbf{1}, \qquad \mathbf{\Phi}^\dagger\mathbf{\Phi} = \mathbf{1} \tag{13g.3}$$

are both unit matrices which are products of Dirac delta functions, the first of the momenta $\delta(\mathbf{p}'-\mathbf{p}'')$, and the second of the coordinates $\delta(\mathbf{q}'-\mathbf{q}'')$.

The unitary matrices $\mathbf{\Phi}$, $\mathbf{\Phi}^{\Gamma}$ transform any hermitian matrix from the coordinate representation to the momenta representation, or vice versa; in particular, for the density matrix $\mathbf{\Omega}$,

$$\mathbf{\Omega}^{(\mathbf{p}',\mathbf{p}'')}=\mathbf{\Phi}\mathbf{\Omega}^{(\mathbf{q}',\mathbf{q}'')}\mathbf{\Phi}^{\dagger}, \qquad \mathbf{\Omega}^{(\mathbf{q}',\mathbf{q}'')}=\mathbf{\Phi}^{\dagger}\mathbf{\Omega}^{(\mathbf{p}',\mathbf{p}'')}\mathbf{\Phi}. \tag{13g.4}$$

Spelled out in terms of the elements the equations are

$$\omega(\mathbf{p}',\mathbf{p}'')=\iint d\mathbf{q}'\,d\mathbf{q}''\,\phi(\mathbf{p}',\mathbf{q}')\omega(\mathbf{q},\mathbf{q}'')\phi^{\dagger}(\mathbf{q}'',\mathbf{p}''),$$

$$\omega(\mathbf{q}',\mathbf{q}'')=\iint d\mathbf{p}'\,d\mathbf{p}''\,\phi^{\dagger}(\mathbf{q}',\mathbf{p}')\omega(\mathbf{p}',\mathbf{p}'')\phi(\mathbf{p}'',\mathbf{q}''). \tag{13g.4'}$$

One can easily check that the Hermitian character, $\omega(\mathbf{q}',\mathbf{q}'')=\omega^*(\mathbf{q}'',\mathbf{q}')$, is preserved in $\omega(\mathbf{p}',\mathbf{p}'')$ but also that, if $\omega(\boldsymbol{q}',\boldsymbol{q}'')$ is real, that is, $\omega^*(\mathbf{q}',\mathbf{q}'')=\omega(\mathbf{q}',\mathbf{q}'')$, then $\omega(\mathbf{p}',\mathbf{p}'')$ is real.

13h. THE WIGNER FUNCTION

The Wigner function is a transformation of the density matrix elements to a function of the coordinates $\mathbf{q}^{(\Gamma)}$ and momenta $\mathbf{p}^{(\Gamma)}$, analogous to a representation $\mathbf{q}'$, $\mathbf{q}''$ or $\mathbf{p}'$, $\mathbf{p}''$ but one in which the Hermitian character of the matrix elements, $\omega(\mathbf{q}',\mathbf{q}'')=\omega^*(\mathbf{q}'',\mathbf{q}')$, is lost. Start with the elements $\omega(\mathbf{q}',\mathbf{q}'')$ in the coordinate representation and introduce for each coordinate q_α, $1\le\alpha\le\Gamma$,

$$q_\alpha=\tfrac{1}{2}(q'_\alpha+q''_\alpha), \qquad q'''_\alpha=q'_\alpha-q''_\alpha, \tag{13h.1}$$

so that

$$q'_\alpha=q_\alpha+\tfrac{1}{2}q'''_\alpha, \qquad q''_\alpha=q_\alpha-\tfrac{1}{2}q'''_\alpha, \tag{13h.1''}$$

and

$$dq'_\alpha\,dq''_\alpha=dq_\alpha\,dq'''_\alpha. \tag{13h.1''}$$

The Wigner function $W_Q(\mathbf{p},\mathbf{q})$, which is the quantum-mechanical equivalent of the classical probability density $W_{cl}(\mathbf{p},\mathbf{q})$, with $\mathbf{q}=\mathbf{q}^{(\Gamma)}\equiv q_1$, $q_2,\ldots,q_\alpha,\ldots,q_\Gamma$ and $\mathbf{p}$ a similar set, is defined by

$$W_Q(\mathbf{p},\mathbf{q})=h^{-\Gamma}\int dq'''\,\omega(\mathbf{q}'+\tfrac{1}{2}\mathbf{q}''',\mathbf{q}-\tfrac{1}{2}\mathbf{q}''')e^{-2\pi ih^{-1}\mathbf{q}'''\cdot\mathbf{p}}, \tag{13h.2}$$

where the convention $\mathbf{q}''' \cdot \mathbf{p} = \sum_{\alpha \geq 1}^{\Gamma} q'''_\alpha p_\alpha$ is used. .The inverse transform is

$$\omega(\mathbf{q}+\tfrac{1}{2}\mathbf{q}''', \mathbf{q}-\tfrac{1}{2}\mathbf{q}''') = h^{-\Gamma} \int d\mathbf{p}\ W_Q(\mathbf{p}, \mathbf{q}) e^{2\pi i h^{-1} \mathbf{q}''' \cdot \mathbf{p}}. \qquad (13h.2')$$

Similarly to eq. (1), we define

$$p_\alpha = \tfrac{1}{2}(p'_\alpha + p''_\alpha), \qquad p'''_\alpha = p'_\alpha - p''_\alpha, \qquad (13h.3)$$

$$p'_\alpha = p_\alpha + \tfrac{1}{2}p'''_\alpha, \qquad p''_\alpha = p_\alpha - \tfrac{1}{2}p'''_\alpha, \qquad (13h.3')$$

$$dp'_\alpha\, dp''_\alpha = dp_\alpha\, dp'''_\alpha, \qquad (13h.3'')$$

so that

$$\mathbf{q}' \cdot \mathbf{p}' - \mathbf{p}'' \cdot \mathbf{q}'' = (\mathbf{q}+\tfrac{1}{2}\mathbf{q}''') \cdot (\mathbf{p}+\tfrac{1}{2}\mathbf{p}''') - (\mathbf{q}-\tfrac{1}{2}\mathbf{q}''') \cdot (\mathbf{p}-\tfrac{1}{2}\mathbf{p}''')$$
$$= \mathbf{q} \cdot \mathbf{p}''' + \mathbf{p} \cdot \mathbf{q}'''. \qquad (13h.3''')$$

We can now write the transformation of eq. (13g.4′) to the coordinate representation element $\omega(\mathbf{q}+\frac{1}{2}\mathbf{q}''', \mathbf{q}-\frac{1}{2}\mathbf{q}''')$ from the momentum representation $\omega(\mathbf{p}'+\frac{1}{2}\mathbf{p}''', \mathbf{p}'-\frac{1}{2}\mathbf{p}''')$ as

$$\omega(\mathbf{q}+\tfrac{1}{2}\mathbf{q}''', \mathbf{q}-\tfrac{1}{2}\mathbf{q}''')$$
$$= h^{-\Gamma} \iint d\mathbf{p}'\, d\mathbf{p}'''\, \omega(\mathbf{p}'+\tfrac{1}{2}\mathbf{p}''', \mathbf{p}'-\tfrac{1}{2}\mathbf{p}''') e^{-2\pi i h^{-1}(\mathbf{p}' \cdot \mathbf{q}''' + \mathbf{q}' \cdot \mathbf{p}''')}. \qquad (13h.4)$$

Use this in eq. (2) for $W_Q(\mathbf{p}, \mathbf{q})$ to find

$$W_Q(\mathbf{p}, \mathbf{q}) = h^{-2\Gamma} \iiint d\mathbf{p}'\, d\mathbf{p}'''\, d\mathbf{q}'''\, \omega(\mathbf{p}'+\tfrac{1}{2}\mathbf{p}''', \mathbf{p}'-\tfrac{1}{2}\mathbf{p}''')$$
$$\times e^{-2\pi i h^{-1}}[\mathbf{q} \cdot \mathbf{p}''' + \mathbf{q}'''(\mathbf{p}'-\mathbf{p})].$$

Integrate over $d\mathbf{q}'''$ first. The integral $h^{-\Gamma} \int d\mathbf{q}''' e^{2\pi i h^{-1}(\mathbf{p}'-\mathbf{p})\mathbf{q}'''} = \delta(\mathbf{p}'-\mathbf{p})$, where $\delta(\mathbf{p}'-\mathbf{p})$ is the product of Dirac delta functions $\prod_{\alpha=1}^{\Gamma} \delta(p'_\alpha - p_\alpha)$, so that subsequent integration over $d\mathbf{p}'$ replaces $\mathbf{p}'$ by $\mathbf{p}$ [eq. (AVIII.5)], and we have

$$W_Q(\mathbf{p}, \mathbf{q}) = h^{-\Gamma} \int d\mathbf{p}'''\ \omega(\mathbf{p}+\tfrac{1}{2}\mathbf{p}''', \mathbf{p}-\tfrac{1}{2}\mathbf{p}''') e^{-2\pi i h^{-1} \mathbf{p}''' \cdot \mathbf{q}}. \qquad (13h.5)$$

The inverse is

$$\omega(\mathbf{p}=\tfrac{1}{2}\mathbf{p}''', \mathbf{p}-\tfrac{1}{2}\mathbf{p}''') = h^{-\Gamma} \int d\mathbf{q}\ W(\mathbf{p}, \mathbf{q}) e^{2\pi i h^{-1} \mathbf{p}'''| \mathbf{q}}. \qquad (13h.5')$$

The conjugate complex $W_q^*(\mathbf{p}, \mathbf{q})$ of $W(\mathbf{p}, \mathbf{q})$ is, from eq. (5), and the fact that $\omega^*(\mathbf{p}', \mathbf{p}) = \omega(\mathbf{p}, \mathbf{p}')$,

$$W_Q^*(\mathbf{p}, \mathbf{q}) = h^{-\Gamma} \int_{-\infty}^{+\infty} d\mathbf{p}''' \, \omega^*(\mathbf{p} + \tfrac{1}{2}\mathbf{p}''', \mathbf{p} - \tfrac{1}{2}\mathbf{p}''') e^{+2\pi i h^{-1} \mathbf{p}''' \cdot \mathbf{q}}$$

$$= h^{-\Gamma} \int_{-\infty}^{+\infty} d(-\mathbf{p}''') \, \omega[\mathbf{p} + \tfrac{1}{2}(-\mathbf{p}'''), \mathbf{p} - \tfrac{1}{2}(-\mathbf{p}''')] e^{-2\pi i h^{-1} \mathbf{q} \cdot (-\mathbf{p}''')}$$

$$= W_Q(\mathbf{p}, \mathbf{q}), \tag{13h.6}$$

and since any function equal to its own conjugate complex is real, it follows that $W_Q(\mathbf{p}, \mathbf{q})$ is real.

From eq. (2′) and the condition $\int d\mathbf{p} e^{2\pi i \mathbf{q}''' \cdot \mathbf{p}} = \delta(\mathbf{q}''')$, we have that

$$\int d\mathbf{p} \, W_Q(\mathbf{p}, \mathbf{q}) = \int d\mathbf{q}''' \, d\mathbf{p} \omega(\mathbf{q} + \tfrac{1}{2}\mathbf{q}''', \mathbf{q} - \tfrac{1}{2}\mathbf{q}''') e^{-2\pi i h^{-1} \mathbf{q}''' \cdot \mathbf{p}} = \omega(\mathbf{q}, \mathbf{q}), \tag{13h.7}$$

and similarly from eq. (5′),

$$\int d\mathbf{q} \, W_Q(\mathbf{p}, \mathbf{q}) = \omega(\mathbf{p}, \mathbf{p}), \tag{13h.7'}$$

so that the integral of $W_Q(\mathbf{p}, \mathbf{q})$ over either $d\mathbf{p}$ or $d\mathbf{q}$ gives the probability density in the conjugate space, respectively. Since for any power, p_α^n of p_α,

$$\langle p_\alpha^n \rangle = \int d\mathbf{p} \, p_\alpha^n \omega(\mathbf{p}, \mathbf{p}),$$

and any real nonsingular analytic function $f(\mathbf{p})$ of the momenta set $\mathbf{p} \equiv p_1, p_2, \ldots, p_\alpha, \ldots, p_\Gamma$ can be developed as a power series in the variables, it follows that for such a function the ensemble average is

$$\langle f(\mathbf{p}) \rangle_{\text{ens}} = \iint d\mathbf{q} \, d\mathbf{p} \, f(\mathbf{p}) W_Q(\mathbf{p}, \mathbf{q}), \tag{13h.8}$$

just as in classical systems we have

$$\langle f(\mathbf{p}) \rangle_{\text{ens}} = \iint d\mathbf{q} \, d\mathbf{p} \, f(\mathbf{p}) W_{\text{cl}}(\mathbf{p}, \mathbf{q}). \tag{13h.8'}$$

Similarly, for a real analytic (nonsingular) function of the coordinates,

$$\langle f(\mathbf{q}) \rangle_{\text{ens}} = \iint d\mathbf{q} \, d\mathbf{p} \, f(\mathbf{q}) W_Q(\mathbf{p}, \mathbf{q}). \tag{13h.9}$$

It also obviously follows for a function $F(\mathbf{p}, \mathbf{q})$ which can be written as a sum, $F = f_p(\mathbf{p}) + f_q(\mathbf{q})$. Actually, it is not difficult to prove, although we do not do so, that if the function $F(\mathbf{p}, \mathbf{q})$ contains no products $p_\alpha q_\alpha$ of a single conjugated coordinate momentum pair,

$$\left\langle \overset{(\text{no } p_\alpha q_\alpha \text{ pair})}{F(\mathbf{p}, \mathbf{q})} \right\rangle_{\text{ens}} = \iint d\mathbf{q}\, d\mathbf{p}\, F(\mathbf{p}, \mathbf{q}) W_Q(\mathbf{p}, \mathbf{q}). \tag{13h.10}$$

That this cannot be so for $p_\alpha q_\alpha$ is obvious from two facts: first, that in the integral of $p_\alpha q_\alpha W_Q$ the order of writing the products $p_\alpha q_\alpha$ or $q_\alpha p_\alpha$ gives identical answers, whereas for the quantum-mechanical operator $p_\alpha q_\alpha \omega = (2\pi i)^{-1}(\partial/\partial q_\alpha) q_\alpha \omega = (2\pi i)^{-1} h\omega + q_\alpha (2\pi i)^{-1} h(\partial/\partial q_\alpha)\omega = q_\alpha p_\alpha \omega + (2\pi i)^{-1} h\omega$; see Sec. 3b.

If $\omega(\mathbf{q}', \mathbf{q}'')$ is real, as it is for an equilibrium ensemble (see Sec. 13i), then $\omega(\mathbf{q}+\frac{1}{2}\mathbf{q}''', \mathbf{q}-\frac{1}{2}\mathbf{q}''') = \omega^*(\mathbf{q}+\frac{1}{2}\mathbf{q}''', \mathbf{q}-\frac{1}{2}\mathbf{q}''')$, and since from eq. (6) $W_Q(\mathbf{p}, \mathbf{q}) = W_Q^*(\mathbf{p}, \mathbf{q})$ is real, we can as well use

$$W_Q(\mathbf{p}, \mathbf{q}) = h^{-\Gamma} \int d\mathbf{q}'''\ \omega(\mathbf{q}+\tfrac{1}{2}\mathbf{q}''', \mathbf{q}-\tfrac{1}{2}\mathbf{q}''') e^{2\pi i h^{-1}\mathbf{q}'''\cdot\mathbf{p}}. \qquad [\omega(\mathbf{q}'\mathbf{q}'') \text{ real}] \tag{13h.11}$$

13i. SEMICLASSICAL EQUILIBRIUM DENSITY MATRIX

We confine our attention to the simplest case, that of an ensemble of systems of fixed V, β, N of one chemical component monatomic molecules without internal degrees of freedom. The general nature of the result is obviously applicable to more complicated systems. We reserve the question of correct symmetrization of the functions used to construct the elements of the matrix until later (Sec. 13 *l*), but use the approximation that $N!$ of the normalized functions with coordinates of numbered molecules must be used to form one linear combination of the correct symmetry.

The $\mathbf{q}'$, $\mathbf{q}''$ elements $\omega(\mathbf{q}', \mathbf{q}'')$ of Ω for an equilibrium petite canonical ensemble of fixed V, β, N is then the coordinate representation of the matrix

$$\mathbf{\Omega} = (N!)^{-1} e^{\beta(A-\mathcal{H})}, \tag{13i.1}$$

with $\mathcal{H}$ the Hamiltonian operator, and with A the Helmholz free energy determined by the condition

$$\text{tr}(\mathbf{\Omega}) = 1. \tag{13i.1'}$$

The matrix $\mathbf{\Omega}$ is diagonal in the $\mathbf{K}'$, $\mathbf{K}''$ representation of the real stationary eigenfunctions $\Psi_\mathbf{K}$ of the operator $\mathcal{H}$ with elements

$$\omega(\mathbf{q}', \mathbf{q}'') = (N!)^{-1} \sum_\mathbf{K} \Psi_\mathbf{K}(\mathbf{q}') e^{\beta(A-E_\mathbf{K})} \Psi_\mathbf{K}^*(\mathbf{q}''), \tag{13i.2}$$

which are real,

$$\omega(\mathbf{q}', \mathbf{q}'') = \omega^*(\mathbf{q}'', \mathbf{q}') = \omega(\mathbf{q}'', \mathbf{q}'). \tag{13i.2'}$$

We wish to avoid evaluation of the functions $\Psi_{\mathbf{K}}(\mathbf{q})$ and of the $E_{\mathbf{K}}$'s. Use cartesian coordinates for the atoms indexed by j, $1 \leq j \leq N$,

$$q_{j1} = x_j, \qquad q_{j2} = y_j, \qquad q_{j3} = z_j,$$

$$\mathbf{q} = \sum_{j\alpha=1}^{3N} q_{j\alpha}. \tag{13i.3}$$

We have that $-\beta$ times the Hamiltonian is a sum of an operator $-\beta\mathcal{K}$ plus the function $-\beta U(q)$ acting as a multiplicative operator $\mathcal{U}$,

$$-\beta\mathcal{H} = -\beta(\mathcal{K} + \mathcal{U}) \tag{13i.4}$$

with $\sqrt{-\beta} = i\beta^{1/2}$,

$$-\beta\mathcal{K} = \frac{-\beta}{2m}\sum_{j\alpha=1}^{3N}\left(\frac{h}{2ni}\frac{\partial}{\partial q_\alpha}\right)^2 = \sum_{j\alpha=1}^{3N}\left[\left(\frac{\beta h^2}{8\pi^2 m}\right)^{1/2}\frac{\partial}{\partial q_\alpha}\right]^2$$
$$= \sum_{j\alpha=1}^{3N}\left(\frac{\lambda}{2\sqrt{\pi}}\frac{\partial}{\partial q_\alpha}\right)^2, \tag{13i.4$'$}$$

where λ is the ubiquitous thermal de Broglie wavelength $\lambda = (h^2\beta/2\pi m)^{1/2}$ (Sec. 3d).

The operator $e^{-\beta\mathcal{H}}$ is then

$$e^{-\beta\mathcal{H}} = \sum_{n\geq 0}\frac{(-1)^n}{n!}\beta^n(\mathcal{K} + \mathcal{U})^n$$
$$= 1 - \beta(\mathcal{K} + \mathcal{U}) + \frac{\beta^2}{2}(\mathcal{K}^2 + \mathcal{K}\mathcal{U} + \mathcal{U}\mathcal{K} + \mathcal{U}^2) + \cdots. \tag{13i.5}$$

The operator $\mathcal{U}$ acting on a function of the coordinates just multiplies the function of $\mathbf{q}'$ by $U(\mathbf{q}')$ and, on a matrix element such as $\Psi_{\mathbf{K}}(\mathbf{q}')\Psi^*_{\mathbf{K}}(\mathbf{q}'')$, by $\frac{1}{2}[U(\mathbf{q}') + U(\mathbf{q}'')]$ [eq. (3b.6)].

We use the representation in which the kinetic energy operator $\mathcal{K}$ is diagonal, namely, that of the functions

$$\Phi(\mathbf{p}\cdot\mathbf{q}) = h^{-(3N/2)}e^{2\pi i h^{-1}\mathbf{p}\cdot\mathbf{q}}, \tag{13i.6}$$

for which

$$\beta\mathcal{K}\Phi(\mathbf{p}\cdot\mathbf{q}) = \beta(2m)^{-1}\mathbf{p}\cdot\mathbf{p}\Phi(\mathbf{p}\cdot\mathbf{q}). \tag{13i.6$'$}$$

In the last line of eq. (5) we have spelled out the operator $(\mathcal{K} + \mathcal{U})^2$. It contains one term $\mathcal{K}\mathcal{U}$ which, when operating on any function $f(q)$, gives

$$\mathcal{K}\mathcal{U}f = -\frac{1}{2m}\frac{h^2}{4\pi^2}\sum_{j\alpha}\left(\frac{\partial^2}{\partial q_{j\alpha}^2}U\right)f + 2\left(\frac{\partial}{\partial q_{j\alpha}}U\right)\left(\frac{\partial}{\partial q_{j\alpha}}f\right) + U\left(\frac{\partial^2}{\partial q_{j\alpha}}f\right). \tag{13i.7}$$

The last term in eq. (7) is $\mathcal{U}\mathcal{K}$. If f is $\Phi(\mathbf{p}\cdot\mathbf{q})$, we have $\sum_{j\alpha}\partial^2/\partial q_{j\alpha}^2\Phi = \mathbf{p}\cdot\mathbf{p}(4\pi^2/h^2)$. The two factors $(h^2/4\pi^2)$ and $(4\pi^2/h^2)$ cancel. In the other terms containing derivatives of the potential energy a factor h remains and will be small if h is small. We rewrite

$$(\mathcal{K}+\mathcal{U})^2=(\mathcal{K}^2+2\mathcal{U}\mathcal{K}+\mathcal{U}^2)+(\mathcal{K}\mathcal{U}-\mathcal{U}\mathcal{K}), \tag{13i.8}$$

where the last term turns out to be a quantum-mechanical correction, and the first term is part of the semiclassical expression for the matrix $e^{-\beta\mathcal{H}}$.

In general, in $(\mathcal{K}+\mathcal{U})^n$ there is one term for every single way that $n-\nu$ powers of $\mathcal{U}$ combined with ν operators $\mathcal{K}$ can be written in order, and altogether $n!/(n-\nu)!\,\nu!$ different permutations $\mathcal{P}_p$ of that order for given n and ν. Each of these produces one term $\mathcal{U}^{n-\nu}\mathcal{K}^{\nu}$. We define, for $\nu\geq 1$, $\mu\geq 1$, an operator

$$\mathcal{G}_{\nu\mu}(U,\mathcal{K})=\left(\sum_{p=1}^{p=(\nu+\mu)!/\nu!\,\mu!}\mathcal{P}_p\mathcal{U}^{\mu}\mathcal{K}^{\nu}\right)-\frac{(\nu+\mu)!}{\nu!\,\mu!}\mathcal{U}^{\mu}\mathcal{K}^{\nu}, \tag{13i.9}$$

which subtracts from each of the $(\nu+\mu)!/\nu!\,\mu!$ permutations of the order of $\mathcal{U}$ and $\mathcal{K}$ operators the operation when the derivatives $(\partial/\partial q_{i\alpha})^2$ in the operators $\mathcal{K}$ carry thorugh the sequence of $\mathcal{U}$'s to operate only on the function on which $(\mathcal{K}+\mathcal{U})^n$ acts. Thus, when $\mathcal{G}_{\nu\mu}$ operates on the function Φ, it produces a sum of terms having derivatives of the potential $U(\mathbf{q})$ multiplied by positive power of Planck's constant h. With this we have

$$(\mathcal{K}+\mathcal{U})^n=\sum_{\nu=0}^{n}\left\{\left[\frac{n!}{\nu!(n-\nu)!}\mathcal{U}^{n-\nu}\mathcal{K}^{\nu}\right]+\mathcal{G}_{\nu(n-\nu)}(\mathcal{U},\mathcal{K})\right\}, \tag{13i.10}$$

$$\sum_{n\geq 0}^{\infty}\frac{(-\beta)^n}{n!}(\mathcal{K}+\mathcal{U})^n=\sum_{\nu\geq 0}^{\infty}\sum_{\mu\geq 0}^{\infty}\frac{(-\beta\mathcal{U})^{\mu}}{\mu!}\frac{(-\beta\mathcal{K})^{\nu}}{\nu!}$$
$$\times\left\{1+\sum_{\nu\geq 1}^{\infty}\sum_{\mu\geq 1}^{\infty}\frac{1}{\nu!}\frac{1}{\mu!}\mathcal{G}_{\nu\mu}(-\beta\mathcal{U},-\beta\mathcal{K}\right\}. \tag{13i.11}$$

Now neglect the terms $\mathcal{G}_{\nu\mu}$ which are indeed small for all but molecules of low molecular weight at low temperatures. We seek the diagonal elements with $\mathbf{q}'=\mathbf{q}''=\mathbf{q}$ for which $\frac{1}{2}[U(\mathbf{q}')+U(\mathbf{q}'')]=U(\mathbf{q})$. The operation of the kinetic energy operator $\frac{1}{2}\mathcal{K}$ on $\Phi(\mathbf{p}\cdot\mathbf{q})$ or on $\Phi^*(\mathbf{q}\cdot\mathbf{p})$ gives the function Φ or Φ^*, respectively, multiplied by $\frac{1}{2}(2m)^{-1}\mathbf{p}\cdot\mathbf{p}$, and the sum is $(2m)^{-1}\mathbf{p}\cdot\mathbf{p}$, the classical kinetic energy at the set $\mathbf{p}$ of momenta. From eq. (6), $\Phi\Phi^*=h^{-3N}$. Summing eq. (11) over ν and μ $(\sum x^{\nu}/\nu!=e^{x})$ we find, in the limit $\mathcal{G}_{\nu\mu}\Rightarrow 0$,

$$\tfrac{1}{2}[\Phi^*(\mathbf{q}\cdot\mathbf{p})e^{-\beta\mathcal{H}}\Phi(\mathbf{p}\cdot\mathbf{q})+\Phi(\mathbf{p}\cdot\mathbf{q})e^{-\beta\mathcal{H}}\Phi^*(\mathbf{q}\cdot\mathbf{p})]$$
$$=h^{-3N}\exp\{-\beta[2m^{-1}\mathbf{p}\cdot\mathbf{p}+U(\mathbf{q})]\}. \tag{13i.12}$$

The Helmholz free energy is determined for the petite canonical ensemble of fixed V, β, N by the condition of eq. (1′), where $\mathrm{tr}(\mathbf{\Omega})=1$, with $\mathbf{\Omega}$ given by eq. (1), which leads for the limit $\mathscr{G}_{\nu\mu} \Rightarrow 0$ to

$$A(V, \beta, N) = \beta^{-1} \ln \int \cdots \int d\mathbf{p}\, d\mathbf{q}\, (N!\, h^{3N})^{-1} e^{-\beta[(2m)^{-1}\mathbf{p}\cdot\mathbf{p}+U(\mathbf{q})]}. \tag{13i.13}$$

This is the semiclassical expression used in Chapters 8 and 9.

We asserted without proof in Chapter 3 that in any volume $\prod_{j\geq 1}^{\Gamma} \Delta p_j\, \Delta q_j = \Delta(\mathbf{p}^{(\Gamma)}, \mathbf{q}^{(\Gamma)})$, for which $\Delta p_j\, \Delta q_j \gg h$ for every degree of freedom j, the number of (unsymmetrized) quantum states included is $\Delta(\mathbf{p}^{(\Gamma)}\mathbf{q}^{(\Gamma)})h^{-\Gamma}$. Setting $\mathscr{G}_{\nu\mu} \Rightarrow 0$ means that the energy difference between neighboring quantum states along every degree of freedom is negligible compared to $\beta^{-1} = kT$, and that integration over $d\mathbf{p}\, d\mathbf{q}\, h^{-\Gamma}$ can be used to replace summation over quantum states. Thus the derivation of eq. (13) in this section proves the assertion made in Chapter 3 and used repeatedly in Chapters 8 and 9.

In Sec. 13j we discuss the inclusion of the quantum corrections for a potential energy function $U(\mathbf{q})$ which depends only on the $\frac{1}{2}N(N-1)$ distances r_{jk}, $1 \leq j < k \leq N$, between atoms j and k, that is, one with no explicit angular dependence of the bond energies such as one normally assumes in polyatomic molecules. We then place a further limitation to the case in which the potential is a sum of pair terms only, with no three-body interactions, which further simplifies the result.

We note here that the Urey-Bradley field approximation has been used with some success to explain the vibrational spectra of simple polyatomic molecules, replacing the explicit bending force constants by an assumed van der Waals–type interaction betwen atoms not bonded to each other. The treatment that follows applies to this case for the internal degrees of freedom, as well as to the potential between the centers of mass.

13j. THE POTENTIAL $U(\mathbf{q})$, A SUM OF PAIRWISE INTERACTIONS

Assume that $U(\mathbf{q})$ is of the form

$$U(\mathbf{q}) = \sum_{1\leq j<k\leq N}\sum u(r_{jk}). \tag{13j.1}$$

For this $U(\mathbf{q})$, or indeed for any function $F(\sum\sum_{1\leq j<k\leq N} f(r_{jk}))$dependent only on the $\frac{1}{2}N(N-1)$ distances, $r_{jk} = [(x_j - x_k)^2 + (y_j - y_k)^2 + (z_j - z_k)^2]^{1/2}$, the *first* derivative and all *odd* derivatives in eq. (13k.7) for the operation of $\mathscr{H}$ on F are zero provided that the dependence of F on every r_{jk} goes to zero more rapidly than r_{jk}^{-3} as $r_{jk} \rightarrow \infty$. That this is so is seen from the

physical meaning of the derivative along any axis α, $\sum_{j=1}(\partial/\partial q_{j\alpha})F$, since this corresponds to the limit $\Delta q_\alpha \to 0$ of $(\Delta q_\alpha)^{-1}\Delta F$, where in ΔF *all* molecules are displaced by Δq_α, leaving their distances unchanged. Since the result of a second derivative of F is also a function of distances,

$$F''(\mathbf{q}) \to F''\left(\sum\sum_{1\le j<k\le N} r_{jk}\right), \tag{13j.2}$$

the third and all subsequent odd derivatives are zero.

There are of course in a closed system of fixed walls surface terms proportional to $V^{2/3}$ due to the wall potential not given in eq. (1), but these are to be disregarded in seeking functions of the bulk properties. In general, surface free energies proportional to surface area need a special treatment.

The details are as follows. We use

$$r_{jk} = \left[\sum_{\alpha=1}^{3} (q_{j\alpha} - q_{k\alpha})^2\right]^{1/2}, \tag{13j.3}$$

so that

$$r_{jk}^{-2} \sum_{\alpha=1}^{3} (q_{j\alpha} - q_{k\alpha})^2 = 1. \tag{13j.3$'$}$$

We then have

$$\frac{\partial r_{jk}}{\partial q_{j\alpha}} = r_{jk}^{-1}(q_{j\alpha} - q_{k\alpha}) = -\frac{\partial r_{jk}}{\partial q_{ka}}, \tag{13j.4}$$

so that, for any function $f(r_{jk})$,

$$\left(\frac{\partial}{\partial q_{j\alpha}} + \frac{\partial}{\partial q_{k\alpha}}\right) f(r_{jk}) = 0. \tag{13j.4$'$}$$

For the second derivative one finds

$$\frac{\partial^2 f}{\partial q_{j\alpha}^2} = \frac{d^2 f}{dr_{jk}^2} r_{jk}^{-2}(q_{j\alpha} - q_{k\alpha})^2 + \frac{df}{dr_{jk}} \frac{\partial}{\partial q_{j\alpha}} r_{jk}^{-1}(q_{j\alpha} - q_{k\alpha})$$

$$= \frac{d^2 f}{dr_{jk}^2} r_{jk}^{-2}(q_{j\alpha} - q_{k\alpha})^2 + r_{jk}^{-1} \frac{df}{dr_{ik}} [1 - r_{jk}^{-2}(q_{j\alpha} - q_{k\alpha})^2] = \frac{\partial^2 f}{\partial q_{k\alpha}^2}. \tag{3j.5}$$

Of the $3N$ second derivative operators in $\mathcal{K}$ only six, $\partial^2/\partial q_{j\alpha}^2 + \partial^2/\partial q_{k\alpha}^2$, $1 \le \alpha \le 3$, operate on $f(r_{jk})$ and, from eq. (4′), the first derivatives are all zero for each α. The second derivatives, from eq. 5, using eq. (3′) for the sum over the α's and both molecules, with eq. (13k.4′) for $-\beta\mathcal{K}$, gives

$$-\beta\mathcal{K} f(r_{jk}) = \frac{\lambda^2}{4\pi}\left(\frac{2d^2 f}{dr_{jk}^2} + 4 r_{jk}^{-1} \frac{dr}{dr_{jk}}\right). \tag{13j.5$'$}$$

With $U(\mathbf{q})$ of eq. (1), $(-\beta\mathcal{K})(-\beta\mathcal{U})$ is a sum of $\frac{1}{2}N(N-1)$ terms $-\beta\mathcal{K}$ operating on $-\beta\mathcal{U}(r_{jk})$, $1 \leq j < k \leq N$, and $(-\beta\mathcal{K})^{\mu}(-\beta\mathcal{U})$ a similar sum of the same number of functional operators each dependent on the pair distances r_{jk} only. We define two dimensionless multiplicative operator functions,

$$-\beta\mathcal{U}_{\kappa}(\mathbf{q}) = -\beta(-\beta\mathcal{K})^{\kappa-1}\mathcal{U}(\mathbf{q}), \tag{13j.6}$$

$$-\beta u_{\kappa}(r) = -\beta(-\beta\mathcal{K})^{\kappa-1}u(r) = -\beta\left[\frac{\lambda^2}{4\pi}\left(\frac{2d^2}{dr^2} + 4r^{-1}\frac{d}{dr}\right)^{\kappa-1}\right]u(r), \tag{13j.6$'$}$$

so that

$$-\beta\mathcal{U}_{\kappa}(\mathbf{q}) = \sum\sum_{1\leq j<k\leq N} -\beta u_{\kappa}(r_{jk}). \tag{13j.6$''$}$$

Because of the very large number of terms in eq. (6″) the sum over n of $\mathcal{H}^n/n!$ increases in numerical importance up to $n \sim N$. The general recursion formula, using the notation of eq. (6) for which $\mathcal{U} = \mathcal{U}_1$, is

$$\mathcal{H}^{n+1} = (\mathcal{U}_1 + \mathcal{K})\mathcal{H}^n. \tag{13j.7}$$

We can, and shall, use this for small n, finding a pattern for $\mathcal{H}^n$ and $\mathcal{H}^n/n!$ which we can *guess* to be general and subsequently prove by induction with eq. (7) that if it holds for n it holds for $n+1$, hence for all finite n.

$$\begin{aligned}
\mathcal{H}^0 &= 1,\\
\mathcal{H}^1 &= \mathcal{U}_1 + \mathcal{K},\\
\mathcal{H}^2 &= (\mathcal{U}_1^2 + \mathcal{U}_2) + 2\mathcal{U}_1\mathcal{K} + \mathcal{K}^2,\\
\mathcal{H}^3 &= (\mathcal{U}_1^3 + 3\mathcal{U}_1\mathcal{U}_2 + \mathcal{U}_3) + 3(\mathcal{U}_1^2 + \mathcal{U}_2)\mathcal{K} + 3\mathcal{U}_1\mathcal{K}^2 + \mathcal{K}^3,\\
\mathcal{H}^4 &= (\mathcal{U}_1^4 + 6\mathcal{U}_1^2\mathcal{U}_2 + 4\mathcal{U}_1\mathcal{U}_3 + 3\mathcal{U}_2^2 + \mathcal{U}_4)\\
&\quad + 4(\mathcal{U}_1^3 + 3\mathcal{U}_1\mathcal{U}_2 + \mathcal{U}_3)\mathcal{K} + 6(\mathcal{U}_1^2 + \mathcal{U}_2)\mathcal{K}^2 + 4\mathcal{U}_1\mathcal{K}^3 + \mathcal{K}^4.
\end{aligned} \tag{13j.8}$$

Use a brace $\{(\mathcal{K}+\mathcal{U})^n\}$ to indicate the operation of $(\mathcal{K}+\mathcal{U})^n$ alone, omitting the terms in eq. (8) followed by $\mathcal{K}^{\mu}$. We see that up to $n=4$, inclusive, we have that $\mathcal{H}^n$ is given by

$$\mathcal{H}^n = \sum_{\mu\geq 0}^{n} \binom{n}{\mu}\{(\mathcal{K}+\mathcal{U}_1)^{n-\mu}\}\mathcal{K}^{\mu} \tag{13j.9}$$

where $\binom{n}{\mu}$ is the binomial coefficient $\binom{n}{\mu} = n!/(n-\mu)!\,\mu!$. The recursion relation (7) for $\mathcal{H}^{n+1}$ that $\mathcal{H}^{n+1} = \mathcal{U}_1\mathcal{H}^n + \{\mathcal{K}\mathcal{H}^n\} + \mathcal{H}^n\mathcal{K}$ leads necessarily to a development of the form

$$\mathcal{H}^n = \sum_{\mu\geq 0}^{n} \mathcal{V}_{n,\mu}\mathcal{K}^\mu, \tag{13j.10}$$

with $\mathcal{V}_{n,\mu}$ a sum of products $\prod_{\kappa=1}^{n-\mu} \mathcal{U}_\kappa^{\nu_\kappa}$ with various coefficients as displayed in eq. (8 for $n \leq 4$. The recursion relation for the $\mathcal{V}_{n,\mu}$'s, from eq. (7), is seen to be

$$\mathcal{V}_{n+1,\mu} = \mathcal{V}_{n,\mu} + \{(\mathcal{K}+\mathcal{U}_1)\mathcal{V}_{n,\mu}\}. \tag{13j.10'}$$

If eq. (9) in which $\mathcal{V}_{n,\mu} = \binom{n}{\mu}\{(\mathcal{K}+\mathcal{U}_1)^{n-\mu}\}$ holds for n, as it does for $n \leq 4$, then from eq. (10′)

$$\begin{aligned} \mathcal{V}_{n+1,\mu} &= \left[\binom{n}{\mu-1} + \binom{n}{\mu}\right]\{(\mathcal{K}+\mathcal{U}_1)^{n+1-\mu}\} \\ &= \left[\binom{n+1}{\mu}\frac{\mu}{n+1} + \binom{n+1}{\mu}\frac{n+1-\mu}{n+1}\right]\{\mathcal{K}+\mathcal{U}_1)^{n+1-\mu}\} \\ &= \binom{n+1}{\mu}\{\mathcal{K}+\mathcal{U}_1)^{n+1-\mu}\}, \end{aligned} \tag{13.j10''}$$

hence for all n.

Examination of eq. (8) shows that, for $n \leq 4$,

$$\frac{\{(\mathcal{K}+\mathcal{U}_1)^n\}}{n!} = \sum_{\nu_1\geq 0} \cdots \sum_{\nu_\kappa \geq 0} \cdots \prod_{\kappa\geq 1}^{n} \frac{(\mathcal{U}_\kappa/\kappa!)^{\nu_\kappa}}{\kappa!}, \qquad \sum_{\kappa\geq 1}^{n} \kappa\nu_\kappa = n. \tag{13j.11}$$

Again we prove by induction that, if eq. (11) holds for n, it holds for $n+1$, hence for all n. The recursion formula is

$$\frac{\{(\mathcal{K}+\mathcal{U}_1)^{n+1}\}}{n+1!} = (n+1)^{-1}\frac{(\mathcal{K}+\mathcal{U}_1)\{(\mathcal{K}+\mathcal{U}_1)^n\}}{n!} \tag{13j.12}$$

From the definition of $\mathcal{U}_\kappa$ implied in eq. (6) we have that $\mathcal{K}\mathcal{U}_{\kappa-1} = \mathcal{U}_\kappa$, from which we find that for $2 \leq \kappa \leq n+1$

$$\begin{aligned} &\frac{(\mathcal{U}_\kappa/\kappa!)^{\nu_\kappa}}{\nu_\kappa!}\mathcal{K}\frac{\mathcal{U}_{\kappa-1}/(\kappa-1)!]^{\nu_{\kappa-1}}}{\nu_{\kappa-1}!} \\ &= \kappa(\nu_\kappa+1)\frac{(\mathcal{U}_\kappa/\kappa!)^{\nu_\kappa+1}}{(\nu_\kappa+1)!}\frac{(\mathcal{U}_{\kappa-1}/\kappa-1)!\ ^{\nu_{\kappa-1}-1}}{(\nu_{\kappa-1}-1)!} \\ &= \kappa\nu'\frac{(\mathcal{U}_\kappa/\kappa!)^{\nu'_\kappa}}{\nu'_\kappa!}\frac{(\mathcal{U}_{\kappa-1}/(\kappa-1)!)^{\nu'_{\kappa-1}-1}}{\nu'_{\kappa-1}}, \\ &\nu'_\kappa = \nu_\kappa+1, \qquad \nu'_{\kappa-1} = \nu_{\kappa-1}-1, \end{aligned} \tag{13j.12'}$$

where, for $\kappa = n+1$, ν_κ is necessarily zero, but $\nu'_{n+1} = 1$ if ν_n is not zero. Operation by $\mathcal{U}_1$ on the right-hand side of eq. (11) increases ν_1 by unity to $\nu'_1 = \nu_1 + 1$, which is consistent with eq. (12′), with $\nu'_0 = \nu'_{1-1} = 0$. Summing the operation of $\mathcal{U}_1$ on the right of eq. (11) plus the operation of $\mathcal{K}$ on each of the single terms $\mathcal{U}_{\kappa-1}$ for $2 \le \kappa \le n+1$, one finds

$$\frac{(\mathcal{K}+\mathcal{U}_1)\{(\mathcal{K}+\mathcal{U}_1)^n\}}{n!}$$
$$= \sum_{\nu'_1 \ge 0} \cdots \sum_{\nu'_\kappa \ge 0} \cdots \sum_{\kappa=1}^{n+1} \kappa\nu'_\kappa \prod_{\kappa \ge 1}^{n+1} \frac{(\mathcal{U}_\kappa/\kappa!)^{\nu_\kappa}}{\nu'_\kappa!}, \qquad \sum \kappa\nu'_\kappa = n+1. \quad (13j.12'')$$

With $\sum \kappa\nu'_\kappa = n+1$ canceling $(n+1)^{-1}$ in eq. (12) one proves the generality of eq. (11) for all n.

We now proceed as we did in deriving eqs. (13i.12) and (13i.13), where the terms $\mathcal{U}_\kappa$, $\kappa \ge 2$, were assumed to be negligible, and where we retained only $\mathcal{U}$ which we now call $\mathcal{U}_1$. As we did there, we restrict ourselves to the diagonal terms, $\mathbf{q}' = \mathbf{q}'' = \mathbf{q}$, for which $\Phi^* e^{-\beta\mathcal{H}}\Phi = \Phi e^{-\beta\mathcal{H}}\Phi^*$ and $\Phi\Phi^* = h^{-3N}$. We have, with eqs. (9) and (11),

$$\tfrac{1}{2}(\Phi^* e^{-\beta\mathcal{H}}\Phi + \Phi e^{-\beta\mathcal{H}}\Phi^*) = \Phi^* e^{-\beta\mathcal{H}}\Phi = \sum_{n\ge 0} \frac{\Phi^*(-\beta\mathcal{H})^n\Phi}{n!}$$
$$= \Phi^* \sum_{n\ge 0}\left[\sum_{\mu\ge 0}\sum_{\nu_1\ge 0}\cdots\sum_{\nu_\kappa \ge 0}\cdots\prod_{\kappa\ge 1}^{n}\frac{(-\beta\mathcal{U}_\kappa/\kappa!)^{\nu_\kappa}}{\nu_\kappa!}\mathcal{K}^\mu\Phi, \quad \mu + \sum_\kappa \kappa\nu_\kappa = n\right]$$
$$= h^{-3N}\sum_{\mu\ge 0}^{\infty}\sum_{\nu_1\ge 0}^{\infty}\cdots\sum_{\nu_\kappa\ge 0}^{\infty}\cdots\prod_{\kappa\ge 1}^{\infty}\frac{(-\beta U_\kappa/\kappa!)^{\nu_\kappa}}{\nu_\kappa!}[(2m)^{-1}\mathbf{p}\cdot\mathbf{p}]^\mu$$
$$= h^{-3N}\exp\left\{-\beta\left[\frac{(2m)^{-1}\mathbf{p}\cdot\mathbf{p} + \sum_{\kappa\ge 1} - \beta U_\kappa(\mathbf{q})}{\kappa!}\right]\right\}$$
$$= h^{-3N}\exp\left\{-\beta\left[\frac{(2m)^{-1}\mathbf{p}\cdot\mathbf{p} + \sum\sum_{1\le j<k\le N}\sum_{\kappa\ge 1} - \beta u_\kappa(r,k)}{\kappa!}\right]\right\}, \quad (13j.13)$$

since the summation over n removes the restriction on the values of μ and ν_κ. The functions U_κ and u_κ correspond to the multiplicative operators $\mathcal{U}_\kappa$ and $\mathscr{u}_\kappa$ of eqs. (6) and (6′) now impressed by their operation on $\Phi(\mathbf{q}\cdot\mathbf{p})$ with the coordinate value $\mathbf{q}$,

$$-\beta u_\kappa = -\beta\left[\frac{\lambda^2}{4\pi}\left(\frac{2d^2}{dr^2} + 4r^{-1}\frac{d}{dr}\right)\right]^{\kappa-1} u(r),$$

$$-\beta U_\kappa(\mathbf{q}) = \sum\sum_{1\le j<k\le N} - \beta u_\kappa(r_{jk}), \quad (13j.14)$$

where we note that λ^2 is proportional to β, so that βu_κ and βU_κ are proportional to β^κ.

By defining

$$-\beta\,\Delta U_Q(\mathbf{q}) = \sum_{\kappa\geq 2} \frac{-\beta U_\kappa(\mathbf{q})}{\kappa!}$$

$$= \sum\sum_{1\leq j<k\leq N} -\beta\,\Delta u_Q(r_{jk}) = \sum\sum_{1\leq j<k\leq N}\sum_{\kappa\geq 2} \frac{-\beta u_\kappa(r_{jk})}{\kappa!}, \qquad (13j.15)$$

so that $\sum_{\kappa\geq 1} U_\kappa/\kappa! = U + \Delta U_Q$, with U the true (classical) potential energy and ΔU_Q a quantum-mechanical correction to the classical probability function, so that $\omega(V, \beta, N;\ \mathbf{q},\ \mathbf{q})$ the (unsymmetrized) diagonal coordinate representation for an equilibrium ensemble of systems with fixed V, β, N is

$$\omega[V, \beta, N; (\mathbf{q}, \mathbf{q})]$$

$$= (N!)^{-1}\int\cdots\int d\mathbf{p}\, h^{-3N} \exp \beta[A - (2m)^{-1}\mathbf{p}\cdot\mathbf{p} - U - \Delta U_Q]. \quad (13j.16)$$

The condition that the trace be unity,

$$\int\int\cdots\int d\mathbf{q}\, \omega(V, \beta, N; \mathbf{q}, \mathbf{q}) = 1, \qquad (13j.16')$$

leads to the equation for the Helmholz free energy $A(V, \beta, N)$ [see eq. (13i.13),

$$A(V, \beta, N) = -\beta^{-1} \ln(N!\, h^{3N})^{-1} \int\int\cdots\int d\mathbf{p}\, d\mathbf{q} \exp\{-\beta[2m)^{-1}\mathbf{p}\cdot\mathbf{p} + U + \Delta U_Q]\}, \qquad (13j.17)$$

differing from the semiclassical only by ΔU_Q. Obviously, the equation can be extended to multicomponent systems $N = N_a$, $N_b, \ldots$ and to grand canonical ensembles (see Chapter 3).

13k. INTERPRETATION OF $\Delta U_Q(\mathbf{q})$ AS $\Delta A_Q(\mathbf{q})$

Many or most of the applications of the semiclassical treatments replacing summation over quantum states by integration over $h^{-\Gamma}\, d\mathbf{p}\, d\mathbf{q}$ for van der Waals–type potentials assume the potential energy $U(\mathbf{q})$ to be a sum of pair terms only [eq. (13j.1)]. Although this is certainly not exact for real molecules, it appears to be numerically very good, and for most problems the specific three-body interactions $u_3(r_{jk}, f_{kl}, r_{lj})$ can be neglected. Since, as shown in Sec. 13j, if U is a sum of pair terms only, so also is ΔU_Q [eq. (13j.15)], and all the simplifications of the pair interaction treatment can be retained in a quantum-mechanical treatment replacing $u(r)$ by the not drastically different $u(r) + \Delta u_Q(r)$ of eq. (13j.15).

Included in the simplification is the *numerical* advantage of the classical interpretation discussed in Sec. 9b, and used in deriving the YKBBG hierarchy of integral equations is the fact that, with $\rho = \langle N/V\rangle$ and

$\rho_n(\mathbf{r}_1, \ldots, \mathbf{r}_j, \ldots, \mathbf{r}_n)$, the reduced probability density for n molecules,

$$\beta^{-1}\frac{\partial}{\partial x_j}\ln \rho^{-n}\rho_n(\mathbf{r}_1, \ldots, \mathbf{r}_1, \ldots, \mathbf{r}_n) = \left\langle\frac{\partial U(\mathbf{r}_1, \ldots, \mathbf{r}_n)}{\partial x_j}\right\rangle, \qquad (13\text{k}.1)$$

where the quantity on the right is minus the average force at fixed $\mathbf{r}_1, \ldots, \mathbf{r}_n$. In the classical case we called $\beta^{-1}\ln \rho^{-n}\rho_n$ the potential of average force. The equation is still valid, with $U+\Delta U_Q$ replacing U in the quantum formulation, and can be used with equally very doubtful validity to close the hierarchy of equations at integration over ρ_3, with the approximation that $\rho_3(r_{23}, r_{31}, r_{12}) = \rho^{-3}\rho_2(r_{23})\rho_2(r_{31})\rho_2(r_{12})$. However, the physical interpretation is now more complicated and is discussed at the end of this section.

The quantity $-\beta[(2m)^{-1}\mathbf{p}\cdot\mathbf{p}+U(\mathbf{q})+\Delta U_Q(\mathbf{q})]$ in the exponent of eq. (13j.16) is a function of variables $\mathbf{p}$ and $\mathbf{q}$ defined for all exact values of both in complete violation of the uncertainty principle that for any single momentum coordinate pair $q_{j\alpha}$, $p_{j\alpha}$ the product of the uncertainties $\Delta p_{j\alpha}\,\Delta q_{j\alpha} = h$. We cannot interpret $\mathbf{p}$ and $\mathbf{q}$ in this expression as momenta and coordinates. The quantity $\mathbf{p}$ has the physical dimensions mlt^{-1} of momentum but is a quantity analogous to a quantum number in the function Φ. Actually, the $p_{j\alpha}/h$'s are quantum numbers of dimension l^{-1} which, in a macroscopic volume L^3 lie so close in energy that integration can be used to replace summation. Information about the p-distribution is given by the Wigner function of a true $\mathbf{p}$ and $\mathbf{q}$ for which the integral over either $\mathbf{p}$ or $\mathbf{q}$ gives us the correct probability density $\omega(\mathbf{q}, \mathbf{q})$ or $\omega(\mathbf{p}, \mathbf{p})$ of the other (Sec. 13h). For the present we can obtain some insight by evaluation of the total average energy $\langle E\rangle$ as a function of the integral of eq. (13j.16) over both $\mathbf{p}$ and $\mathbf{q}$.

One way to do this is to use the method discussed in Sec. 5a: Since the integral of a probability density must be unity, the derivative of its integral with respect to any parameter in the function must be zero, namely, that

$$\begin{aligned}
&\beta\frac{\partial}{\partial\beta}(N!\,h^{3N})^{-1}\int\!\!\int\cdots\int d\mathbf{p}\,d\mathbf{q}\exp\beta[A-(2m)^{-1}\mathbf{p}\cdot\mathbf{p}-U-\Delta U_Q]\\
&= (N!\,h^{3N})^{-1}\int\!\!\int\cdots\int d\mathbf{p}\,d\mathbf{q}\exp\beta[A-(2m)^{-1}\mathbf{p}\cdot\mathbf{p}-U-\Delta U_Q]\\
&\quad\times\frac{\beta\,\partial}{\partial\beta}\beta[A-(2m)^{-1}\mathbf{p}\cdot\mathbf{p}-U-\Delta U_Q]\\
&= \beta A+\beta^2\frac{\partial A}{\partial\beta}-\beta\langle(2m)^{-1}\mathbf{p}\cdot\mathbf{p}+U\rangle-\left\langle\beta\frac{\partial}{\partial\beta}\beta\,\Delta U_Q\right\rangle\\
&= \beta E-\beta E_{\mathrm{cl}}-\left\langle\beta\frac{\partial}{\partial\beta}\beta\Delta U_Q\right\rangle = 0, \qquad (13\text{k}.2)
\end{aligned}$$

where $E_{\rm cl}$ is the classical expression in terms of the $\mathbf{p}$ and $\mathbf{q}$, which occurs in the equations,

$$E_{\rm cl} = \langle (2\mathrm{m})^{-1}\mathbf{p}\cdot\mathbf{p} + U(\mathbf{q})\rangle, \tag{13k.2$'$}$$

and E is the total thermodynamic energy,

$$E = A + \beta\frac{\partial A}{\partial\beta} = A - T\frac{\partial A}{\partial T} = A + TS. \tag{13k.2$''$}$$

Since $\beta\,\Delta U_Q$ [eq. (13j.15)] is a sum of terms, $\beta\,\Delta U_Q = \sum_{\kappa\geq 2}\beta U_\kappa/\kappa!$, and βU_κ is proportional to β^κ, we find that if we define

$$\begin{aligned}\beta\,\Delta E_Q(\beta,\mathbf{q}) &= \beta\frac{\partial}{\partial\beta}\beta\,\Delta U_Q = \sum_{\kappa\geq 2}\frac{\kappa\beta U_\kappa}{\kappa!} = \sum_{\kappa\geq 2}\frac{\beta U_\kappa}{(\kappa-1)!}\\ &= \sum\sum_{1\leq k<k\leq N}\sum_{\kappa\geq 2}\frac{\beta u_\kappa(r_{jk})}{(\kappa-1)!},\end{aligned} \tag{13k.3}$$

it is a function of the coordinates $\mathbf{q}$ whose average value is an additive term in the true total thermodynamic energy at β in addition to the classical expression, $E_{\rm cl} = (2m)^{-1}\mathbf{p}\cdot\mathbf{p} + U(\mathbf{q})$,

$$E = E_{\rm cl} + \langle\Delta E_Q(\beta,\mathbf{q})\rangle. \tag{13k.4}$$

One has, then, that

$$\beta\,\Delta U_Q(\beta,\mathbf{q}) = \int_0^\beta d\beta'\,\Delta E_Q(\beta',\mathbf{q}) + B \tag{13k.5}$$

with B a constant, is the general solution to eq. (3). Since the initial term, $\kappa = 2$, in both $\beta\,\Delta U_Q$ and $\beta\,\Delta E_Q$ are proportional to β^2 and the higher terms to β^κ, $\kappa > 2$ we have that

$$\lim_{\beta\to 0}[\beta\,\Delta U_Q(\beta,\mathbf{q}) - \beta\,\Delta E_Q(\beta,\mathbf{q})] = B = 0. \tag{13k.5$'$}$$

Integrate eq. (5) by parts using $\Delta E_Q = 0$ at $\beta = 0$ to find

$$\beta\,\Delta U_Q(\beta,\mathbf{q}) = \beta\,\Delta E_Q - \int_0^\beta \beta'\left[\frac{d\,\Delta E_Q}{d\beta'}\right]d\beta'. \tag{13k.6}$$

Now ΔE_Q is an energy term and its derivative with respect to T is a heat capacity which we call ΔC_Q,

$$\Delta C_Q(\beta,\mathbf{q}) = \frac{d\,\Delta E_Q(T,\mathbf{q})}{dT} \tag{13k.7}$$

and

$$\beta \frac{d\,\Delta E_Q}{d\beta} = -T\frac{d\,\Delta E_Q}{dT} = -T\,\Delta C_Q \tag{13k.7$'$}$$

$$\int_0^\beta d\beta' = \int_T^\infty d(kT)^{-1} = -\int_T^\infty k^{-1}T^{-2}\,dT,$$

so that, for the systems of constant volume,

$$\begin{aligned}\beta\,\Delta U_Q(\beta,\mathbf{q}) &= \beta\,\Delta E_Q + k^{-1}\int^\infty (T')^{-1}\,\Delta C_Q(T')\,dT' \\ &= \beta\,\Delta E_Q - k^{-1}\,\Delta S_Q(\beta,\mathbf{q}) \\ &= \beta\,\Delta A_Q(\beta,\mathbf{q}),\end{aligned} \tag{13k.7$''$}$$

where $\Delta S_Q(\beta,\mathbf{q})$ and $\Delta A_Q(\beta,\,\mathbf{q})$, like $\Delta E_Q(\beta,\,\mathbf{q})$, are functions of the coordinate positions $\mathbf{q}$ whose average values over the coordinate space are quantum corrections to the thermodynamic functions entropy, Helmholtz free energy, and energy, respectively.

These functions all obey, at each value $\mathbf{q}$ of the coordinates, the standard thermodynamic relations, namely, as functions of the more familiar $T=(k\beta)^{-1}$,

$$\begin{aligned}\Delta A_Q(V,T,N;\mathbf{q}) &= \Delta E_Q(\mathbf{q}) - T\,\Delta S_Q(\mathbf{q}) \\ \frac{\partial A_Q(V,T,N;\mathbf{q})}{\partial T} &= -\Delta S_Q(\mathbf{q}),\end{aligned} \tag{13k.8}$$

so that

$$\frac{\beta\partial}{\partial\beta}\,\beta\,\Delta A_Q(\beta,\mathbf{q}) = \Delta A_Q(\beta,\mathbf{q}) + \frac{\beta A(\beta,\mathbf{q})}{d\beta} = \Delta E_Q(\beta,\mathbf{q}) \tag{13k.8$'$}$$

The functions are

$$\begin{aligned}\Delta E_Q(\mathbf{q}) &= \sum_{\kappa\ge 2}\frac{U_\kappa(\mathbf{q})}{(\kappa-1)!}, \\ \Delta A_Q(\mathbf{q}) &= \sum_{\kappa\ge 2}\frac{U_\kappa(\mathbf{q})}{\kappa!}, \\ T\Delta S_Q(\mathbf{q}) &= \sum_{\kappa\ge 2}\frac{(\kappa-1)U_\kappa(\mathbf{q})}{\kappa!}, \\ T\Delta C_Q(\mathbf{q}) &= \sum_{\kappa\ge 2}\frac{U_\kappa(\mathbf{q})}{(\kappa-2)!},\end{aligned} \tag{13k.9}$$

and their averaged values in the ensemble are the quantum corrections to the semiclassical values of the corresponding functions for which U_κ, $\kappa\ge 2$, are assumed to be zero.

We return to the physical meaning of $\beta^{-1}(\partial/\partial x_j)\ln \rho^{-n}\rho_n(\mathbf{r}_1, \ldots, \mathbf{r}_j, \ldots, \mathbf{r}_n)$, which in eq. (1) we identified as the average in the ensemble $\langle \partial U(\mathbf{r}_1, \ldots, \mathbf{r}_j, \ldots, \mathbf{r}_n/\partial x_j)\rangle$, namely, that $\beta^{-1} \ln \rho^{-n}\rho_n(\mathbf{r}_1, \ldots, \mathbf{r}_n)$ of the dimensions of energy is the potential of average force. This interpretation arises for the classical case because in cartesian coordinates, $\mathbf{q} = \sum q_{j\alpha}$, and their conjugated momenta, $\mathbf{p} = \sum p_{j\alpha}$, the Hamiltonian is separable into the potential function of the coordinates alone $U(\mathbf{q})$ plus the kinetic energy $(2m)^{-1}\mathbf{p}\cdot\mathbf{p}$ dependent on $\mathbf{p}$ alone. At equilibrium the average $\langle (2m)^{-1}\mathbf{p}\cdot\mathbf{p}\rangle = \frac{3}{2}NkT$ is independent of $\mathbf{q}$. In quantum mechanics the separability no longer exists. We must replace U by $U + \Delta U_Q = U + \Delta A_Q$, so that

$$\beta^{-1}\frac{\partial}{\partial x_j}\ln \rho^{-n}\rho_n(\mathbf{r}_1, \ldots, \mathbf{r}_j, \ldots, \mathbf{r}_n)$$

$$= \langle \partial[U(\mathbf{r}_1, \ldots, \mathbf{r}_n, \ldots, \mathbf{r}_n) + \Delta A_Q(\mathbf{r}_1, \ldots, \mathbf{r}_j, \ldots, \mathbf{r}_n]\, \partial x_j\rangle. \quad (13k.10)$$

Now $U + \Delta A_Q$ is the only part of the Helmholtz free energy explicitly dependent on $\mathbf{q}$, so that eq. (10) can be written as

$$\beta^{-1}\frac{\partial}{\partial x_j}\ln \rho^{-n}\rho_n(\mathbf{r}_1, \ldots, \mathbf{r}_j, \ldots, \mathbf{r}_n)$$

$$= \left\langle \frac{\partial A(\mathbf{r}_1, \ldots, \mathbf{r}_j, \ldots, \mathbf{r}_n)}{\partial x_j}\right\rangle \quad (13k.10')$$

namely, the ensemble average of the derivative of the local Helmholtz free energy at fixed $\mathbf{r}_1, \ldots, \mathbf{r}_n$. But for a system of fixed V, T, N the Helmholtz free energy, often called the work function, is the energy function whose derivative with respect to any virtual parameter other than V, T, N gives the reversible work. We therefore find that $-kT \ln \rho^{-n}\rho_n(\mathbf{r}_1, \ldots, \mathbf{r}_n)$ is the quantum-mechanical solution for n molecules at equilibrium in the averaged force due to the work function at fixed positions $\mathbf{r}_1, \ldots, \mathbf{r}_n$.

The coordinate position $\mathbf{q}$ imposed on $U(\mathbf{q})$ by operation of the operator $\mathcal{U}$ on $\Phi(\mathbf{p}, \mathbf{q})$ is the true coordinate position in the Wigner function, so that $U(\mathbf{q})$ is the true potential energy at fixed $\mathbf{q}$. The remaining term, $\Delta E_Q(\mathbf{q})$ of eq. (9), is then extra quantum-mechanical energy above the classical value $\langle U(\mathbf{q} + \mathbf{p}\cdot\mathbf{p}/2m)\rangle$ and represents an averaged contribution to the kinetic energy at coordinate position $\mathbf{q}$. It is a sum of terms for each pair coordinate r_{kj}, $1 \le k < j \le N$, associated with vibration along the bond coordinate r_{jk} and related to the $\frac{1}{2}h\nu$ energy above minimum potential of the lowest vibrational state. It is positive at r_{jk} values near the minimum of potential for which the second derivative of $u(r_{jk})$ is positive, which is the position most highly weighted in the exponential $\exp[-\beta u(r_{jk})]$. It gives a negative contribution at larger r_{jk} values, where $d^2u(r_{jk})/dr_{jk}^2$ becomes negative.

Of course, we must emphasize again that any interpretation of an *exact* coordinate value r_{jk} and an *exact* conjugated momentum p_{jk} violates the uncertainty principle and can be interpreted only in a vague average sense, as evidenced by the fact that the Wigner function, which is the nearest analogue of a probability density, although real and not complex, can have negative values.

Unfortunately, as of this writing we have been unable to produce a simply interpretable self-consistent equation for the Wigner function.

13*l*. SYMMETRIZATION

In Sec. 3e we discussed Bose-Einstein and Fermi-Dirac systems. The state function $\Psi_i(\mathbf{q})$ of any system must change sign, that is, be antisymmetric in exchange of position of two identical elementary fermions: protons, neutrons, or electrons. If we use coordinates of atoms or molecules, as we do, and if the unit atom or molecule contains an even number of fermions, the exchange of coordinates of two such units requires the function to be multiplied by -1 raised to an even power, or $+1$. The function must be symmetric, that is, unchanged, by such an exchange; the molecular units are said to be bosons. If the number of elementary fermions in the atom or molecule is odd, then we call the unit a fermion, and the function $\Psi_i(\mathbf{q})$ must be antisymmetric in exchange of identical pairs of such units. If a system is composed of only one kind of atom, and one kind means of course one isotope of one element, then the system is either pure Fermi-Dirac or pure Bose-Einstein, but a multicomponent system may well be mixture; the symmetrization is slightly more complicated to discuss in detail.

If $\mathcal{P}_p$ is a permutation operator which permutes the order of writing the coordinates $\mathbf{q}^{(N)} = \mathbf{q}_1, \ldots, \mathbf{q}_j, \ldots, q_N$ of molecules j, $1 \le j \le N$, in a function $f(\mathbf{q}^{(N)})$ of the coordinates, the function

$$f_s(\mathbf{q}) = (N!)^{-1} \sum_{p=1}^{N!} \mathcal{P}_P f(\mathbf{q}) \tag{13l.1}$$

is symmetric in all pair exchanges. We number the index p on the permutation $\mathcal{P}_p$ so that it is even if an even number of pairs have been exchanged in the permutation $\mathcal{P}_p$ and odd if the number of pair exchanges is odd. If this is done, then

$$f_a(\mathbf{q}) = (N!)^{-1} \sum_{p=1}^{N!} (-)^p \mathcal{P}_p f(\mathbf{q}) \tag{13l.1'}$$

is antisymmetric in all pair exchanges.

Any physically meaningful operator acting on a function $\omega(\mathbf{q}', \mathbf{q}'')$ of the coordinates of N *identical* molecules must lead to the same integral if the artificial labeling indices of *both* $\mathbf{q}'$ and $\mathbf{q}''$ are exchanged in the same way, or the molecules have distinguishable properties and therefore are not identical. A permutation $\mathcal{P}_{p'}(\mathbf{q}')$ on $\mathbf{q}'$ followed by another $\mathcal{P}_p''(\mathbf{q}'')$ on $\mathbf{q}''$ can always be made by first operating with $\mathcal{P}_{p''}$ on both $\mathbf{q}'$ and $\mathbf{q}''$ followed by another permutation $\mathcal{P}_{p'''}$ on $\mathbf{q}'$ alone. If $\omega_u(\mathbf{q}', \mathbf{q}'')$ is the unsymmetrized density matrix element, which we have already arbitrarily divided by $N!$, we have for an ensemble of bosons, constituting a Bose-Einstein ensemble,

$$\omega_{BE}(\mathbf{q}'\mathbf{q}'') = \sum_{p=1}^{N!} \mathcal{P}_p(\mathbf{q}')\omega_u(\mathbf{q}', \mathbf{q}'') = \sum_{p=1}^{N!} \mathcal{P}_p(\mathbf{q}'')\omega_u(\mathbf{q}', \mathbf{q}''). \tag{13l.2}$$

For an ensemble of fermions, namely, of Fermi-Dirac systems,

$$\omega_{FD}(\mathbf{q}', \mathbf{q}'') = \sum_{p=1}^{N!} (-)^P \mathcal{P}_p(\mathbf{q}')\omega_u(\mathbf{q}', \mathbf{q}''), \tag{13l.2'}$$

which of course can also be written in the other forms of eq. (2).

We anticipate that pair exchanges in eq. (2) or (2′) will be unimportant if $\lambda^{-1}(|\mathbf{r}_j - \mathbf{r}_k| \gg 1$. For instance, the simplest model for a crystal is the pure harmonic oscillator model of atoms in different lattice sites oscillating in their own lattice site with an infinite potential separating two neighboring lattice sites. In this case the two oscillator functions, $\Psi_\nu(\mathbf{r}_j)$ for molecule j at lattice site ν and $\Psi_\mu(\mathbf{r}_j)$ at site μ, are orthogonal, $\int d\mathbf{r}_j \psi_\nu(\mathbf{r}_j)\psi_\mu^*(\mathbf{r}_j) = \delta(\nu - \mu)$. Only the identical permutation which permutes nothing contributes to the thermodynamic properties, and we could well use $\omega_{BE} = \omega_{FD} = \omega_u$ in this case.

The same equations, (2) and (2′), symmetrize the unsymmetrized functions $\omega_u(\mathbf{p}', \mathbf{p}'')$ in the momentum representation.

13m. PERFECT BOSE GAS CLUSTER FUNCTION

In Sec. 11b we found that, with λ the de Broglie wavelength, $\lambda = (\beta h^2/2\pi m)^{1/2}$ (Sec. 7d), and b_n defined as

$$b_n = n^{-5/2}\lambda^{3(n-1)}, \tag{13m.1}$$

we could write βPV for the perfect gas of bosons with the same equation used for an imperfect classical gas,

$$\beta PV = \sum_{n\geq 1} b_n z^n. \tag{13m.2}$$

in this, z is the absolute activity, $z=\lambda^{-1}e^{\beta\mu}$, with μ the chemical potential (Sec. 8c), and b_n [eq. (8g.8)] is an integral in the coordinate space $\mathbf{r}^{(n)}$ of n molecules of a cluster function $g_n(\mathbf{r}^{(n)})$,

$$b_n = \lim_{V\to\infty}[V^{-1}(n!)^{-1}\textstyle\int_V d\mathbf{r}^{(n)}\, g_n(\mathbf{r}^{(n)})]. \qquad (13m.3)$$

In Sec. 11b we promised that we would show the form of eq. (2) not to be fortuitous but that a definition [eq. (3)] could be used for b_n in which the cluster function attraction arose from the symmetrization of $\omega(\mathbf{q}, \mathbf{q})$.

For a perfect gas $U(\mathbf{q})\equiv 0$. With

$$\phi^\dagger(x'p)\phi(px'') = h^{-1}e^{2\pi i h^{-1}p(x''-x')} \qquad (13m.4)$$

for a single coordinate momentum pair and J defined by

$$J(x'', x') = h^{-1}\textstyle\int dp\, e^{-\beta(2m)^{-1}p^2+2\pi i h^{-1}p(x''-x')}, \qquad (13m.5)$$

we can write eq. (13i.13) for $\omega(\mathbf{q}', \mathbf{q}'')$ as

$$\omega(\mathbf{q}', \mathbf{q}'') = (N!)^{-1}\; e^{-\beta A}\textstyle\prod_{j\alpha=1}^{3N} J(x''_{j\alpha}, x'_{j\alpha}), \qquad (13m.6)$$

for $1\le j\le N$ and $\alpha=1, 2, 3$ for the three axes x, y, z. To evaluate J use a dummy variable for each single xp conjugate pair,

$$\zeta = \lambda h^{-1}p - i\lambda^{-1(x''-x')^2} \qquad (13m.7)$$

to write the exponent in the integrand of J as

$$-[\beta(2m)^{-1}p^2 - 2\pi i h^{-1}p(x''-x')] = -\pi[\zeta^2 + \pi\lambda^{-2}(x''-x')^2]. \qquad (13m.7')$$

With $h^{-1}\,dp = \lambda^{-1}\,d\zeta$ we integrate ζ along the line $i\lambda^{-1}(x''-x')$ from $-\infty$ to $+\infty$ and, with $\int e^{-\pi\zeta^2}\,d\zeta = 1$, find

$$J(x'', x') = \lambda^{-1}e^{-\pi\lambda^{-2}(x''-x')^2}. \qquad (13m.7'')$$

and

$$\prod_{\alpha=x,y,z} J(x_{j\alpha}, x_{j\alpha}) = \lambda^{-3}e^{-\pi\lambda^{-1}(|\mathbf{r}_j-\mathbf{r}_j|)2}, \qquad (13m.7''')$$

so that

$$\omega(\mathbf{q}',\mathbf{q}'') = (N!)^{-1}\lambda^{-3N}e^{-\beta A}\prod_{j=1} e^{-\pi\lambda^{-2}(|\mathbf{r}_j''-\mathbf{r}_j'|)^2}. \qquad (13m.8)$$

The effect of a permutation of one of the coordinate pair $\omega(\mathbf{q}, \mathbf{q})$ of the *diagonal* element $\omega(\mathbf{q},\mathbf{q})$ is to produce an off-diagonal element $\omega[(\mathcal{P}_p\mathbf{q}), \mathbf{q}]$ of the unsymmetrized element $\omega_u(\mathbf{q}, \mathbf{q})$. For every pair permuted it introduces a factor

$$f(r_{jk}) = e^{-\pi\lambda^{-2}(|\mathbf{r}_i-\mathbf{r}_j|)^2}. \qquad (13m.9)$$

Every permutation of N molecules can be analyzed as a product of cycle permutations. A cycle permutation of n numbered objects is one in which, say, object 1 replaces object 2, object 2 replaces object 3, ..., object $n-1$ replaces object n, and finally object n replaces object 1. The factor introduced into the diagonal element of $\omega(\mathbf{q}, \mathbf{q})$ by such a single ordered cycle is

$$C_n(\mathbf{r}_1, \mathbf{r}_2, \ldots, \mathbf{r}_n) = f(\mathbf{r}_{12})f(\mathbf{r}_{23}) \cdots f(\mathbf{r}_{n-1,\,n})f(\mathbf{r}_{n,1}), \qquad (13\text{m}.10)$$

which is a function that for finite n approaches zero value unless all n molecules are close together. We define the cluster function $g_n(\mathbf{r}^{(n)})$ of eq. (3) as the sum of the $(n-1)!$ different ways of cycling the same n molecules,

$$g_n(\mathbf{r}^{(n)}) = \sum_{\alpha=1}^{(n-1)!} C_{n\alpha}(\mathbf{r}_1, \ldots, \mathbf{r}_n). \qquad (13\text{m}.11)$$

With b_n defined by eq. (3) it is now more facile to go to the grand canonical ensemble to derive eq. (2). The procedure exactly follows that of Chapter 8.

The evaluation of b_n is now done by folding. With

$$\begin{aligned} g(t) &= \int 4\pi r^2\, dr\, e^{-\pi\lambda^2 r^2}(2\pi t r)^{-1} \sin 2\pi t r \\ &= \lambda^{-3} e^{-\pi\lambda^2 t^2}, \end{aligned} \qquad (13\text{m}.12)$$

the factor of $(n-1)!$ different permutations in eq. (11) and $(n!)^{-1}$ in eq. (3) gives

$$\begin{aligned} b_n &= n^{-1} 4\pi t^2\, dt [g(t)]^n \\ &= n^{-5/2} \lambda^{3(n-1)}, \end{aligned} \qquad (13\text{m}.13)$$

in agreement with eq. (1).

APPENDIX

AI. SOME DEFINITE INTEGRALS

$$\int_0^\infty x^n e^{-ax}\,dx = \frac{1}{a^{n+1}}\int_0^\infty z^n e^{-z}\,dz = \frac{n!}{a^{n+1}}.$$

$$\int_0^\infty x^{2k+1} e^{-ax^2}\,dx = \frac{1}{2a^{k+1}}\int_0^\infty y^k e^{-y}\,dy = \frac{k!}{2a^{k+1}}.$$

$$\int_{-\infty}^{+\infty} e^{-ax^2}\,dx = \left(\frac{\pi}{a}\right)^{1/2} = \frac{1}{a^{1/2}}1.772{,}453.$$

$$\int_{-\infty}^{+\infty} x^2 e^{-ax^2}\,dx = \frac{1}{2a}\left(\frac{\pi}{a}\right)^{1/2}.$$

$$\int_{-\infty}^{+\infty} x^4 e^{-ax^2}\,dx = \frac{3}{4a^2}\left(\frac{\pi}{a}\right)^{1/2}.$$

$$\int_{-\infty}^{+\infty} x^6 e^{-ax^2}\,dx = \frac{15}{8a^3}\left(\frac{\pi}{a}\right)^{1/2}.$$

$$\int_{-\infty}^{+\infty} x^{2n} e^{-ax^2}\,dx = \frac{(2n-1)!!}{2^n a^n}\left(\frac{\pi}{a}\right)^{1/2},$$

where

$$(2n-1)!! = (2n-1)(2n-3)\cdots 5\cdot 3\cdot 1.$$

AII. THE EULER–MACLAURIN SUMMATION FORMULA

The sum of a function of some variable j, for integral values of the variable between two limits m and n, is symbolized by

$$\sum_{j=m}^{j=n} f(j) = f(m) + f(m+1) + \cdots + f(n-1) + f(n). \tag{AII.1}$$

If the function f is definable for nonintegral values of the variable $f(x)$, the sum may be approximated in terms of the integral of $f(x)$ between the limits m and n and the values of the function and its derivatives at the two limits. The approximation formula is known as the Euler–Maclaurin summation formula. The symbol

$$f(a)^r = \left(\frac{d^r f(x)}{dx^r}\right)_{x=a} \tag{AII.2}$$

is used for the rth derivative of the function at the value $x = a$. Then,

$$\sum_{j=m}^{j=n} f(j) = \int_m^n f(x)\,dx + \tfrac{1}{2}[f(m)+f(n)] + \sum_{k\geq 1} (-1)^k \frac{B_k}{(2k)!}[f(m)^{(2k-1)} - f(n)^{(2k-1)}], \tag{AII.3}$$

where the numbers B_k are the Bernoulli numbers, $B_1 = \frac{1}{6}$, $B_2 = \frac{1}{30}$, $B_3 = \frac{1}{42}$, $B_4 = \frac{1}{30}$, $B_5 = \frac{5}{66}$.

The first few terms of this development may be readily checked geometrically. In Fig. AII.1 a function $f(x)$ is plotted against x between the limits m and n. The integral is the area under the smooth curve. The values of the function at integral values of the variable x are shown as perpendicular lines, so that the sum is the sum of the heights of these lines. Since the lines are a unit distance apart, the sum of the heights of all but the last, $f(n)$, is given by the shaded area under the stepwise figure. If the points $f(j)$ and $f(j+1)$ are connected by straight lines, the areas under

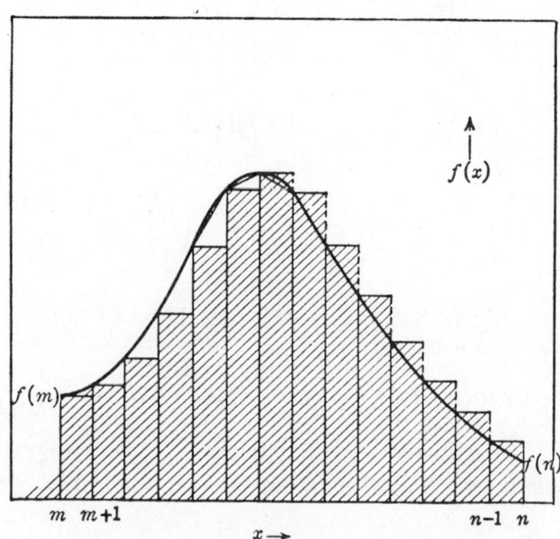

Figure AII.1

the triangles so formed above the steps form the first correction to the difference between the area of the stepwise figure and that under the smooth curve. The area of each triangle is $\frac{1}{2}[f(j)-f(j+1)]$, and the sum of their areas is $\frac{1}{2}[f(m)-f(n)]$, if account is taken of the fact that the triangles above the steps must be subtracted from the integral and those below must be added to the integral in order to approximate the sum. Adding the integrals, the areas of the triangles, and the last term of the sum, $f(n)$, one obtains the approximation eq. (3) up to the terms containing the derivatives.

The correction of the first derivative can also be seen geometrically rather simply, but we omit the argument.

If, however, it is assumed that a general equation like eq. (3) *can* be obtained, that is, one that expresses the sum between two limits in terms of the integral between these limits, and the values of the function and its derivatives at the limits, then the coefficients of the first terms may be obtained readily by consideration of a special case.

The integral of the function e^{-ax} between zero and infinity is a^{-1}. The value of the function and of all its derivatives at $x=\infty$ is zero. The value of the function at $x=0$ is unity, and of its rth derivative is $(-a)^r$, at this value of x. The sum e^{-aj} from $j=0$ to $j=\infty$ can be summed in closed form and the expression expanded as a power series in a containing powers minus one, zero, and all positive powers. The coefficients of this series give the coefficients in eq. (AII.3).

The steps are simple enough algebraically. The quantity $1/(1-e^{-a})$ can be seen to be the desired sum if the indicated division is carried out. The exponential e^{-a} may be expanded in a power series in a, and the analytic expression obtained may be divided into unity, obtaining

$$\sum_{j=0}^{j=\infty} e^{-aj} = \frac{1}{1-e^{-a}} = \frac{1}{a} + \frac{1}{2} + \frac{1}{12}a - \frac{1}{720}a^3 + \frac{1}{30{,}240}a^5 - \cdots. \tag{AII.4}$$

The coefficients in eq. (4) agree with those in eq. (3).

AIII. THE FACTORIAL AND THE STIRLING APPROXIMATION

The product of N factors $N(N-1)(N-2)(N-3)\cdots 3\cdot 2\cdot 1$ is called N factorial and written $N!$. The convention is adopted that zero factorial is unity. The logarithm of $N!$,

$$\ln N! = \sum_{j=1}^{j=N} \ln j, \tag{AIII.1}$$

may be approximated by the Euler–Maclaurin summation formula [eq. (AII.3)]. The integral of $\ln x$ is $x \ln x - x$, and between the limits 1 and N gives $N \ln N - N + 1$. The value of the function at $x = N$ is $\ln N$, and at $x = 1$ is zero. The derivatives of $\ln x$ are inverse powers of x, so that at the limit $x = N$ the derivatives may all be neglected for large values of N. At $x = 1$ the derivatives are independent of N. One obtains

$$\ln N! \cong N \ln N - N + \tfrac{1}{2} \ln N + c, \tag{AIII.2}$$

where c is a term containing a constant and inverse powers of N. The constant term in c is actually $\frac{1}{2} \ln 2\pi$, so that

$$N! \cong N^N e^{-N} (2\pi N)^{1/2} \tag{AIII.3}$$

for large values of N.

AIV. THE VOLUME OF AN N-DIMENSIONAL SPHERE

In an N-dimensional space of orthogonal coordinates $x_1, x_2, \ldots, x_N$, the part of the space for which

$$x_1^2 + x_2^2 + x_3^2 + \cdots + x_N^2 \leq r^2$$

constitutes the inside of an N-dimensional sphere of radius r. The volume of this sphere is

$$V_N = \frac{\pi^{N/2}}{(\frac{1}{2}N)!} r^N \qquad \text{for } N \text{ even}, \tag{AIV.1}$$

and

$$V_N = \frac{2^N \pi^{(N-1)/2} (\frac{1}{2}N - \frac{1}{2})!}{N!} r^N \qquad \text{for } N \text{ odd}. \tag{AIV.1$'$}$$

Using the Stirling approximation for the factorial, the asymptotic expression for $\ln V_N$ is the same for both cases,

$$\ln V_N \cong \frac{N}{2} \ln\left(\frac{2\pi e r^2}{N}\right). \tag{AIV.2}$$

AV. THE METHOD OF UNDETERMINED MULTIPLIERS

Suppose one seeks the maximum (or minimum) value of a function F of N variables $x_1, x_2, \ldots, x_N$. This value is determined by the condition that the variation

$$\delta F = \sum_{\nu=1}^{N} \frac{\partial F}{\partial x_\nu} \delta x_\nu = 0 \qquad \text{(AV.1)}$$

in F for any conceivable small variation $\delta x_1, \ldots, \delta x_N$ of the variables be zero. This condition can be satisfied for all possible variations of the variables only if every partial derivative is zero,

$$\frac{\partial F}{\partial x_\nu} = 0. \qquad \text{(AV.2)}$$

Whether the extremum so found is a maximum or not must be determined by the values of the second derivatives.

However, it may be that not all the variables x are independent, but that some condition,

$$G(x) = 0, \qquad \text{(AV.3)}$$

where G is a function of all the x_ν's, must be obeyed.

One may then seek the maximum value of F, subject to the condition in eq. (3).

In this case, eq. (1), where $\delta F = 0$, must still hold at the maximum, although not for all arbitrary variations of the variables, but only for those that are such that the variation in G is zero,

$$\delta G = \sum_{\nu=1}^{N} \frac{\partial G}{\partial x_\nu} \delta x_\nu = 0, \qquad \text{(AV.4)}$$

so that eq. (3) is maintained. One can, then, not conclude that eq. (2) must hold.

Since, for the allowed variations of the variables, the quantity δG [eq. (4)] is zero, and of course also $z\,\delta G = 0$ for any value of z, one may subtract $z\,\delta G$ from eq. (1), without altering its value, and obtain

$$\delta F = \sum_{\nu=1}^{N} \left(\frac{\partial F}{\partial x_\nu} - z \frac{\partial G}{\partial x_\nu} \right) \delta x_\nu = 0. \qquad \text{(AV.5)}$$

Equation (4) gives us one condition by means of which the variation δx_μ of one of the variables x_μ is determined if all the other variations

$$\delta x_1, \ldots, \delta x_{\mu-1}, \delta x_{\mu+1}, \ldots, \delta x_N$$

are arbitrarily assigned, that is, the condition in eq. (3) may be maintained with any arbitrary variation in the values of $N-1$ of the N variables. The numerical value of z may be so chosen that, if $\partial G / \partial x_\mu$ is not zero,

$$\frac{\partial F}{\partial x_\mu} - z \frac{\partial G}{\partial x_\mu} = 0, \qquad \text{(AV.6)}$$

and in the sum of eq. (5) the μth term is zero whatever the variation in x_μ is. However, the sum over the other $N-1$ terms must be zero if F is to be a maximum, and since all conceivable variations of the $N-1$ terms are consistent with eq. (4), one finds that, for all values of ν,

$$\frac{\partial F}{\partial x_\nu} - z\frac{\partial G}{\partial x_\nu} = 0. \tag{AV.7}$$

This is a necessary condition for the position of the maximum of F subject to the condition in eq. (3) that G be held constant. The numerical value of z must be determined in such a way that the function G has the particular value zero if the condition in eq. (3) is to be obeyed.

From eq. (7) it is seen that

$$z = \frac{dF}{dG}. \tag{AV.8}$$

The extension to two conditions is fairly obvious and is not discussed in detail.

AVI. COMBINATORY PROBLEMS

Problems frequently arise in statistical mechanics which may be shown to be identical to some problem concerning the number of ways in which objects can be arranged in order or placed in piles or boxes. The answers to some of these problems follow.

1. The number of ways in which N distinguishable objects can be arranged in order is

$$N!$$

The object to occupy the first place may be selected out of the N different objects in N different ways. Independently of the choice for first place, the selection for second place may be made from the remaining $N-1$ objects in $N-1$ ways, so that the first two places may be filled in $N(N-1)$ ways. Similarly, the object for the kth place may be chosen in $N-k-1$ ways after the first $k-1$ places have been filled. The product

$$N(N-1)(N-2)\cdots 3\cdot 2\cdot 1 \text{ is } N!$$

This number $N!$ is also the number of permutations of N objects.

2. The number of different ways in which M objects may be selected from N distinguishable objects, irrespective of the order of choice, is

$$\frac{N!}{(N-M)!\,M!}.$$

The objects may be arranged in order in $N!$ different ways, and the objects in the first M places always chosen. However, many orderings lead to a choice of the same M objects, namely, all those that differ from each other only by permutations of the objects in the first M places among each other, and permutations of the remaining $N-M$ objects among each other. The number of these permutations is $M!(N-M)!$, so that every choice of the first M objects corresponds to this many different arrangements of all the particles in order. The number of choices one can make is then $N!$ divided by the product $M!(N-M)!$.

3. The number of ways in which N objects may be arranged in two piles of, respectively, M and $N-M$ objects each is

$$\frac{N!}{(N-M)!\,M!}.$$

The problem is the same as the preceding one since, after the selection of M objects, the remaining $N-M$ are uniquely determined.

4. The number of ways in which N objects may be placed in P piles, m_1 in the first pile, m_2 in the second pile, ..., m_i in the ith, with

$$\sum_{i=1}^{P} m_i = N,$$

is

$$\frac{N!}{\prod_{i=1}^{P} m_i!}.$$

The N objects are ordered, and those in the first m_1 places assigned to the first pile, those in the places m_1+1 to m_1+m_2, inclusive, assigned to the second pile, and so on. There are $N!$ different ways of arranging the N objects in order, but any one assignment of m_1 definite objects to the first pile, m_2 definite objects to the second, and so on, corresponds to $\prod m_i!$ different arrangements of the objects in order, since all permutations of the m_i objects with each other, within the one pile i, lead to a new arrangement of the N objects but not to a new assignment to piles. The total number of ways the objects can be assigned to the piles, with given values of the m_i's, is the total number of arrangements in order, $N!$, divided by the number of arrangements per assignment to piles, $\prod m_i!$, or $N!/\prod m_i!$.

5. The multinomial coefficient, the coefficient of $\prod z_i^{m_i}$ in the sum $\sum x_i$ raised to the power N is

$$\frac{N!}{\prod_i m_i!},$$

with $\sum_{i=1}^{P} m_i = N$ if there are P members x_i in the sum. This problem can be shown to be identical to problem 4. One writes the Nth power of the sum $\sum x_i$ as N consecutive distinguishable factors. Each term in the product contains one and only one x from each factor. If from one factor the x chosen is x_i, that factor will be assigned to the pile i. The number of ways we can make selections from the numbered factors so as to have m_i objects in the pile i for all values of i is the above expression, and this is just the number of terms in the expansion having the given powers m_i of x_i.

A special case of problem 5 is that there are just two values of i, so that

$$\frac{N!}{(N-M)!\,M!}$$

is the binomial coefficient, the coefficient of $x^M y^{N-M}$ in the expansion of $(x+y)^N$.

From problem 5 the values of certain sums may be ascertained. For instance,

$$\sum_{M=0}^{N} \frac{N!}{(N-M)!\,M!} = (1+1)^N = 2^N,$$

$$\sum_{\substack{m_i=0 \\ \sum_{i=1}^{i=P} m_i = N}}^{m_i=N} \frac{N!}{\prod_{i=1}^{i=P} m_i!} = P^N,$$

$$\sum_{M=0}^{N} \frac{N!}{(N-M)!\,M!}(-1)^M = (1-1)^N = 0.$$

An example of a more complicated trick which may be employed to evaluate a sum is

$$\sum_{M=0}^{N} \frac{N!}{(N-M)!\,M!}(A+BM)^k(-1)^M = \left(\frac{\partial^k}{\partial x^k}\right)_{x=0} e^{Ax}(1-e^{Bx})^N.$$

$$= 0 \qquad \text{if } k < N.$$

6. The number of ways in which N distinguishable objects may be assorted into piles with m_1 piles of one object each, m_2 piles of two objects each, . . . , m_k piles of k objects each, so that

$$\sum_{k\geq 1} k m_k = N,$$

is

$$\frac{N!}{\prod m_k!\,(k!)^{m_k}}.$$

This problem differs from problem 4 in that the piles are unnumbered. In problem 4, if the first pile contained three objects and the second pile the same number of objects, the arrangement would be different if the objects in the two piles were all exchanged, whereas in this problem only the number of piles of three objects each is specified. With numbered piles the number of arrangements would be $N!/\prod_i k_i!$, which must be divided by $\prod_k m_k!$ for the numbers of permutations of piles with identical numbers of objects.

AVII. MATRICES

In this appendix we explain the operations and notation used in this book where matrices occur. We do so without proof, and limit ourselves completely to the two special classes of matrices that we use. A clear and concise more general discussion is given in Charles L. Perrin, *Mathematics for Chemists*, Wiley-Interscience, New York, 1970, Chapters 8 and 9.

A finite $n \times m$ matrix $\mathbf{A}$ is a two-dimensional array of numbers a_{ij}, $1 \leq i \leq n$, $1 \leq j \leq m$,

$$\mathbf{A} \equiv \begin{matrix} a_{11} & a_{12} & \cdots & a_{1j} & \cdots & a_{1m} \\ a_{21} & a_{22} & \cdots & a_{2j} & \cdots & a_{2m} \\ \cdot & \cdot & & \cdot & & \cdot \\ \cdot & \cdot & & \cdot & & \cdot \\ \cdot & \cdot & & \cdot & & \cdot \\ a_{i1} & a_{i2} & \cdots & a_{ij} & \cdots & a_{im} \\ \cdot & \cdot & & \cdot & & \cdot \\ \cdot & \cdot & & \cdot & & \cdot \\ \cdot & \cdot & & \cdot & & \cdot \\ a_{n1} & a_{n2} & \cdots & a_{nj} & \cdots & a_{nm}. \end{matrix} \qquad \text{(AVII.1)}$$

Three fundamental operations are defined:

1. Addition. $\mathbf{A}+\mathbf{B}=\mathbf{B}+\mathbf{A}=\mathbf{D}$, where if $\mathbf{A}$ and $\mathbf{B}$ have elements a_{ij}, b_{ij}, $1 \leq i \leq n$, $1 \leq j \leq m$, then $\mathbf{D}$ is an $n \times m$ matrix of elements

$$d_{ij} = a_{ij} + b_{ij}. \qquad \text{(AVII.2)}$$

2. Multiplication. When $\mathbf{B}$ is a matrix of elements b_{jk}, $1 \leq j \leq m$,

$$\mathbf{C} = \mathbf{AB}, \text{ having elements } c_{ik} = \sum_{j=1}^{j=m} a_{ij} b_{jk}, \qquad \text{(AVII.3)}$$

is not generally the same as **BA**. The order of multiplication is important. If **AB** = **BA**, the matrices **A** and **B** are said to commute.

3. Multiplication of a matrix by a number c is simply defined by

$$c\mathbf{A} \text{ has elements } ca_{ij}, \tag{AVII.4}$$

including of course $c=-1$, so that $-\mathbf{A}$ has elements $-a_{ij}$.

Matrix algebra for finite matrices, including the proof of many theorems, had been well developed by mathematicians before Born and Heisenberg introduced the use of infinite matrices, $1 \le i \le \infty$, $1 \le j \le \infty$, in quantum mechanics. The infinite range caused some dismay among mathematicians until John von Neumann proved that, for the matrices used in quantum mechanics, the theorems for finite matrices employed were indeed valid. The introduction of infinite matrices permits a natural extension to continuous indices, usually coordinates q or momenta p, in which case the summation of eq. (3) for multiplication is replaced by integration,

$$\mathbf{C}=\mathbf{AB}, \qquad c_{ik}=\int dq\, a_{iq} b_{qk}. \tag{AVII.3$'$}$$

We present without proof some of these theorems for two classes of matrices, *Hermitian* matrices which correspond to physical observables, and *unitary* matrices which transform Hermitian matrices from one representation to another, rather analogously to two different functional representations of one physical quantity in, say, cartesian coordinates x, y, z and in polar coordinates r, θ, ϕ. We particularly stress some characteristics of their use in physical problems not often emphasized or even mentioned in mathematical texts. The elements a_{ij} of our Hermitian matrices are physical quantities, usually having physical dimensions $m^{\nu} l^{\mu} t^{\kappa}$. Also, the indices i, j have physical significance; when they are discrete numbers, they usually are quantum numbers indexing the eigenfunctions of a particular Hermitian operator $\mathscr{F}$, the quantum-mechanical operator corresponding to the physically observable function of coordinates and momenta $f(\mathbf{q}, \mathbf{p})$ (see Sec. 3b).

An operator $\mathscr{F}$ is said to be *Hermitian* if, for any (reasonably behaved)[(1)] complex functions $\Phi(x)$, $\Psi(x)$ whose complex conjugates are $\Phi^*(x)$, $\Psi^*(x)$, formed by replacing the imaginary $i=\sqrt{-1}$ by $-i$ in Φ and Ψ,

$$\int dx\, \Psi^*(x)\mathscr{F}\Phi(x)=\left[\int dx\, \Phi^*(x)\mathscr{F}\Psi(x)\right]^*. \tag{AVII.5}$$

[(1)] Very often $\mathscr{F}$ is Hermitian only if the functions $\Phi(x)$, $\Psi(x)$ obey boundary conditions such as being zero at walls of a volume V.

An eigenfunction $\Phi_\nu(x)$ of a Hermitian operator $\mathscr{F}$ is one for which

$$\mathscr{F}\Phi_\nu(x) = f_\nu \Phi_\nu(x), \tag{AVII.6}$$

with f_ν a real (not complex) number with physical dimensions of $f(\mathbf{q}, \mathbf{p})$. From eq. (5), with eq. (6), it follows that, if $f_\nu \neq f_{\nu'}$, then Φ_ν and $\Phi_{\nu'}$ are *orthogonal*,

$$\int dx\, \Phi_{\nu'}^* \Phi_\nu = 0, \qquad f_{\nu'} \neq f_\nu, \tag{AVII.7}$$

and, since eq. (6) is still valid if Φ_ν is multiplied by a constant, it can be normalized,

$$\int dx\, \Phi_\nu^*(x)\Phi_\nu(x) = 1. \tag{AVII.7′}$$

If there are two or more different normalized solutions of eq. (6) with the same numerical value of f, they are said to be *degenerate* and there are linear combinations of them that are orthogonal; any linear combination of them satisfies eq. (6). Equations (7) and (7′) can be combined, for the infinite number, $1 \leq \nu \leq \infty$, of solutions of eq. (6) for which all the *orthonormal* functions $\Phi_\nu(x)$ are said to form a *complete set* (of unit vectors in Hilbert space), into

$$\int dx\, \Phi_{\nu'}^*(x)\Phi_\nu(x) = \delta(\nu' - \nu), \tag{AVII.7″}$$

where the Kronecker delta $\delta(\nu' - \nu)$ is

$$\begin{aligned} \delta(\nu' - \nu) &= 1 \qquad \text{if } \nu' = \nu, \\ &= 0 \qquad \text{if } \nu' \neq \nu. \end{aligned} \tag{AVII.7‴}$$

The complete set of functions Φ_ν obeys an analogous relation,

$$\sum_\nu^\infty \Phi_\nu^*(x')\Phi_\nu(x) = \delta(x' - x), \tag{AVII.8}$$

where we use the same symbol δ for the Dirac delta function, which is zero if $x' \neq x$ but infinite if $x' = x$, in such a way that

$$\int dx'\, \delta(x' - x) = 1, \tag{AVII.8′}$$

so that, for any function $f(x')$ that is continuous at $x' = x$,

$$\int dx'\, f(x')\delta(x' - x) = f(x). \tag{AVII.8″}$$

It is sometimes helpful in visualizing $\delta(x'-x)$ to regard it as the limit, when α goes to infinity, of $(\alpha/\sqrt{\pi})e^{-\alpha^2(x'-x)^2}$

To any Hermitian operator $\mathcal{G}$ [eq. (5)] there corresponds an infinite Hermitian matrix **G** of elements,

$$g_{\nu'\nu} = \int dx\, \Phi^*_{\nu'}(x)\mathcal{G}\Phi_\nu(x) = g^*_{\nu\nu'} = \left[\int dx\, \Phi^*_\nu(x)\mathcal{G}\Phi_\nu(x)\right]^*, \tag{AVII.9}$$

where the functions $\Phi_{\nu'}$ and Φ_ν are members of an infinite complete orthonormal set of functions, normally eigenfunctions [eq.(6)] of a Hermitian operator. In the general case the elements $g_{\nu'\nu}$ are said to be the elements of the matrix **G** in the representation of the quantum numbers ν of the functions Φ_ν. From eq. (5) the diagonal elements, $\nu'=\nu$, namely, $g_{\nu\nu}$, are real (not complex). If the functions Φ_ν are eigenfunctions of the same operator $\mathcal{G}$, then, from eqs. (6) and (7), only the diagonal elements are nonzero, and the matrix is said to be in its diagonal representation.

A Hermitian matrix is square in character, that is, the two indices $\nu'\nu$ or ij have the same significance and the order in which they are written must be the same. Both addition and multiplication have no physical meaning unless the indices of the two matrices **A** and **B** of eqs. (2) and (3) are in the same representation, and addition is meaningless unless the physical dimensions of the elements a_{ij} and b_{ij} are identical.

We can now summarize the properties of a quantum-mechanical Hermitian matrix **F** of elements f_{ij}, adding some definitions and theorems.

The elements f_{ij} represent some physical observable corresponding to a classical function $f(p, q)$ which has a Hermitian operator $\mathcal{F}$.

The indices i, j have the same physical significance. They may be *quantum numbers* of an orthonormal set of state functions Ψ_i which are eigenstates of an operator $\mathcal{G}$, or continuous variables such as q or p. If i, j are quantum states of $\mathcal{F}\Psi_i = f_i\Psi_i$, then **F** is diagonal, $f_{ij} = f_i\delta(i-j)$.

We have necessarily for Hermitian matrices:

$$f_{ij} = f_{ji}{}^*, \qquad f_{ii} \text{ real.}$$

if $f_{ij}=0$ for all $i \neq j$, the matrix **F** is said to be *diagonal.* If **F** and **G** are both diagonal in the same representation, then **FG** = **GF** is diagonal of elements $f_{ii}g_{ii}$ from eq. (3), and **F** and **G** are said to *commute.* The unit matrix **1**, if in a discrete representation, is diagonal of elements unity, and if in a continuous representation the elements are the Dirac delta $\delta(q'-q)$. From eq. (3) or (3′), $\mathbf{1F} = \mathbf{F1} = \mathbf{F}$. The trace of a Hermitian matrix **F** is written as $\operatorname{tr}\mathbf{F} = \sum_i f_{ii}$ if the indices i are discrete, and as $\operatorname{tr}\mathbf{F} = \iint dq\, dq'\, f(q, q')\, \delta(q-q')$ if the representation qq is continuous. The trace is *invariant in the representation,* that is, it is the same in all representations. The determinant of a matrix **F** is written as $|\mathbf{F}|$ and is also invariant in the representation. If $|\mathbf{F}|=0$, the matrix $\mathbf{F}$ is said to be singular. If $|\mathbf{F}|\neq 0$, the matrix $\mathbf{F}$ has a reciprocal $\mathbf{F}^{-1}$ such that $\mathbf{F}^{-1}\mathbf{F}=\mathbf{F}\mathbf{F}^{-1}=\mathbf{1}$, the unit matrix. If $|\mathbf{F}|\neq 0$, the matrix $\mathbf{F}$ can always be *diagonalized*, that is, brought into a diagonal representation. If $\mathbf{F}$ is in a diagonal representation with elements f_{ii}, its reciprocal $\mathbf{F}^{-1}$ from eq. (3) is diagonal of elements $f_{ii}^{-1}=(f_{ii})^{-1}$. If $\mathcal{F}$ and $\mathcal{G}$ are two Hermitian operators, the elements $c_{n'n}$ of the product matrix $\mathbf{C}=\mathbf{FG}$, from eqs. (9), (3) and (3′), are

$$c_{n'n}=\iint dx'\,dx\sum_{n''}^{\infty}\Phi_{n'}^{*}(x)\mathcal{F}\Phi_{n''}(x)\Phi_{n''}^{*}(x')\mathcal{G}\Phi_{n}(x').$$

Sum over n'' first, using eq. (8) to obtain $\delta(x'-x)$. Integrate over x' using eq. (8″), which simply makes $x=x'$, and one finds

$$c_{n'n}=\int dx\,\Phi_{n'}(x)\mathcal{F}\mathcal{G}\Phi_{n}(x), \tag{AVII.10}$$

namely, the product matrix $\mathbf{C}$ in any representation is the matrix of the product operator $C=\mathcal{F}\mathcal{G}$, not generally the same as $\mathcal{G}\mathcal{F}$.

Any function $f(x)$ obeying the same boundary conditions, such as being single-valued and continuous for $a\leq x\leq b$ and zero for $x\leq a$, $x\geq b$, as those of a complete orthonormal set $\Psi_n(x)$, $1\leq n\leq\infty$, can be expanded as a linear combination,

$$f(x)=\sum_{n=1}^{\infty}a_n\Psi_n(x), \tag{AVII.11}$$

this being really the criterion of completeness of the set. Multiply both sides by $\Psi_{n'}^{*}(x)$ and integrate using eq. (7″) to find

$$\int dx\,f(x)\Psi_{n'}^{*}(x)=\sum_{n=1}^{\infty}a_n\int dx\,\Psi_n(x)\Psi_{n'}^{*}(x)=\sum_{n}a_n\delta(n'-n)=a_{n'}. \tag{AVII.11′}$$

If $f(x)\equiv\Phi_\nu(x)$ is one of another complete orthonormal set, we can expand for each ν using eq. (11′),

$$\Phi_\nu^{*}(x)=\sum_{n=1}^{\infty}u_{\nu n}\Psi_n^{*}(x),\qquad u_{\nu n}=\int dx\,\Phi_\nu^{*}(x)\Psi_n(x). \tag{AVII.12}$$

Alternatively, we may define coefficients $u_{n\nu}^{\dagger}$ by

$$\Psi_n^{*}(x)=\sum_{\nu=1}^{\infty}u_{n\nu}^{\dagger}\Phi_\nu^{*}(x),\qquad u_{n\nu}^{\dagger}=\int dx\,\Psi_n^{*}(x)\Phi_\nu(x)=u_{\nu n}^{*}. \tag{AVII.12′}$$

If we use eq. (12) for Φ_ν^* in eq. (12′), multiply both sides by $\Psi_{n'}(x)$ and integrate using the orthonormality condition in eq. (7″),

$$\int dx\,\Psi_{n'}(x)\Psi_n^*(x) = \sum_{\nu=1}^{\infty} u_{n'\nu}^\dagger u_{\nu n} = \delta(n'-n), \qquad \text{(AVII.13)}$$

and symmetrically

$$\sum_{n=1} u_{\nu' n} u_{n\nu}^\dagger = \delta(\nu'-\nu). \qquad \text{(AVII.13′)}$$

A unitary matrix **U** with its adjoint $\mathbf{U}^\dagger$ transforms a Hermitian matrix **F** from one representation to another representation. In the elements $u_{\nu n}$ of a unitary matrix **U** the two indices ν and n refer to *different entities*, if discrete, one is normally the quantum number of one complete set of eigenfunctions of some Hermitian operator and the other those of a different set of a different operator, or one may be a quantum number and the other a continuous variable such as a coordinate, or both may be different continuous variables such as coordinate and momentum.

To any matrix **A** of elements $a_{\nu n}$ one can define an *adjoint* matrix $\mathbf{A}^\dagger$ of elements,

$$a_{n\nu}^\dagger \equiv (\mathbf{A}^\dagger)_{n\nu} = a_{\nu n}^* \equiv (\mathbf{A}^*)_{\nu n}. \qquad \text{(AVII.14)}$$

A Hermitian matrix is self-adjoint. A matrix is unitary if

$$\mathbf{UU} = \mathbf{U}^\dagger\mathbf{U} = \mathbf{1},$$
$$\sum_n u_{\nu' n} u_{n\nu}^\dagger = \delta(\nu'-\nu), \quad \sum_\nu u_{n'\nu} u_{\nu n} = \delta(n'-n), \qquad \text{(AVII.15)}$$

where, if the middle index in either case is continuous, summation is replaced by integration and, if the outer index is continuous, the $\delta(x'-x)$ is the Dirac delta function rather than the Kronecker delta. If $\mathbf{F}^{(n)}$ is the Hermitian matrix **F** in the n', n reresentation, then

$$\mathbf{F}^{(\nu)} = \mathbf{U}\mathbf{F}^{(n)}\mathbf{U}^\dagger,$$
$$f_{\nu'\nu} = \sum_n \sum_{n'} u_{\nu' n'} f_{n'n} u_{n\nu}, \qquad \text{(AVII.16)}$$

is the matrix **F** in the $\nu'\nu$ representation if the elements $u_{\nu n}$ and $u_{n\nu}$ are given by eqs. (12) and (12′).

Any complete set of orthonormal functions $\Phi_\nu(x)$ can be regarded as a unitary matrix $\mathbf{\Phi}$ of elements, $\phi_{x\nu} \equiv \Phi_\nu(x)$, whose unitary adjoint $\mathbf{\Phi}^\dagger$ has elements $\phi_{\nu x} = \phi_{x\nu}^* \equiv \Phi_\nu^*(x)$. Equations (7″) and (8), written in matrix notation, are those of eq. (15) defining unitary matrices,

$$\mathbf{\Phi\Phi}^\dagger = \mathbf{\Phi}^\dagger\mathbf{\Phi} = \mathbf{1} \qquad \text{(AVII.17)}$$

The functions $\mathscr{F}\Phi_\nu(x)$ resulting from the operator $\mathscr{F}$ acting on $\Phi_\nu(x)$ can be regarded as the elements $(\mathscr{F}\Phi)_{x\nu} \equiv \mathscr{F}\Phi_\nu(x)$ of a matrix, and the matrix $\mathbf{F}^{(\nu)}$ in the $\nu'\nu$ representation may be written symbolically as

$$\mathbf{F}^{(\nu)} = \mathbf{\Phi}^\dagger(\mathscr{F}\mathbf{\Phi}). \tag{AVII.18}$$

The matrix $\mathbf{F}^{(x)}$ in the coordinate representation is then Hermitian, if $\mathscr{F}$ is a Hermitian operator, and is

$$\mathbf{F}^{(x)} = \mathbf{\Phi}\mathbf{\Phi}^\dagger(\mathscr{F}\mathbf{\Phi})\mathbf{\Phi}^\dagger \tag{AVII.19}$$

of elements

$$\begin{aligned} f_{x'x} &= \sum_{\nu'\nu}\sum \int dx''\, \Phi_{\nu'}(x')\Phi_{\nu'}(x'')[\mathscr{F}\Phi_\nu(x'')]\Phi^*(x) \\ &= \sum_{\nu'\nu}\sum \phi_{x\nu'} f_{\nu'\nu}\phi_{\nu x}, \end{aligned} \tag{AVII.19$'$}$$

and is independent of the representation $\nu'\nu$.

In the previous discussion we assumed that n', n were single quantum numbers and implicitly assumed that, if i is a continuous variable such as q, it is a single coordinate. Just as one can extend the notion of a function $f(x)$ to one of many variables $f(\mathbf{q}^{(\Gamma)})$, so we can, and do, extend our i, j's to be a set $\mathbf{i}$, $\mathbf{j}$ of numbers, even as horrible a set as $\mathbf{K}'$, $\mathbf{K}$, the quantum numbers of a system of 10^{24} molecules, or of their coordinates $\mathbf{q}'^{(\Gamma)}$, $\mathbf{q}^{(\Gamma)}$ or momenta $\mathbf{p}'^{(\Gamma)}$, $\mathbf{p}^{(\Gamma)}$.

AVIII. FOURIER INTEGRAL TRANSFORMS, MOMENTUM FUNCTIONS

For a single coordinate x, a single-valued continuous function $f(x)$ defined in the range $-\frac{1}{2}l \le x \le +\frac{1}{2}l$ can be expanded as linear combinations of the complete set (see Appendix A VII) of orthonormal functions, with k an integer, $-\infty \le k \le \infty$,

$$\varphi_k(x) = \frac{1}{\sqrt{l}}e^{2\pi ikx/l}, \qquad \varphi_k^*(x) = \frac{1}{\sqrt{l}}e^{-2\pi ikx/l}, \tag{AVIII.1}$$

$$\int dx\, \varphi_{k'}^*(x)\varphi_k(x) = \delta(k'-k), \qquad \sum_k \varphi_k^*(x')\varphi_k(x) = \delta(x'-x), \tag{AVIII.2}$$

where $k'-k$ is the Kronecker delta, and $x'-x$ is the Dirac delta function.

The first condition of eq. (2) permits us to write, for any function $f(k')$, that

$$\int dx \left(\sum_{k'} f(k')\varphi_{k'}^*(x)\right]\varphi_k(x) = f(k). \tag{AVIII.3}$$

Suppose that the range of integration over x is not limited, in other words, that l approaches infinity, so that with h Planck's constant the quantity $p = hkl^{-1}$, which has the dimension of momentum conjugate to x, becomes continuous. Also, then $k = lph^{-1}$ may be regarded as continuous, and summation in eq. (3) over k' may be replaced by integration over $dk' = h^{-1}l\,dp'$. Rewriting $\varphi_k(x)$ of eq. (1) as a function $\phi(p, x)$ with a changed normalization,

$$\phi(p, x) = h^{-1/2}e^{(2\pi i/h)px} \tag{AVIII.4}$$

we have, instead of eq. (3),

$$\int dx\left[\int dp' f(p')\phi(p', x)\right]\phi(p, x) = f(p)$$
$$= dp' f(p', x)\delta(x' - x). \tag{AVIII.5}$$

In place of eq. (2) we now have

$$\int dx\,\phi^*(p'x)\phi(p'x) = \delta(p' - p), \qquad \int dp\,\phi^*(p, x)\phi(p, x) = \delta(x' - x), \tag{AVIII.6}$$

with both δ's the Dirac function. The functions $\phi(p, x)$ of eq. (4) form a complete orthonormal set. They are eigenfunctions of the momentum operator:

$$\mathcal{P} = \left(\frac{h}{2\pi i}\right)\frac{\partial}{\partial x},$$
$$\mathcal{P}\phi(p, x) = p\phi(p, x), \qquad -\infty \le p \le \infty. \tag{AVIII.7}$$

Let $\Psi_\nu(x)$ and $\Psi_\mu(x)$ be two of an orthonormal complete set of state functions in the coordinate space. Develop each as a linear combination of the doubly continuous functions $\phi(p, x)$, for instance,

$$\Psi_\nu(x) = \int dp'\,\Phi_\nu(p')\phi(p'x), \tag{AVIII.8}$$

with a similar equation relating Ψ_ν, Ψ_μ. Multiply both sides of eq. (8) by $\phi^*(px)$ and integrate over x, using eq. (7) and the property of the complete set given in eq. (AVII.8). One finds

$$\int dx\,\Psi_\nu(x)\phi^*(px) = \int dp'\,\Phi_\nu(p')\int dx\phi\,(p'x)\phi^*(px)$$
$$= \int dp'\,\Phi_\nu(p')\delta(p' - p) = \Phi_\nu(p), \tag{AVIII.8$'$}$$

with of course a corresponding equation for Φ_μ. Now examine $\int dp\,\Phi_\nu(p)\Phi_\mu^*(p)$ using eq. (8$'$) and its equivalent for Φ_μ^*, with eq. (6) and the orthonormality condition for Ψ_ν, Ψ_μ. One has

$$\int dp\, \Phi_\nu(p)\Phi_\mu^*(p) = \iint dx\, dx'\, \Psi_\nu(x')\left[\int dp\, \phi^*(px')\phi(px)\right]\Psi_\mu^*(x)$$
$$= \iint dx\, dx'\, \Psi_\nu(x')\delta(x'-x)\Psi_\mu(x)$$
$$= \int dx\, \Psi_\nu(x)\Psi_\mu(x) = \delta(\nu-\mu), \qquad \text{(AVIII.9)}$$

namely, the functions Φ_ν, $\Phi_\mu, \ldots$ form an infinite orthonormal set of functions of the momentum space having one-to-one correspondence with the infinite complete set Ψ_ν, $\Psi_\mu, \ldots$ of the coordinate space. Equations (8) and (8′) give $\Psi_\nu(x)$ as a Fourier integral transform of $\Phi_\nu(p)$ and $\Phi_\nu(p)$ as the transform of $\Psi_\nu(x)$, respectively.

We now prove that just as the probability density $w_\nu(x)$ of the position x in the state $\Psi_\nu(x)$ is $w_\nu(x) = \Psi_\nu(x)\Psi_\nu^*(x)$, so also the probability density $w(p)$ in momentum is $\Phi_\nu(p)\Phi_\nu^*(p)$. We note that with eq. (3) for the momentum operator $\mathcal{P} = (h/2\pi i)(\partial/\partial x)$, and eq. (6) for $\phi(px)$, $\mathcal{P}\phi(px) = p\phi(px)$, so that, with eq. (8) for Ψ_ν,

$$\mathcal{P}\Psi_\nu(x) = \int dp'\, \Phi_\nu(p')p'\phi(p'x). \qquad \text{(AVIII.10)}$$

Now the probability density of momentum having the numerical value p is

$$w(p) = \int dx\, \Psi_\nu^*(x)[\delta(p-\mathcal{P})]\Psi_\nu(x). \qquad \text{(AVIII.10')}$$

From eq. (8) write $\Psi_\nu^* = \int dp''\, \Phi_\nu^*(p'')\phi^*(p''x)$ in eq. (10′), with eq. (10) for Ψ_ν, to obtain

$$w(p) = \iiint dx\, dp'\, dp''\, \Phi_\nu^*(p'')\phi(p'')\phi(p''x)[\delta(p-p']\Psi_\nu(p')\phi(p'x). \qquad \text{(AVIII.10'')}$$

Integrate over dx using eq. (7) to find

$$w(p) = \iint dp'\, dp''\, \Phi_\nu^*(p'')\delta[p''-p']\delta(p-p')\Phi_\nu(p') = \Phi_\nu^*(p)\Phi_\nu(p) \qquad \text{(AVIII.11)}$$

AIX. DIMENSIONLESS THERMODYNAMIC OSCILLATOR FUNCTIONS

Numerical values of $(RT)^{-1}$ times energy E, R^{-1} times heat capacity C, $(RT)^{-1}$ times Gibbs free energy G, and R^{-1} times entropy S (all per mole) are listed for a monochromatic harmonic oscillator. (For values of $u = \theta/T = h\nu/kT$ for which extrapolations of certain functions are inaccurate more easily extrapolated functions are given, that is, $(F/RT) + \ln u$ and $(S/R) + \ln u$.)

Table AIX2

$u = \frac{\theta}{T} = \frac{h\nu}{kT}$	$\frac{E}{RT} = \frac{u}{e^u - 1}$	$-\frac{\Delta f}{\Delta u}$	$\frac{C}{R} = \frac{u^2 e^u}{(e^u - 1)^2}$	$-\frac{\Delta f}{\Delta u}$	$-\frac{G}{RT} = [-\ln (1 - e^{-u})]$	$-\frac{G}{RT} + \ln u$	$\frac{S}{R}$	$\frac{S}{R} + \ln u$
1	2	2a	3	3a	4	4a	5	5a
0.001	0.9995		1.0000		6.9083	0.0005	7.9078	1.0000
		0.50						
0.005	0.9975		1.0000		5.3008	0.0025	6.2983	1.0000
		0.50						
0.010	0.9950		1.0000		4.6102	0.0050	5.6052	1.0000
		0.49		0.00				
0.050	0.9752		0.9998		3.0206	0.0249	3.9957	1.0001
		0.49		0.01				
0.10	0.9508		0.9992		2.3522	0.0496	3.3030	1.0004
		0.48		0.02				
0.15	0.9269		0.9981		1.9711	0.0740	2.8981	1.0010
		0.47		0.03				
0.20	0.9033		0.9967		1.7077	0.0983	2.6111	1.0017
		0.46		0.04				
0.25	0.8802		0.9948		1.5087	0.1224	2.3889	1.0026
		0.45		0.04				
0.30	0.8575		0.9925		1.3502	0.1462	2.2078	1.0038
		0.44		0.05				
0.35	0.8352		0.9898		1.2197	0.1699	2.0549	1.0051
		0.44		0.06				
0.40	0.8133		0.9868		1.1096	0.1933	1.9230	1.0067
		0.43		0.07				
0.45	0.7919		0.9832		1.0150	0.2165	1.8070	1.0085
		0.42		0.07				
0.50	0.7707		0.9794		0.9327	0.2396	1.7035	1.0104
		0.41		0.08		$-\Delta f/\Delta u$		
0.55	0.7501		0.9752		0.8602		1.6104	1.0126
		0.40		0.09		1.29		
0.60	0.7295		0.9705		0.7958		1.5257	1.0149
		0.39		0.10		1.15		
0.65	0.7100		0.9655		0.7383		1.4482	1.0174
		0.39		0.10		1.04		
0.70	0.6905		0.9602		0.6864		1.3769	1.0202
		0.38		0.11		0.94		
0.75	0.6715		0.9544		0.6393		1.3109	1.0232
		0.37		0.12		0.85		
0.80	0.6528		0.9484		0.5965		1.2494	$-\Delta f/\Delta u$
		0.36		0.13		0.78		1.15

Table AIX2 (contd.)

$u = \frac{\theta}{T} = \frac{h\nu}{kT}$	$\frac{E}{RT}$ $\frac{u}{e^u - 1}$	$-\frac{\Delta f}{\Delta u}$	$\frac{C}{R} = \frac{u^2 e^u}{(e^u - 1)^2}$	$-\frac{\Delta f}{\Delta u}$	$-\frac{G}{RT} = [-\ln (1 - e^{-u})]$	$-\frac{G}{RT}$ $+\ln u$	$\frac{S}{R}$	$\frac{S}{R}$ $+\ln u$
1	2	2a	3	3a	4	4a	5	5a
0.85	0.6345		0.9420		0.5576		1.1920	
		0.35		0.13		0.71		1.07
0.90	0.6166		0.9353		0.5218		1.1385	
		0.35		0.14		0.655		1.01
0.95	0.5991		0.9282		0.4890		1.0881	
		0.34		0.15		0.60		0.95
1.00	0.5820		0.9207		0.4587		1.0406	
		0.34		0.15		0.56		0.89
1.05	0.5652		0.9130		0.4307		0.9959	
		0.33		0.16		0.52		0.84
1.10	0.5489		0.9050		0.4047		0.9536	
		0.32		0.16		0.48		0.80
1.15	0.5329		0.8967		0.3807		0.9136	
		0.31		0.17		0.45		0.76
1.20	0.5172		0.8882		0.3584		0.8756	
		0.30		0.17		0.415		0.72
1.25	0.5019		0.8795		0.3376		0.8395	
		0.30		0.18		0.39		0.68
1.30	0.4870		0.8706		0.3182		0.8052	
		0.29		0.18		0.36		0.65
1.35	0.4725		0.8613		0.3001		0.7726	
		0.28		0.19		0.34		0.62
1.40	0.4582		0.8516		0.2831		0.7413	
		0.28		0.20		0.31		0.59
1.45	0.4444		0.8417		0.2673		0.7117	
		0.27		0.20		0.29		0.57
1.50	0.4308		0.8318		0.2525		0.6833	
		0.26		0.20		0.28		0.54
1.55	0.4176		0.8218		0.2386		0.6562	
		0.26		0.20		0.26		0.52
1.60	0.4048		0.8115		0.2255		0.6303	
		0.25		0.21		0.24		0.49
1.65	0.3922		0.8010		0.2133		0.6055	
		0.24		0.21		0.23		0.47
1.70	0.3800		0.7903		0.2017		0.5817	
		0.24		0.21		0.22		0.46
1.75	0.3681		0.7796		0.1909		0.5587	
		0.23		0.21		0.20		0.44

Table AIX2 (contd.)

$u = \frac{\theta}{T} = \frac{h\nu}{kT}$	$\frac{E}{RT} = \frac{u}{e^u - 1}$	$-\frac{\Delta f}{\Delta u}$	$\frac{C}{R} = \frac{u^2 e^u}{(e^u - 1)^2}$	$-\frac{\Delta f}{\Delta u}$	$-\frac{G}{RT} = [-\ln (1 - e^{-u})]$	$-\frac{G}{RT} + \ln u$	$\frac{S}{R}$	$\frac{S}{R} + \ln u$
1	2	2a	3	3a	4	4a	5	5a
1.80	0.3564		0.7688		0.1807		0.5368	
		0.23		0.22		0.19		0.42
1.85	0.3451		0.7578		0.1711		0.5159	
		0.22		0.22		0.18		0.40
1.90	0.3342		0.7467		0.1620		0.4960	
		0.21		0.22		0.17		0.38
1.95	0.3235		0.7354		0.1535		0.4770	
		0.21		0.22		0.16		0.37
2.00	0.3130		0.7241		0.1454		0.4584	
		0.20		0.23		0.15		0.35
2.10	0.2931		0.7013		0.1303		0.4234	
		0.19		0.23		0.13		0.32
2.20	0.2743		0.6783		0.1172		0.3915	
		0.18		0.23		0.12		0.30
2.30	0.2565		0.6553		0.1054		0.3619	
		0.17		0.23		0.11		0.27
2.40	0.2397		0.6320		0.0948		0.3346	
		0.16		0.23		0.10		0.25
2.50	0.2236		0.6089		0.0854		0.3092	
		0.15		0.23		0.09		0.24
2.60	0.2085		0.5859		0.0769		0.2855	
		0.14		0.23		0.08		0.22
2.70	0.1944		0.5630		0.0692		0.2637	
		0.13		0.23		0.07		0.20
2.80	0.1813		0.5404		0.0624		0.2439	
		0.12		0.22		0.06		0.19
2.90	0.1689		0.5182		0.0562		0.2253	
		0.11		0.22		0.06		0.17
3.00	0.1572		0.4963		0.0507		0.2179	
		0.11		0.22		0.05		0.16
3.10	0.1462		0.4747		0.0458		0.1920	
		0.10		0.21		0.04		0.15
3.20	0.1360		0.4536		0.0413		0.1773	
		0.09		0.21		0.04		0.14
3.30	0.1264		0.4329		0.0373		0.1637	
		0.09		0.20		0.04		0.13
3.40	0.1173		0.4128		0.0336		0.1509	
		0.08		0.19		0.03		0.12

Table AIX2 (contd.)

$u = \frac{\theta}{T} = \frac{h\nu}{kT}$	$\frac{E}{RT} = \frac{u}{e^u - 1}$	$-\frac{\Delta f}{\Delta u}$	$\frac{C}{R} = \frac{u^2 e^u}{(e^u - 1)^2}$	$-\frac{\Delta f}{\Delta u}$	$-\frac{G}{RT} = [-\ln (1 - e^{-u})]$	$-\frac{G}{RT} + \ln u$	$\frac{S}{R}$	$\frac{S}{R} + \ln u$
1	2	2a	3	3a	4	4a	5	5a
3.50	0.0190		0.3933		0.0304		0.1393	
		0.08		0.19		0.03		0.11
3.60	0.1011		0.3743		0.0275		0.1286	
		0.07		0.18		0.03		0.10
3.70	0.0938		0.3559		0.0248		0.1187	
		0.07		0.18		0.02		0.10
3.80	0.0870		0.3381		0.0223		0.1093	
		0.06		0.17		0.02		0.09
3.90	0.0806		0.3208		0.0200		0.1006	
		0.06		0.17		0.02		0.08
4.00	0.0746		0.3041		0.0180		0.0925	
		0.05		0.16		0.02		0.07
4.20	0.0640		0.2726		0.0148		0.0787	
		0.05		0.14		0.01		0.06
4.40	0.0547		0.2437		0.0119		0.0666	
		0.04		0.13		0.01		0.05
4.60	0.0467		0.2169		0.0097		0.0564	
		0.04		0.12		0.01		0.04
4.80	0.0398		0.1927		0.0079		0.0477	
		0.03		0.11		0.01		0.04
5.00	0.0339		0.1707		0.0063		0.0403	
		0.02		0.10				
5.20	0.0289		0.1507		0.0052		0.0341	
		0.02		0.09				
5.40	0.0245		0.1328		0.0042		0.0287	
		0.02						
5.60	0.0208		0.1168		0.0034		0.0242	
		0.01						
5.80	0.0178		0.1024		0.0027		0.0205	
		0.01						
6.00	0.0149		0.0898		0.0022		0.0171	
		0.01						
6.50	0.0098		0.0636		0.0010		0.0107	
7.00	0.0064		0.0446		0.0003		0.0067	
7.50	0.0042		0.0310					

AX. PHYSICAL CONSTANTS

These values are from those listed in the *American Institute of Physics Handbook*, 3d ed., compiled by B. N. Taylor, W. H. Parker, and D. N. Langenberg. Plus-minus values are standard deviation uncertainties in the last digits of the given value.

Velocity of light	$c = 2.9979250 \pm 10 \times 10^{10}$ cm sec^{-1}
Charge on electron	$e = 1.6021917 \pm 70 \times 10^{19}$ C
Planck's constant	$h = 6.626196 \pm 50 \times 1\bar{0}^{27}$ erg sec
	$\hbar = \dfrac{h}{2\pi} = 1.0545919 \pm 80 \times 10^{-27}$ erg sec
Mass of electron	$m_e = 9.109558 \pm 54 \times 10^{-28}$ g
Mass of hydrogen atom	$M_H = 1.673525 \pm 11 \times 10^{-24}$ g
Ratio	$\dfrac{M_H}{m_e} = 1\,837.109 \pm 11$
Ratio, charge to mass of electron	$\dfrac{e}{m_e} = \begin{matrix} 1.7588028 \pm 54 \times 10^{7} \text{ emu g}^{-1} \\ 5.272759 \pm 16 \times 10^{17} \text{ esu g}^{-1} \end{matrix}$
Boltzmann constant	$k = 1.380662 \pm 59 \times 10^{-16}$ erg K^{-1}
Avogadro's number	$N_0 = 6.022169 \pm 40 \times 10^{23}$ mol^{-1}
Volume of perfect gas at 0°C and 1 atm	$= 2.24136 \pm 94 \times 10^{4}$ cm^3 mol^{-1}
Gas constant per mole	$R = 8.31434 \pm 34 \times 10^{7}$ erg mol^{-1} K^{-1}
	$= 1.98588^{(1)}$ cal mol^{-1} K^{-1}
Faraday constant	$F = 9.648670 \pm 54 \times 10^{4}$ C mol^{-1}
	$= 2.892599 \pm 16 \times 10^{14}$ esu mol^{-1}
Normal acceleration of gravity at equator	$g = 978.049$ (defined) cm sec^{-2}
Normal atmosphere	$= 1.013250$ (defined) $\times 10^{6}$ dyne cm^2
Temperature of melting ice, 0°C	$= 273.16$K
Density of mercury (0°C and 1 atm)	$= 13.595 \pm 5$ g cm^{-3}
Radius of first Bohr orbit	$a_0 = 5.2917715 \pm 81 \times 10^{-9}$ cm
Reciprocal of find structure constant	$\alpha^{-1} = 137.03602 \pm 21$
Bohr magneton $\mu = \dfrac{eh}{4\pi m_e c}$	$= 9.274096 \pm 64 \times 10^{-21}$ erg G^{-1}

(1) The calorie is defined as the quantity of heat necessary to raise 1 g of water 1°C. Its precise value depends on the interval of temperature used in the measurement. We use a value adopted at the International Steam Table Conference in 1929 modified by the absolute system of electrical units, namely, 1 cal = 4.18674 J.

AXI. CONVERSION OF ENERGY UNITS

The following are all used as energy units: erg, the fundamental energy unit of the cgs system; joule, 10^7 ergs; calorie, the heat required to raise

	Ergs/molecule	Seconds^{-1}	Centimeters^{-1}	Electronvolts	Joules mole	Kilocalories mole	Atomic units	Kelvins
1 erg per molecule equals	1	1.50916×10^{26}	5.03402×10^{15}	6.24145×10^{11}	6.02217×10^{16}	1.43840×10^{13}	2.29367×10^{10}	7.24311×10^{15}
Unit frequency (in sec^{-1}) equals	6.62620×10^{-27}	1	3.33564×10^{-11}	4.13571×10^{-15}	3.99041×10^{-10}	9.53109×10^{-14}	1.51983×10^{-16}	4.79943×10^{-11}
Unit wave number (in cm^{-1}) equals	1.98648×10^{-16}	2.99792×10^{10}	1	1.23985×10^{-4}	11.9623	2.85720×10^{-3}	4.55633×10^{-6}	1.43883
1 absolute electron volt equals	1.60219×10^{-12}	2.41797×10^{14}	8.06546×10^{3}	1	9.64866×10^{4}	23.0459	3.67489×10^{-2}	1.16049×10^{4}
1 joule per mole equals	1.66053×10^{-17}	2.50601×10^{9}	8.35914×10^{-2}	1.03641×10^{-5}	1	2.3885×10^{-4}	3.80871×10^{-7}	0.1202741
1 kilocalorie per mole equals	6.95217×10^{-14}	1.04920×10^{13}	3.49993×10^{2}	4.33917×10^{-2}	4.18674×10^{3}	1	1.59460×10^{-3}	5.03555×10^{-2}
1 atomic unit equals	4.35983×10^{-11}	6.57968×10^{15}	2.19475×10^{5}	27.2116	2.62556×10^{6}	6.27115×10^{2}	1	3.15787×10^{5}
kT at 1 kelvin equals	1.38062×10^{-16}	2.08358×10^{10}	0.695008	8.61707×10^{-5}	8.31434	1.98588	3.16669×10^{-6}	1

1 g of water 1°C at 15°C; kilocalorie, 10^3 cal; frequency (in sec^{-1}), the energy of a photon of that frequency; wave number (in cm^{-1}), the energy of a photon of that wave number; volt, the energy of an electron accelerated through that number of volts; degree, the energy divided by k; atomic units, the energy in units e^2/a_0, twice the ionization energy of a hypothetical hydrogen isotope of infinite mass.

AXII. GREEK ALPHABET

Α α	Alpha		Ν ν	Nu
Β β	Beta		Ξ ξ	Xi
Γ γ	Gamma		Ο o	Omicron
Δ δ	Delta		Π π	Pi
Ε ϵ	Epsilon		Ρ ρ	Rho
Ζ ζ	Zeta		Σ σ	Sigma
Η η	Eta		Τ τ	Tau
Θ ϑ θ	Theta		Υ υ	Upsilon
Ι ι	Iota		Φ φ ϕ	Phi
Κ κ	Kappa		Χ χ	Chi
Λ λ	Lambda		Ψ ψ	Psi
Μ μ	Mu		Ω ω	Omega

INDEX

References are to pages; **boldface type** indicates definitions or more important references.